BASIC PULSE CIRCUITS

basic pulse circuits

RICHARD BLITZER
RCA Institutes, Inc.

McGRAW-HILL BOOK COMPANY
New York · Toronto · London

BASIC PULSE CIRCUITS

Library of Congress Catalog Card Number: 63-15107

4 5 6 7 8 9 – MP – 9 8

dedication

To the thousands of students whom I helped train, and who helped me. Without them this book never would have been possible.

Preface

This book is intended as the next step for the technician who is completing or has completed a basic course in electronics. It should also prove useful as a study and reference source for the serviceman desiring to upgrade himself and for the electronic technician and technical writer who, though experienced in some particular phase, require information on certain advanced circuits.

The book is based on extensive notes which the author has used in his work as instructor at RCA Institutes for the past sixteen years and as a consultant technical writer on radar and missile projects. The topics covered are mainly those which comprise the subject matter taught in the senior term of the electronic technician course at RCA Institutes.

The circuits described are those used extensively in radars, computers, and missiles. Each circuit is completely explained and treated independently of other similar circuits. Because the material is presented in this manner, the user need read of only that particular circuit to gain a knowledge of it.

As in practically all technician courses, the current flow used is the movement of electrons from negative to positive. Since most circuits can be explained satisfactorily by this method, it is employed throughout the text. Although no calculus is used, some skill in basic mathematics is necessary. The mathematics used in a discussion is explained in detail in that particular section. Some electronics fundamentals are also briefly reviewed where desirable. Numerous examples are worked out with the same detail as would be presented at the chalkboard. Problems at the end of each chapter are mainly those the author has used in quizzes and are arranged to become progressively more challenging. Answers to most problems are included at the end of the text.

The author wishes to thank his colleagues on the faculty at RCA Institutes for their cooperation. Faculty members Patrick Clinton, Hyman Gellman, Alvin Liff, Aram Ehramjian, and Henry Warner were particularly helpful. In addition to the many computer, radar, and missile engineers and technicians who aided the author, Alfred Cook and

Walter Neiman of Grumman Aircraft Engineering Corporation offered many invaluable suggestions.

Grateful thanks are offered also to my daughter Jane and my son Robert for their understanding, and especially to my wife Connie for her encouragement and patience, her excellent typing of the manuscript, and her endless hours of help with the proofreading.

Richard Blitzer

Contents

CHAPTER 1

NETWORK CIRCUITS

1·1 General

Network circuits, also called *mesh circuits,* are simply more complicated arrangements of series-parallel combinations. Pulse circuitry often includes these networks. Explanations of these circuits require the ability to understand and solve them. Usually, solutions may be possible in any of several different methods. In this chapter, many examples of network diagrams are given together with several suggested methods of solution for each. The mathematics involved uses only simple algebra and, in one of the methods of solution, simultaneous equations. The next section of this chapter, therefore, reviews the solution of simultaneous equations using two methods: (1) the elimination of all unknowns except one and (2) the use of *determinants.* If the reader is completely familiar with *both* of these methods of solution of simultaneous equations, he may choose to skip Secs. 1·2 and 1·3 and go right into the circuitry and examples of the sections following these.

1·2 Simultaneous Equations

One method of solving mesh circuits involves *simultaneous equations.* These are simple algebraic equations using *two* or more unknowns. Equations (1·1) and (1·2), as shown in Example 1·1, have two unknowns, x and y.

Example 1·1 Solve for x and y in the following:

$$x + 3y = 7 \tag{1·1}$$
$$x - 2y = -3 \tag{1·2}$$

To solve for the values of x and y, the two equations must be combined by either addition or subtraction so that one of the unknowns is eliminated, yielding a new equation containing only a single unknown. To have one unknown term (either x or y) drop out, it is necessary that the *coefficient* of the x term or the y term be identical in both equations. In Eqs. (1·1) and (1·2), the coefficient of x is $+1$, while the coefficients of y are $+3$ in Eq. (1·1) and -2 in Eq. (1·2). Since the coefficient of x is identical $(+1)$ in both equations, if the two equations are subtracted, the x term becomes $0x$, and the x term is eliminated as shown in the following:

$$x + 3y = 7 \qquad (1\cdot1)$$

$$\text{Subtract} \quad x - 2y = -3 \qquad (1\cdot2)$$

$$+1x - (+1x) + 3y - (-2y) = +7 - (-3)$$

$$+1x - 1x + 3y + 2y = +7 + 3$$

$$0x + 5y = +10 \qquad (1\cdot3)$$

$$y = 2$$

Note that when algebraically *subtracting* Eq. (1·2) from Eq. (1·1), the signs of the *subtrahend* [Eq. (1·2)] are changed. Then the two equations are *added* algebraically, resulting in Eq. (1·3). The new Eq. (1·3) contains only the single unknown y since the x term has been eliminated. Solving for y in Eq. (1·3) yields $y = 2$.

After finding the value of one unknown, $y = 2$ in this case, substitute this value for y in either Eq. (1·1) or (1·2), and solve for x. For example:

$$x + 3y = 7 \qquad (1\cdot1)$$

Substitute for y:

$$x + 3(2) = 7$$

$$x + 6 = 7$$

$$x = 7 - 6$$

$$= 1$$

or, again solving for x,

$$x - 2y = -3 \qquad (1\cdot2)$$

Substitute for y:

$$x - 2(2) = -3$$

$$x - 4 = -3$$

$$x = 4 - 3$$

and again $\quad x = 1$

In the foregoing, the x term was eliminated. The y term could be eliminated by first multiplying Eqs. (1·1) and (1·2) by *any* numbers which produce equal coefficients of y in both resulting equations. A simple procedure is to multiply each equation by the y coefficient of the *other* equation. For example, multiply Eq. (1·1) by −2 [the y coefficient of Eq. (1·2)], and multiply Eq. (1·2) by +3 [the y coefficient of Eq. (1·1)]. This procedure is as follows:

$$x + 3y = 7 \tag{1·1}$$
$$x - 2y = -3 \tag{1·2}$$

Multiply Eq. (1·1) by −2, giving

$$-2x - 6y = -14 \tag{1·4}$$

and multiply Eq. (1·2) by +3, giving

$$3x - 6y = -9 \tag{1·5}$$

Algebraically subtract Eq. (1·5) from Eq. (1·4) by changing the signs of all terms in Eq. (1·5), and adding the two equations. This gives

$$\begin{aligned} -5x + 0y &= -5 \qquad (1·6)\\ -5x &= -5\\ x &= \frac{-5}{-5}\\ &= 1 \end{aligned}$$

Note that this value, $x = 1$, is, of course, the same as was found previously. After finding the value of x, solve for y by substituting for x in one of the original equations, either (1·1) or (1·2).

$$x + 3y = 7 \tag{1·1}$$

Substitute for x:

$$\begin{aligned} 1 + 3y &= 7\\ 3y &= 7 - 1\\ &= 6\\ y &= \frac{6}{3}\\ &= 2 \end{aligned}$$

or, again solving for y,

$$x - 2y = -3 \tag{1·2}$$

Substitute for x:

$$\begin{aligned} 1 - 2y &= -3\\ -2y &= -3 - 1\\ -2y &= -4\\ y &= \frac{-4}{-2}\\ &= 2 \end{aligned}$$

This value, $y = 2$, is, of course, the same as was found previously.

Simultaneous equations, three unknowns. The previous discussion covered equations with *two* unknowns. When there are *three* unknowns, three simultaneous equations must be employed. The method of solution is similar to the elimination of an unknown term described previously. Since there are now three unknowns, two of them must be eliminated in order to solve for the third. The method is as follows.

Example 1·2 Solve for x, y, and z in the following.

$$x + y + z = 6 \tag{1·7}$$
$$2x + 3y - 2z = 2 \tag{1·8}$$
$$-3x - 2y + 3z = 2 \tag{1·9}$$

The procedural steps to solve for x, y, and z are as follows:

1. First eliminate one term, say the x, by combining two of the three equations, say Eqs. (1·7) and (1·8).
2. This results in a new equation containing only y and z terms.
3. Then eliminate the x term again by combining one of the original equations, say (1·7), with the third equation, (1·9).
4. This again results in a new equation containing only y and z terms.
5. Finally, eliminate one of these terms, say, the y, by combining the two *new* equations containing the y and z terms.
6. Solve for z.
7. Then substitute the value of z in either of the equations which contain only the y and z terms, and solve for y.
8. Finally, substitute the values of y and z in any of the original equations, say (1·7), and solve for x.

Following the steps of the given procedures and using Eqs. (1·7) to (1·9), solve for x, y, and z as follows:

1. Eliminate the x term by making the coefficients of x equal, using Eqs. (1·7) and (1·8). To do this, multiply Eq. (1·7) by 2 [the x coefficient of Eq. (1·8)], giving

$$2x + 2y + 2z = 12 \tag{1·10}$$

Rewrite Eq. (1·8):

$$2x + 3y - 2z = 2 \tag{1·8}$$

Subtract Eq. (1·8) from (1·10), giving

$$-y + 4z = 10 \tag{1·11}$$

2. Equation (1·11) contains only y and z terms and is the result of combining Eqs. (1·7) and (1·8).
3. Again eliminate the x term, this time using Eqs. (1·7) and (1·9), by making the coefficients of x equal. To do this, multiply Eq. (1·7) by -3

[the x coefficient of Eq. (1·9)], giving

$$-3x - 3y - 3z = -18 \qquad (1\cdot12)$$

Rewrite Eq. (1·9):

$$-3x - 2y + 3z = 2 \qquad (1\cdot9)$$

Subtract Eq. (1·9) from (1·12), giving

$$-y - 6z = -20 \qquad (1\cdot13)$$

4. Equation (1·13) contains only y and z terms and is the result of combining Eqs. (1·7) and (1·9).

5. Combine Eqs. (1·11) and (1·13) containing only the y and z terms, to eliminate a second unknown, say the y term. To do this, the y coefficients must be made equal. In this particular example, these coefficients happen to be equal, as shown in the following Eqs. (1·11) and (1·13).

$$-y + 4z = 10 \qquad (1\cdot11)$$
$$-y - 6z = -20 \qquad (1\cdot13)$$

Subtracting Eq. (1·13) from (1·11) gives

$$10z = 30 \qquad (1\cdot14)$$

6. Solving Eq. (1·14) for z gives

$$\begin{aligned} 10z &= 30 \qquad (1\cdot14) \\ z &= \frac{30}{10} \\ &= 3 \end{aligned}$$

7. Substitute this value of z in either Eq. (1·11) or (1·13), and solve for y.

$$\begin{aligned} -y + 4z &= 10 \qquad (1\cdot11) \\ -y + 4(3) &= 10 \\ -y + 12 &= 10 \\ -y &= 10 - 12 \\ &= -2 \\ y &= 2 \end{aligned}$$

8. Substitute the z value (3) and the y value (2) in any of the original Eqs. (1·7), (1·8), or (1·9), containing the x, y, and z terms, and solve for x.

$$\begin{aligned} x + y + z &= 6 \qquad (1\cdot7) \\ x + 2 + 3 &= 6 \\ x &= 6 - 2 - 3 \\ &= 1 \end{aligned}$$

The final answers of Example 1·2 are therefore $x = 1$, $y = 2$, and $z = 3$.

1·3 Determinants

Another method of solving simultaneous equations is by the use of *determinants*. This system will be employed later in this chapter in solving network or mesh circuits. Using the two equations of Example 1·1 from the previous section, x and y are solved with the determinant method as follows:

$$x + 3y = 7 \tag{1·1}$$
$$x - 2y = -3 \tag{1·2}$$

To solve for x and for y, first list the coefficients of x and y in Eq. (1·1) in their order of appearance reading from left to right as $+1$ and $+3$.

Repeat, for Eq. (1·2) as $+1$ and -2. These numbers (the coefficients of x and y) are now called the *elements* of the determinant.

Write these elements in horizontal rows one under the other:

$$\begin{matrix} +1 & +3 \\ +1 & -2 \end{matrix}$$

These elements, as shown directly above, are called the *determinant of the system* and become the *denominator* for both the x term and the y term in this method of solution.

The *numerator* for the x term is similar, except that in place of the x coefficient, the number at the right side of the equation is used. In Eq. (1·1), this number is 7. Writing this number and then the y coefficient in Eq. (1·1) gives $+7$ and $+3$.

Repeat for Eq. (1·2), where the number -3 from the right side replaces the x coefficient, followed by the y coefficient, giving -3 and -2.

The numerator for the x term then consists of these last four digits written in two horizontal rows one under the other:

$$\begin{matrix} +7 & +3 \\ -3 & -2 \end{matrix}$$

The complete expression for the x term is then

$$x = \frac{\text{numerator}}{\text{denominator}}$$

$$= \frac{\begin{vmatrix} +7 & +3 \\ -3 & -2 \end{vmatrix}}{\begin{vmatrix} +1 & +3 \\ +1 & -2 \end{vmatrix}}$$

The vertical lines simply indicate that the above expression is in the determinant form. The complete expression for the y term has exactly the same denominator as the x. The numerator for y consists of the x coefficients followed by the numbers $+7$ and -3 [the right side of Eqs. (1·1) and (1·2)] in place of the y coefficients.

$$y = \frac{\begin{vmatrix} +1 & +7 \\ +1 & -3 \end{vmatrix}}{\begin{vmatrix} +1 & +3 \\ +1 & -2 \end{vmatrix}}$$

The solution for x, after writing the elements in the numerator and denominator, as repeated below, is continued in the following manner:

$$x = \frac{\text{numerator}}{\text{denominator}} = \frac{\begin{vmatrix} +7 & +3 \\ -3 & -2 \end{vmatrix}}{\begin{vmatrix} +1 & +3 \\ +1 & -2 \end{vmatrix}}$$

In the numerator, multiply the diagonally opposite numbers, starting with the upper left number, $+7$. Multiply $+7$ and -2, giving -14. From this product of -14, subtract the next product consisting of -3 and $+3$ or -9. The numerator for the x term is then

$$\begin{aligned} x &= \frac{\text{numerator}}{\text{denominator}} \\ &= \frac{(+7)(-2) - (-3)(+3)}{\text{denominator}} \\ &= \frac{(-14) - (-9)}{\text{denominator}} \\ &= \frac{-14 + 9}{\text{denominator}} \\ &= \frac{-5}{\text{denominator}} \end{aligned}$$

The denominator for the x term is found in the same manner as the numerator. That is, multiply the diagonally opposite numbers, starting with the upper left number, $+1$, times the lower right one, -2. From this product, subtract the next product of the lower left number, $+1$, and the upper right one, $+3$. The denominator for the x term is then

$$x = \frac{\text{numerator}}{\text{denominator}}$$
$$= \frac{-5}{(+1)(-2) - (+1)(+3)}$$
$$= \frac{-5}{(-2) - (+3)}$$
$$= \frac{-5}{-2 -3}$$
$$= \frac{-5}{-5}$$

Finally, solving for x gives

$$x = \frac{-5}{-5}$$
$$= +1$$

Solving for y and repeating the determinant expression for y gives

$$y = \frac{\begin{vmatrix} +1 & +7 \\ +1 & -3 \end{vmatrix}}{\begin{vmatrix} +1 & +3 \\ +1 & -2 \end{vmatrix}}$$
$$= \frac{(+1)(-3) - (+1)(+7)}{(+1)(-2) - (+1)(+3)}$$
$$= \frac{(-3) - (+7)}{(-2) - (+3)}$$
$$= \frac{-3 -7}{-2 -3}$$
$$= \frac{-10}{-5}$$
$$= +2$$

The denominators for the x and y terms, of course, are identical.

Note that these answers for Example 1·1, of $x = 1$ and $y = 2$, of course, agree with the answers found previously in Sec. 1·2.

Determinants for three unknowns. When there are three equations with three unknowns, the determinant method is similar to that just discussed for two unknowns. The work, however, is more lengthy and is shown in the following for the three equations of Example 1·2 from Sec. 1·2.

$$x + y + z = 6 \qquad (1 \cdot 7)$$
$$2x + 3y - 2z = 2 \qquad (1 \cdot 8)$$
$$-3x - 2y + 3z = 2 \qquad (1 \cdot 9)$$

Solving for x, list the coefficients of x, y, and z in the denominator as shown.

$$x = \frac{\text{numerator}}{\text{denominator}}$$

$$= \left| \frac{\text{numerator}}{\begin{matrix} +1 & +1 & +1 \\ +2 & +3 & -2 \\ -3 & -2 & +3 \end{matrix}} \right|$$

This denominator is the same for the y and z terms.

In the numerator for the x term, replace the x coefficients with the numbers from the right sides of Eqs. (1·7), (1·8), and (1·9), and list the y and z coefficients, as shown.

$$x = \frac{\begin{vmatrix} +6 & +1 & +1 \\ +2 & +3 & -2 \\ +2 & -2 & +3 \end{vmatrix}}{\begin{vmatrix} +1 & +1 & +1 \\ +2 & +3 & -2 \\ -3 & -2 & +3 \end{vmatrix}}$$

To solve for x, first repeat the first two vertical columns, placing these to the right of the third column as shown in Fig. 1·1. Solve for the *numerator* by performing six multiplications. Starting at the upper left number,

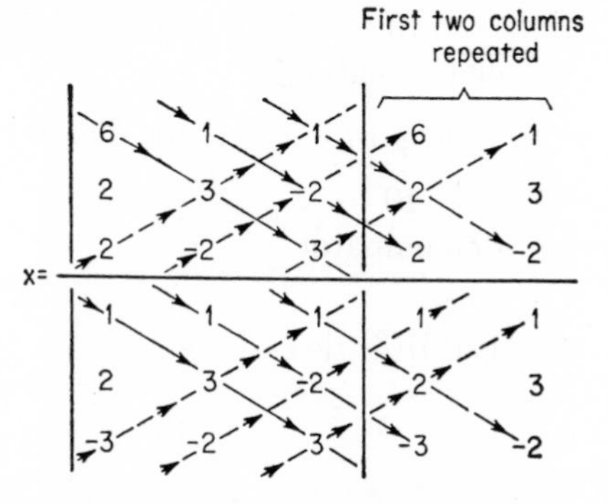

Figure 1·1 **Solving determinants.**

+6, in Fig. 1·1 (the + sign is omitted for simplicity), follow the solid-line arrow diagonally downward to the right and multiply the three numbers 6, 3, and 3, giving a product of 54. To this product, add the next product which consists of the three digits joined by the center solid-line arrow. These digits are 1, -2, and 2, and the product is -4. To this second product add a third one which consists of the third diagonal row of numbers joined by the right-hand solid-line arrow. These digits are 1, 2, and -2 and have a product of -4.

From the sum of these three products (formed by the diagonal solid-line arrows) there must now be *subtracted* three more products. These three new products consist of the diagonal row of numbers joined by the dotted-line arrows of Fig. 1·1 pointing diagonally upward to the right. The first of these products which are to be subtracted consists of the number starting at the lower left (in the numerator), 2, multiplied by the digits 3 and 1, producing a resultant of −6. The next product to be subtracted consists of the digits which are joined by the middle dotted-line arrows. These digits (in the numerator) are −2, −2, and 6, and the product to be subtracted is −24. The last product consists of the digits which are joined by the third dotted-line arrow. These are (in the numerator) 3, 2, and 1, giving the product to be subtracted of −6.

The numerator of the x term is therefore (referring to the above discussion and to Fig. 1·1)

$$
\begin{aligned}
x &= \frac{\text{numerator}}{\text{denominator}} \\
&= \frac{(6)(3)(3) + (1)(-2)(2) + (1)(2)(-2) - (2)(3)(1) - (-2)(-2)(6) - (3)(2)(1)}{\text{denominator}} \\
&= \frac{(54) + (-4) + (-4) - (6) - (24) - (6)}{\text{denominator}} \\
&= \frac{54 - 4 - 4 - 6 - 24 - 6}{\text{denominator}} \\
&= \frac{10}{\text{denominator}}
\end{aligned}
$$

Referring to Fig. 1·1, the denominator of the x term is as shown below:

$$
\begin{aligned}
x &= \frac{10}{\text{denominator}} \\
&= \frac{10}{(1)(3)(3) + (1)(-2)(-3) + (1)(2)(-2) - (-3)(3)(1) - (-2)(-2)(1) - (3)(2)(1)} \\
&= \frac{10}{(9) + (6) + (-4) - (-9) - (4) - (6)} \\
&= \frac{10}{9 + 6 - 4 + 9 - 4 - 6} \\
&= 10/10
\end{aligned}
$$

The x term is therefore

$$
\begin{aligned}
x &= \frac{10}{10} \\
&= 1
\end{aligned}
$$

This value, $x = 1$, of course, agrees with the answer of Example 1·2 in Sec. 1·2.

Solving for y, and referring to the three equations of Example 1·2, (1·7) to (1·9), gives the following determinant. The denominator is the same as in the x term. The numerator has the x and z coefficients, but the y coefficient is replaced by the digits from the right side of the equations.

$$y = \frac{\text{numerator}}{\text{denominator (same as } x \text{ denominator)}}$$

$$= \left| \frac{\begin{matrix} 1 & 6 & 1 \\ 2 & 2 & -2 \\ -3 & 2 & 3 \end{matrix}}{10} \right|$$

Repeating the first two columns of the numerator and placing these to the right of the third column, similar to Fig. 1·1, gives the following determinant and solution.

$$y = \left| \frac{\begin{matrix} 1 & 6 & 1 \\ 2 & 2 & -2 \\ -3 & 2 & 3 \end{matrix}}{10} \right| \quad \begin{matrix} 1 & 6 \\ 2 & 2 \\ -3 & 2 \end{matrix}$$

$$= \frac{(1)(2)(3) + (6)(-2)(-3) + (1)(2)(2) - (-3)(2)(1) - (2)(-2)(1) - (3)(2)(6)}{10}$$

$$= \frac{(6) + (36) + (4) - (-6) - (-4) - (36)}{10}$$

$$= \frac{6 + 36 + 4 + 6 + 4 - 36}{10}$$

$$= \frac{20}{10}$$

$$= 2$$

This value, $y = 2$, of course, agrees with the answer of Example 1·2 in Sec. 1·2.

Knowing the values of x and y (1 and 2 respectively), z can now be easily found by substituting the values of x and y in any one of the original three equations, (1·7), (1·8), or (1·9). However, for the sake of practice, z will be solved by again using the determinant method. The *numerator* for the z term contains the x and y coefficients, while the z coefficients are replaced by the digits from the right sides of Eqs. (1·7) to (1·9). The following also shows the first two columns repeated, similar to that of Fig. 1·1.

$$z = \frac{\text{numerator}}{\text{denominator (same as } x \text{ denominator)}}$$

$$= \frac{\begin{vmatrix} 1 & 1 & 6 \\ 2 & 3 & 2 \\ -3 & -2 & 2 \end{vmatrix} \begin{matrix} 1 & 1 \\ 2 & 3 \\ -3 & -2 \end{matrix}}{10}$$

$$= \frac{(1)(3)(2) + (1)(2)(-3) + (6)(2)(-2) - (-3)(3)(6) - (-2)(2)(1) - (2)(2)(1)}{10}$$

$$= \frac{(6) + (-6) + (-24) - (-54) - (-4) - (4)}{10}$$

$$= \frac{6 - 6 - 24 + 54 + 4 - 4}{10}$$

$$= \frac{30}{10}$$

$$= 3$$

This value, $z = 3$, of course, agrees with the answer of Example 1·2 in Sec. 1·2.

The discussion presented here of the determinant method of solution is by no means a complete one, simply indicating the mechanics of manipulation. The reader should consult any good college algebra text for a more comprehensive presentation.

1·4 Simple Network Circuits

A simple mesh or network circuit is shown in Fig. 1·2. This circuit consists of two *loops* acting as independent series circuits. Loop 1 is made up

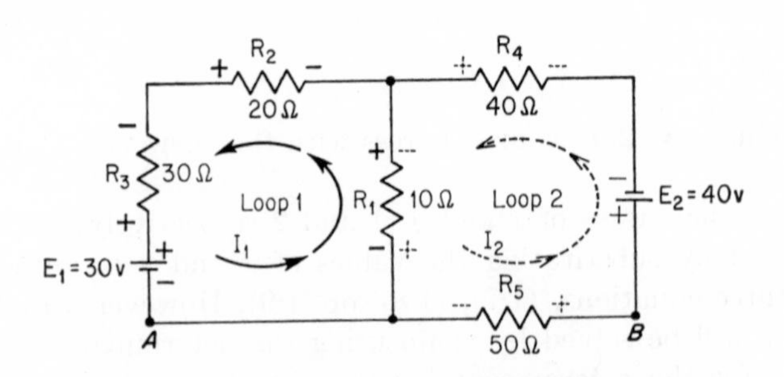

Figure 1·2 Two-loop mesh circuit of Example 1·3.

of the applied d-c voltage E_1, producing the current I_1 which flows through R_1, R_2, and R_3. Loop 2 comprises the other applied voltage E_2, causing current I_2 to flow through R_4, R_1, and R_5. The direction of these currents (electron movement) may be assumed to be as shown, with I_1 indicated by the solid-line arrows and I_2 by the dashed-line arrows. In many network circuits it may be impossible to correctly estimate the current *direction*. This is not important since when solving for the current, a *positive* answer will denote the fact that the assumed direction is the correct one. A *negative*-current answer indicates that the true current direction is backwards to that assumed. Of the many methods of mesh-circuit solutions, several of the more commonly used ones are reviewed here.

Example 1·3 In the circuit of Fig. 1·2, find the currents I_1 and I_2 and also find the voltage across R_1.

Solution. Assume current directions in each loop to be as shown in the diagram. Starting at *any* point in loop 1, "walk" completely around the loop in the same direction as the assumed current, adding up all the voltages therein. This total sum, of course, from Kirchhoff's voltage law, is equal to zero. When "walking" around the loop, note the polarity of each voltage. Indicate a *negative* sign for a voltage if that voltage (across either a resistor or a battery) is "entered" at the negative end. Indicate a *positive* sign for a voltage where the entering end, as the walk is taken, is the positive side. As an example, start at point A and walk up through R_1. The lower end of R_1, as shown by the solid-line signs, is negative because of the assumed direction of current I_1. This voltage (or IR) is then called $-I_1R_1$. However, the lower end of R_1 is also the positive side due to I_2, as shown by the dashed-line signs, and this voltage is called $+I_2R_1$. Since R_1 is 10 ohms, the total voltage across R_1 as the walk up through R_1 is taken is $-10I_1 + 10I_2$.

For loop 1, Kirchhoff's voltage law results in the following equation, starting at point A and walking completely around this loop.

$$-10I_1 + 10I_2 - 20I_1 - 30I_1 + 30 = 0 \qquad (1{\cdot}15)$$

$$-60I_1 + 10I_2 = -30 \qquad (1{\cdot}16)$$

For loop 2, Kirchhoff's voltage law results in the following equation, starting at point B and proceeding around this loop in the direction of the assumed current I_2.

$$+40 - 40I_2 - 10I_2 + 10I_1 - 50I_2 = 0 \qquad (1{\cdot}17)$$

$$+10I_1 - 100I_2 = -40 \qquad (1{\cdot}18)$$

For simplicity, reduce Eqs. (1·16) and (1·18) to simpler terms by dividing each by 10. Equation (1·16) then becomes

$$-6I_1 + I_2 = -3 \tag{1·19}$$

and Eq. (1·18) becomes

$$I_1 - 10I_2 = -4 \tag{1·20}$$

Solving these simultaneous equations, (1·19) and (1·20), for I_1 and I_2, multiply Eq. (1·20) by 6 [the coefficient of I_1 in Eq. (1·19)] to make both I_1 coefficients the same. Equation (1·20) multiplied by 6 now becomes

$$6I_1 - 60I_2 = -24 \tag{1·21}$$

Rewriting Eq. (1·19),

$$-6I_1 + I_2 = -3 \tag{1·19}$$

Add Eqs. (1·21) and (1·19) to remove one of the unknowns, I_1. This gives

$$-59I_2 = -27 \tag{1·22}$$
$$I_2 = 0.457 \text{ amp}$$

Solve for I_1 by substituting for I_2 in Eq. (1·19).

$$\begin{aligned} -6I_1 + I_2 &= -3 \\ -6I_1 + 0.457 &= -3 \\ -6I_1 &= -3 \quad -0.457 \\ &= -3.457 \\ I_1 &= 0.576 \text{ amp} \end{aligned}$$

Since both currents, I_1 and I_2, result in positive answers and not negative ones, the directions of the assumed currents are correct.

To find the voltage across R_1, note that I_1 and I_2 are flowing in opposite directions through this resistor. The actual current through R_1 is therefore the difference between the two currents, or

$$\begin{aligned} I \text{ through } R_1 &= I_1 - I_2 \\ &= 0.576 - 0.457 \\ &= 0.119 \text{ amp} \\ E_{R_1} &= IR_1 \\ &= (0.119)(10) \\ &= 1.19 \text{ volts} \end{aligned}$$

Another solution for I_1 and I_2 from Eqs. (1·19) and (1·20) is through the use of *determinants*, as explained in Sec. 1·3. Solving for I_1 gives

$$I_1 = \frac{\text{numerator}}{\text{denominator}}$$

$$= \frac{\begin{vmatrix} -3 & 1 \\ -4 & -10 \end{vmatrix}}{\begin{vmatrix} -6 & 1 \\ 1 & -10 \end{vmatrix}}$$

$$= \frac{(-3)(-10) - (-4)(1)}{(-6)(-10) - (1)(1)}$$

$$= \frac{30 + 4}{60 - 1}$$

$$= \frac{34}{59}$$

$$I_1 = 0.576 \text{ amp}$$

Solving for I_2, also by determinants, from Eqs. (1·19) and (1·20) gives

$$I_2 = \frac{\text{numerator}}{\text{denominator, same as } I_1 \text{ denominator}}$$

$$= \frac{\begin{vmatrix} -6 & -3 \\ 1 & -4 \end{vmatrix}}{59}$$

$$= \frac{(-6)(-4) - (1)(-3)}{59}$$

$$= \frac{24 + 3}{59}$$

$$= \frac{27}{59}$$

$$= 0.457 \text{ amp}$$

Note that these answers for I_1 and I_2 using determinants agree, of course, with the previous results. A good method of checking the results is to add up the voltages around the *outer* loop. This sum should equal zero. The voltage sum in the large outside loop of Fig. 1·2, starting at point A and going around this loop counterclockwise, is

$$E_{R_5} + E_2 + E_{R_4} + E_{R_2} + E_{R_3} + E_1$$

Observing the polarities of these voltages (IR products) as we move counterclockwise from point A gives

$$-(I_2R_5) + E_2 - (I_2R_4) - (I_1R_2) - (I_1R_3) + E_1$$

Substituting the values $I_1 = 0.576$ and $I_2 = 0.457$ in the above produces

$$-(0.457)(50) + 40 - (0.457)(40) - (0.576)(20) - (0.576)(30) + 30$$
$$-22.85 + 40 - 18.28 - 11.52 - 17.28 + 30$$
$$-69.93 + 70$$

For a perfect check, the negative and positive voltages should be equal, producing zero volts total. However, because of the dropping of digits beyond the third decimal place, a slight variation from zero is present. It is usually a satisfactory check of the answers if the difference between the negative sum voltage above (-69.93) and the positive one ($+70$), which is 0.07, is within 1 per cent of both the negative and the positive sum voltages. Since this is true in this case, the answers for I_1 and I_2 are correct.

This method of solution using loop currents is known as *Maxwell's cyclic currents.*

When setting up the equations for the various loops in this Maxwell cyclic-current method, two considerations must prevail. These are known as *Helmholtz's criteria* and are as follows: (*a*) Each equation must contain at least one *additional* circuit element not appearing in a previous equation, and (*b*) *every* circuit element must appear in at least one of the equations.

Three-loop network. The circuit of Fig. 1·3 consists of three separate loops, with currents I_1, I_2, and I_3 flowing, as shown by the solid-line, dashed-line, and dotted-line arrows respectively.

Example 1·4 In the circuit of Fig. 1·3, find the values of the currents I_1, I_2, and I_3 and also the voltage between points B and D.

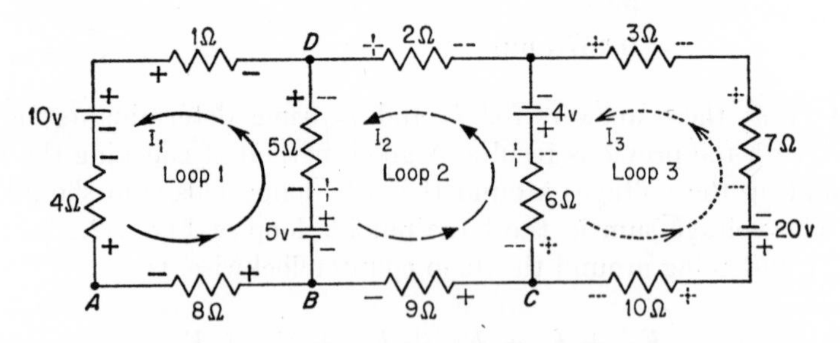

Figure **1·3** **Three-loop mesh circuit of Example 1·4.**

Solution. Assume current flow in each loop in any direction. If the assumed direction is incorrect, it will be indicated as such by a negative answer for that particular current. As shown, the loop currents are assumed to be, in this problem, all counterclockwise. Totaling the voltages in each loop, from Kirchhoff's voltage law, equals zero. Starting at point A in loop 1 and walking completely around this loop in the same direction as the assumed current, observing the polarities of the entering end of each voltage, gives

$$\textit{Loop } 1 \quad -8I_1 - 5 - 5I_1 + 5I_2 - 1I_1 + 10 - 4I_1 = 0$$

$$-18I_1 + 5I_2 = -5 \tag{1·23}$$

Starting at point B in loop 2 and repeating the procedure produces

$$\textit{Loop } 2 \quad -9I_2 - 6I_2 + 6I_3 + 4 - 2I_2 - 5I_2 + 5I_1 + 5 = 0$$

$$5I_1 - 22I_2 + 6I_3 = -9 \tag{1·24}$$

In loop 3, start at point C, and repeat the walk around, giving:

$$\textit{Loop } 3 \quad -10I_3 + 20 - 7I_3 - 3I_3 - 4 - 6I_3 + 6I_2 = 0$$

$$6I_2 - 26I_3 = -16 \tag{1·25}$$

Rewriting Eqs. (1·23) to (1·25) for convenience,

$$\begin{aligned} -18I_1 + 5I_2 \qquad\quad &= -5 & (1{\cdot}23) \\ 5I_1 - 22I_2 + 6I_3 &= -9 & (1{\cdot}24) \\ 6I_2 - 26I_3 &= -16 & (1{\cdot}25) \end{aligned}$$

Note that the coefficient of I_3 is zero in Eq. (1·23) and also zero for I_1 in Eq. (1·25).

Solving for I_1 by the use of determinants gives the following:

$$I_1 = \frac{\left|\begin{array}{rrr|rr} -5 & 5 & 0 & -5 & 5 \\ -9 & -22 & 6 & -9 & -22 \\ -16 & 6 & -26 & -16 & 6 \end{array}\right|}{\left|\begin{array}{rrr|rr} -18 & 5 & 0 & -18 & 5 \\ 5 & -22 & 6 & 5 & -22 \\ 0 & 6 & -26 & 0 & 6 \end{array}\right|}$$

$$= \frac{(-5)(-22)(-26) + (5)(6)(-16) + (0)(-9)(6) - (-16)(-22)(0) - (6)(6)(-5) - (-26)(-9)(5)}{(-18)(-22)(-26) + (5)(6)(0) + (0)(5)(6) - (0)(-22)(0) - (6)(6)(-18) - (-26)(5)(5)}$$

$$= \frac{-2860 + (-480) + 0 - 0 - (-180) - (1170)}{-10296 + 0 + 0 - 0 - (-648) - (-650)}$$

$$= \frac{-2860 - 480 + 180 - 1170}{-10296 + 648 + 650}$$

$$= \frac{-4330}{-8998}$$

$$= 0.481 \text{ amp}$$

I_2 could now be found by substituting the I_1 value in Eq. (1·23). However, I_2 will be solved here by using determinants again, producing

$$I_2 = \frac{\left|\begin{array}{ccc|cc} -18 & -5 & 0 & -18 & -5 \\ 5 & -9 & 6 & 5 & -9 \\ 0 & -16 & -26 & 0 & -16 \end{array}\right|}{\text{same denominator as for } I_1,\ -8998}$$

$$= \frac{(-18)(-9)(-26) + (-5)(6)(0) + (0)(5)(-16) - (0)(-9)(0) - (-16)(6)(-18) - (-26)(5)(-5)}{-8998}$$

$$= \frac{-4212 + 0 + 0 - 0 - 1728 - 650}{-8998}$$

$$= \frac{-6590}{-8998}$$

$$= 0.732 \text{ amp}$$

I_3 could now be found by substituting the I_2 value in Eq. (1·25), or by substituting both I_1 and I_2 values in Eq. (1·24). However, I_3 will be solved here by using determinants. This produces

$$I_3 = \frac{\left|\begin{array}{ccc|cc} -18 & 5 & -5 & -18 & 5 \\ 5 & -22 & -9 & 5 & -22 \\ 0 & 6 & -16 & 0 & 6 \end{array}\right|}{\text{same denominator as for } I_1,\ -8998}$$

$$= \frac{(-18)(-22)(-16) + (5)(-9)(0) + (-5)(5)(6) - (0)(-22)(-5) - (6)(-9)(-18) - (-16)(5)(5)}{-8998}$$

$$= \frac{-6336 + 0 + (-150) - 0 - (972) - (-400)}{-8998}$$

$$= \frac{-6336 - 150 - 972 + 400}{-8998}$$

$$= \frac{-7058}{-8998}$$

$$= 0.784 \text{ amp}$$

Note that the values found, $I_1 = 0.481$, $I_2 = 0.732$, and $I_3 = 0.784$, are all positive. This means that these currents are in the directions that were assumed, as shown in Fig. 1·3. The last part of Example 1·4 is to find the voltage between points B and D. This includes the 5 volts applied d-c and the voltage across the 5-ohm resistor. As shown in Fig.

1·4a, I_1 flows *up* through the resistor with a voltage polarity indicated by the solid-line signs. I_2 flows *down* through this resistor with the opposite polarity voltage across the resistor, as indicated by the dashed-line signs. Since the downward current I_2 is larger than upward I_1, the net current through the 5-ohm resistor is downward and is the difference between the two. Voltage across the resistor is then

$$\begin{aligned} E &= I_{\text{net}}R \\ &= (I_2 - I_1)R \\ &= (0.732 - 0.481)5 \\ &= (0.251)5 \\ &= 1.255 \text{ volts} \end{aligned}$$

Since the net current is downward through this resistor, then the 1.255 volts across the resistor has the polarity as indicated by the dashed-line

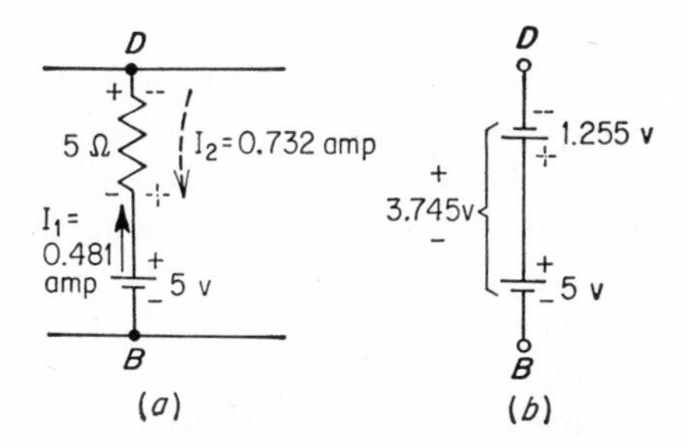

Figure 1·4 Voltage between two points, part of a mesh circuit.

signs of Fig. 1·4a and b, negative at the top. This 1.255 volts is opposite in polarity to the 5-volt battery, as shown in Fig. 1·4b. The voltage between points B and D is simply the difference between 5 and 1.255 volts, or 3.745 volts, with point D the positive side and point B the negative end with respect to each other, as shown in Fig. 1·4b.

A check of the answers of Example 1·4 involves taking the sum of all voltages around the large outer loop of Fig. 1·3. Starting at point A and going around this outside loop in a counterclockwise direction, the voltages are

$$E_{8\text{ ohms}} + E_{9\text{ ohms}} + E_{10\text{ ohms}} + 20 \text{ volts} + E_{7\text{ ohms}} + E_{3\text{ ohms}} + E_{2\text{ ohms}} + E_{1\text{ ohms}} + 10 \text{ volts} + E_{4\text{ ohms}}$$

Taking the polarities into account, as shown in Fig. 1·3, these voltages are

$$-8I_1 + (-9I_2) + (-10I_3) + 20 + (-7I_3) + (-3I_3) + (-2I_2) + (-1I_1) + 10 + (-4I_1)$$

Substituting the values of $I_1(0.481)$, $I_2(0.732)$, and $I_3(0.784)$ gives the following:

$$\begin{aligned}
-(8)(0.481) &- (9)(0.732) - (10)(0.784) + 20 - (7)(0.784) \\
&- (3)(0.784) - (2)(0.732) - (1)(0.481) + 10 - (4)(0.481) \\
&= -3.848 - 6.588 - 7.84 + 20 - 5.488 - 2.352 - 1.464 \\
&\qquad - 0.481 + 10 - 1.924 \\
&= -29.985 + 30
\end{aligned}$$

For a perfect check, the negative and positive voltages should be equal, producing zero volts total. However, due to the dropping of digits beyond the third decimal place, a slight variation is usually present. It is a satisfactory check of the answers if the difference between the sum of the negative voltages above (-29.985) and the positive ones ($+30$), which is 0.015, is within 1 per cent of both the negative and positive sums. Since this is true in this proof of Example 1·4, the answers for I_1, I_2, and I_3 are correct.

Superposition. Another method of solving mesh circuits is known as *superposition.* This involves finding the various currents due to *one* voltage source at a time, with all *other* applied voltages replaced with their internal resistances only. Then, the total currents through the various resistors are simply the sum of the currents due to *each* of the applied voltages. In the simple two-loop mesh circuit of Fig. 1·5*a*, the currents through the various resistors will be found first because of the applied voltage E_1, while E_2 is shorted across. Then the currents due to E_2 alone will be found, while E_1 is short-circuited. Finally, the currents through each resistor, because of *both* of these applied voltages, are found by *adding* or *subtracting* the currents, depending upon whether these currents are in the same or opposing directions through each component.

Example 1·5 In the circuit of Fig. 1·5*a*, find the current through each resistor, the direction of each current, and the voltage between points A and B using the *superposition* method.

Solution. Short out E_2, and redraw the circuit as shown in Fig. 1·5*b*. R_3 and R_2 are now in parallel, and this combination is in series with R_1. The parallel circuit resistance between points A and B as shown in Fig. 1·5*b* and *c* is

$$\begin{aligned}
R'_{A-B} &= \frac{R_3R_2}{R_3 + R_2} \\
&= \frac{6(3)}{6 + 3} \\
&= 2 \text{ ohms}
\end{aligned}$$

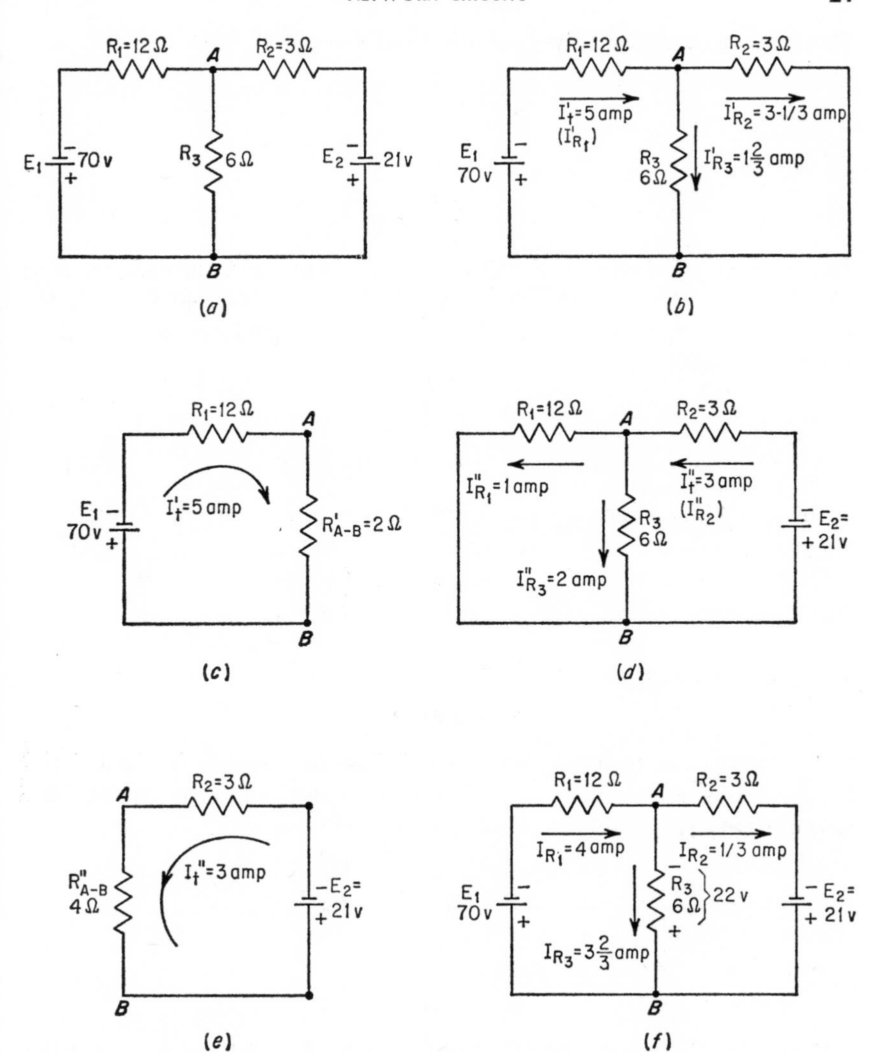

Figure 1·5 Simple two-loop mesh circuit and superposition of Example 1·5. (a) Original circuit; (b) circuit with E_2 shorted out; (c) simplified circuit with E_2 shorted out; (d) circuit with E_1 shorted out; (e) simplified circuit with E_1 shorted out; (f) answers to Example 1·5.

The total resistance R'_T that applied voltage E_1 "sees" is the series circuit R_1 and R'_{A-B}, as shown in Fig. 1·5*c*, and is

$$\begin{aligned} R'_T &= R_1 + R'_{A-B} \\ &= 12 + 2 \\ &= 14 \text{ ohms} \end{aligned}$$

The total current flow, due to E_1, is I'_T and is

$$I'_T = \frac{E_1}{R'_T} = \frac{70}{14} = 5 \text{ amp}$$

This current, as shown in Fig. 1·5*b*, flows left to right through R_1, and *down* through R_3, and also left to right through R_2. Current through R_1, due to E_1, I'_{R1}, is the same as I'_T, 5 amp. Voltage between points A and B, due to E_1, called E'_{A-B} is

$$E'_{A-B} = I'_{R_1}R'_{A-B} = 5(2) = 10 \text{ volts}$$

With 10 volts between points A and B, due to E_1, current through R_2, called I'_{R_2} is

$$I'_{R_2} = \frac{E'_{A-B}}{R_2} \quad \text{or} \quad \frac{E'_{R_2}}{R_2} = \frac{10}{3} = 3\tfrac{1}{3} \text{ amp}$$

The difference between the total current (I'_T), due to E_1, of 5 amp and the $3\frac{1}{3}$ amp flowing through R_2 is $1\frac{2}{3}$ amp and must be flowing through R_3. This current, I'_{R_3}, could also have been found from

$$I'_{R_3} = \frac{E'_{A-B}}{R_3} \quad \text{or} \quad \frac{E'_{R_3}}{R_3} = \frac{10}{6} = 1\tfrac{2}{3} \text{ amp}$$

Now, after having found the currents in each resistor due to E_1, the process must be repeated, this time shorting out E_1 and using E_2 (Fig. 1·5*a*) as the only applied source voltage. With E_1 shorted, the circuit appears as shown in Fig. 1·5*d*. E_2 "sees" R_2 in series with the A-to-B parallel combination. The parallel circuit resistance between points A and B is R''_{A-B} and is now

$$R''_{A-B} = \frac{R_1R_3}{R_1 + R_3} = \frac{12(6)}{12 + 6} = 4 \text{ ohms}$$

The total resistance R''_T that E_2 "sees" is R_2 added to R''_{A-B}, as shown in Fig. 1·5*e*, and is

$$\begin{aligned} R''_T &= R_2 + R''_{A-B} \\ &= 3 + 4 \\ &= 7 \text{ ohms} \end{aligned}$$

The total current flow, due to E_2, is I''_T and is

$$\begin{aligned} I''_T &= \frac{E_2}{R''_T} \\ &= \frac{21}{7} \\ &= 3 \text{ amp} \end{aligned}$$

This total current due to E_2 flows through R_2, and $I''_T = I''_{R_2} = 3$ amp. This current, as shown in Fig. 1·5*d*, flows right to left through R_2 (opposite to the current through R_2 due to E_1), down through R_3 (same as the current due to E_1), and also right to left through R_1 (opposite to the current due to E_1). Voltage between points A and B, due to E_2, called E''_{A-B} is

$$\begin{aligned} E''_{A-B} &= I''_T R''_{A-B} \\ &= 3(4) \\ &= 12 \text{ volts} \end{aligned}$$

With 12 volts between points A and B, due to E_2, current through R_1, called I''_{R_1}, is

$$\begin{aligned} I''_{R_1} &= \frac{E''_{A-B}}{R_1} \quad \text{or} \quad \frac{E''_{R_1}}{R_1} \\ &= \frac{12}{12} \\ &= 1 \text{ amp} \end{aligned}$$

The difference between the total current (I''_T), due to E_2, of 3 amp, and the 1 amp (I''_{R_1}) flowing through R_1 is 2 amp and must be flowing through R_3. This current, I''_{R_3}, could also have been found as follows:

$$\begin{aligned} I''_{R_3} &= \frac{E''_{A-B}}{R_3} \quad \text{or} \quad \frac{E''_{R_3}}{R_3} \\ &= \frac{12}{6} \\ &= 2 \text{ amp} \end{aligned}$$

After finding these currents and their directions through each resistor due to E_1 and E_2 *alone*, the *total* currents due to *both* E_1 and E_2 can now be determined as follows.

Current through R_1 due to E_1 is $I'_{R1} = 5$ amp and flows, as shown in Fig. 1·5*b*, left to right. Current through this same R_1 due to E_2 is $I''_{R1} = 1$ amp and flows, as shown in Fig. 1·5*d*, right to left. The total current through R_1, I_{R_1}, due to both E_1 and E_2, is therefore the difference between these bucking currents and is

$$\begin{aligned} I_{R_1} &= I'_{R1} - I''_{R1} \\ &= 5 - 1 \\ &= 4 \text{ amp} \end{aligned}$$

This 4-amp I_{R_1} flows through R_1 left to right, in the same direction as the larger current, I'_{R1}.

Current through R_2 due to E_1 is $I'_{R_2} = 3\frac{1}{3}$ amp and flows, as shown in Fig. 1·5*b*, left to right. At the same time, current through this same resistor R_2 due to E_2 is $I''_{R_2} = 3$ amp and flows, as shown in Fig. 1·5*d*, right to left. Total current through R_2, I_{R_2}, due to both E_1 and E_2, is therefore the *difference* between these bucking currents and is

$$\begin{aligned} I_{R_2} &= I'_{R2} - I''_{R2} \\ &= 3\tfrac{1}{3} - 3 \\ &= \tfrac{1}{3} \text{ amp} \end{aligned}$$

This $\frac{1}{3}$-amp I_{R_2} flows left to right through R_2 in the same direction as the larger current I'_{R2}.

Current through R_3 due to E_1 is $I'_{R_3} = 1\frac{2}{3}$ amp and flows down, as shown in Fig. 1·5*b*. Current through this same resistor, R_3, because of E_2 is $I''_{R_3} = 2$ amp and also flows down, as shown in Fig. 1·5*d*. Total current through R_3, I_{R_3}, due to both E_1 and E_2, is therefore the sum of these aiding (same direction) currents and is

$$\begin{aligned} I_{R_3} &= I'_{R3} + I''_{R3} \\ &= 1\tfrac{2}{3} + 2 \\ &= 3\tfrac{2}{3} \text{ amp} \end{aligned}$$

To find the *total* voltage between points A and B (as requested in the last part of Example 1·5),

$$\begin{aligned} E_{A-B} &= I_{A-B}R_{A-B} \\ &= I_{R_3}R_3 \\ &= 3\tfrac{2}{3}(6) \\ &= 22 \text{ volts} \end{aligned}$$

Since the current I_{R_3} flows down through R_3, point A (top of R_3) is negative with respect to point B (lower end of R_3). This total voltage between A and B of 22 volts could also have been found by simply adding the voltage between these points due to E_1, E'_{A-B}, to that because of E_2, E''_{A-B}:

$$\begin{aligned} E_{A-B} &= E'_{A-B} + E''_{A-B} \\ &= 10 + 12 \\ &= 22 \text{ volts} \end{aligned}$$

The answers to Example 1·5 are shown in the diagram of Fig. 1·5*f*.

Nodal currents. A method of solution for mesh circuits which sometimes requires fewer equations than the Maxwell's cyclic-current method (loop currents) is called the *nodal-current* method. The diagram in Fig. 1·6 is the same circuit as Fig. 1·5*a*. The following example and solution will use the nodal-current method.

Example 1·6 In the circuit of Fig. 1·6 (same as Fig. 1·5*a*) find the current magnitudes and directions through each resistor, and also find the voltage between points A and B using the nodal-current method.

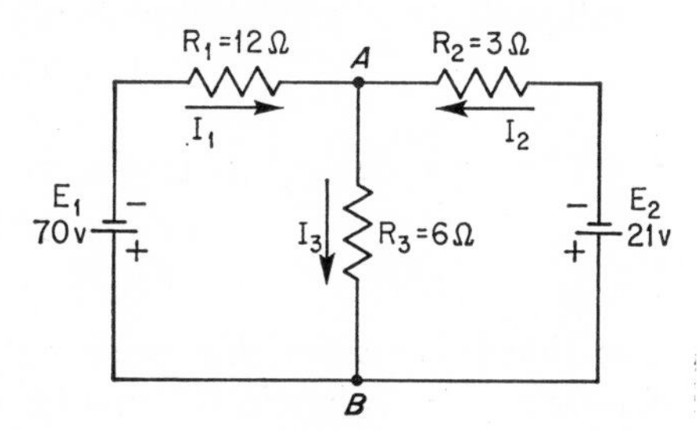

Figure 1·6 Simple two-loop mesh, nodal currents of Example 1·6. (I_1, I_2, and I_3 directions are assumed.)

Solution. If the loop-current method were used, two equations would be required to solve Fig. 1·6. In the nodal-current system, only one equation will be required for this circuit as will be shown. Assuming a current flow, I_1, from the negative terminal of E_1, this current flows left to right through R_1 to point A as indicated by the arrow of Fig. 1·6. Also assume a current flow, I_2, from the negative terminal of E_2 right to left through R_2 to point A, also indicated in Fig. 1·6. At point A, these two currents join and flow down through R_3, the combined current being I_3.

Kirchhoff's *current* law states that the sum of all currents flowing into a point is equal to the sum of all currents flowing away from that point. I_1 and I_2 flow into point A, while I_3 flows away from this point. Therefore,

$$I_3 = I_1 + I_2 \qquad (1·26)$$

From Kirchhoff's *voltage* law,

$$E_1 = E_{R_1} + E_{R_3} \qquad (1·27)$$

$$E_1 - E_{R_3} = E_{R_1} \qquad (1·28)$$

and also from Kirchhoff's *voltage* law,

$$E_2 = E_{R_2} + E_{R_3} \tag{1·29}$$

$$E_2 - E_{R_3} = E_{R_2} \tag{1·30}$$

From Eq. (1·26) and Ohm's law ($I = E/R$),

$$\frac{E_{R_3}}{R_3} = \frac{E_{R_1}}{R_1} + \frac{E_{R_2}}{R_2} \tag{1·31}$$

Substituting for E_{R_1} [from Eq. (1·28)] and for E_{R_2} [from Eq. (1·30)] in Eq. (1·31) gives

$$\frac{E_{R_3}}{R_3} = \frac{E_1 - E_{R_3}}{R_1} + \frac{E_2 - E_{R_3}}{R_2} \tag{1·32}$$

From the values given in Fig. 1·6, substitute in Eq. (1·32), giving

$$\frac{E_{R_3}}{6} = \frac{70 - E_{R_3}}{12} + \frac{21 - E_{R_3}}{3} \tag{1·33}$$

Multiply by 12 to clear fractions from Eq. (1·33), giving

$$\begin{aligned} 2E_{R_3} &= (70 - E_{R_3}) + 4(21 - E_{R_3}) \\ &= 70 - E_{R_3} + 84 - 4E_{R_3} \\ 2E_{R_3} + E_{R_3} + 4E_{R_3} &= 70 + 84 \\ 7E_{R_3} &= 154 \\ E_{R_3} &= 22 \text{ volts} \end{aligned}$$

Note that this answer, $E_{R_3} = 22$ volts, agrees, of course, with the result of Example 1·5, wherein the voltage drop across R_{A-B} or R_3 is 22 volts.

The various currents are found as follows:

$$I_1 = \frac{E_{R_1}}{R_1}$$

Substitute for E_{R_1} from Eq. (1·28):

$$\begin{aligned} I_1 &= \frac{E_1 - E_{R_3}}{R_1} \\ &= \frac{70 - 22}{12} \\ &= \frac{48}{12} \\ &= 4 \text{ amp} \end{aligned}$$

Note that this result of $I_1 = 4$ amp agrees with the current flowing through R_1, I_{R_1}, found in Example 1·5 and shown in Fig. 1·5*f*.

$$I_2 = \frac{E_{R_2}}{R_2}$$

Substituting for E_{R_3} from Eq. (1·30) gives

$$I_2 = \frac{E_2 - E_{R_3}}{R_2} = \frac{21 - 22}{3} = \frac{-1}{3} = -\tfrac{1}{3} \text{ amp}$$

The negative sign for I_2 simply indicates that the assumed direction of I_2, shown in Fig. 1·6 as going *right to left* through R_2, is really just the reverse, actually going *left to right* through R_2. The current magnitude is ⅓ amp, and the *left-to-right direction* agrees with the result found in Example 1·5 and shown in Fig. 1·5*f*.

$$I_3 = \frac{E_{R_3}}{R_3} = \frac{22}{6} = 3\tfrac{2}{3} \text{ amp}$$

This result of $I_3 = 3\tfrac{2}{3}$ amp also agrees with the answer for the current through $R_3(I_{R_3})$ found in Example 1·5 and shown in Fig. 1·5*f*.

Thévenin theorem. A common method used to find the voltage and current between two points in a mesh circuit makes use of *Thévenin's theorem*. This theorem, in effect, states that between any two points in a mesh circuit there exists an *equivalent* voltage and resistance (impedance). The voltage is that which a voltmeter would read, while the resistance is that indicated by an ohmmeter if all the applied voltages in the circuit were shorted out. The following example illustrates this method.

Example 1·7 In the simple network of Fig. 1·7*a* (the same used in Examples 1·5 and 1·6) find the voltage and current between points *A* and *B*.

Solution. First open the circuit just to the left of point *A*, as shown in Fig. 1·7*b*, disconnecting R_1 and E_1. Solve for the resistance, R'_{A-B} of this partial circuit between points *A* and *B* with E_2 shorted out. This places R_3 and R_2 in parallel.

$$R'_{A-B} = \frac{R_3R_2}{R_3 + R_2} = \frac{6(3)}{6 + 3} = 2 \text{ ohms}$$

Solve for the voltage, E'_{A-B}, of this partial circuit consisting of R_2, R_3, and E_2 as a simple series circuit.

$$E'_{A-B} = \frac{R_3}{R_3 + R_2} E_2$$

$$= \frac{6}{6+3} \times 21$$

$$= \frac{6}{9} \times 21$$

$$= 14 \text{ volts}$$

In this partial circuit of Fig. 1·7*b*, current due to E_2 flows down through R_3, making point A (the top of R_3) *negative* with respect to point B (the

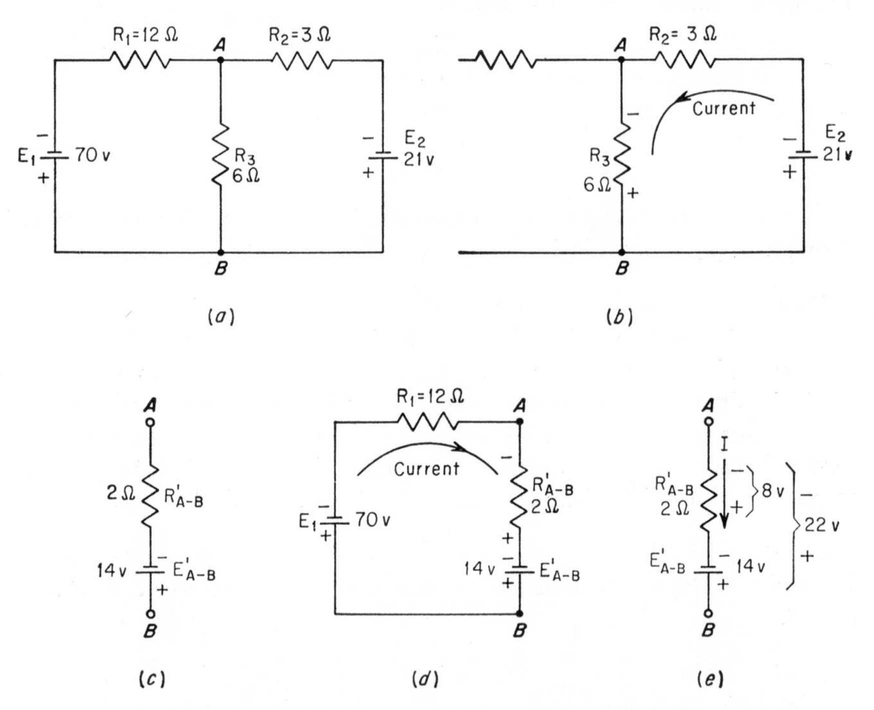

Figure 1·7 **Simple two-loop mesh, Thévenin equivalent method of Example 1·7.**

lower end of R_3). Therefore, between these two terminals there exists an equivalent 2 ohms (R'_{A-B}) and 14 volts (E'_{A-B}), as shown in Fig. 1·7*c*.

Now complete the remainder of the circuit by reconnecting R_1 and E_1 as shown in Fig. 1·7*d*. The circuit is now a simple series circuit consisting of R_1, R'_{A-B}, E'_{A-B}, and E_1. Since E_1 and E'_{A-B} are series opposing, the net applied voltage in this circuit is the difference between the two.

$$\begin{aligned} E_{\text{net}} &= E_1 - E'_{A-B} \\ &= 70 - 14 \\ &= 56 \text{ volts} \end{aligned}$$

Since E_1 is larger than E'_{A-B}, current flows down through R'_{A-B}, as shown in Fig. 1·7*d* with the polarity across the resistor as shown, negative at the top. Voltage across R'_{A-B} is

$$\begin{aligned} E'_{RA-B} &= \frac{R'_{A-B}}{R'_{A-B} + R_1} E_{\text{net}} \\ &= \frac{2}{2 + 12} \times 56 \\ &= \frac{2}{14} \times 56 \\ &= 8 \text{ volts} \end{aligned}$$

The total voltage between points A and B, E_{A-B}, as shown in Fig. 1·7*e*, is the sum of E'_{RA-B} and E'_{A-B}, or

$$\begin{aligned} E_{A-B} &= E'_{RA-B} + E'_{A-B} \\ &= 8 + 14 \\ &= 22 \text{ volts} \end{aligned}$$

This result of 22 volts between A and B agrees with that found previously in Example 1·6.

Current through R_3 from point A to point B is simply

$$\begin{aligned} I_{A-B} &= \frac{E_{A-B}}{R_3} \\ &= \frac{22}{6} \\ &= 3\tfrac{2}{3} \text{ amp} \end{aligned}$$

This result also agrees with the answer in Example 1·6. The other currents, although not requested in Example 1·7, could be found also. Current through R_1 in Fig. 1·7*a* could be found by referring to Fig. 1·7*d*. In this simple series circuit, the net applied voltage is 56 volts, as determined previously. Current is therefore

$$\begin{aligned} I &= \frac{E_{\text{net}}}{R_1 + R'_{A-B}} \\ &= \frac{56}{12 + 2} \\ &= 4 \text{ amp} \end{aligned}$$

With 4 amp flowing through R_1 *to* point A in Fig. 1·7*a* and $3\tfrac{2}{3}$ amp flowing *away* from point A down through R_3 (found previously), the remainder ($4 - 3\frac{2}{3}$) of $\frac{1}{3}$ amp flows *away* from point A through R_2. These values of currents agree with those found previously in Example 1·6.

Incidentally, the equivalent resistance between points A and B of Fig. 1·7*a*, such as an ohmmeter would read with E_1 and E_2 short circuited, consists of R_1, R_3, and R_2 in parallel. The total *equivalent* resistance between A and B is then

$$\begin{aligned} R_{A-B_{\text{equiv}}} &= \frac{1}{\dfrac{1}{R_1} + \dfrac{1}{R_3} + \dfrac{1}{R_2}} \\ &= \frac{1}{\dfrac{1}{12} + \dfrac{1}{6} + \dfrac{1}{3}} \\ &= \frac{1}{\dfrac{1 + 2 + 4}{12}} \\ &= \frac{1}{\dfrac{7}{12}} \\ &= \frac{12}{7} \\ &= 1\tfrac{5}{7} \text{ ohms} \end{aligned}$$

A three-loop mesh circuit using the Thévenin equivalent method of solution is shown in the following example.

Example 1·8 In the circuit of Fig. 1·8*a*, find the voltage between points D and B.

Solution. This circuit is the same as that shown in Fig. 1·3, Example 1·4. First open the circuit just to the left of point Y, as shown in Fig. 1·8*b*, and solve for the following in the right-hand loop: The *equivalent* resistance between points Y and C, R'_{Y-C}, with the 4- and 20-volt sources shorted, consists of the 6-ohm resistor in parallel with the series resistors 3, 7, and 10 ohms.

$$\begin{aligned} R'_{Y-C} &= \frac{6(3 + 7 + 10)}{6 + (3 + 7 + 10)} \\ &= \frac{6(20)}{6 + 20} \\ &= \frac{120}{26} \\ &= 4.62 \text{ ohms} \end{aligned}$$

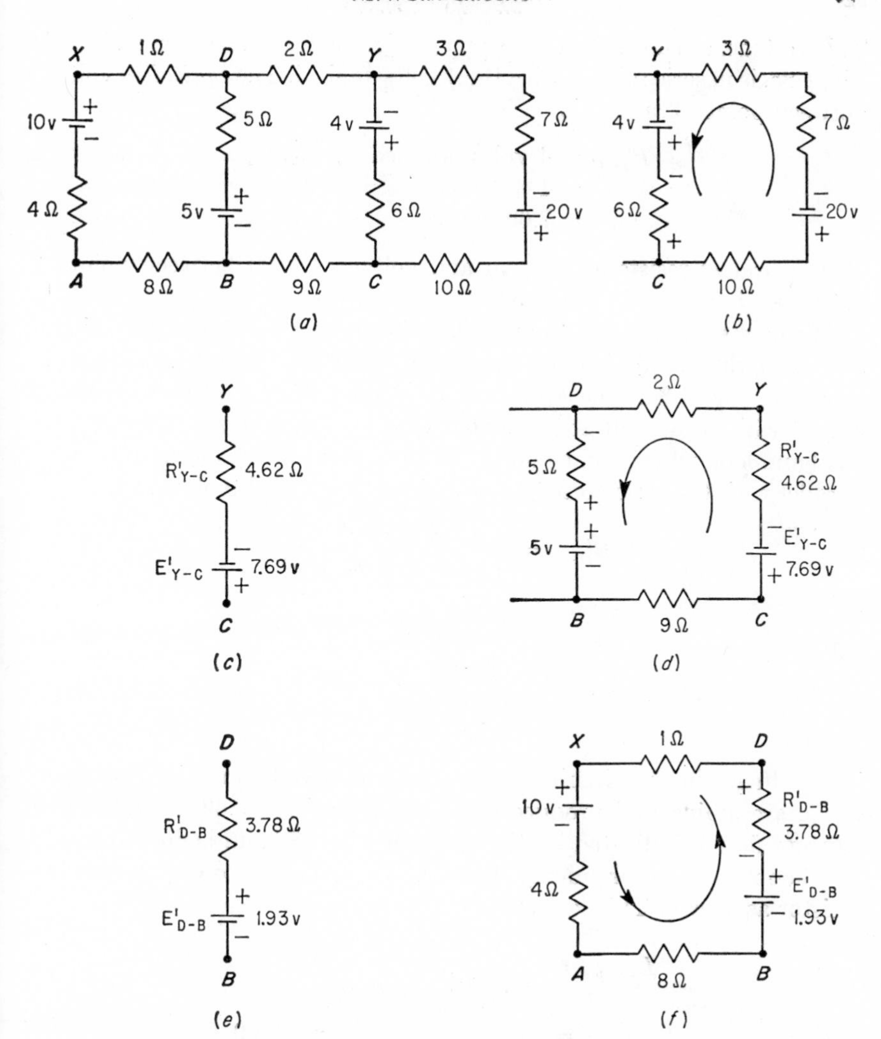

Figure 1·8 Three-loop mesh, Thévenin equivalent method of Example 1-8.

Current flow, as indicated by the arrow in Fig. 1·8*b*, is down through the 6-ohm resistor, with the polarities as shown. The 4 and 20 volts applied are opposing, giving a *net* applied voltage of 16 volts. Voltage across the 6-ohm resistor is

$$E_{6\,\text{ohms}} = \frac{6}{6 + 10 + 7 + 3} E_{\text{net}}$$

$$= \frac{6}{26} \times 16$$

$$= 3.69 \text{ volts}$$

This voltage, as shown by the − and + signs across the 6-ohm resistor in Fig. 1·8*b*, is in series aiding with the 4 volts applied. The *equivalent* voltage between points Y and C is

$$\begin{aligned} E'_{Y-C} &= 4 \text{ volts (applied)} + E_{6\,\text{ohms}} \\ &= 4 + 3.69 \\ &= 7.69 \text{ volts} \end{aligned}$$

The *equivalent* resistance (R'_{Y-C}) and voltage (E'_{Y-C}) between points Y and C are indicated in Fig. 1·8*c*.

The next step is to add the middle loop to the Y-to-C *equivalent* circuit and open the circuit just to the left of point D. The circuit is now as shown in Fig. 1·8*d*. The *equivalent* resistance, R'_{D-B}, between points D and B with the 5-volt source and E'_{Y-C} shorted (but *not* including the left-hand loop which will be added later) consists of the 5-ohm resistor in parallel with the series resistors of 2 ohms, R'_{Y-C}, and 9 ohms.

$$\begin{aligned} R'_{D-B} &= \frac{5(2 + 4.62 + 9)}{5 + (2 + 4.62 + 9)} \\ &= \frac{5(15.62)}{5 + 15.62} \\ &= \frac{78.1}{20.62} \\ &= 3.78 \text{ ohms} \end{aligned}$$

The voltages in the circuit of Fig. 1·8*d*, the 5 volts and the 7.69 volts, E'_{Y-C}, are aiding and produce a *net* voltage of $5 + 7.69 = 12.69$ volts. Current flow is indicated by the arrow of Fig. 1·8*d* and is down through the 5-ohm resistor, producing the − and + polarity voltage across this resistor as shown. This voltage is

$$\begin{aligned} E_{5\,\text{ohms}} &= \frac{5}{5 + 9 + R'_{Y-C} + 2} E_{\text{net}} \\ &= \frac{5}{5 + 9 + 4.62 + 2} \times 12.69 \\ &= \frac{5}{20.62} \times 12.69 \\ &= 3.07 \text{ volts} \end{aligned}$$

As shown in Fig. 1·8*d*, this voltage is opposing the 5 volts applied, and the equivalent voltage, E'_{D-B} (not including the left-hand loop to be added later), is the difference of the two:

$$\begin{aligned} E'_{D-B} &= 5 - E_{5\,\text{ohms}} \\ &= 5 - 3.07 \\ &= 1.93 \text{ volts} \end{aligned}$$

This *temporary* (since it does not include the last loop yet) equivalent voltage (E'_{D-B}) and resistance (R'_{D-B}) are shown in Fig. 1·8*e*.

The last procedure is to add the left-hand loop of Fig. 1·8*a* to the *D*-to-*B* equivalent circuit. This is shown in Fig. 1·8*f*. The applied 10 volts is opposing the E'_{D-B} of 1.93 volts, producing a *net* voltage of 8.07 volts, and current flows, as indicated by the arrow, up through the 3.78-ohm resistor, R'_{D-B}. The voltage across this resistor is

$$\begin{aligned} E'_{RD-B} &= \frac{R'_{D-B}}{R'_{D-B} + 1 + 4 + 8} E_{\text{net}} \\ &= \frac{3.78}{3.78 + 1 + 4 + 8} \times 8.07 \\ &= \frac{3.78}{16.78} \times 8.07 \\ &= 1.82 \text{ volts} \end{aligned}$$

As indicated by the $-$ and $+$ voltage polarities in Fig. 1·8*f*, E'_{RD-B} is aiding E'_{D-B}, and the *total* equivalent voltage (E_{D-B}) between points *D* and *B* is

$$\begin{aligned} E_{D-B} &= E'_{RD-B} + E'_{D-B} \\ &= 1.82 + 1.93 \\ &= 3.75 \text{ volts} \end{aligned}$$

This final result checks within slide-rule tolerance with the voltage between points *D* and *B* found in Example 1·4.

The *total* equivalent resistance between points *D* and *B* (not asked for in Example 1·8, however) consists of R'_{D-B} in parallel with the series resistors 1, 4, and 8 ohms and is

$$\begin{aligned} R_{D-B_{\text{equiv}}} &= \frac{3.78(1 + 4 + 8)}{3.78 + (1 + 4 + 8)} \\ &= \frac{3.78(13)}{16.78} \\ &= \frac{49.1}{16.78} \\ &= 2.93 \text{ ohms} \end{aligned}$$

1·5 More Complex Network Circuits

The mesh circuits shown previously were, of course, more complicated than simple series-parallel circuits. The networks of Fig. 1·9*a*, *b*, and *c* are, similarly, more difficult than the previous ones of Sec. 1·4. Circuits such as those of Fig. 1·9 can best be solved by changing some sections of the circuits into their equivalents which then allows the circuit to be redrawn into a simple series-parallel circuit. The groups or sections to be changed and the method of transformation are explained in the following discussion.

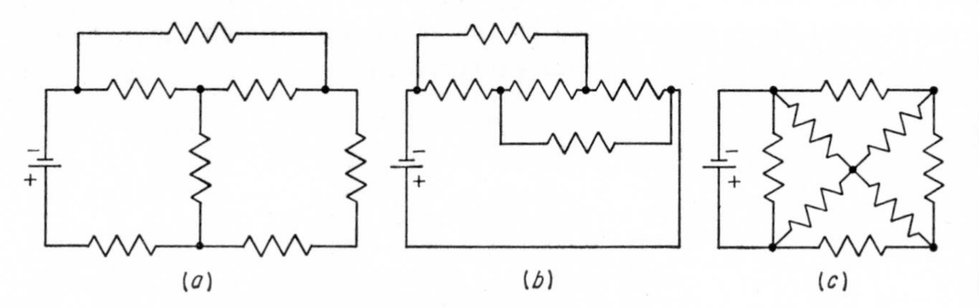

Figure 1·9 More complex mesh circuits.

Delta or Pi circuit transformed to wye or tee. The circuit of Fig. 1·10*a* illustrates resistors in a delta (Δ) arrangement. The name *delta* describes the *shape* of the circuit drawing, with R_1, R_2, and R_3 acting as the sides of the triangular-shaped Greek letter *delta*. Note that there are *three* junction points, A, B, and C. The circuit of Fig. 1·10*b* is exactly the same diagram except that the lower ends of R_2 and R_3 are connected to a common wire instead of the common junction point C. This, of course, does

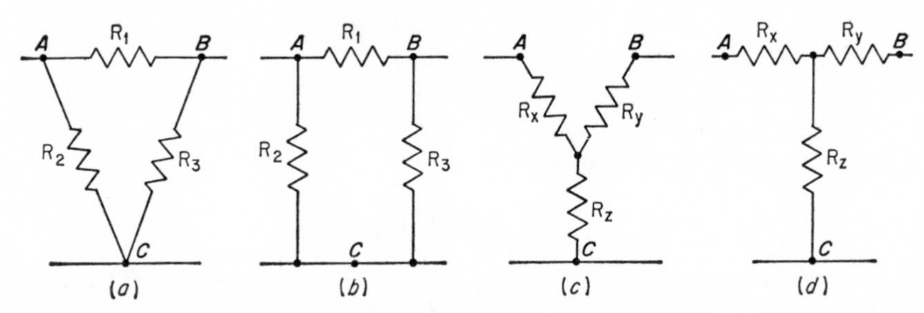

Figure 1·10 Delta, pi, wye, and tee circuits. (a) Δ arrangement; (b) π arrangement; (c) Y arrangement; (d) T arrangement.

not change the circuit electrically. The arrangement in Fig. 1·10*b* simply simulates the shape of the Greek letter *pi* (π). Circuits (*a*) and (*b*) are identical electrically.

In the diagram of Fig. 1·10*c*, the three resistors R_X, R_Y, and R_Z simulate the shape of the letter Y, while in Fig. 1·10*d*, the identical circuit electrically, the three resistors form the letter T.

In many complex mesh circuits, the solution may be simplified by changing a pi or delta circuit into an equivalent wye or tee. The circuit of

Fig. 1·11a is a simple series-parallel one and is used here only to show the validity of the delta-to-wye transformation equations.

Example 1·9 Find the *total* resistance between points A and B in Fig. 1·11a, using first the simple series-parallel circuit solution and then the transformation equations which are given later.

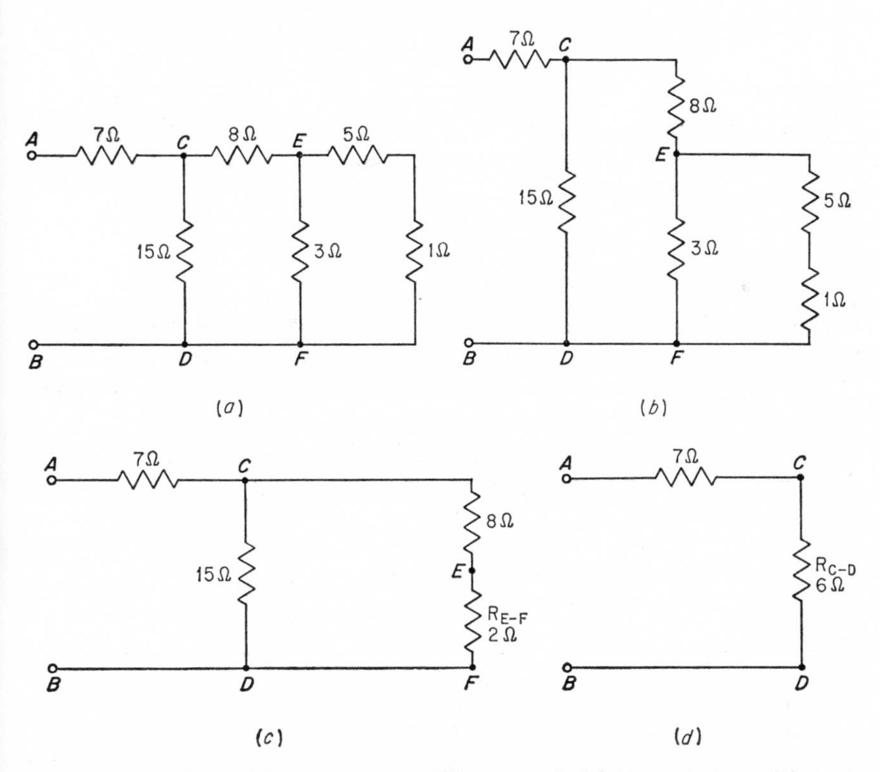

Figure 1·11 **Series-parallel combination of Example 1·9. (*a*) Original circuit; (*b*) circuit redrawn; (*c*) circuit simplified; (*d*) circuit simplified further.**

Solution 1. Using the simple series-parallel solution, the circuit is redrawn, as shown in Fig. 1·11b. Between points E and F, it can be readily seen that the 3-ohm resistor is in parallel with the series resistors, 5 ohms and 1 ohm. The resistance between E and F is

$$R_{E-F} = \frac{3(5+1)}{3+(5+1)} = \frac{18}{9} = 2 \text{ ohms}$$

Redrawing the circuit in more simplified form, as shown in Fig. 1·11*c*, it is apparent that between points C and D, the 15-ohm resistor is in parallel with the series circuit made up of the 8 ohms and the 2 ohms (R_{E-F}). The resistance between points C and D is

$$R_{C-D} = \frac{15(8 + 2)}{15 + (8 + 2)} = \frac{150}{25} = 6 \text{ ohms}$$

Finally, redrawing the circuit in its simplest form, the series circuit of Fig. 1·11*d* shows that the *total* resistance between points A and B consists of the 7-ohm resistor in series with the 6 ohms (R_{C-D}).

$$R_{\text{total}}, \text{ or } R_{A-B} = 7 + 6 = 13 \text{ ohms}$$

Before solving Example 1·9 for the total resistance using *delta-to-wye transformation*, the required equations for this method must be given. In Fig. 1·12*a*, a delta circuit is shown with the resistors, R_1, R_2, and R_3,

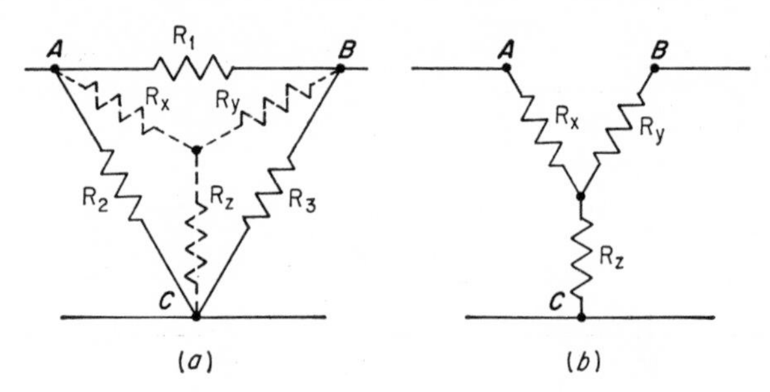

Figure 1·12 Δ and Y equivalent circuits. (*a*) Δ arrangement; (*b*) Y arrangement.

drawn with solid lines. The wye circuit, resistors R_X, R_Y, and R_Z, is shown using dotted-line components and also separately in Fig. 1·12*b*.

Resistor R_X takes the place of R_1 and R_2, and is equal to the *product* of the two (R_1 and R_2) divided by the *sum* of all three resistors of the delta (R_1, R_2, and R_3), or

$$R_X = \frac{R_1R_2}{R_1 + R_2 + R_3} \tag{1·34}$$

Similarly,

$$R_Y = \frac{R_1R_3}{R_1 + R_2 + R_3} \tag{1·35}$$

and

$$R_Z = \frac{R_2R_3}{R_1 + R_2 + R_3} \tag{1·36}$$

Note that the denominators are identical.

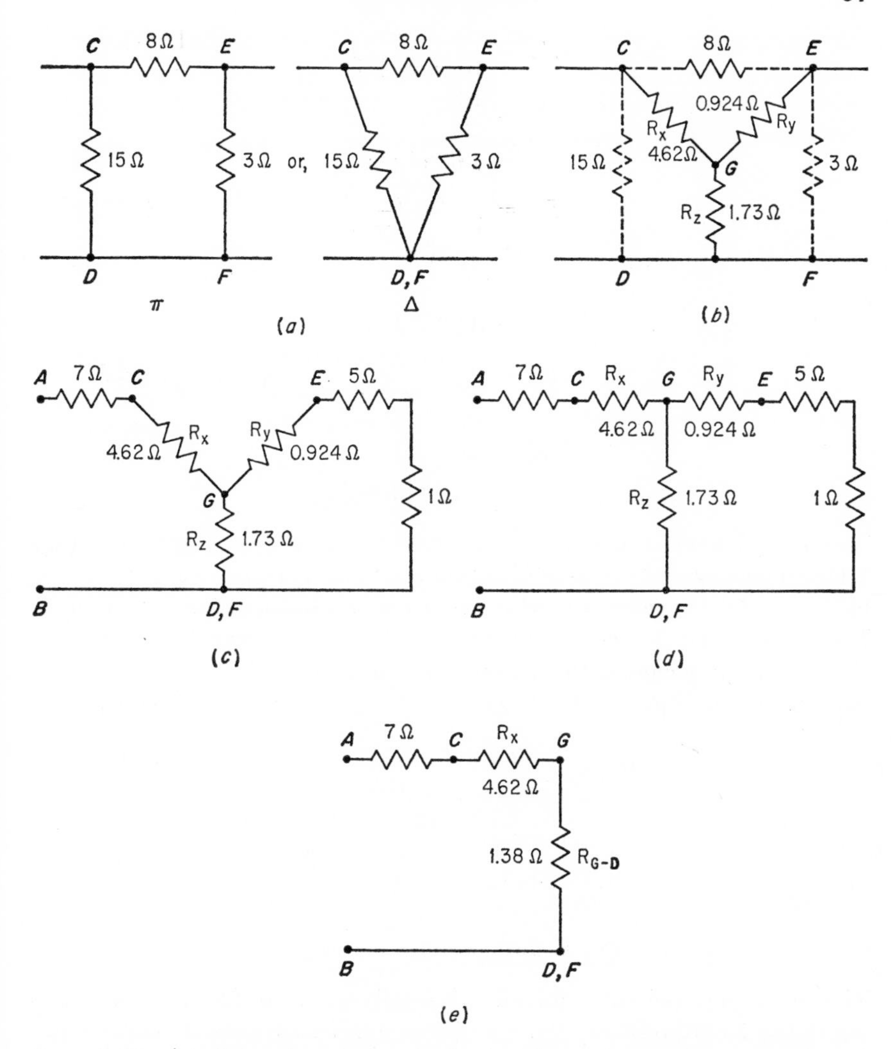

Figure 1·13 Δ-to-Y transformation method of solving series-parallel combination of Fig. 1·11*a*, Example 1·9. (*a*) The π or Δ section removed from the complete circuit; (*b*) π (or Δ) to T (or Y) transformation; (*c*) the T (or Y) section placed back in the complete circuit; (*d*) circuit redrawn; (*e*) circuit simplified.

Equations (1·34) to (1·36) are the delta-to-wye transformation formulas. Their validity will be checked in the following.

Solution 2. Referring to the original diagram of Fig. 1.11*a*, the 15-, 8-, and 3-ohm resistors comprise a pi or delta circuit and are shown separately in Fig. 1·13*a*. The wye equivalent circuit is shown by the solid-line resistors in Fig. 1·13*b*, and their values are

$$R_X = \frac{15(8)}{15 + 8 + 3} \tag{1·34}$$
$$= \frac{120}{26}$$
$$= 4.62 \text{ ohms}$$

$$R_Y = \frac{8(3)}{15 + 8 + 3} \tag{1·35}$$
$$= \frac{24}{26}$$
$$= 0.924 \text{ ohm}$$

$$R_Z = \frac{15(3)}{15 + 8 + 3} \tag{1·36}$$
$$= \frac{45}{26}$$
$$= 1.73 \text{ ohms}$$

The Y-equivalent circuit is now inserted in the original diagram of Fig. 1·11*a*, replacing the 15-, 8-, and 3-ohm resistors. The circuit is now shown in Fig. 1·13*c*. This circuit is redrawn in Fig. 1·13*d* for simplicity. Now, in Fig. 1·13*d*, it can be seen as a simpler series-parallel circuit consisting of the series components, the 7-ohm resistor and R_X, in series with the parallel circuit from point G to point D or F. This parallel circuit is

$$R_{G-D} = \frac{R_Z(R_Y + 5 + 1)}{R_Z + (R_Y + 5 + 1)}$$
$$= \frac{1.73(0.924 + 5 + 1)}{1.73 + (0.924 + 5 + 1)}$$
$$= \frac{11.97}{8.654}$$
$$= 1.38 \text{ ohms}$$

The simple series-circuit equivalent is finally shown in Fig. 1·13*e*, consisting of the 7-ohm resistor, R_X, and R_{G-D}. The *total* resistance is therefore

$$R_{\text{total}} \text{ or } R_{A-B} = 7 + R_X + R_{G-D}$$
$$= 7 + 4.62 + 1.38$$
$$= 13 \text{ ohms}$$

Note that this is the same result as was found in solution 1 of this Example 1·9, proving the validity of the delta-to-wye transformation equations (1·34) to (1·36).

Wye (or tee) to delta (or pi) transformation. Just as it is sometimes useful to change a delta arrangement into an equivalent wye circuit by the method just previously described, it is also helpful to do just the reverse. A wye or tee can be transformed into an equivalent delta or pi circuit. Referring to Fig. 1·12*a* and *b*, the *wye*-circuit components are R_X, R_Y, and R_Z, while the *delta*-circuit components are R_1, R_2, and R_3. To transform from wye to delta, each delta-circuit component (R_1 in Fig. 1·12*a*, for example) is equal to the sum of the products of each pair of wye components ($R_XR_Y + R_YR_Z + R_XR_Z$) divided by the wye component that is opposite the delta component (R_Z is opposite R_1, in Fig. 1·12*a*). Note that in the following wye-to-delta transformation equations, the numerators are identical.

$$R_1 = \frac{R_XR_Y + R_YR_Z + R_XR_Z}{R_Z} \tag{1·37}$$

$$R_2 = \frac{R_XR_Y + R_YR_Z + R_XR_Z}{R_Y} \tag{1·38}$$

$$R_3 = \frac{R_XR_Y + R_YR_Z + R_XR_Z}{R_X} \tag{1·39}$$

Example 1·10 Change the wye arrangement of Fig. 1·14 (R_X, R_Y, and R_Z) into its delta equivalent (dotted-line resistors R_1, R_2, and R_3).

Solution. The values of the wye components (R_X, R_Y, and R_Z) in Fig. 1·14 are from Fig. 1·13*b*. Solving for the delta or pi circuit component values gives

$$\begin{aligned} R_1 &= \frac{R_XR_Y + R_YR_Z + R_XR_Z}{R_Z} && (1·37) \\ &= \frac{(4.62)(0.924) + (0.924)(1.73) + (4.62)(1.73)}{1.73} \\ &= \frac{4.26 + 1.6 + 8}{1.73} \\ &= \frac{13.86}{1.73} \\ &= 8 \text{ ohms} \\ R_2 &= \frac{\text{same numerator as for } R_1 \text{ above}}{R_Y} && (1·38) \\ &= \frac{13.86}{0.924} \\ &= 15 \text{ ohms} \\ R_3 &= \frac{\text{same numerator as for } R_1 \text{ above}}{R_X} && (1·39) \\ &= \frac{13.86}{4.62} \\ &= 3 \text{ ohms} \end{aligned}$$

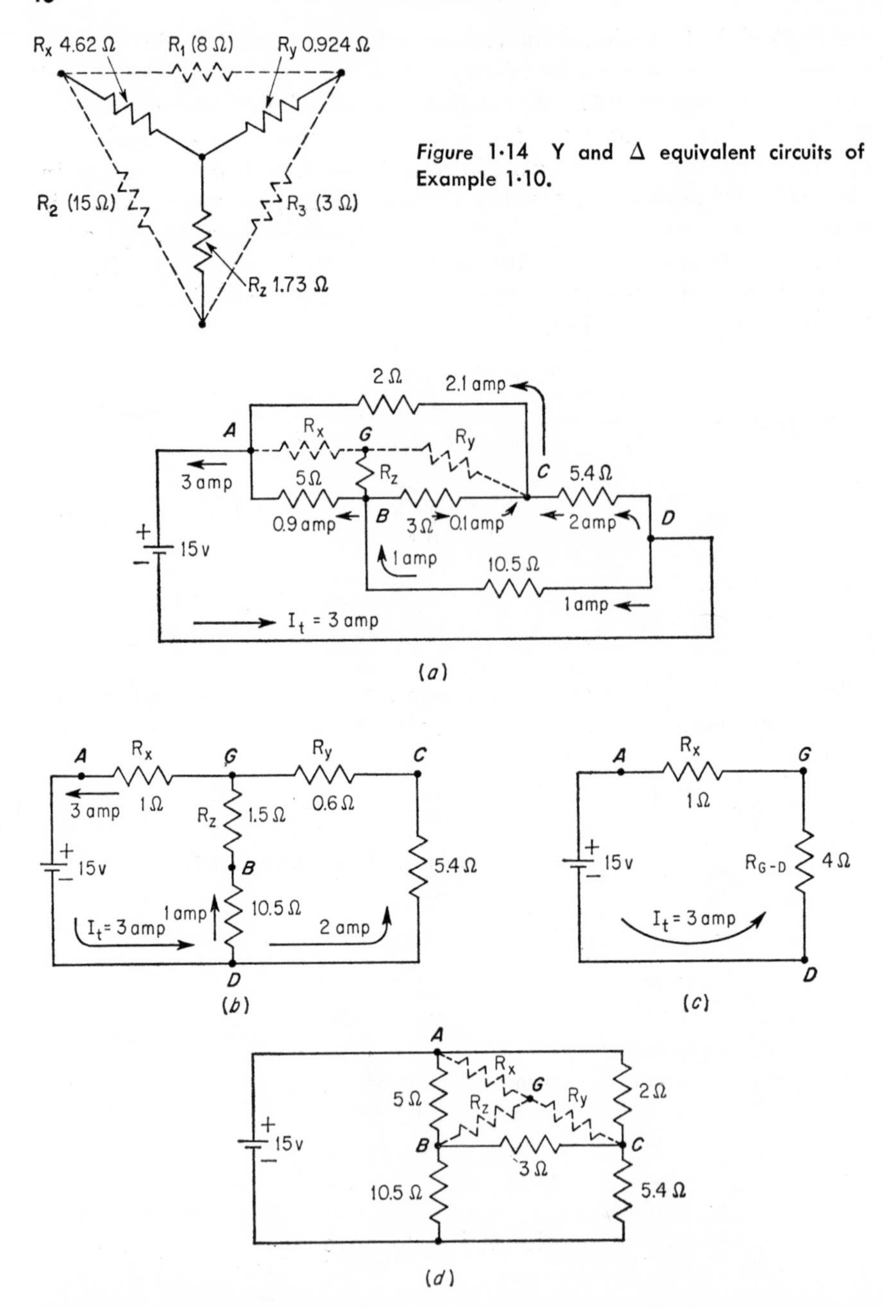

Figure 1·14 Y and Δ equivalent circuits of Example 1·10.

Figure 1·15 Mesh circuit of Example 1·11. (*a*) Original circuit; (*b*) circuit simplified by Δ-to-Y transformation; (*c*) circuit further simplified to a series type; (*d*) original circuit redrawn as a bridge type.

These results ($R_1 = 8$ ohms, $R_2 = 15$ ohms, and $R_3 = 3$ ohms) agree with the original values of the delta-circuit resistors shown in Fig. 1·13*a*. This proves the validity of the wye-to-delta transformation Eqs. (1·37) to (1·39).

Multiloop circuits. The following three Examples (1·11 to 1·13) are some of the many possible complex mesh circuits, three of which were referred to previously in Fig. 1·9*a*, *b*, and *c*. In each of the following, the delta-to-wye transformation method (or wye-to-delta) is employed. The reader is urged to solve these examples on his own, using one or two of the other methods of solution.

Example 1·11 In the resistor-mesh circuit of Fig. 1·15*a*, find the total resistance, and also find the current values through each resistor.

Solution. The diagram of Fig. 1·15*a* can be simplified into a series-parallel circuit by first finding the wye equivalent of the delta circuit made up of the three resistors, 2-, 5-, and 3-ohm. Although at first glance these resistors may not appear to be in a delta arrangement, a more careful analysis, noting the junction points *A*, *B*, and *C*, should show the delta characteristics. The dotted-line resistors, R_X, R_Y, and R_Z, are the wye equivalent.

$$R_X = \frac{2(5)}{2+5+3} \qquad [\Delta \text{ to Y transformation from Eqs. (1·34) to (1·36)}]$$

$$= \frac{10}{10}$$

$$= 1 \text{ ohm}$$

$$R_Y = \frac{2(3)}{2+5+3} \qquad [\Delta \text{ to Y transformation, Eqs. (1·34) to (1·36)}]$$

$$= \frac{6}{10}$$

$$= 0.6 \text{ ohms}$$

$$R_Z = \frac{5(3)}{2+5+3} \qquad [\Delta \text{ to Y transformation, Eqs. (1·34) to (1·36)}]$$

$$= \frac{15}{10}$$

$$= 1.5 \text{ ohms}$$

The original circuit of Fig. 1·15*a* is shown simplified in (*b*) with the Y or T arrangement of R_X, R_Y, and R_Z replacing the delta circuit components, the 2-, 5-, and 3-ohm resistors. The circuit of Fig. 1·15*b* is now a simple series-parallel one, with the 1-ohm resistor, R_X, in series with the parallel

G to D circuit of 12 ohms (1.5 ohms, R_Z + 10.5 ohms) and 6 ohms (0.6 ohm, R_Y, + 5.4 ohms). The G-to-D parallel-circuit resistance is

$$R_{G-D} = \frac{(1.5 + 10.5)(0.6 + 5.4)}{(1.5 + 10.5) + (0.6 + 5.4)}$$
$$= \frac{12(6)}{12 + 6}$$
$$= 4 \text{ ohms}$$

This 4 ohms, R_{G-D}, is in series with the 1 ohm, R_X, as shown in Fig. 1·15c, making a total resistance, R_T, of

$$R_T = R_X + R_{G-D}$$
$$= 1 + 4$$
$$= 5 \text{ ohms}$$

Total current I_T is

$$I_T = \frac{E_T}{R_T}$$
$$= \frac{15}{5}$$
$$= 3 \text{ amp}$$

Voltage between points G and D in Fig. 1·15c is

$$E_{G-D} = I_T R_{G-D}$$
$$= 3(4)$$
$$= 12 \text{ volts}$$

Current flow, in Fig. 1·15b, in the left branch of the G–D parallel circuit (through the 10.5 ohms + 1.5 ohms) is

$$I_{G-D\text{left}} = \frac{E_{G-D}}{R_{G-D\text{left}}}$$
$$= \frac{12}{1.5 + 10.5}$$
$$= 1 \text{ amp}$$

From Kirchhoff's current law, with 3 amp total current flowing into point D of Fig. 1·15b and 1 amp flowing away from point D through the left branch of the parallel circuit, the current through the right branch must then be 2 amp as shown in the diagram.

To find the current flow through the 3-ohm resistor of Fig. 1·15a, let us first determine the voltage at point B with respect to a reference point and also at point C with respect to this same reference. Using the negative end of the applied 15-volt source, point D, as a reference point, then points B and C in Fig. 1·15b are both positive with respect to reference point D. Voltage at point B with respect to D is the voltage across the 10.5-ohm resistor and is

$$\begin{aligned} E_B &= I_{G-D_{\text{left}}} R_{10.5\,\Omega} \\ &= 1(10.5) \\ &= +10.5 \text{ volts} \end{aligned}$$

Voltage at point C with respect to D is the voltage across the 5.4-ohm resistor and is

$$\begin{aligned} E_C &= I_{G-D_{\text{right}}} R_{5.4\,\Omega} \\ &= 2(5.4) \\ &= +10.8 \text{ volts} \end{aligned}$$

The difference between the voltage at point C, +10.8 volts, and that at point B, +10.5 volts, is 0.3 volt and is the voltage across the 3-ohm resistor. Since point C is more positive, electron movement or current flow is from point B, left to right through the 3-ohm resistor to point C, as shown in Fig. 1·15*a*. This current is

$$\begin{aligned} I_{3\text{ ohms}} &= \frac{E_{B-C}}{R_{3\text{ ohms}}} \\ &= \frac{0.3}{3} \\ &= 0.1 \text{ amp} \end{aligned}$$

To find the current through the 5-ohm resistor in Fig. 1·15*a*, use is again made of Kirchhoff's current law. In Fig. 1·15*a*, 1 amp flows *into* point B, and 0.1 amp flows *away* from this point to the right through the 3-ohm resistor. The remainder, or 0.9 amp, flows to the left, *away* from point B, through the 5-ohm resistor.

Current through the last resistor, the 2 ohms, can be calculated as follows from Kirchhoff's current law again. In Fig. 1·15*a*, the currents flowing *into* point C are 2 amp, flowing through the 5.4-ohm resistor, and 0.1 amp, flowing through the 3-ohm resistor. These currents add so that the current flowing *away* from point C is 2.1 amp through the 2-ohm resistor. This 2.1 amp and the 0.9 amp flowing through the 5-ohm resistor, flow into point A (Fig. 1·15*a*). The sum of these currents $(2.1 + 0.9 = 3)$ is the total current, I_T, which checks with the value of I_T found previously.

The circuit of Fig. 1·15*a* could also have been redrawn as a bridge circuit, as shown in Fig. 1·15*d*, with the Y or T transformation circuit shown by the dotted-line resistors R_X, R_Y, and R_Z.

Example 1·12 The mesh circuit of Fig. 1·16*a* is another case involving a delta arrangement. In this circuit, find the total resistance and the current flow through each resistor.

Solution. The three resistors, 10, 50, and 40 ohms, connected to the junction points A, B, and C form a delta arrangement. These resistors could be replaced in the circuit by the wye or tee transformation resistors, R_X, R_Y, and R_Z, which are indicated by the dotted-line components in

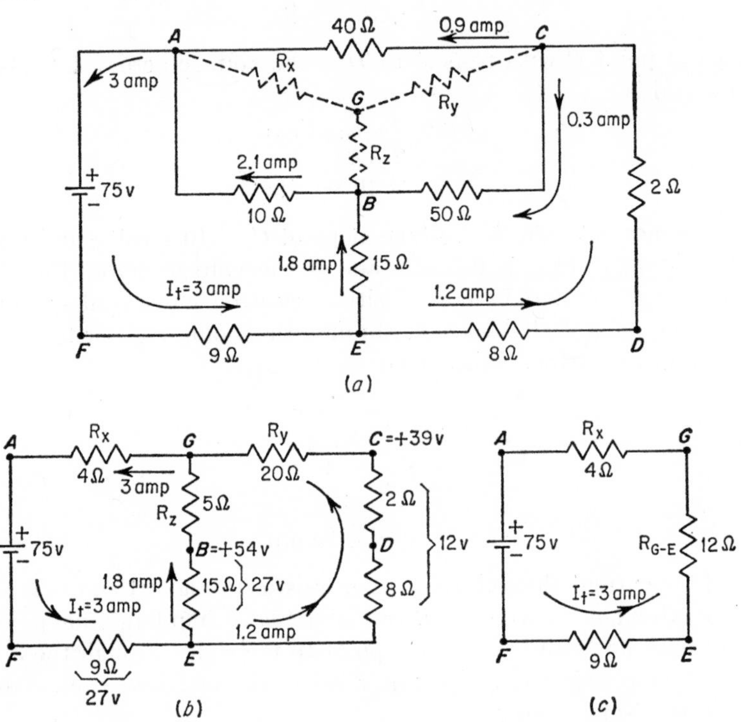

Figure **1·16** **Mesh circuit of Example 1·12. (a) Original circuit; (b) circuit simplified by Δ-to-Y transformation; (c) circuit further simplified.**

Fig. 1·16*a* and are also shown in the simplified circuit of Fig. 1·16*b*. Changing the *delta* arrangement of the 10-, 50-, and 40-ohm resistors into the *wye* circuit of R_X, R_Y, and R_Z yields the following:

$$R_X = \frac{10(40)}{10 + 40 + 50} \qquad [\Delta \text{ to Y transformation, Eqs. (1·34) to (1·36)}]$$

$$= \frac{400}{100}$$

$$= 4 \text{ ohms}$$

$$R_Y = \frac{50(40)}{10 + 40 + 50} \qquad [\Delta \text{ to Y transformation, Eqs. (1·34) to (1·36)}]$$

$$= \frac{2000}{100}$$

$$= 20 \text{ ohms}$$

$$R_Z = \frac{10(50)}{10 + 40 + 50} \qquad [\Delta \text{ to Y transformation, Eqs. (1·34) to (1·36)}]$$
$$= \frac{500}{100}$$
$$= 5 \text{ ohms}$$

These Y-arrangement resistors, R_X, R_Y, and R_Z, are inserted in the circuit of Fig. 1·16*a*, in place of the delta circuit. The new circuit is shown in Fig. 1·16*b* and is a simple series-parallel type. R_X is in series with the parallel G to E circuit and in series with the 9-ohm resistor. The parallel G to E circuit, as shown, is made up of the left branch consisting of R_Z and the 15-ohm resistor in series and the right branch comprising R_Y, and the 2- and 8-ohm resistors in series. The parallel circuit is

$$R_{G-E} = \frac{(R_Z + 15)(R_Y + 2 + 8)}{(R_Z + 15) + (R_Y + 2 + 8)}$$
$$= \frac{(5 + 15)(20 + 2 + 8)}{(5 + 15) + (20 + 2 + 8)}$$
$$= \frac{20(30)}{20 + 30}$$
$$= \frac{600}{50}$$
$$= 12 \text{ ohms}$$

The G-to-E resistance of 12 ohms is shown in Fig. 1·16*c* in series with R_X and the 9-ohm resistor, as a simple series circuit. The total resistance, R_T, is

$$R_T = R_X + R_{G-E} + 9$$
$$= 4 + 12 + 9$$
$$= 25 \text{ ohms}$$

Total current I_T is

$$I_T = \frac{E_T}{R_T}$$
$$= \frac{75}{25}$$
$$= 3 \text{ amp}$$

Voltage across the 12-ohm R_{G-E} in Fig. 1·16*c* is

$$E_{G-E} = I_T R_{G-E}$$
$$= 3(12)$$
$$= 36 \text{ volts}$$

Current through the *left* branch of the parallel G to E circuit in Fig. 1·16*b* is then

$$I_{G-E_{\text{left}}} = \frac{E_{G-E}}{R_{G-E_{\text{left}}}} = \frac{36}{R_Z + 15} = \frac{36}{5 + 15} = 1.8 \text{ amp}$$

From Kirchhoff's current law, with 3 amp flowing *into* point E of Fig. 1·16*a* and *b* and with 1.8 amp *leaving* this point and flowing through the *left* branch, the remainder ($3 - 1.8 = 1.2$ amp) *leaves* point E and flows through the *right* branch. These currents are shown in the circuits of Fig. 1·16*a* and *b* and account for the current through the 9-ohm resistor ($I_T = 3$ amp), through the 15-ohm resistor (1.8 amp), and through the 8- and 2-ohm resistors (1.2 amp).

To find the current through the 50-ohm resistor of Fig. 1.16*a*, first determine the voltage at point B and at point C with respect to a *reference* point. Using the negative end of the applied 75-volt source, point F, as a *reference* point, B and C are both positive since the electron movement is from E toward B and also from E toward C. Voltage at point B with respect to F is the sum of the voltages across the 9- and 15-ohm resistors. These are

$$E_{9\text{ ohms}} = I_T R_{9\text{ ohms}} = 3(9) = 27 \text{ volts}$$

$$E_{15\text{ ohms}} = I_{G-E_{\text{left}}} R_{15\text{ ohms}} = 1.8(15) = 27 \text{ volts}$$

Point B with respect to point F is then $+54$ volts ($27 + 27$). Similarly, point C with respect to point F is the sum of the voltages across the 9-ohm resistor (27 volts) and those across the 8- and 2-ohm resistors. These latter ones are

$$E_{8\text{ ohms}} + E_{2\text{ ohms}} = (I_{G-E_{\text{right}}})(R_{8\text{ ohms}} + R_{2\text{ ohms}}) = 1.2(8 + 2) = 1.2(10) = 12 \text{ volts}$$

Point C with respect to F is then 27 + 12 or +39 volts. In Fig. 1·16a, point B (+54 volts) is at the left end of the 50-ohm resistor, with point C (+39 volts) at the right end of the resistor. Voltage across this resistor is the difference between these two points (+54 volts and +39 volts), or 15 volts. Current through this resistor is then

$$\begin{aligned} I_{50\text{ ohms}} &= \frac{E_{50\text{ ohms}}}{R_{50\text{ ohms}}} \\ &= \frac{15}{50} \\ &= 0.3 \text{ amp} \end{aligned}$$

This current is indicated in Fig. 1·16a. Note that the *direction* of electron movement is from point C (+39 volts) toward the higher voltage at point B (+54 volts).

Current through the 40-ohm resistor of Fig. 1·16a can be ascertained by first finding the voltage across this resistor, or between points A and C. Point A, the positive end of the applied voltage source, is +75 volts with respect to the reference point F. Point C, as already established, is +39 volts. The voltage across the 40-ohm resistor is therefore 36 volts (the difference between the +75 and the +39 voltage points). Current through this resistor is then

$$\begin{aligned} I_{40\text{ ohms}} &= \frac{E_{40\text{ ohms}}}{R_{40\text{ ohms}}} \\ &= \frac{36}{40} \\ &= 0.9 \text{ amp} \end{aligned}$$

This current is indicated in Fig. 1·16a.

The final current to be found is that which flows through the 10-ohm resistor of Fig. 1·16a. This can be found by again using Kirchhoff's current law. Currents flowing into point B are the 1.8 amp ($I_{G-E_{\text{left}}}$) and the 0.3 amp flowing through the 50-ohm resistor of Fig. 1·16a. Current *leaving* point B in Fig. 1·16a is the sum of these two *entering* currents, or 1.8 + 0.3 = 2.1 amp as indicated in the diagram. This 2.1 amp can be checked by solving for this current through the 10-ohm resistor by first finding the voltage across the resistor. This, as shown, is between points A and B. As has already been stated, A is +75 volts, while B is +54 volts. The difference between them is 21 volts, and this is across the 10-ohm resistor.

Current through it is therefore

$$I_{10\text{ ohms}} = \frac{E_{10\text{ ohms}}}{R_{10\text{ ohms}}} = \frac{21}{10} = 2.1 \text{ amp}$$

This checks with the answer found previously. A final check can be seen from Kirchhoff's current law at point A. The *sum* of the *two* currents entering A, in Fig. 1·16a, the 0.9 amp through the 40-ohm resistor, and the 2.1 amp through the 10-ohm resistor, gives a total current, I_T, of 3 amp as previously established.

Example 1·13 In the mesh circuit of Fig. 1·17a, find the total resistance between points A and D.

Solution. This circuit will be solved by first changing the delta arrangements of R_1, R_2, and R_3 into a wye, as shown by the dotted-line resistors, $R_{1,2}$, $R_{1,3}$, $R_{2,3}$, in Fig. 1·17a. The Y-circuit resistors are

$$R_{1,2} = \frac{R_1R_2}{R_1 + R_2 + R_3} \qquad [\Delta \text{ to Y transformation, Eqs. (1·34) to (1·36)}]$$

$$= \frac{20(30)}{20 + 30 + 50} = \frac{600}{100} = 6 \text{ ohms}$$

$$R_{1,3} = \frac{R_1R_3}{R_1 + R_2 + R_3} \qquad [\Delta \text{ to Y transformation, Eqs. (1·34) to (1·36)}]$$

$$= \frac{20(50)}{20 + 30 + 50} = \frac{1{,}000}{100} = 10 \text{ ohms}$$

$$R_{2,3} = \frac{R_2R_3}{R_1 + R_2 + R_3} \qquad [\Delta \text{ to Y transformation, Eqs. (1·34) to (1·36)}]$$

$$= \frac{30(50)}{20 + 30 + 50} = \frac{1{,}500}{100} = 15 \text{ ohms}$$

These Y- or T-circuit resistors, $R_{1,2}$, $R_{1,3}$, and $R_{2,3}$, are shown in Fig. 1·17*b* replacing the original delta resistors, R_1, R_2, and R_3 of Fig. 1·17*a*.

The next step in the solution is to transform the lower delta arrangement of R_7, R_5, and R_8 into a Y circuit, as shown by the dotted-line

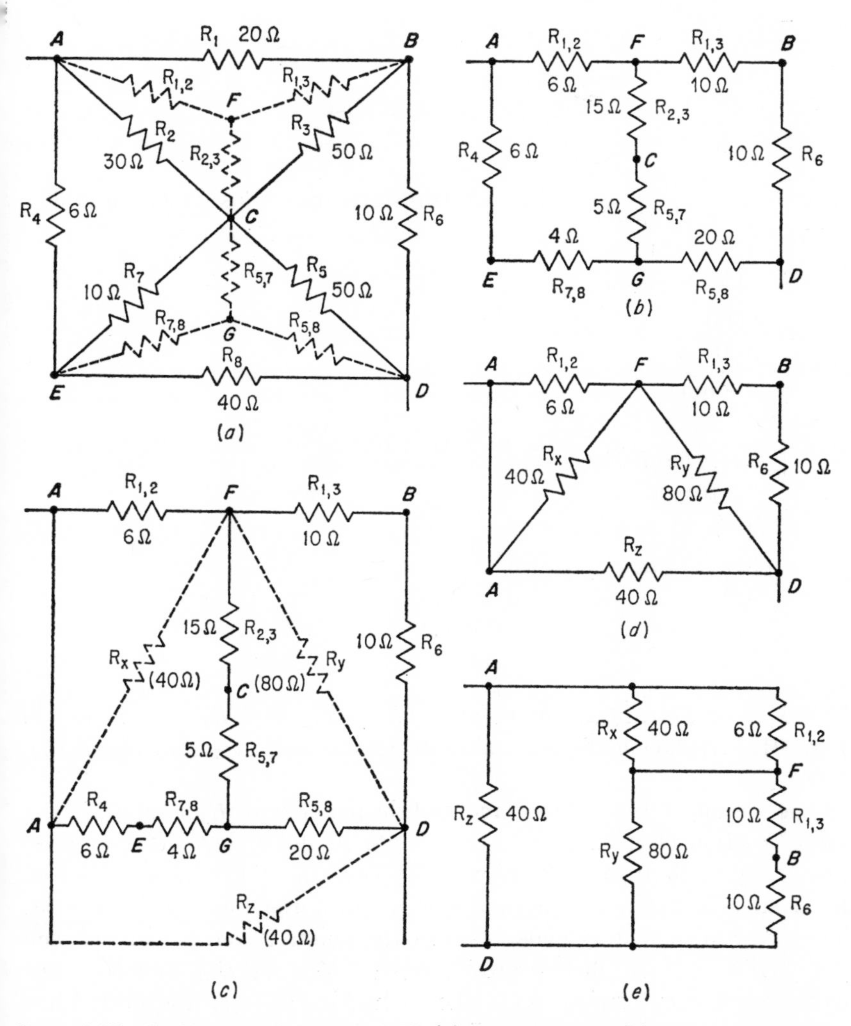

Figure 1·17 Mesh circuit of Example 1·13. (*a*) Original circuit; (*b*) circuit simplified by Δ-to-Y transformations; (*c*) circuit redrawn; (*d*) circuit further simplified by Y-to-Δ transformation; (*e*) circuit redrawn.

Y-circuit components, $R_{5,7}$, $R_{7,8}$, and $R_{5,8}$, of Fig. 1·17*a*. These Y-circuit resistors are

$$R_{5,7} = \frac{R_5R_7}{R_5 + R_7 + R_8} \qquad \text{[}\Delta \text{ to Y transformation, Eqs. (1·34) to (1·36)]}$$

$$= \frac{50(10)}{50 + 10 + 40}$$

$$= \frac{500}{100}$$

$$= 5 \text{ ohms}$$

$$R_{5,8} = \frac{R_5R_8}{R_5 + R_7 + R_8} \qquad \text{[}\Delta \text{ to Y transformation, Eqs. (1·34) to (1·36)]}$$

$$= \frac{50(40)}{50 + 10 + 40}$$

$$= \frac{2{,}000}{100}$$

$$= 20 \text{ ohms}$$

$$R_{7,8} = \frac{R_7R_8}{R_5 + R_7 + R_8} \qquad \text{[}\Delta \text{ to Y transformation, Eqs. (1·34) to (1·36)]}$$

$$= \frac{10(40)}{50 + 10 + 40}$$

$$= \frac{400}{100}$$

$$= 4 \text{ ohms}$$

These Y- or T-arrangement resistors, $R_{5,7}$, $R_{5,8}$, and $R_{7,8}$, are shown in Fig. 1·17*b*, replacing the original delta resistors, R_5, R_7, and R_8, of Fig. 1·17*a*.

The circuit of Fig. 1·17*b* is redrawn in part *c* with R_4, 6 ohms, in the same position electrically between points A and E but simply moved physically to that shown in Fig. 1·17*c*. Now, the T arrangement (or an inverted T as it is in the diagram) will be transformed into a delta circuit. This T arrangement extends between the points F and G and between points A and D and includes resistors $R_{2,3}$, $R_{5,7}$, R_4, $R_{7,8}$, and $R_{5,8}$. The delta-arrangement resistors are shown in Fig. 1·17*c* as the dotted-line components, R_X, R_Y, and R_Z. Resistors $R_{2,3}$ and $R_{5,7}$, between points F and G, will be referred to as R_{F-G} for simplicity. Resistors R_4 and $R_{7,8}$, between points A and G, will be referred to as R_{A-G} for simplicity. Similarly, $R_{5,8}$, between points G and D, is referred to as R_{G-D}. The values of the delta resistors are

$$R_X = \frac{R_{F-G}R_{G-D} + R_{G-D}R_{A-G} + R_{A-G}R_{F-G}}{R_{G-D}}$$ [Y to Δ transformation, Eqs. (1·37) to (1·39)]

$$= \frac{20(20) + 20(10) + 10(20)}{20}$$

$$= \frac{400 + 200 + 200}{20}$$

$$= 40 \text{ ohms}$$

$$R_Y = \frac{R_{F-G}R_{G-D} + R_{G-D}R_{A-G} + R_{A-G}R_{F-G}}{R_{A-G}}$$ [Y to Δ transformation, Eqs. (1·37) to (1·39)]

$$= \frac{800}{10}$$

$$= 80 \text{ ohms}$$

$$R_Z = \frac{R_{F-G}R_{G-D} + R_{G-D}R_{A-G} + R_{A-G}R_{F-G}}{R_{F-G}}$$ [Y to Δ transformation, Eqs. (1·37) to (1·39)]

$$= \frac{800}{20}$$

$$= 40 \text{ ohms}$$

The delta resistors R_X, R_Y, and R_Z are shown in the diagram of Fig. 1·17*d* connected between the junction points *A*, *F*, and *D* and are shown replacing the T-circuit resistors of Fig. 1·17*c* between these same points. The circuit of Fig. 1·17*d* is shown redrawn in *e* for simplicity purposes. In Fig. 1·17*e*, it can be seen that the circuit is a parallel-series combination. Resistor R_X is in parallel with $R_{1,2}$, and R_Y is in parallel with the series combination of $R_{1,3}$ and R_6. The two parallel circuits are in series with each other and in parallel with R_Z. The resistance of R_X and $R_{1,2}$ in parallel from point *A* to *F* is

$$R_{A-F} = \frac{R_X R_{1,2}}{R_X + R_{1,2}}$$

$$= \frac{40(6)}{40 + 6}$$

$$= 5.22 \text{ ohms}$$

The resistance of the parallel circuit, between points *F* and *D*, of R_Y and $R_{1,3}$ and R_6 is

$$R_{F-D} = \frac{R_Y(R_{1,3} + R_6)}{R_Y + (R_{1,3} + R_6)}$$

$$= \frac{80(10 + 10)}{80 + 10 + 10}$$

$$= \frac{1{,}600}{100}$$

$$= 16 \text{ ohms}$$

The *total* resistance, R_T, of the entire circuit, between points A and D, consists of R_Z in parallel with the series circuit of R_{A-F} and R_{F-D} and is

$$\begin{aligned} R_T &= \frac{R_Z(R_{A-F} + R_{F-D})}{R_Z + (R_{A-F} + R_{F-D})} \\ &= \frac{40(5.22 + 16)}{40 + 5.22 + 16} \\ &= \frac{40(21.22)}{61.22} \\ &= \frac{848.8}{61.22} \\ &= 13.85 \end{aligned}$$

PROBLEMS

1·1 Using loop currents, find the voltage across resistor R_2 in the accompanying diagram.

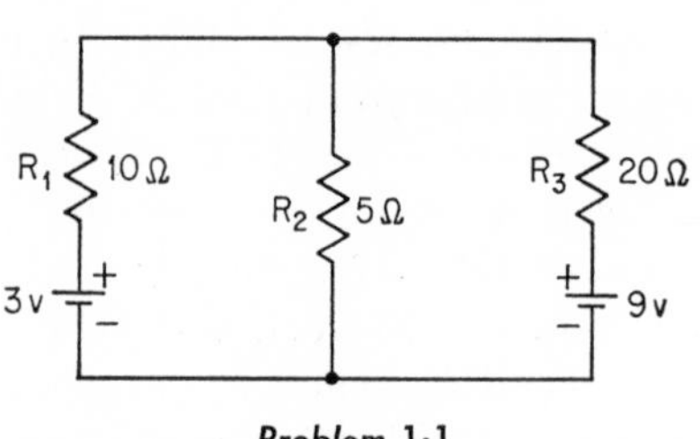

Problem 1·1

1·2 Using Thévenin equivalent, check the above result.

1·3 Using loop currents in the circuit shown, find (*a*) I_1; (*b*) I_2; (*c*) I_3.

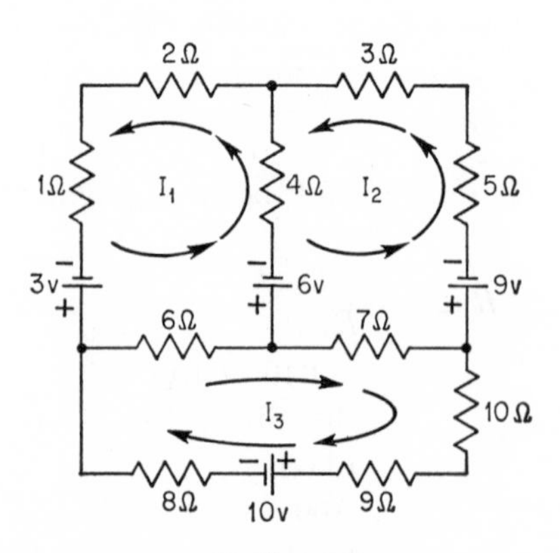

Problem 1·3

1·4 In the circuit of Prob. 1·3, find the voltage across the 4-ohm resistor.

1·5 In the circuit shown, calculate the Thévenin equivalent resistance and voltage between points A and B.

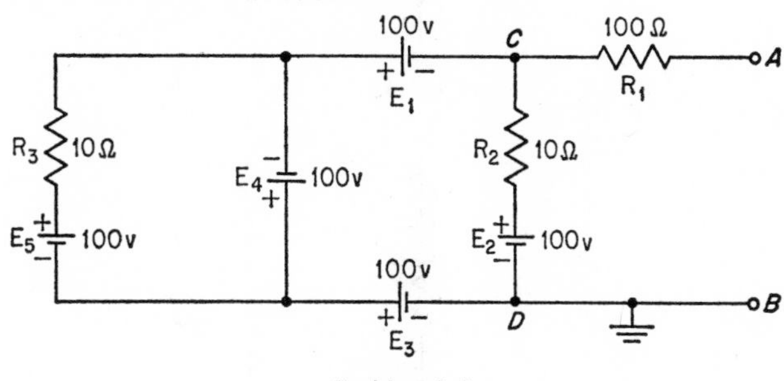

Problem 1·5

MORE CHALLENGING PROBLEMS

1·6 Find the total resistance between points A and B in the accompanying diagram.

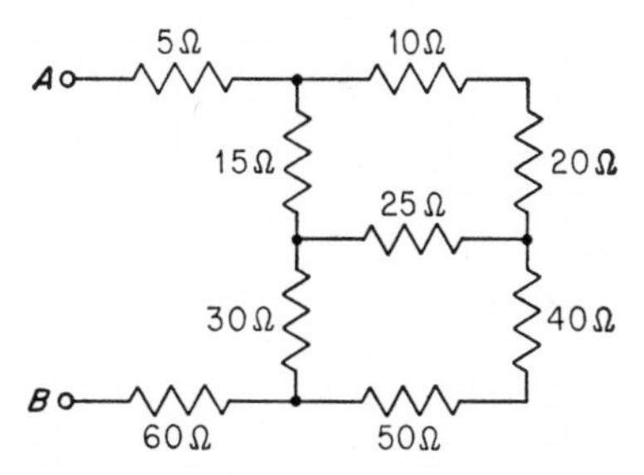

Problem 1·6

1·7 Find the voltage across resistor R_5 (between point X and ground) in the circuit shown.

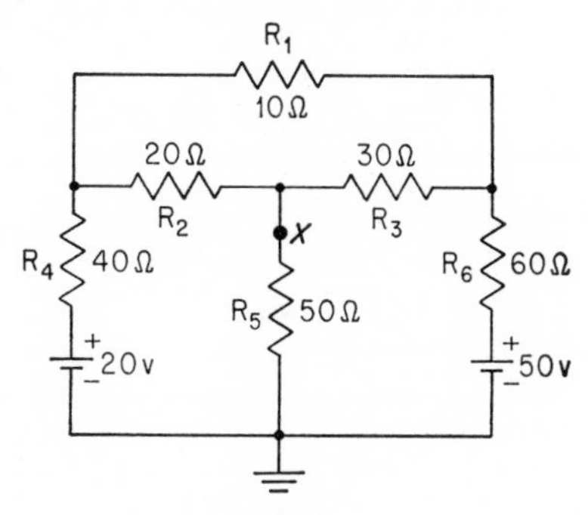

Problem 1·7

1·8 In diagram of Prob. 1·7, find the voltage across R_2.

1·9 In the accompanying diagram, find the Thévenin equivalent resistance between points X and Y.

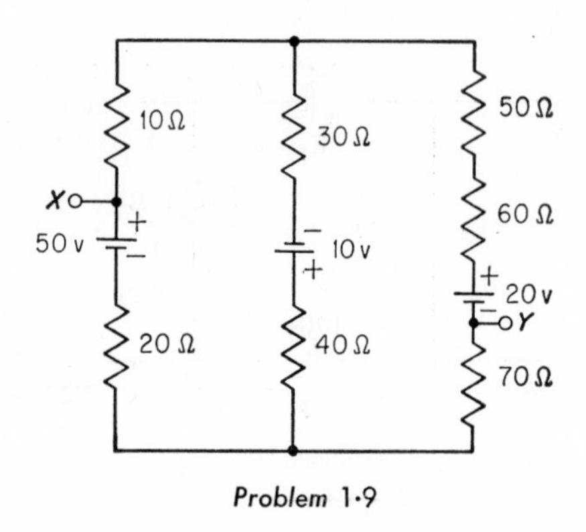

Problem 1·9

1·10 In the diagram of Prob. 1·9, find the voltage between points X and Y.

CHAPTER 2

PULSE AMPLIFIERS

2·1 Pulse Amplifiers, General

A complex waveshape such as a square wave or a pulse contains a fundamental sine wave plus one or more harmonics. The fundamental is a sine-wave signal at the frequency of the complex wave, while the harmonics are simply other sine-wave signals at multiples of the fundamental frequency. The complex wave shown in Fig. 2·1*a* consists of the

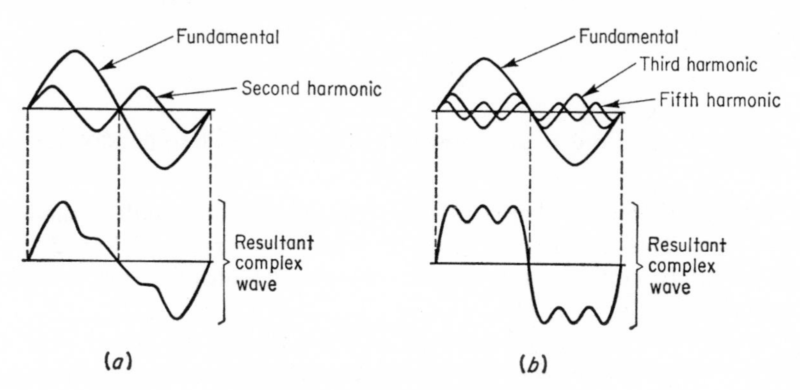

Figure 2·1 Complex wave content.

fundamental sine wave and the second harmonic as shown. When the fifth harmonic is also added, as shown in Fig. 2·1*b*, the resultant complex wave becomes more of a square wave. Actually, a symmetrical square wave is made up of the fundamental and numerous *odd* harmonics. An unbalanced or asymmetrical square wave such as a pulse, on the other hand, contains the fundamental sine wave plus many odd and even harmonics. An amplifier, therefore, that must handle pulse signals must be capable of producing equal amplification to *all* the frequencies contained in the pulse. Such an amplifier is called a *pulse amplifier* or a *wideband amplifier* and is also commonly called a *video amplifier*.

The various capacitors in an amplifier circuit prevent equal gain at all frequencies, the gain decreasing for the extreme low- and high-frequency signals. Additional components must be added to the amplifier circuit in order to compensate for this smaller amplification at these extreme frequencies.

2·2 Amplifier High-frequency Response

The voltage gain or amplification of the pentode and the transistor circuits of Fig. 2·2 depends on, among other things, the load impedance. The amplification for the pentode circuit of Fig. 2·2*a* is approximately

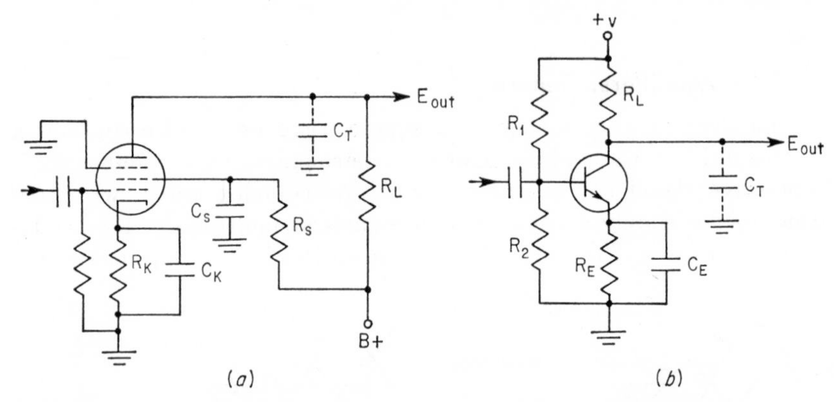

Figure **2·2 Amplifier circuits, uncompensated. (*a*) Pentode amplifier; (*b*) NPN transistor amplifier.**

equal to G_M times Z_L, where G_M is the transconductance and Z_L the load impedance. The voltage gain for the transistor circuit of Fig. 2·2*b* is approximately equal to the current gain or β multiplied by the ratio Z load/Z input, or $\beta Z_L/Z_{\text{in}}$.

The load impedance Z_L consists of the load resistor R_L and the capacitor C_T. This capacitor is shown in Fig. 2·2 using dotted-line connections since it is not a physical component that is wired into the circuit. C_T consists of the stray capacitance existing between the wiring of the pentode anode (or the transistor collector) and chassis ground and also the interelectrode capacitance of the tube (or the capacitance of the transistor output impedance).

Where the output from the plate of one amplifier stage is coupled to the grid of the next succeeding stage, the interelectrode capacitance of this second stage becomes an important part of C_T and is usually called the *input capacitance*, C_{in}. This consists of grid-to-cathode interelectrode capacitance, C_{GK}, plus grid-to-plate capacitance, C_{GP}, increased by a factor called the *Miller effect*. This is due to the *gain* and *phase inversion* characteristics of an amplifier where input signal is applied to the grid, output is taken off at the plate, and the load is resistive. As an example, if an amplifier stage has a gain of 30 and the input signal drives the grid 1 volt in a negative direction, then the plate voltage rises or goes 30 volts in a positive direction. This has the effect of changing the charge on the grid-to-plate interelectrode capacitor by 30 + 1 (gain + 1) or 31 volts for every 1 volt input signal to the grid. This Miller effect therefore results in an effective capacitance between grid and plate which is C_{GP} multiplied by the factor gain + 1. Input capacitance C_{in} is therefore $C_{GK} + C_{GP}$ times gain + 1.

The load impedance Z_L consists of R_L essentially in parallel with the reactance of C_T, since R_L connects to $B+$, and $B+$ is grounded (to the a-c signal) through the filter capacitor. The value of C_T is quite small, about 20 to 40 pf. At low and medium frequencies, X_{C_T} is very much larger than R_L and presents no problem. At these frequencies Z_L is just about equal to R_L, as shown in the following two examples.

Example 2·1 If C_T is 30 pf, and R_L is 10,000 ohms (10 kilohms), and the frequency is 100 cps, then X_{C_T} is about 50 megohms

$$X_{C_T} = \frac{1}{2\pi f C_T} = \frac{1}{6.28(100)30 \times 10^{-12}} = 50 \times 10^6 \text{ (approx.)}$$

The 50-megohm X_{C_T} in *parallel* with the 10-kilohm R_L produces a total Z_L of

$$\begin{aligned} Z_L &= \frac{\text{product}}{\text{sum}} \\ &= \frac{X_{C_T} R_L}{\sqrt{X_{C_T}^2 + R_L^2}} \\ &= \frac{50 \times 10^6 (10 \times 10^3)}{\sqrt{(50 \times 10^6)^2 + (10 \times 10^3)^2}} \\ &= \frac{50 \times 10^6 (10 \times 10^3)}{50 \times 10^6 \text{ (approx.)}} \\ Z_L &= 10 \times 10^3 \\ &= 10 \text{ kilohms} \end{aligned}$$

Example 2·2 At a medium frequency of 20,000 cycles per second (20 Kc), X_{C_T} is still quite large, about 260 kilohms

$$X_{C_T} = \frac{1}{2\pi f C_T} = \frac{1}{6.28(20 \times 10^3)30 \times 10^{-12}} = 260 \text{ kilohms (approx.)}$$

The 260-kilohm X_{C_T} in *parallel* with the 10-kilohm R_L produces a total Z_L of

$$\begin{aligned} Z_L &= \frac{X_{C_T} R_L}{\sqrt{X_{C_T}^2 + R_L^2}} \\ &= \frac{260 \times 10^3(10 \times 10^3)}{\sqrt{(260 \times 10^3)^2 + (10 \times 10^3)^2}} \\ &= \frac{260 \times 10^3(10 \times 10^3)}{260 \times 10^3 \text{ (approx.)}} \\ &= 10 \times 10^3 \\ &= 10 \text{ kilohms} \end{aligned}$$

In the previous two examples, Z_L is about equal to R_L (10 kilohms) at the low frequency (100 cps) and at the medium frequency (20 kc). As the frequency goes higher, X_{C_T} decreases. The following two examples illustrate the effect that a decreasing X_{C_T} has on Z_L.

Example 2·3 At a higher frequency of 500 kc, X_{C_T} falls to 10 kilohms

$$X_{C_T} = \frac{1}{2\pi f C_T} = \frac{1}{6.28(5 \times 10^5)30 \times 10^{-12}} = 10 \text{ kilohms (approx.)}$$

The 10-kilohm X_{C_T} in *parallel* with the 10-kilohm R_L produces a total Z_L of

$$\begin{aligned} Z_L &= \frac{X_{C_T} R_L}{\sqrt{X_{C_T}^2 + R_L^2}} \\ &= \frac{10 \times 10^3(10 \times 10^3)}{\sqrt{(10 \times 10^3)^2 + (10 \times 10^3)^2}} \\ &= \frac{100 \times 10^6}{\sqrt{100 \times 10^6 + 100 \times 10^6}} \\ &= \frac{100 \times 10^6}{\sqrt{200 \times 10^6}} \\ &= \frac{100 \times 10^6}{14.1 \times 10^3} \\ &= 7.07 \times 10^3 \\ &= 7.07 \text{ kilohms} \end{aligned}$$

Example 2·4 At a much higher frequency of 5,000,000 cycles per second (5 Mc), X_{C_T} decreases to 1 kilohm,

$$X_{C_T} = \frac{1}{2\pi f C_T} = \frac{1}{6.28(5 \times 10^6)30 \times 10^{-12}} = 1 \text{ kilohm (approx.)}$$

The 1-kilohm X_{C_T} in *parallel* with the 10-kilohm R_L produces a total Z_L of

$$\begin{aligned}
Z_L &= \frac{X_{C_T} R_L}{\sqrt{X_{C_T}^2 + R_L^2}} \\
&= \frac{1 \times 10^3(10 \times 10^3)}{\sqrt{(1 \times 10^3)^2 + (10 \times 10^3)^2}} \\
&= \frac{10 \times 10^6}{\sqrt{1 \times 10^6 + 100 \times 10^6}} \\
&= \frac{10 \times 10^6}{\sqrt{101 \times 10^6}} \\
&= \frac{10 \times 10^6}{10 \times 10^3 \text{ (approx.)}} \\
&= 1 \times 10^3 \text{ (approx.)} \\
&= 1 \text{ kilohm}
\end{aligned}$$

Since the gain of an amplifier is directly proportional to the value of Z_L, then the gain decreases at the higher frequencies due to the decreasing value of X_{C_T}, causing Z_L to fall. This is illustrated in Fig. 2·3*a* and *b*.

Freq	X_{C_T} (approx)	R_L	Z_L
(example 1) 100 cps	50 Meg	10 K	10 K
(example 2) 20 KC	260 K	10 K	10 K
(example 3) 500 KC	10 K	10 K	7.07 K
(example 4) 5 MC	1 K	10 K	1 K

(*a*)

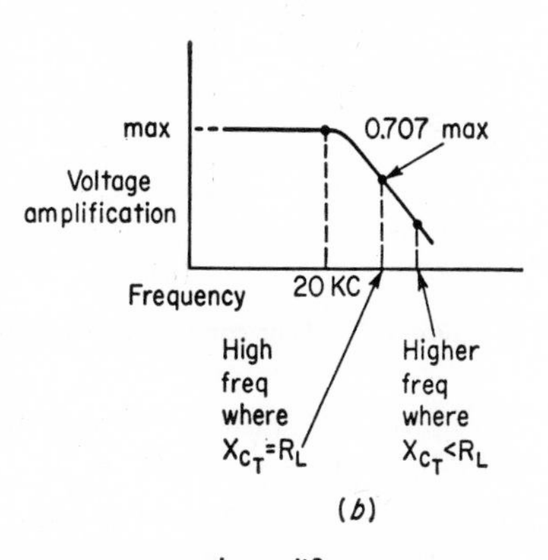

(*b*)

Figure 2·3 **High-frequency effect on uncompensated amplifier.**

In Fig. 2·3*a*, the results of the four previous examples are tabulated, and in Fig. 2·3*b*, the frequency-response curve (the graph of gain versus frequency) of the amplifier is shown. As shown, Z_L remains at 10 kilohms over a band of frequencies (Examples 2·1 and 2·2), and the gain remains constant or flat for these frequencies. At the particular higher frequency where X_{C_T} becomes equal to R_L (Example 2·3, 500 kc), Z_L drops to 0.707

of its former constant value (10 kilohms × 0.707 = 7.07 kilohms), and as a result, the gain decreases proportionally. As shown in Fig. 2·3, the gain drops considerably at the still higher frequency of 5 Mc (Example 2·4) where X_{C_T} is much smaller than R_L (1 kilohm and 10 kilohms respectively).

By having a frequency-response curve which falls off at the higher frequencies, an amplifier could not properly reproduce a complex wave such as a pulse or a square wave since the higher-frequency harmonic components would not be adequately amplified. To prevent this type of distortion, called *frequency distortion*, the amplifier must be compensated by using peaking coils as explained in the next section. Another type of distortion, called *phase distortion*, is discussed later in Secs. 2·5 and 2·6.

2·3 High-frequency Compensation

By adding an inductance called a *peaking* coil in the plate circuit of an amplifier, the load impedance Z_L and the gain can be prevented from decreasing at the higher frequencies even though X_{C_T} continues to fall. The effect of the peaking coil compensates for the decreasing value of X_{C_T} at the higher frequencies.

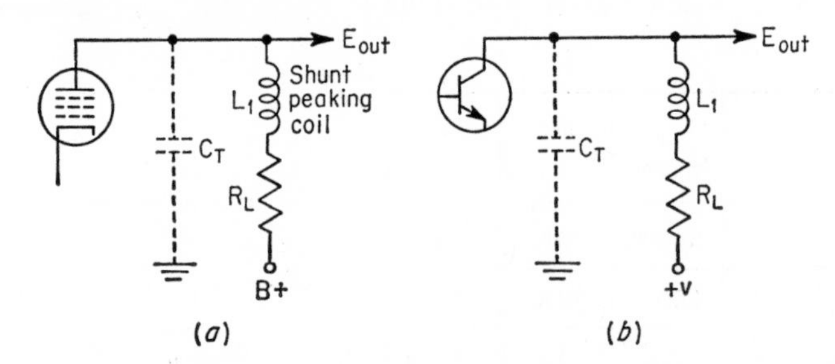

Figure 2·4 Shunt compensation. (a) Vacuum-tube amplifier; (b) transistor amplifier.

Figure 2·4 shows a simple method of high-frequency compensation called *shunt compensation*. The values of this coil L_1, and the load resistor R_L, depend on the amount of C_T in the circuit and the highest desirable frequency (f_0) to be amplified. The values are

$$R_L = X_{C_T} \text{ at } f_0 \tag{2·1}$$

where R_L is in ohms, C_T in farads, and f_0 in cycles per second.

$$L_1 = \tfrac{1}{2} C_T (R_L)^2 \tag{2·2}$$

where L_1 is in henrys, C_T in farads, and R_L in ohms.

The load impedance Z_L now consists of X_{C_T} in *parallel* with the *series* circuit of L_1 and R_L and is

$$Z_L = \frac{\text{product}}{\text{sum}} = \frac{X_{C_T}\sqrt{X_{L_1}{}^2 + R_L{}^2}}{\sqrt{R_L{}^2 + (X_{L_1} - X_{C_T})^2}} \tag{2·3}$$

To see the results of adding L_1 to the circuit, the following four examples (2·5 to 2·8) use the same frequencies and the same value of C_T as shown previously in Examples 2·1 to 2·4 of Fig. 2·3*a*. C_T is 30 pf, and the frequencies are 100 cps, 20 kc, 500 kc, and 5 Mc.

Example 2·5 At the 100-cps frequency and using a value for C_T of 30 pf, determine the required values of R_L and L_1 if the highest frequency (f_0) is 5 Mc, and find Z_L.

R_L is found as follows:

$$\begin{aligned} R_L &= X_{C_T} \text{ at } f_0 &(2·1)\\ &= \frac{1}{2\pi f_0 C_T} \\ &= \frac{1}{6.28(5 \times 10^6)30 \times 10^{-12}} \\ &= 1 \text{ kilohm (approx.)} \end{aligned}$$

The shunt peaking coil L_1 is found from

$$\begin{aligned} L_1 &= \tfrac{1}{2}C_T R_L{}^2 &(2·2)\\ &= \tfrac{1}{2}(30 \times 10^{-12})(1 \times 10^3)^2 \\ &= 15 \times 10^{-12}(1 \times 10^6) \\ &= 15 \times 10^{-6} \text{ henry} \\ L_1 &= 15\ \mu\text{h} \end{aligned}$$

The load impedance Z_L at the 100-cycle frequency can be found after determining X_{C_T} and X_{L_1} at this frequency.

$$\begin{aligned} X_{C_T} \text{ at 100 cps} &= \frac{1}{2\pi f C_T} \\ &= \frac{1}{6.28(100)30 \times 10^{-12}} \\ &= 50 \text{ megohms} \\ X_{L_1} \text{ at 100 cps} &= 2\pi f L \\ &= 6.28(100)15 \times 10^{-6} \\ &= 0.01 \text{ ohms} \end{aligned}$$

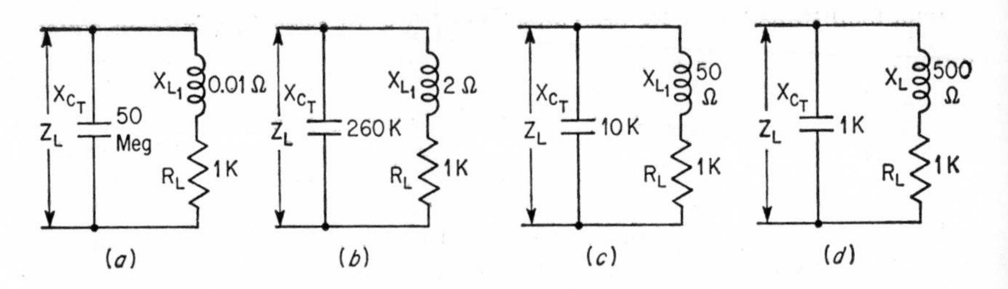

Freq	X_{C_T} (approx)	X_{L_1} (approx)	R_L	Z_L
(example 5) 100 cps	50 Meg	0.01 Ω	1K	1K
(example 6) 20 KC	260 K	2 Ω	1K	1K
(example 7) 500 KC	10 K	50 Ω	1K	1K
(example 8) 5 MC	1 K	500 Ω	1K	1K

(e)

Figure 2·5 High-frequency effect on shunt-compensated amplifier. (a) At 100 cps; (b) at 20 kc; (c) at 500 kc; (d) at 5 Mc.

These values for X_{C_T} and X_{L_1} are shown in Fig. 2·5*a* and *e*.

$$Z_L \text{ at 100 cps} = \frac{X_{C_T}\sqrt{X_{L_1}{}^2 + R_L{}^2}}{\sqrt{R_L{}^2 + (X_{L_1} - X_{C_T})^2}} \tag{2·3}$$

$$= \frac{50 \times 10^6 \sqrt{0.01^2 + (1 \times 10^3)^2}}{\sqrt{(1 \times 10^3)^2 + (0.01 - 50 \times 10^6)^2}}$$

$$= \frac{50 \times 10^6(1 \times 10^3) \text{ (approx.)}}{50 \times 10^6 \text{ (approx.)}}$$

$$= 1 \times 10^3$$

$$Z_L = 1 \text{ kilohm}$$

Example 2·6 At a frequency of 20 kc, find Z_L after determining the values of X_{C_T} and X_{L_1}.

$$X_{C_T} \text{ at 20 kc} = \frac{1}{2\pi f C_T}$$

$$= \frac{1}{6.28(20 \times 10^3)30 \times 10^{-12}}$$

$$= 260 \text{ kilohms (approx.)}$$

$$X_{L_1} \text{ at 20 kc} = 2\pi f L_1$$

$$= 6.28(20 \times 10^3)15 \times 10^{-6}$$

$$= 2 \text{ ohms (approx.)}$$

These values for X_{C_T} and X_{L_1} are shown in Fig. 2·5*b* and *e*.

$$Z_L \text{ at 20 kc} = \frac{X_{C_T}\sqrt{X_{L_1}{}^2 + R_L{}^2}}{\sqrt{R_L{}^2 + (X_{L_1} - X_{C_T})^2}} \qquad (2\cdot3)$$

$$= \frac{260 \times 10^3 \sqrt{2^2 + (1 \times 10^3)^2}}{\sqrt{(1 \times 10^3)^2 + [2 - (260 \times 10^3)]^2}}$$

$$= \frac{260 \times 10^3(1 \times 10^3)}{260 \times 10^3 \text{ (approx.)}}$$

$$= 1 \times 10^3$$

$$= 1 \text{ kilohm}$$

Example 2·7 At a frequency of 500 kc, find Z_L after determining the values of X_{C_T} and X_{L_1}.

$$X_{C_T} \text{ at 500 kc} = \frac{1}{2\pi f C_T}$$

$$= \frac{1}{6.28(500 \times 10^3)30 \times 10^{-12}}$$

$$= 10 \text{ kilohms}$$

$$X_{L_1} \text{ at 500 kc} = 2\pi f L_1$$

$$= 6.28(500 \times 10^3)15 \times 10^{-6}$$

$$= 50 \text{ ohms (approx.)}$$

These values for X_{C_T} and X_{L_1} are shown in Fig. 2·5*c* and *e*.

$$Z_L \text{ at 500 kc} = \frac{X_{C_T}\sqrt{X_{L_1}{}^2 + R_L{}^2}}{\sqrt{R_L{}^2 + (X_{L_1} - X_{C_T})^2}} \qquad (2\cdot3)$$

$$= \frac{(10 \times 10^3)\sqrt{50^2 + (1 \times 10^3)^2}}{\sqrt{(1 \times 10^3)^2 + [50 - (10 \times 10^3)]^2}}$$

$$= \frac{10 \times 10^3(1 \times 10^3)}{10 \times 10^3 \text{ (approx.)}}$$

$$= 1 \times 10^3$$

$$= 1 \text{ kilohm}$$

Example 2·8 At a frequency of 5 Mc, find Z_L after determining the values of X_{C_T} and X_{L_1}.

$$X_{C_T} \text{ at 5 Mc} = \frac{1}{2\pi f C_T}$$

$$= \frac{1}{6.28(5 \times 10^6)10 \times 10^{-12}}$$

$$= 1 \text{ kilohm}$$

$$X_{L_1} \text{ at 5 Mc} = 2\pi f L$$

$$= 6.28(5 \times 10^6)15 \times 10^{-6}$$

$$= 500 \text{ ohms (approx.)}$$

These values for X_{C_T} and X_{L_1} are shown in Fig. 2·5*d* and *e*.

$$Z_L \text{ at 5 Mc} = \frac{X_{C_T}\sqrt{X_{L_1}{}^2 + R_L{}^2}}{\sqrt{R_L{}^2 + (X_{L_1} - X_{C_T})^2}} \tag{2·3}$$

$$= \frac{1 \times 10^3 \sqrt{500^2 + (1 \times 10^3)^2}}{\sqrt{(1 \times 10^3)^2 + [500 - (1 \times 10^3)]^2}}$$

$$= 1 \times 10^3$$

$$= 1 \text{ kilohm}$$

The results of Examples 2·5 to 2·8, tabulated in Fig. 2·5*e*, show that Z_L remains a constant value as the frequency increases despite the fact that X_{C_T} *decreases*. Since the voltage amplification depends on Z_L, then the gain also remains constant as the frequency is increased.

By varying the value of the shunt-peaking coil L_1 of Fig. 2·4 from the size given in Eq. (2·2) ($L = \frac{1}{2}C_T R_L{}^2$) the gain at the high frequency (f_0) changes. A larger coil increases this high-frequency gain, while not affecting it at the medium frequencies. This rise at f_0 is called overcompensation and is shown in the frequency-response curves of Fig. 2·7.

Similarly, varying R_L results in a change of gain at f_0. A larger R_L decreases the gain at f_0, while a smaller R_L causes a rise at f_0 with the overcompensation shown in the curves of Fig. 2·7. Note that, as shown in these curves, the frequency response of the shunt-compensated amplifier is also not completely uniform but rises slightly at f_0. By making *both* R_L and L_1 smaller, where R_L is made only $0.85X_{C_T}$ at f_0 (instead of equal to X_{C_T}) and L_1 is $0.42C_T R_L{}^2$ (instead of $0.5C_T R_L{}^2$), a smaller gain at *all* frequencies results, without any rise or overcompensation at f_0.

Another and more complex method of high-frequency compensation, shown in Fig. 2·6*a*, is called *series compensation*. The series peaking coil L_2 separates C_T into two parts, C_{out} and C_{in}. C_{out} is the tube capacitance and wiring capacitance between the plate and ground, while C_{in} is the capacitance between the grid and ground of the next stage as well as the stray wiring capacitance at this point. This includes the *Miller effect* as explained in Sec. 2·2.

The series peaking coil L_2 and C_{in} act as a series resonant circuit at the highest desirable frequency f_0. At this frequency, the output signal voltage across C_{in} rises to a maximum value. This, of course, is true of any series resonant circuit where a constant voltage input, varying in frequency, produces a maximum voltage across the capacitor at the resonant frequency. A 1-volt a-c input produces, at resonance, a larger a-c voltage across the capacitor. The amplitude of the capacitor voltage at resonance depends upon the input signal and the Q of the circuit. A series resonant circuit having a Q of 10 steps up the capacitor voltage by a factor of 10 so that a 1-volt a-c input produces 10 volts across the capacitor. This is often called the *resonant rise* of voltage or the *gain* of a resonant circuit.

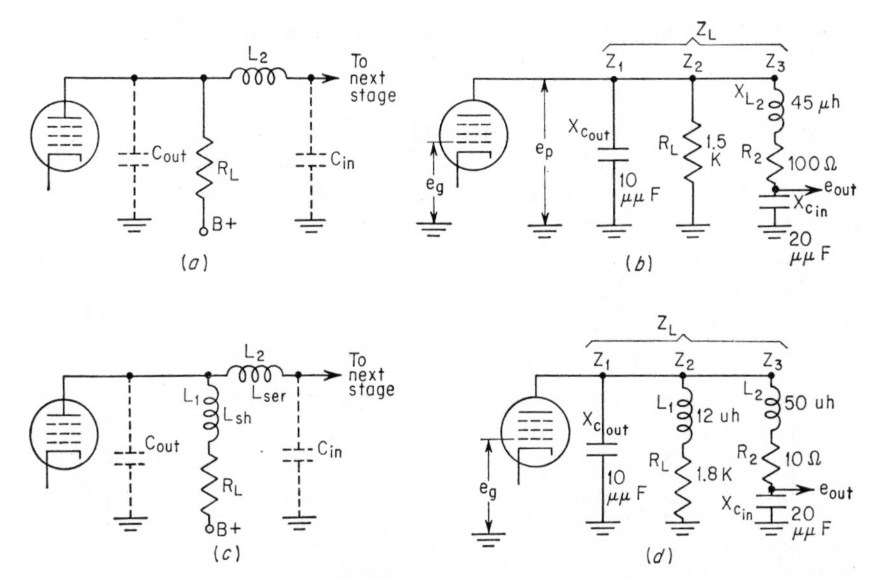

Figure 2·6 **High-frequency compensation. (*a*) Series compensation; (*b*) series compensation example; (*c*) shunt-series compensation; (*d*) shunt-series compensation example.**

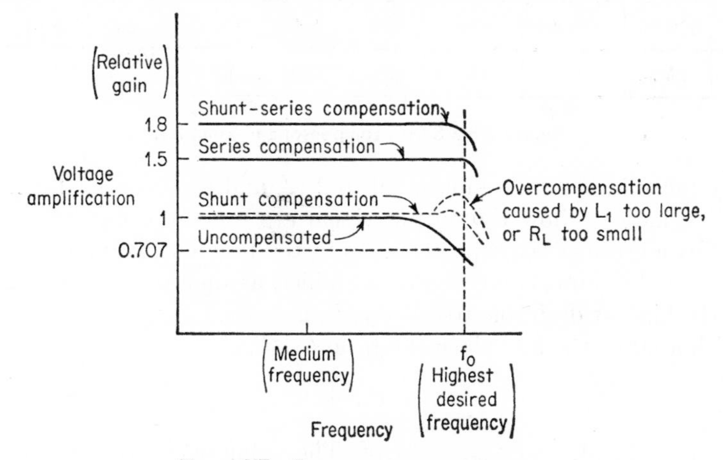

Figure 2·7 **Frequency-response curves.**

The two parts of C_T, C_{in} and C_{out}, separated by the series peaking coil L_2, should have a 2:1 ratio in the following discussions. That is, $\dfrac{C_{\text{in}}}{C_{\text{out}}} = \dfrac{2}{1}$. By the proper positioning of the coil and other parts, this ratio can be attained. The values of the peaking coil L_2 and the load resistor R_L, similar to those of the *shunt* compensation circuit discussed previously, depend on C_T and the highest desirable frequency f_0. These values are

$$R_L = 1.5X_{C_T} \text{ at } f_0 \tag{2·4}$$

where R_L is in ohms, C_T in farads, and f_0 in cycles per second.

$$L_2 = 0.67C_T(R_L)^2 \tag{2·5}$$

where L_2 is in henrys, C_T in farads, and R_L in ohms.

The series compensation circuit of Fig. 2·6a can be redrawn to that of Fig. 2·6b. Here, it can be more easily seen that the load impedance, Z_L, of the tube consists of three parallel branches, Z_1, Z_2, and Z_3. Z_1 is simply $X_{C_{out}}$, Z_2 is simply R_L, while Z_3 consists of X_{L_2}, R_2 (the resistance of the peaking coil), and $X_{C_{in}}$. The a-c signal voltage applied to the tube is shown as e_G, while the plate-voltage signal is called e_P. The output signal e_{out}, available to the next stage, is the voltage across $X_{C_{in}}$.

	Freq	Z_1: $X_{C_{out}}$ (approx) 10 μμf	X_{C_t} (approx) 30 μμf	Z_2: R_L	Z_3: $X_{C_{in}}$ (approx) 20 μμf	Z_3: X_{L_2} (approx) 45 uh	Z_L (approx)	e_p	e_{out} $E_{X_{C_{in}}}$
ex 9	100 cps	160 Meg	50 Meg	1.5 K	80 Meg	0.03 Ω	1.5 K	7.5 v	7.5 v
ex 10	20 KC	800 K	260 K	1.5 K	400 K	6 Ω	1.5 K	7.5 v	7.5 v
ex 11	500 KC	30 K	10 K	1.5 K	15 K	150 Ω	1.5 K	7.5 v	7.5 v
ex 12	5 MC	3 K	1 K	1.5 K	1.5 K	1.5 K	100 Ω (R_2)	0.5 v	7.5 v

Figure 2·8 **Series compensation examples.**

The following four examples, 2·9 to 2·12, and the tabulated results of Fig. 2·8, indicate how the output signal e_{out} remains constant over the desired wide range of frequencies. As in the previous examples and discussion of the shunt compensated amplifier, assume that C_T is 30 pf and that the highest desirable frequency, f_0, is 5 Mc.

With a 30-pf C_T, and since the ratio $C_{in}/C_{out} = 2/1$, and

$$C_T = C_{in} + C_{out}$$

then $C_{in} = 20$ pf, and $C_{out} = 10$ pf. The value of R_L required is, from Eq. (2·4),

$$\begin{aligned} R_L &= 1.5X_{C_T} \text{ at } f_0 \\ &= (1.5)\frac{1}{2\pi f_0 C_T} \\ &= (1.5)\frac{1}{6.28(5 \times 10^6)30 \times 10^{-12}} \\ &= 1.5(1{,}000) \\ &= 1.5 \text{ kilohms} \end{aligned}$$

The value of the series peaking coil, L_2, required is, from Eq. (2·5),

$$
\begin{aligned}
L_2 &= 0.67 C_T R_L^2 \\
&= 0.67(30 \times 10^{-12})(1.5 \times 10^3)^2 \\
&= 20 \times 10^{-12}(2.25 \times 10^6) \\
&= 45 \times 10^{-6} \text{ henry} \\
&= 45\ \mu\text{h}
\end{aligned}
$$

The values $C_{\text{out}} = 10$ pf, $C_{\text{in}} = 20$ pf, $L_2 = 45\ \mu$h, and $R_L = 1.5$ kilohms are shown in Fig. 2·6*b*. Also shown is an assumed 100 ohms of resistance, R_2, in the coil L_2.

Example 2·9 In the circuit of Fig. 2·6*b* at a frequency of 100 cps, find the output voltage, e_{out}, after finding all the items shown in the table of Fig. 2·8, $X_{C_{\text{out}}}$, $X_{C_{\text{in}}}$, X_{C_T}, X_{L_2}, Z_L and plate signal e_P. Assume that the input signal, e_G, is 1 volt and that the tube transconductance, G_M, is 5,000 μmhos.

$$
\begin{aligned}
X_{C_{\text{out}}} \text{ at 100 cycles} &= \frac{1}{2\pi f C_{\text{out}}} \\
&= \frac{1}{6.28(100)10 \times 10^{-12}} \\
&= 160 \text{ megohms (approx.)} \\
X_{C_{\text{in}}} \text{ at 100 cycles} &= \frac{1}{6.28(100)20 \times 10^{-12}} \\
&= 80 \text{ megohms (approx.)} \\
X_{C_T} \text{ at 100 cycles} &= \frac{1}{2\pi f C_T} \\
&= \frac{1}{6.28(100)30 \times 10^{-12}} \\
&= 50 \text{ megohms (approx.)} \\
X_{L_2} \text{ at 100 cycles} &= 2\pi f L \\
&= 6.28(100)45 \times 10^{-6} \\
&= 0.03 \text{ ohm (approx.)}
\end{aligned}
$$

Z_L at 100 cycles consists of Z_1, Z_2, and Z_3 in parallel. Z_1 is simply $X_{C_{\text{out}}}$, which, at 100 cycles, is 160 megohms. Z_2 is simply R_L, which is 1.5 kilohms. Z_3 is X_{L_2}, R_2 (the coil resistance of 100 ohms), and $X_{C_{\text{in}}}$ in series.

$$
\begin{aligned}
Z_3 &= \sqrt{R_2^2 + (X_{L_2} - X_{C_{\text{in}}})^2} \\
&= \sqrt{100^2 + (0.03 - 80 \text{ megohms})^2} \\
&= 80 \text{ megohms}
\end{aligned}
$$

Since Z_2 (or R_L) is only 1.5 kilohms and is much smaller than Z_1 (160 megohms) or Z_3 (80 megohms), then Z_L is just about equal to R_L alone. The plate signal E_P is the product of the input signal e_G and the voltage amplification (VA).

$$\begin{aligned} e_P &= e_G \text{VA} \\ &= e_G G_M Z_L \\ &= 1(5{,}000 \times 10^{-6})1.5 \times 10^3 \\ &= 7.5 \text{ volts} \end{aligned}$$

e_{out} is that part of e_P which appears across $X_{C_{\text{in}}}$

$$\begin{aligned} e_{\text{out}} &= e_P \frac{X_{C_{\text{in}}}}{Z_3} \\ &= \left(7.5\right)\frac{80 \times 10^6}{80 \times 10^6} \\ &= 7.5 \text{ volts} \end{aligned}$$

Example 2·10 In the circuit of Fig. 2·6*b*, find e_{out} at a frequency of 20 kc, after finding $X_{C_{\text{out}}}$, $X_{C_{\text{in}}}$, X_{C_T}, X_{L_2}, and Z_L as listed in Fig. 2·8.

$$\begin{aligned} X_{C_{\text{out}}} \text{ at 20 kc} &= \frac{1}{2\pi f C_{\text{out}}} \\ &= \frac{1}{6.28(20 \times 10^3)10 \times 10^{-12}} \\ &= 800 \text{ kilohms (approx.)} \\ X_{C_{\text{in}}} \text{ at 20 kc} &= \frac{1}{2\pi f C_{\text{in}}} \\ &= \frac{1}{6.28(20 \times 10^3)20 \times 10^{-12}} \\ &= 400 \text{ kilohms (approx.)} \\ X_{C_T} \text{ at 20 kc} &= \frac{1}{2\pi f C_T} \\ &= \frac{1}{6.28(20 \times 10^3)30 \times 10^{-12}} \\ &= 260 \text{ kilohms (approx.)} \\ X_{L_2} \text{ at 20 kc} &= 2\pi f L_2 \\ &= 6.28(20 \times 10^3)45 \times 10^{-6} \\ &= 6 \text{ ohms (approx.)} \end{aligned}$$

Z_3 consists of X_{L_2}, R_2, and $X_{C_{in}}$ in series.

$$\begin{aligned} Z_3 &= \sqrt{R_2{}^2 + (X_{L_2} - X_{C_{in}})^2} \\ &= \sqrt{100^2 + (6 - 400 \text{ kilohms})^2} \\ &= 400 \text{ kilohms} \end{aligned}$$

Z_L consists of Z_1, Z_2, and Z_3 in parallel. Since Z_2 (or R_2) is 1.5 kilohms and is much smaller than Z_1 (or $X_{C_{out}}$), which is 800 kilohms, or Z_3, which is 400 kilohms, then Z_L is just about equal to R_2.

$$\begin{aligned} e_P &= e_G \text{VA} \\ &= e_G G_M Z_L \\ &= 1(5{,}000 \times 10^{-6})1.5 \times 10^3 \\ &= 7.5 \text{ volts} \end{aligned}$$

$$\begin{aligned} e_{\text{out}} &= e_P \frac{X_{C_{in}}}{Z_3} \\ &= (7.5) \frac{400 \times 10^3}{400 \times 10^3} \\ &= 7.5 \text{ volts} \end{aligned}$$

Example 2·11 At a frequency of 500 kc in the circuit of Fig. 2·6*b*, find e_{out} after finding $X_{C_{out}}$, $X_{C_{in}}$, X_{C_T}, X_{L_2}, and Z_L.

$$\begin{aligned} X_{C_{out}} \text{ at 500 kc} &= \frac{1}{2\pi f C_{\text{out}}} \\ &= \frac{1}{6.28(500 \times 10^3)10 \times 10^{-12}} \\ &= 30 \text{ kilohms (approx.)} \end{aligned}$$

$$\begin{aligned} X_{C_{in}} \text{ at 500 kc} &= \frac{1}{2\pi f C_{\text{in}}} \\ &= \frac{1}{6.28(500 \times 10^3)20 \times 10^{-12}} \\ &= 15 \text{ kilohms (approx.)} \end{aligned}$$

$$\begin{aligned} X_{C_T} \text{ at 500 kc} &= \frac{1}{2\pi f C_T} \\ &= \frac{1}{6.28(500 \times 10^3)30 \times 10^{-12}} \\ &= 10 \text{ kilohms (approx.)} \end{aligned}$$

$$\begin{aligned} X_{L_2} \text{ at 500 kc} &= 2\pi f L \\ &= 6.28(500 \times 10^3)45 \times 10^{-6} \\ &= 150 \text{ ohms (approx.)} \end{aligned}$$

$$\begin{aligned} Z_3 &= \sqrt{R_2{}^2 + (X_{L_2} - X_{C_{in}})^2} \\ &= \sqrt{100^2 + (150 - 15 \text{ kilohms})^2} \\ &= 15 \text{ kilohms (approx.)} \end{aligned}$$

Z_L consists of Z_1, Z_2, and Z_3 in parallel. Since Z_2 (or R_L) is 1.5 kilohms and is much smaller than Z_1 (or $X_{C_{out}}$), which is 30 kilohms, or Z_3, which is 15 kilohms, then Z_L is just about equal to Z_2 (or R_L).

$$\begin{aligned} e_P &= e_G \text{VA} \\ &= e_G G_M Z_L \\ &= 1(5{,}000 \times 10^{-6})1.5 \times 10^3 \\ &= 7.5 \text{ volts} \\ e_{\text{out}} &= e_P \frac{X_{C_{\text{in}}}}{Z_3} \\ &= 7.5 \times \frac{15 \text{ kilohms}}{15 \text{ kilohms}} \\ &= 7.5 \text{ volts} \end{aligned}$$

Example 2·12 At a frequency of 5 Mc (f_0) in the circuit of Fig. 2·6*b*, find e_{out} after solving for $X_{C_{\text{out}}}$, $X_{C_{\text{in}}}$, X_{C_T}, X_{L_2}, and Z_L.

$$\begin{aligned} X_{C_{\text{out}}} \text{ at 5 Mc} &= \frac{1}{2\pi f C_{\text{out}}} \\ &= \frac{1}{6.28(5 \times 10^6)10 \times 10^{-12}} \\ &= 3 \text{ kilohms (approx.)} \\ X_{C_{\text{in}}} \text{ at 5 Mc} &= \frac{1}{2\pi f C_{\text{in}}} \\ &= \frac{1}{6.28(5 \times 10^6)20 \times 10^{-12}} \\ &= 1.5 \text{ kilohms (approx.)} \\ X_{C_T} \text{ at 5 Mc} &= \frac{1}{2\pi f C_T} \\ &= \frac{1}{6.28(5 \times 10^6)30 \times 10^{-12}} \\ &= 1 \text{ kilohm (approx.)} \\ X_{L_2} \text{ at 5 Mc} &= 2\pi f L_2 \\ &= 6.28(5 \times 10^6)45 \times 10^{-6} \\ &= 1.5 \text{ kilohms (approx.)} \end{aligned}$$

Note that at this 5-Mc frequency, $X_{C_{\text{in}}}$ is equal to X_{L_2} (both 1.5 kilohms), making L_2 and C_{in} a series resonant circuit. The impedance, Z_3, of this series-resonant circuit is

$$\begin{aligned} Z_3 &= \sqrt{R_2{}^2 + (X_{L_2} - X_{C_{in}})^2} \\ &= \sqrt{100^2 + (1.5 \text{ kilohms} - 1.5 \text{ kilohms})^2} \\ &= 100 \text{ ohms} \end{aligned}$$

Z_L consists of Z_1, Z_2, and Z_3 in parallel. At a frequency of 5 Mc, Z_3 (the series-resonant circuit) is only 100 ohms and is much smaller than Z_1 (or $X_{C_{out}}$), which is 3 kilohms, or Z_2 (or R_L), which is 1.5 kilohms. Z_L is therefore equal to Z_3.

$$\begin{aligned} e_P &= e_G \text{VA} \\ &= e_G G_M Z_L \\ &= 1(5{,}000 \times 10^{-6})100 \\ &= 0.5 \text{ volt} \end{aligned}$$

Note that at this 5-Mc frequency, the series-resonant circuit (Z_3) reduces Z_L, causing a smaller gain (VA). This produces a plate-voltage signal (e_P) of only 0.5 volt, which is much smaller than the previous values of e_P (7.5 volts).

$$\begin{aligned} e_{out} &= e_P \frac{X_{C_{in}}}{Z_3} \\ &= 0.5 \left(\frac{1{,}500}{100}\right) \\ &= 7.5 \text{ volts} \end{aligned}$$

Another method of finding e_{out} (which is the voltage across C_{in}) of the series resonant circuit Z_3 is $e_{C_{in}} = E_{applied} \times Q$ of the resonant circuit

or

$$\begin{aligned} e_{out} &= e_P \frac{X_{L_2}}{R_2} \\ &= 0.5 \left(\frac{1{,}500}{100}\right) \\ &= 7.5 \text{ volts} \end{aligned}$$

Note that e_{out} at 5 Mc is the same amplitude (7.5 volts) as at the lower frequencies despite the fact that the load impedance (Z_L) and the gain (VA) decrease considerably. This, of course, is due to the resonant rise of voltage across C_{in} at the resonant frequency.

From Examples 2·9 to 2·12 of the *series-compensated* circuit of Fig. 2·6*b*, shown tabulated in Fig. 2·8, it can be seen that as the *frequency* of the 1-volt input signal (e_G) is increased, the output signal (e_{out}) across C_{in} remains a constant value of 7.5 volts. The *series*-peaking coil L_2 does not reduce the effect of C_T such as the *shunt*-peaking coil (discussed previously) did. L_2 and C_{in} actually lower Z_L, causing a reduction of plate signal (e_P), but they provide the voltage step up of a series-resonant circuit in order to maintain a constant amplitude output signal (e_{out}).

Comparing the values of the load resistors for shunt compensation ($R_L = X_{C_T}$ at f_0), and series compensation ($R_L = 1.5X_{C_T}$ at f_0), Eqs. (2·1) and (2·4) respectively, it is seen that for the same values of C_T and f_0, R_L in series compensation is 50 per cent larger than for shunt peaking. This produces a 50 per cent greater gain in series compensation, as shown in the frequency-response curves of Fig. 2·7. To produce the proper amount of voltage rise by the series resonant L_2 and C_{in}, the Q of this circuit must be correct. In Examples 2·9 to 2·12, it was assumed that L_2 had a resistance, R_2, of 100 ohms. An additional resistor, not shown, is often placed in parallel with L_2 to produce the required Q and the resulting correct resonant rise of voltage across C_{in}. Phase shift or time delay, which is discussed more fully in the topic of low-frequency compensation, is also better with series compensation than with the shunt method. This phase shift, caused by the reactances in the circuit, ideally should be linear and proportional to the frequency. That is, if a fundamental frequency is shifted 10°, then the second harmonic should be shifted twice this amount, or 20°, while the third harmonic should undergo a 30° delay, etc. Series peaking more nearly approaches this ideal linear phase shift than the shunt-compensated circuit.

A third type of high-frequency compensation, and the most commonly used, is a *combination shunt-series* compensation circuit, the diagram of which is shown in Fig. 2·6c. This circuit, as shown, uses a shunt-peaking coil, L_1, and a series coil, L_2. As in series compensation, L_2 separates C_T into the two components C_{out} and C_{in}. The following relationships assume a 2:1 ratio of C_{in} to C_{out}, as was used in the previous *series* compensation. The values of the various components are

$$R_L = 1.8X_{C_T} \text{ at } f_0 \tag{2·6}$$

where R_L is in ohms, C_T is in farads, and the frequency is in cycles per second.

$$L_1 \text{ (the shunt coil)} = 0.12C_T R_L^2 \tag{2·7}$$

$$L_2 \text{ (the series coil)} = 0.52C_T R_L^2 \tag{2·8}$$

where L_1 and L_2 are in henrys, C_T is in farads, and R_L in ohms.

The following four Examples, 2·13 to 2·16, shown tabulated in Fig. 2·9, illustrate the method whereby the shunt-series compensation circuit of Fig. 2·6c maintains a constant output signal, e_{out}, at all frequencies up to the highest desirable frequency, f_0. The values of C_{out} and C_{in} are the same as those assumed for the series peaking discussion, namely 10 pf and 20 pf respectively. C_T is the sum of the two, or 30 pf. Using, as before,

5 Mc as f_0, then the value of R_L is

$$R_L = 1.8X_{C_T} \text{ at } f_0 \tag{2·6}$$
$$= 1.8\left(\frac{1}{2\pi f_0 C_T}\right)$$
$$= 1.8\left[\frac{1}{6.28(5 \times 10^6)30 \times 10^{-12}}\right]$$
$$= 1.8 \text{ kilohms (approx.)}$$

The value of the shunt coil L_1 is

$$L_1 = 0.12C_T R_L^2 \tag{2·7}$$
$$= 0.12(30 \times 10^{-12})(1.8 \times 10^3)^2$$
$$= 12\ \mu\text{h (approx.)}$$

The value of the series peaking coil L_2 is

$$L_2 = 0.52C_T R_L^2 \tag{2·8}$$
$$= 0.52(30 \times 10^{-12})(1.8 \times 10^3)^2$$
$$= 50\ \mu\text{h (approx.)}$$

A value of resistance, R_2, for the L_2 coil will be assumed to be 10 ohms and is shown in Fig. 2·6*d* along with the other components. The load impedance, Z_L, consists of Z_1, Z_2, and Z_3 in parallel. Z_1 is simply $X_{C_{out}}$. Z_2 is made up of X_{L_1} and R_L in series, while Z_3 consists of X_{L_2}, R_2, and $X_{C_{in}}$ in series. The following discussion and examples use an input signal, e_G, of 1 volt, and a tube transconductance, G_M, of 5,000 μmhos. Plate signal is e_P, and the output signal, e_{out}, is across C_{in}, as shown in Fig. 2·6*d*.

		Z_1	Z_2		Z_3		Z_L		
	FREQ	$X_{C_{out}}$ (approx) 10 $\mu\mu$F	$X_{L_1(shunt)}$ (approx) 12 uh	R_L	$X_{L_2(series)}$ (approx) 50 uh	$X_{C_{in}}$ (approx) 20 $\mu\mu$F	(approx)	e_P	e_{out} (approx) $E_{X_{C_{in}}}$
ex 13	100 cps	160 Meg	0.008 Ω	1.8 K	0.032 Ω	80 Meg	1.8 K	9 v	9 v
ex 14	20 KC	800 K	1.6 Ω	1.8 K	6.4 Ω	400 K	1.8 K	9 v	9 v
ex 15	500 KC	30 K	40 Ω	1.8 K	160 Ω	15 K	1.8 K	9 v	9 v
ex 16	5 MC	3 K	400 Ω	1.8 K	1.6 K	1.5 K	100 Ω Z_3	0.5 v	7.5 v

Figure 2·9 **Shunt-series compensation examples.**

Example 2·13 In the shunt-series compensation circuit of Fig. 2·6*c* and *d*, find the output signal, e_{out}, at a frequency of 100 cps after solving for Z_1, Z_2, Z_3, etc., as shown in the tabulation of Fig. 2·9.

$$
\begin{aligned}
Z_1 &= X_{C_{\text{out}}} = \frac{1}{2\pi f C_{\text{out}}} \\
&= \frac{1}{6.28(100)10 \times 10^{-12}} \\
&= 160 \text{ megohms} \\
Z_2 &= \sqrt{R_L{}^2 + X_{L_1}{}^2} \\
&= \sqrt{(1.8 \times 10^3)^2 + (2\pi f L_1)^2} \\
&= \sqrt{(1.8 \times 10^3)^2 + [6.28(100)12 \times 10^{-6}]} \\
&= \sqrt{(1.8 \times 10^3)^2 + 0.008^2} \\
&= 1.8 \text{ kilohms} \\
Z_3 &= \sqrt{R_2{}^2 + (X_{L_2} - X_{C_{\text{in}}})^2}
\end{aligned}
$$

First solving for X_{L_2} and $X_{C_{\text{in}}}$ yields

$$
\begin{aligned}
X_{L_2} &= 2\pi f L_2 \\
&= 6.28(100)50 \times 10^{-6} \\
&= 0.032 \text{ ohms (approx.)}
\end{aligned}
$$

and

$$
\begin{aligned}
X_{C_{\text{in}}} &= \frac{1}{2\pi f C_{\text{in}}} \\
&= \frac{1}{6.28(100)20 \times 10^{-12}} \\
&= 80 \text{ megohms (approx.)}
\end{aligned}
$$

Therefore

$$
\begin{aligned}
Z_3 &= \sqrt{R_2{}^2 + (X_{L_2} - X_{C_{\text{in}}})^2} \\
&= \sqrt{10^2 + (0.032 - 80 \text{ megohms})^2} \\
&= 80 \text{ megohms (approx.)}
\end{aligned}
$$

Z_L consists of Z_1, Z_2, and Z_3 in parallel. Since Z_2 is only 1.8 kilohms and is much smaller than Z_1 (160 megohms), or Z_3 (80 megohms), then Z_L is just about equal to Z_2, or 1.8 kilohms.

The gain or voltage amplification is

$$
\begin{aligned}
\text{VA} &= G_M Z_L \\
&= 5{,}000 \times 10^{-6}(1.8 \times 10^3) \\
&= 9
\end{aligned}
$$

The plate signal e_P is the gain multiplied by the grid signal e_G of 1 volt.

$$
\begin{aligned}
e_P &= \text{VA}(e_G) \\
&= 9(1) \\
&= 9 \text{ volts}
\end{aligned}
$$

The output signal e_{out} is across $X_{C_{\text{in}}}$ and is a part of e_P, that part which is the ratio of $X_{C_{\text{in}}}$ to Z_3.

$$\begin{aligned} e_{out} &= e_P \frac{X_{C_{in}}}{Z_3} \\ &= 9\left(\frac{80 \text{ megohms}}{80 \text{ megohms}}\right) \\ &= 9 \text{ volts} \end{aligned}$$

Example 2·14 At a frequency of 20 kc in the circuit of Fig. 2·6*c* and *d*, solve for the output signal e_{out}, in the same way as the previous example. The solutions for all necessary values of $X_{C_{out}}$, X_{L_1}, Z_2, etc., are shown below, and the results are tabulated in Fig. 2·9.

$$\begin{aligned} Z_1 = X_{C_{out}} &= \frac{1}{2\pi f C_{out}} \\ &= \frac{1}{6.28(20 \times 10^3)10 \times 10^{-12}} \\ &= 800 \text{ kilohms (approx.)} \end{aligned}$$

$$\begin{aligned} Z_2 &= \sqrt{R_L{}^2 + X_{L_1}{}^2} \\ &= \sqrt{(1.8 \times 10^3)^2 + (2\pi f L_1)^2} \\ &= \sqrt{(1.8 \times 10^3)^2 + [6.28(20 \times 10^3)12 \times 10^{-6}]^2} \\ &= \sqrt{(1.8 \times 10^3)^2 + 1.6^2} \\ &= 1.8 \text{ kilohms} \end{aligned}$$

$$\begin{aligned} X_{L_2} &= 2\pi f L_2 \\ &= 6.28(20 \times 10^3)50 \times 10^{-6} \\ &= 6.4 \text{ ohms (approx.)} \end{aligned}$$

$$\begin{aligned} X_{C_{in}} &= \frac{1}{2\pi f C_{in}} \\ &= \frac{1}{6.28(20 \times 10^3)20 \times 10^{-12}} \\ &= 400 \text{ kilohms (approx.)} \end{aligned}$$

$$\begin{aligned} Z_3 &= \sqrt{R_2{}^2 + (X_{L_2} - X_{C_{in}})^2} \\ &= \sqrt{10^2 + (6.4 - 400 \text{ kilohms})^2} \\ &= 400 \text{ kilohms (approx.)} \end{aligned}$$

Z_L consists of Z_1, Z_2, and Z_3 in parallel. Since, as in the previous example, Z_2 is 1.8 kilohms and is much smaller than Z_1 (800 kilohms), or Z_3 (400 kilohms), then Z_L is equal to Z_2, or 1.8 kilohms.

The gain or voltage amplification is

$$\begin{aligned} \text{VA} &= G_M Z_L \\ &= 5{,}000 \times 10^{-6}(1.8 \times 10^3) \\ &= 9 \end{aligned}$$

Plate signal e_P is the gain multiplied by the 1-volt grid signal, e_G.

$$\begin{aligned} e_P &= \mathrm{VA}e_G \\ &= 9(1) \\ &= 9 \text{ volts} \end{aligned}$$

The output signal e_{out} is across $X_{C_{\text{in}}}$ and is a part of e_P; that part which is the ratio of $X_{C_{\text{in}}}$ to Z_3.

$$\begin{aligned} e_{\text{out}} &= e_P \frac{X_{C_{\text{in}}}}{Z_3} \\ &= 9 \left(\frac{400 \text{ kilohms}}{400 \text{ kilohms}} \right) \\ &= 9 \text{ volts} \end{aligned}$$

Example 2·15 At a frequency of 500 kc in the circuit of Fig. 2·6*c* and *d*, find the output signal, e_{out}, as in Examples 2·13 and 2·14. The solutions for all necessary values of $X_{C_{\text{out}}}$, X_{L_1}, etc., are shown in the following and are tabulated in Fig. 2·9.

$$\begin{aligned} Z_1 = X_{C_{\text{out}}} &= \frac{1}{2\pi f C_{\text{out}}} \\ &= \frac{1}{6.28(500 \times 10^3)10 \times 10^{-12}} \\ &= 30 \text{ kilohms (approx.)} \end{aligned}$$

$$\begin{aligned} Z_2 &= \sqrt{R_L{}^2 + X_{L_1}{}^2} \\ &= \sqrt{(1.8 \times 10^3)^2 + (2\pi f L_1)^2} \\ &= \sqrt{(1.8 \times 10^3)^2 + [6.28(500 \times 10^3)12 \times 10^{-6}]^2} \\ &= \sqrt{(1.8 \times 10^3)^2 + 40^2} \text{ (approx.)} \\ &= 1.8 \text{ kilohms} \end{aligned}$$

$$\begin{aligned} X_{L_2} &= 2\pi f L_2 \\ &= 6.28(500 \times 10^3)50 \times 10^{-6} \\ &= 160 \text{ ohms (approx.)} \end{aligned}$$

$$\begin{aligned} X_{C_{\text{in}}} &= \frac{1}{2\pi f C_{\text{in}}} \\ &= \frac{1}{6.28(500 \times 10^3)20 \times 10^{-12}} \\ &= 15 \text{ kilohms (approx.)} \end{aligned}$$

$$\begin{aligned} Z_3 &= \sqrt{R_2{}^2 + (X_{L_2} - X_{C_{\text{in}}})^2} \\ &= \sqrt{10^2 + (160 - 15 \text{ kilohms})^2} \\ &= 15 \text{ kilohms (approx.)} \end{aligned}$$

Z_L consists of Z_1, Z_2, and Z_3 in parallel. Since Z_2 is 1.8 kilohms and is much smaller than Z_1 (30 kilohms), or Z_3 (15 kilohms), then Z_L is just about equal to Z_2, or 1.8 kilohms.

The gain or voltage amplification is

$$\begin{aligned} \text{VA} &= G_M Z_L \\ &= 5{,}000 \times 10^{-6}(1.8 \times 10^3) \\ &= 9 \end{aligned}$$

Plate signal e_P is the gain multiplied by the grid signal, e_G, of 1 volt.

$$\begin{aligned} e_P &= \text{VA}e_G \\ &= 9(1) \\ &= 9 \text{ volts} \end{aligned}$$

The output signal e_{out} is across $X_{C_{\text{in}}}$ and is a part of e_P; that part which is the ratio of $X_{C_{\text{in}}}$ to Z_3.

$$\begin{aligned} e_{\text{out}} &= e_P \frac{X_{C_{\text{in}}}}{Z_3} \\ &= 9\left(\frac{15 \text{ kilohms}}{15 \text{ kilohms}}\right) \\ &= 9 \text{ volts} \end{aligned}$$

Example 2·16 At a frequency of 5 Mc in the circuit of Fig. 2·6*c* and *d*, find the output signal, e_{out}, as in Examples 2·13 to 2·15. The solutions for all the necessary values of $X_{C_{\text{out}}}$, X_{L_1}, etc., are shown in the following and are tabulated in Fig. 2·9.

$$\begin{aligned} Z_1 = X_{C_{\text{out}}} &= \frac{1}{2\pi f C_{\text{out}}} \\ &= \frac{1}{6.28(5 \times 10^6)10 \times 10^{-12}} \\ &= 3 \text{ kilohms (approx.)} \end{aligned}$$

$$\begin{aligned} Z_2 &= \sqrt{R_L{}^2 + X_{L_1}{}^2} \\ &= \sqrt{(1.8 \times 10^3)^2 + (2\pi f L_1)^2} \\ &= \sqrt{(1.8 \times 10^3)^2 + [6.28(5 \times 10^6)12 \times 10^{-6}]^2} \\ &= 1.85 \text{ kilohms (approx.)} \end{aligned}$$

$$\begin{aligned} X_{L_2} &= 2\pi f L_2 \\ &= 6.28(5 \times 10^6)50 \times 10^{-6} \\ &= 1.6 \text{ kilohms (approx.)} \end{aligned}$$

$$
\begin{aligned}
X_{C_{\text{in}}} &= \frac{1}{2\pi f C_{\text{in}}} \\
&= \frac{1}{6.28(5 \times 10^6)20 \times 10^{-6}} \\
&= 1.5 \text{ kilohms (approx.)} \\
Z_3 &= \sqrt{R_2{}^2 + (X_{L_2} - X_{C_{\text{in}}})^2} \\
&= \sqrt{10^2 + (1.6 \text{ kilohms} - 1.5 \text{ kilohms})^2} \\
&= \sqrt{10^2 + 100^2} \\
&= 100 \text{ ohms (approx.)}
\end{aligned}
$$

Z_L consists of Z_1, Z_2, and Z_3 in parallel. Since Z_3 is now only 100 ohms, it is much smaller than Z_1 (3 kilohms) or Z_2 (1.85 kilohms). Z_L is therefore just about equal to Z_3.

The gain or voltage amplification is

$$
\begin{aligned}
\text{VA} &= G_M Z_L \\
&= 5{,}000 \times 10^{-6}(100) \\
&= 0.5
\end{aligned}
$$

Plate signal e_P is the gain multiplied by the 1-volt grid input signal e_G.

$$
\begin{aligned}
e_P &= \text{VA}(e_G) \\
&= 0.5(1) \\
&= 0.5 \text{ volt}
\end{aligned}
$$

Output signal, e_{out}, is larger than e_P since L_2 (the series coil) and C_{in} are *almost* a series-resonant circuit at this 5-Mc frequency. A series-resonant circuit produces a rise of voltage across the capacitor at the resonant frequency. The rise or gain of the series-resonant circuit is proportional to the Q of the circuit. This output signal, e_{out} is also (as shown in Examples 2·13 to 2·15)

$$
\begin{aligned}
e_{\text{out}} &= e_P \frac{X_{C_{\text{in}}}}{Z_3} \\
&= 0.5 \left(\frac{1{,}500}{100}\right) \\
&= 7.5 \text{ volts}
\end{aligned}
$$

At resonance, X_{L_2} and $X_{C_{\text{in}}}$ are equal, and therefore cancel. Z_3 is therefore only 10 ohms or R_2, and Z_L is equal to 10 ohms also. The gain is then

$$
\begin{aligned}
\text{VA} &= G_M Z_L \\
&= 5{,}000 \times 10^{-6}(10) \\
&= 0.05
\end{aligned}
$$

Plate signal at series resonance is then

$$e_P = \text{VA}(e_G)$$
$$= 0.05(1)$$
$$= 0.05 \text{ volt}$$
$$e_{\text{out}} = Qe_P$$

The 5-Mc frequency is almost the resonant frequency, and an approximate value of e_{out} is

$$e_{\text{out}} = Qe_P$$
$$= \frac{X_{L_2}}{R_2} e_P$$
$$= \left[\frac{1{,}600 \text{ (approx.)}}{10}\right] 0.05$$
$$= 8 \text{ volts}$$

The results of Examples 2·13 to 2·16 are shown in Fig. 2·9. From this compilation of values, it can be seen that e_{out} remains fairly constant over the desired wide range of frequencies. The shunt-series compensation circuit of Fig. 2·6*c* combines the characteristics of the shunt peaked and the series-peaked amplifiers of Figs. 2·4 and 2·6*a* with, however, a greater gain. Since R_L is larger ($1.8X_{C_T}$) in a shunt-series combination than that of the series circuit ($R_L = 1.5X_{C_T}$), or that of the shunt peaking circuit ($R_L = X_{C_T}$), then the combination circuit has the largest gain. This is shown in the frequency-response curves of Fig. 2·7. The phase shift of the shunt-series combination is better than that of the shunt-peaked amplifier and just about as linear as the series-compensated circuit.

2·4 Distributed Amplifier

The frequency-response curve of several identical conventional amplifiers in cascade is narrower than that of *one* of these amplifiers. This can be seen from Fig. 2·10. The lower response curve for the single amplifier has a maximum voltage gain of 10. The bandwidth is measured at a level which is 3 decibels (db) down from the maximum voltage gain. This is the equivalent of 0.707 of the maximum, or approximately 7 volts. The bandwidth, as shown, is therefore from the low-frequency f_1 to the high-frequency f_2.

If a second identical amplifier were added in cascade with the first, the voltage amplification or gain of the two stages would multiply. Where the gain of *each* is a maximum of 10, the gain of the two in cascade becomes 10 times 10 or 100, as shown in the upper-response curve of Fig. 2·10. At frequency f_1 and also at f_2, where each stage *alone* has a gain of about 7, the gain of the two stages in cascade becomes 7 times 7 or about 50.

The bandwidth of the upper curve for the two amplifiers in cascade is similarly measured between the points where the voltage gain has decreased 3 db, or 0.707 of the maximum of 100 to a value of about 70. The bandwidth, as shown, is from low frequency f_3 to high frequency f_4. Note that this bandwidth is narrower than that of the single stage which has a width from f_1 to f_2.

In order to increase the bandwidth or frequency response of an amplifier, the load resistor R_L must be decreased, producing a smaller gain. As discussed in the previous section on high-frequency compensation, the value of R_L depends upon C_T, the amount of stray wiring capacitance and tube interelectrode capacitance, and the highest desirable frequency f_0. This frequency, f_0, must be increased if a wider bandwidth is desired. In shunt compensation, R_L should be equal to X_{C_T} at f_0 [from Eq. (2·1)]; in

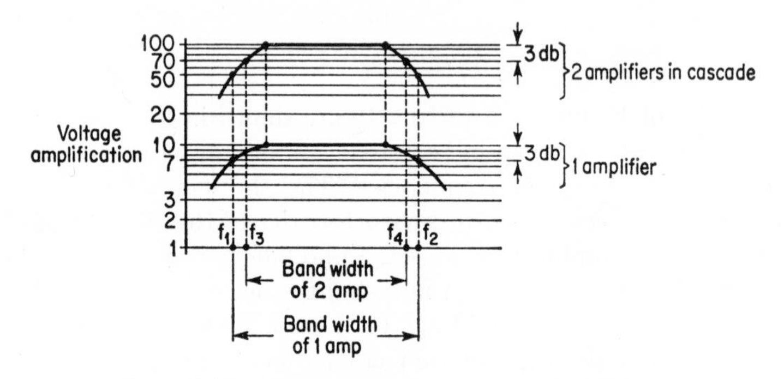

Figure 2·10 Frequency-response curves of amplifiers.

series compensation, $R_L = 1.5X_{C_T}$ at f_0 [from Eq. (2·4)]; and in shunt-series compensation, $R_L = 1.8X_{C_T}$ at f_0 [from Eq. (2·6)]. Therefore, to increase the bandwidth, f_0 is increased and X_{C_T} is decreased, requiring a smaller value of R_L, thus lowering the gain of the stage. Also, as has just been shown, adding additional stages in cascade produces greater gain but narrower bandwidth. A circuit that has a much broader bandwidth than the previous high-frequency compensated amplifier is called a *distributed amplifier*. A tube circuit and one of a transistor are shown in Fig. 2·11*a* and *b*.

A transmission line, such as a coaxial cable, is used in conjunction with a vacuum tube or semiconductor to produce amplification. A coaxial cable, like any transmission line, contains a series of inductors and capacitors along its entire length. These are shown in Fig. 2·11*a* as L_G and C_G in the line which connects to the tubes control grids and also as L_P and C_P in the plate line. In Fig. 2·11*b* these are L_B and C_B in the line connected to the transistor bases and L_C and C_C in the collector line. As shown in Fig. 2·11*a*, an input signal is applied across one end of the grid

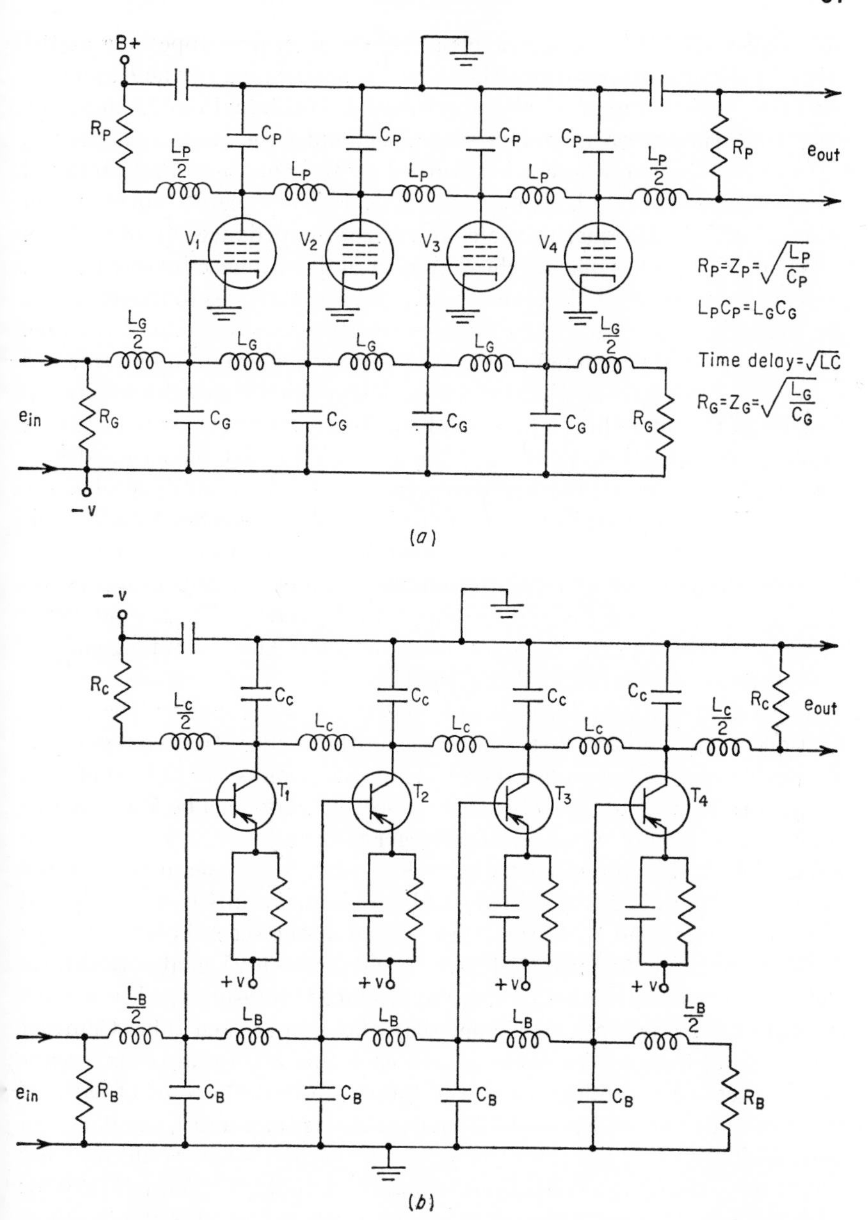

Figure 2·11 Distributed amplifiers. (a) Vacuum-tube amplifier stage; (b) PNP-transistor amplifier stage.

line, across R_G. This signal travels along the line, first appearing at the grid of V_1. Then, after a time delay which is determined by the L and C of the line, the signal appears at the grid of V_2. The signal moves along the line, appearing at each grid in succession, until the signal reaches the right end of the line where it is absorbed by resistor R_G so that no signal remains to be reflected back down the line. This complete absorption will take place if R_G is equal to the characteristic impedance, Z_0, of the line.

When the input signal reaches the grid of V_1, a phase-inverted signal appears at the plate of V_1. Half of this plate signal moves to the left along the plate line and is absorbed by resistor R_P at the B+ end of the line. The other half of the signal at V_1 plate moves to the right along this line and reaches the plate of V_2 after a time delay which is determined by the L and C of this plate line. The L_P and C_P product of the plate line is made equal to the L_G and C_G product of the grid line so that their time delays are equal. As a result, the inverted signal at the plate of V_1 reaches the plate of V_2 at the same instant that the input signal reaches V_2 grid. This signal at the grid of V_2 produces an inverted signal at V_2 plate which *adds* to the signal arriving at V_2 plate from the plate of V_1. This process is repeated by each of the succeeding amplifiers, V_3 and V_4. The output signal is taken from the right end of the plate line, R_P. This output signal, e_{out}, is the *sum* of the plate signals on V_1, V_2, V_3, and V_4.

A distributed amplifier, therefore, has a unique property: The amplification of each tube *adds* to that of the other tubes. In a conventional amplifier (stages in cascade), the total amplification is the *product* of each stage. As a result, the gain of each tube section may even be *less* than one (unity), and the output will still be larger than the input. For example, a 10-mv input signal applied to six tubes, each with a gain of only 0.5 (actually a loss), in a distributed amplifier circuit results in a 5-mv signal at the first plate (10×0.5); a 10-mv signal at the second plate ($5 + 5$); a 15-mv signal at the third ($5 + 5 + 5$); with a 30-mv signal output from the plate of the sixth tube. In order to produce this result, the loss of each section of the line must be smaller than the gain of each tube. Contrast this result of the *addition* of the gain of each tube with that of six *conventional* cascaded amplifiers where the total gain is the *product* of each. If these cascaded amplifiers each had a gain of *less* than unity, say 0.5, then with an input of 10 mv the plate signal of the first stage would be 5 mv (10×0.5); the plate signal of the second tube would be 2.5 mv (5×0.5); with only 0.156 mv at the plate of the sixth tube ($10 \times 0.5 \times 0.5 \times 0.5 \times 0.5 \times 0.5 \times 0.5$, or 10×0.5^6). This, as expected, shows that *conventional cascaded amplifiers*, each with a gain less than unity, are useless as an amplifier, while six similar tubes each with a gain less than unity in the *distributed-amplifier* circuit do produce amplification.

Furthermore, since the total amplification of the distributed amplifier is the *sum* of the gain of each tube, then, adding additional sections, V_5, V_6, V_7, etc., to the circuit of Fig. 2·11*a* will increase the gain without narrowing the bandwidth. These amplifiers have a frequency response that is fairly uniform up to several hundred megacycles, the high frequency being limited by the transit time of electrons within the tube. Each tube, V_1, V_2, V_3, etc., is usually called a *section*, with several such sections constituting a complete *stage* of a distributed amplifier. The transmission line or delay line is often made up of a series of actual coils and capacitors instead of the coaxial type of line. Capacitors C_G and C_P in Fig. 2·11*a* may be the C_{in} and C_{out} of each tube. Often, capacitors are added in shunt to make the tube capacitances less significant. Replacing a tube then poses no problem.

A *transistor distributed amplifier* is shown in Fig. 2·11*b*. The input signal is applied at the lower left corner at the input to the base delay line, across resistor R_B. This signal travels to the right along the line reaching the base of transistor T_1 first. An inverted signal now appears at the collector of T_1. Half of this signal travels to the left on the collector line and is absorbed by the resistor R_C, at the upper left. The remaining half of T_1 collector signal travels to the right and, after a delay which is determined by the L_C and C_C of this line, reaches the collector of T_2. At this same instant, the input signal has reached the base of T_2 since the time delay of the base-line section between T_1 and T_2 is equal to the delay of the collector-line section. The input signal at T_2 base produces an inverted signal at the collector of T_2. This signal *adds* to that from the collector of T_1. This process is repeated as the input signal moves further along to the base of T_3, arriving there as the signal from T_2 collector reaches the collector of T_3. The signal produced by the collector of T_3 *adds* to that which arrived there from the T_2 collector. An output signal, e_{out}, is developed across the upper-right-corner resistor R_C, which is the *sum* of all the collector signals.

When the input signal reaches the right end of the base delay line, it is absorbed by resistor R_B, preventing reflections back down the line. Two resistors, R_C, one at each end of the collector line, also absorb signals from the collectors to prevent reflections back along this line.

2·5 Amplifier Low-frequency Response

The gain or voltage amplification *decreases* at the lower frequencies just as it did at the higher frequencies (as described previously in Sec. 2·2), but for other reasons. This is called *frequency distortion*. The lower response curve of Fig. 2·10 shows that at some low frequency, f_1, the voltage amplification has dropped from a maximum value of 10, at some medium or higher frequency, to a gain of 7. At still lower frequencies than f_1, the gain decreases even further.

This decrease of gain at the lower frequencies is due to the various capacitors in the conventional amplifier circuit, a typical example of which is shown in Fig. 2·12. The bypass capacitors, C_K in the cathode circuit and C_S in the screen circuit, and the coupling capacitor C_C produce distortion at the lower frequencies.

Another type of distortion, besides the decrease of signal at the lower frequencies caused by these capacitors, is called *phase distortion*. Each of

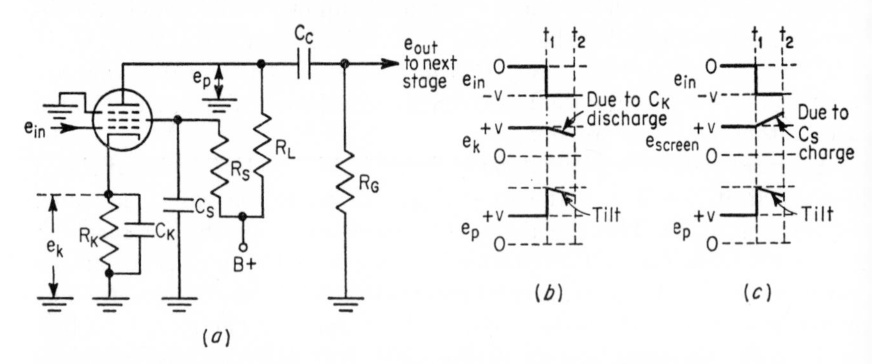

Figure 2·12 Conventional uncompensated amplifier and resulting phase distortions. (a) Conventional uncompensated amplifier; (b) e_P shows a leading phase shift; (c) e_P shows a leading phase shift.

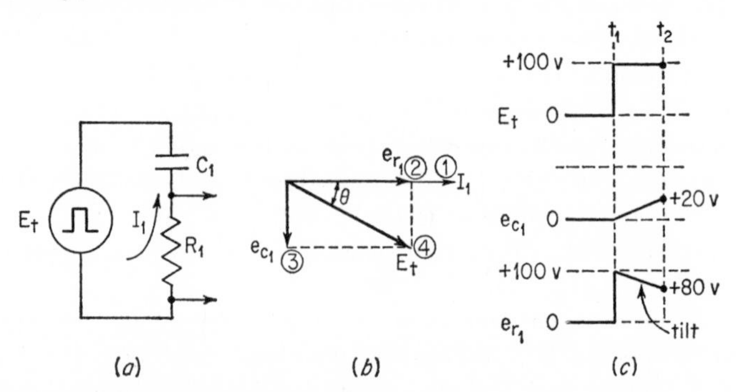

Figure 2·13 Leading phase-shift explanation. (a) Circuit; (b) e_{R_1} leads E_T by angle θ; (c) e_{R_1} denotes a leading phase shift.

the capacitors in Fig. 2·12 produces a *leading* phase shift. That is, the output signal voltage, e_{out}, leads the applied input signal, e_{in}, as will be shown.

An example of a circuit producing a leading phase shift is shown in Fig. 2·13*a*. The explanation is given below and uses the *vectors* or *phasors* shown in Fig. 2·13*b*.

1. As a result of the applied voltage, E_T, a current I_1 flows through R_1 and C_1. I_1 is shown as vector 1.

2. The voltage across resistor R_1, e_{R_I}, as is true of all resistors, is *in phase* with the current I_1. This is shown as vector 2, e_{R_I}, in the same direction as I_1 vector.

3. Voltage across capacitor C_1, e_{C_I}, like all capacitors, *lags* the current I_1 by 90°. Vectors are always assumed to rotate in a counterclockwise direction. Therefore, as shown, vector 3, e_{C_I}, is *lagging* vector 1, I_1, by 90°.

4. The total voltage applied, E_T, is the vector sum of the two voltages, e_{R_I} and e_{C_I}. To add two vectors, a parallelogram is constructed, as shown, by drawing a line from the arrowhead of one vector parallel to the other vector, and repeating this for the other vector arrowhead. These lines are shown as the dotted ones in Fig. 2·13*b*. The vector sum E_T is the diagonal of the parallelogram, extending from the junction (or point of origin) of the original two vectors, e_{R_I} and e_{C_I}, to the junction of the two dotted lines. This sum E_T is shown as vector 4.

5. Comparing e_{R_I} (vector 2) with E_T (vector 4), and recalling that vectors rotate counterclockwise, it can be seen that e_{R_I} is ahead or *leading* E_T by the phase angle, θ.

Another example of a leading phase shift is shown by the voltage waveshapes of Fig. 2·13*c*. Here it is assumed that the generator in the circuit of Fig. 2·13*a* is applying a +100-volt pulse, E_T, to the series circuit of C_1 and R_1. The following discussion explains the waveforms of Fig. 2·13*c*.

1. Before instant t_1, no voltage is yet being applied, and no current flows.

2. There is zero volts across R_1 and C_1.

3. At instant t_1, the applied voltage E_T is +100 volts.

4. Current flows, but at this first instant the capacitor C_1 has not had a chance to start charging yet, and e_{C_I} is still zero.

5. From Kirchhoff's voltage law, the sum of e_{C_I} and e_{R_I} must be equal to the applied +100 volts. At instant t_1, e_{C_I} is still zero, and e_{R_I} is therefore +100 volts.

6. During the period of time from instant t_1 to t_2, capacitor C_1 charges up somewhat. Assume that C_1 reaches a charge of 20 volts.

7. Since $e_{C_I} + e_{R_I} = 100$ volts, then, at instant t_2, when e_{C_I} is +20 volts, e_{R_I} decreases to +80 volts, as shown.

8. Note that with a flat-topped applied pulse E_T, the voltage across R_1, e_{R_I}, has a downward-sloping top. This type of waveshape denotes a *leading* phase shift.

Effect of R_K and C_K. The cathode-bias resistor R_K and its bypass capacitor C_K, shown in Fig. 2·12*a*, produce both *frequency* and *phase* distortion. C_K is such a value that its reactance X_{C_K}, at the lowest desired frequency, is much smaller than R_K. The parallel combination of the two is the cathode impedance Z_K and is just about equal to X_{C_K}. At the lower frequencies X_{C_K} increases, making Z_K larger. The voltage amplification or gain of a pentode using cathode bias with R_K and C_K is

$$\mathrm{VA} = \frac{G_M R_L}{1 + G_M Z_K} \tag{2·9}$$

With a large value of Z_K, at the lower frequencies, the gain decreases, producing frequency distortion. At the medium and higher frequencies, X_{C_K} becomes practically zero, making Z_K the same. The denominator of Eq. (2·9) then becomes 1, and the gain, at the medium and higher frequencies, becomes the usual equation: $\mathrm{VA} = G_M R_L$.

Phase distortion is produced by R_K and C_K as shown by the downward tilt of the plate signal, e_P, in Fig. 2·12*b*. This is brought about as follows:

1. At instant t_1, the input signal e_{in} drives the grid negative, causing plate current to decrease.

2. The decreased plate current through R_K causes the cathode voltage C_K to attempt to become less positive.

3. C_K prevents this voltage from decreasing immediately.

4. The smaller plate current through the load resistor R_L, at instant t_1, causes the plate signal voltage, e_P, to rise.

5. Just after instant t_1, during the period t_1 to t_2, the input signal, e_{in}, remains at a steady negative voltage.

6. Capacitor C_K now starts discharging somewhat through R_K, and the cathode voltage e_K decreases slightly. R_K and C_K constitute a large time constant so that C_K cannot discharge very much, and a fairly constant cathode voltage is maintained.

7. The falling cathode voltage (negative-going), due to C_K discharging, has the same effect as if the grid were going positive.

8. During the time period t_1 to t_2, the falling cathode voltage therefore causes plate *current* to increase slightly, resulting in the falling of the plate voltage signal, e_P.

9. This falling e_P (during t_1 to t_2 time) has a downward tilt, denoting a leading phase shift, as was also shown in Fig. 2·13*c*.

Effect of R_S and C_S. In Fig. 2·12*a*, resistor R_S is a voltage-dropping resistor for the screen. It simply makes the screen the desired positive voltage, where screen voltage is the difference between the B+ voltage and the voltage across R_S (screen current multiplied by R_S). Capacitor C_S, like C_K in the cathode circuit, is a bypass capacitor. Its purpose is to prevent screen voltage from fluctuating when an input signal is applied to the control grid. If the screen voltage were permitted to vary because of an input signal, degeneration would result with resultant decrease of gain. For example, if an input signal to the control grid went positive, plate current and screen current would *increase*. This larger screen current, flowing through R_S, produces a larger voltage *across* R_S, and the screen voltage becomes less positive. This would tend to cause plate current to *decrease*, which opposes the action that the control grid has, resulting in degeneration and smaller gain. C_S is in the circuit in order to help prevent this. By being part of a large *R-C* time constant, C_S keeps the screen-grid voltage fairly constant. However, screen voltage will still vary slightly, especially at the lower frequencies. Here, at the lower frequencies, an alternation or half cycle is a longer time period than at the medium and higher frequencies, and C_S is given more time in which to charge up or to discharge.

Phase distortion due to a leading phase shift is also produced by C_S, as shown in Fig. 2·12*c*. This is similar to that described for the cathode capacitor C_K, and is described in the following:

1. At instant t_1, the input signal e_{in} goes negative, and plate current and screen current decrease.
2. With less plate current flowing through resistor R_L, a smaller voltage appears *across* this resistor, and as expected, plate voltage e_P rises toward B+.
3. Screen-grid voltage, at instant t_1, also tries to rise but is prevented from immediately doing so by C_S.
4. During the period from t_1 to t_2, e_{in} is a steady negative voltage, and the plate voltage should also remain steady. C_S now starts charging, and the screen voltage rises slightly.
5. This causes plate current to increase, causing plate voltage e_P to decrease instead of remaining constant.
6. The downward tilt of e_P is a phase distorted signal, denoting a leading phase shift.

Effect of C_C and R_G. C_C, usually called the *blocking* capacitor and also the *coupling* capacitor, does both these jobs. The signal voltage at the plate of the amplifier in Fig. 2·12*a* is a fluctuating d-c voltage: This is made up of a d-c-voltage component (the average) and an a-c-voltage component. Capacitor C_C blocks the positive d-c-voltage component at the plate from getting to the grid of the next stage. At the same time, C_C permits the a-c-voltage component to get to the next grid. C_C has a reactance (X_{C_C}) to this a-c-voltage component which varies *inversely* with frequency ($X_{C_C} = 1/2\pi f C_C$). X_{C_C} and resistor R_G act as a voltage divider for the a-c signal voltage. At medium and higher frequencies, X_{C_C} is very small compared to R_G, and practically the entire a-c voltage at the plate appears across R_G and is applied to the next stage grid. At the lower frequencies, X_{C_C} increases and is no longer insignificant compared to R_G. Now, a larger proportion of the a-c signal voltage at the plate appears across X_{C_C}, causing that which appears across R_G to decrease. This has the same effect as if the gain on the stage were decreased at the lower frequencies and is another example of *frequency distortion.*

Phase distortion is also produced by C_C and R_G. This can be seen in Fig. 2·13*a*, *b*, and *c* by simply calling C_1 and R_1 C_C and R_G. E_T would be the a-c plate-signal voltage e_P. From Fig. 2·13*b*, e_{R_G} (shown as e_{R_1}) *leads* e_P (shown as E_T) by the phase angle θ.

A flat-topped pulse appearing at the plate, e_P (shown as E_T, in Fig. 2·13*c*), results in a downward-tilted voltage wave across the resistor, corresponding to e_{R_G} (shown as e_{R_1}). This tilted waveshape, instead of a flat-topped wave, is a phase distortion caused by the leading phase shift.

2·6 Low-frequency Compensation and Correction

In the previous section, the effects of cathode bypass capacitor C_K, screen bypass C_S, and coupling capacitor C_C on the gain at lower frequencies and on the phase shift were explained. To reduce this *frequency distortion* and *phase distortion,* the conventional amplifier circuit of Fig. 2·12*a* must be changed somewhat.

Cathode bypass capacitor C_K could be omitted. This, of course, would produce degeneration and a smaller gain. However, the smaller gain would be uniform for all frequencies and would not decrease at the lower ones only. If the decrease in gain due to the degeneration could not be tolerated, then some bias other than cathode bias should be used.

Screen bypass capacitor C_S may be omitted if the screen voltage dropping resistor R_S is deleted too in Fig. 2·12*a*. In this case, the screen grid is then connected directly to B+, and the B+ circuit itself must have a low impedance if degeneration is to be avoided.

Coupling capacitor C_C can also be deleted by connecting the plate of one amplifier to the control grid of the next stage. This, of course, is a direct-coupled amplifier, called the Loftin-White amplifier. A direct-coupled amplifier has the disadvantage that each succeeding stage requires progressively more positive voltages. This is due to the fact that the grid of a second stage is at, or near, the potential on the previous plate. To produce bias on this second stage means that the cathode must be made more positive than its grid. With a positive grid and a positive cathode, the plate of this second stage must connect to a higher B+ voltage than the previous stage. As a result of this, a direct-coupled amplifier circuit requires a larger power supply with tapped connections for each stage.

Low-frequency compensation components R_F *and* C_F, as shown in Fig. 2·14*a*, are added in the plate circuit in order to boost the gain at the lower frequencies and, even more important, to compensate for the leading phase shift produced by either C_C, C_K, or C_S.

At high frequencies, the phase shift should be linear and directly proportional to the frequency. For example, if the phase shift at 100 kc is

1°, then at 200 kc, the phase shift should be 2°; at 500 kc, 5°; and at 5 Mc, 50°. This is not possible at the low frequencies where a 1° phase shift at 30 cps would require 100° at 3,000 cycles and 200° at 6,000 cycles, etc. These are not possible since the *maximum* phase shift due to any capacitor is only 90°.

In order to prevent phase distortion, the phase shift must either be zero over a band of frequencies, as shown by graph 1 in Fig. 2·14*b* or must be linear and proportional to frequency, as shown by graph 2 in Fig. 2·14*b*. Another way of expressing *phase shift* is as *time delay*. The time delay should be uniform or *constant* over a band of frequencies if phase distortion is to be avoided. Time-delay graphs are shown in Fig. 2·14*c*, where graph 1 shows a zero and constant *time delay* which corresponds to the

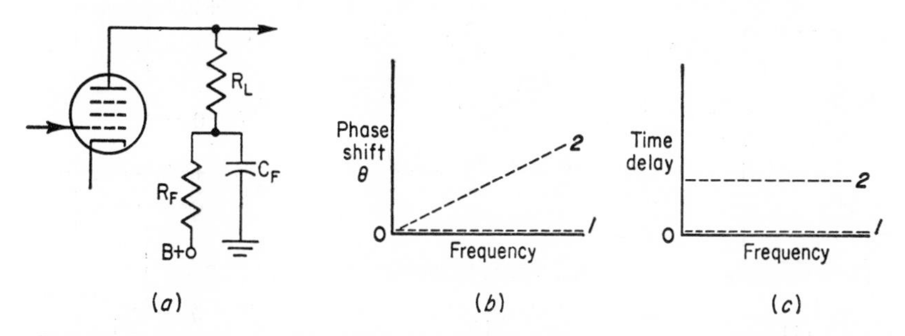

Figure 2·14 Low-frequency compensation. (a) Low-frequency compensation circuit.

zero phase shift of graph 1 in Fig. 2·14*b*. Time-delay graph 2 of Fig. 2·14*c* shows a *constant* time delay with a value other than zero. This corresponds to *phase-shift* graph 2 of Fig. 2·14*b* which is linear and proportional to frequency.

The effect of R_F and C_F in Fig. 2·14*a* is to produce a greater gain at the lower frequencies and to produce a *lagging* phase shift in order to compensate for the *leading* phase shift caused by the capacitors.

The increase of gain at the lower frequencies is due to the following: R_F is very much larger than X_{C_F}, and the impedance Z_F of these two, which are essentially in parallel to the a-c signal, is practically X_{C_F} alone. As the frequency decreases, X_{C_F} increases, increasing Z_F. The load impedance Z_L of the tube is the vector sum of R_L and Z_F (or actually X_{C_F}). At the lower frequencies, therefore, X_{C_F} increases, causing Z_L to increase. Since the gain or voltage amplification of the stage is directly proportional to Z_L ($VA = G_M Z_L$) then the gain increases at the lower frequencies. This compensates for the effect of the various capacitors (C_K, C_S, and C_C) which tend to lower the gain.

The *compensation* for the *leading* phase shift caused by the various capacitors (C_K, C_S, and C_C) is produced by C_F and is explained by the following: Fig. 2·15*a* shows a generator with a voltage E_T and an internal resistance R_P connected to a series resistor, R_L, and capacitor C_F. This circuit is basically that of the vacuum-tube amplifier of Fig. 2·14*a*, using load resistor R_L and low-frequency compensating capacitor C_F. Since R_F is very much larger than X_{C_F} and these two are in parallel, R_F may be neglected. The circuit of Fig. 2·15*a* produces a *lagging* phase shift shown by the vectors of Fig. 2·15*b*, and the discussion follows.

1. The applied voltage E_T from the generator causes a current, I, to flow through R_P, R_L, and C_F. This current I is shown as vector 1.

2. Voltage across the resistor R_L, e_{R_L}, is in phase with the current and is shown as vector 2.

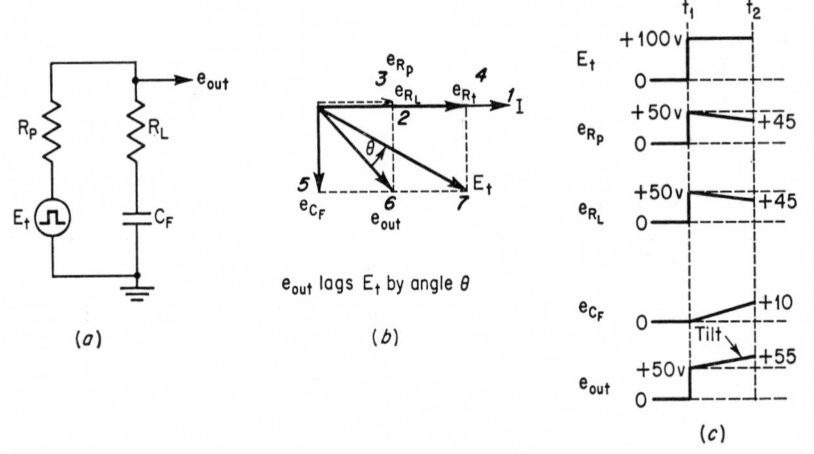

Figure 2·15 **Lagging phase shift. (a) Circuit; (b) e_{out} lags E_T by angle θ; (c) waveshapes.**

3. Voltage across internal resistor R_P, e_{R_P}, is also in phase with the current and is shown as vector 3.

4. Total resistor voltage e_{R_T} is the vector sum of e_{R_P} and e_{R_L} and is shown as vector 4.

5. Voltage across the capacitor C_F, e_{C_F}, *lags* the current I by 90° and is shown as vector 5, remembering that vectors are assumed to be rotating counterclockwise.

6. The output voltage, e_{out}, of Fig. 2·15*a* is across R_L and C_F and is the vector sum of e_{R_L} (vector 2) and e_{C_F} (vector 5) and is shown as vector 6.

7. The applied voltage E_T is the vector sum of both resistor voltages, e_{R_T} (vector 4) and the capacitor voltage, e_{C_F} (vector 5), and is shown as vector 7.

8. From the vector diagram of Fig. 2·15*b*, it can be seen that e_{out} (vector 6) is *lagging* the applied voltage E_T (vector 7) by the angle θ. The circuit of Fig. 2·15*a* therefore produces a lagging phase angle. The lower the frequency, the larger X_{C_F} becomes, and the larger e_{C_F} becomes. As a result, e_{out} lags E_T by a larger angle.

The *lagging* phase angle produced by C_F is just the reverse of the *leading* phase shift produced by the capacitors (C_C, C_K, and C_S) of Fig. 2·12*a*. The effect of a lagging phase angle on a pulse is shown in Fig. 2·15*c* and is explained below. The output voltage e_{out} has a tilt which is just the opposite of that of the distorted leading phase-shift pulse e_{R_1} of Fig. 2·13*c*.

The lagging phase-shift circuit of Fig. 2·15*a* distorts a pulse signal, as shown in Fig. 2·15*c* and explained in the following:

1. At instant t_1, the applied voltage E_T rises from zero to 100 volts.
2. At this first instant, the capacitor C_F has not charged up yet, and e_{C_F} is still zero, and the applied 100 volts (E_T) will appear across R_P and R_L in series. Assuming that these resistors are equal, e_{R_P} and e_{R_L} are both 50 volts at the first instant.
3. During the time interval t_1 to t_2, capacitor C_F charges up somewhat. Assume that at instant t_2, e_{C_F} has become 10 volts. Then the resistor voltages, e_{R_P} and e_{R_L}, each decreases to 45 volts, making $e_{R_P} + e_{R_L} + e_{C_F} = E_T$ or $45 + 45 + 10 = 100$.
4. The output signal voltage e_{out}, as shown in the circuit of Fig. 2·15*a*, is across R_L and C_F and is $e_{R_L} + e_{C_F}$ or $45 + 10 = 55$ volts at instant t_2.
5. The output-signal pulse e_{out}, of Fig. 2·15*c*, tilts *upward* from a value of 50 volts at instant t_1 to a value of 55 volts at instant t_2. This upward tilt is a phase-distorted wave indicative of a *lagging* phase shift. Note that the tilt due to a *leading* phase shift (e_{R_1} of Fig. 2·13*c*) is in the reverse direction or *downward*.

By using the correct value C_F in the compensated circuit of Fig. 2·14*a*, the lagging phase angle produced can compensate for the leading phase shift caused by the other capacitors (Fig. 2·12*a*) resulting in an output signal having *zero phase shift* and *zero time delay*, as shown in graphs 1 of Fig. 2·14*b* and *c*.

The values of the low-frequency compensating components, R_F and C_F of Fig. 2·14*a*, depend upon whether compensation for C_K or C_C of Fig. 2·12*a* is required. If compensation for C_C is desired, which is more common, then the *R-C* time constants are made equal.

$$R_L C_F = R_G C_C \tag{2·10}$$

where the value of R_L is determined by *high-frequency compensation* (discussed in Sec. 2·3); R_G should be as large as possible and is recommended by the tube manufacturer, its size being limited by ionization inside the next tube causing possible grid emission. An excessively large R_G could damage that tube. C_C likewise should be as large as possible to have as little X_C as possible; its value is limited to its physical dimensions increasing the stray capacitance, thus effecting C_T and high-frequency compensation. Equation (2·10) then gives

$$C_F = \frac{R_G C_C}{R_L} \tag{2·11}$$

The resistor R_F is made much larger than X_{C_F}, and its presence in the circuit causes a lowering of the d-c plate voltage, requiring therefore a higher value of B+. The value of R_F is

$$R_F = (10 \text{ to } 20)(X_{C_F} \text{ at the } \textit{lowest} \text{ desired frequency}) \qquad (2\cdot12)$$

Where it is desired to compensate for the *cathode bypass capacitor* C_K, the values of R_F and C_F are

$$\frac{R_F}{R_K} = \text{VA} \qquad (2\cdot13)$$

where R_K is determined by the cathode bias desired and by the current through R_K (plate + screen current). Equation (2·13) gives

$$R_F = R_K\text{VA} \qquad (2\cdot14)$$

The value of C_F is, similarly

$$\frac{C_K}{C_F} = \text{VA} \qquad (2\cdot15)$$

where C_K is determined by the value of R_K; X_{C_K} at the lowest frequency must be much less, about $\frac{1}{10}$ of R_K. From Eqs. (2·13) and (2·15),

$$\frac{R_F}{R_K} = \frac{C_K}{C_F} \qquad (2\cdot16)$$

Cross multiplication yields

$$R_FC_F = R_KC_K \qquad (2\cdot17)$$

PROBLEMS

2·1 In the uncompensated pentode amplifier of Fig. 2·2*a*, which components effect the gain at (*a*) the higher frequencies and (*b*) the lower frequencies?

2·2 (*a*) Draw a PNP transistor amplifier using series compensation. (*b*) Draw an NPN transistor amplifier using shunt-series compensation.

2·3 In the shunt-compensated pentode amplifier of Fig. 2·4*a*, if the highest desirable frequency is 2 Mc and C_T is 20 pf, find (*a*) R_L; (*b*) L_1.

2·4 Using the values of Prob. 2·3, (*a*) calculate the load impedance at the 2-Mc frequency. (*b*) How should this compare with the impedance at frequencies less than 2 Mc?

2·5 Of the following possibilities: uncompensated amplifier, shunt-compensated amplifier, series compensation, and shunt-series compensation, which produces (*a*) the least gain at the highest desirable frequency; (*b*) the most gain at the highest desirable frequency; (*c*) the most gain possible at some medium frequency?

2·6 Explain the effect of capacitor C_F in Fig. 2·14*a* on low-frequency signals.

2·7 In the accompanying diagram, identify each output, E_{out} (*a*) and E_{out} (*b*), as indicative of a *leading* or *lagging* phase shift.

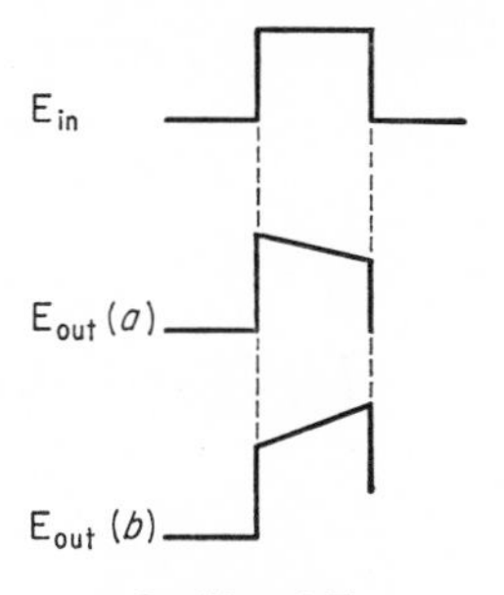

Problem 2·7

2·8 How does the bandwidth of a single amplifier compare to that of this stage in cascade with a second identical stage?

2·9 From the circuit and vectors shown for this problem, identify each of the vectors (*a*, *b*, *c*, etc.) by the names shown in the circuit.

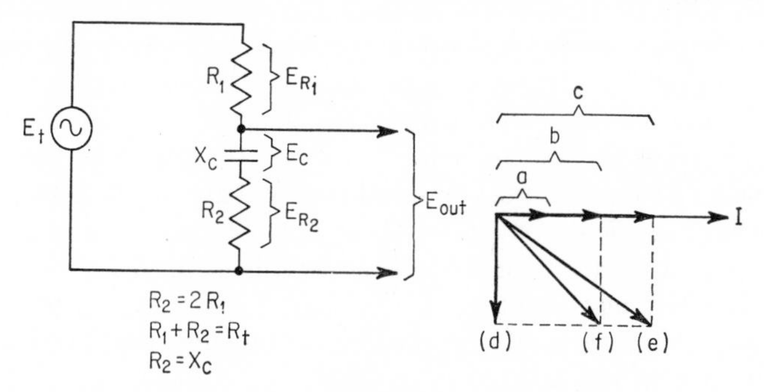

Problem 2·9

2·10 Briefly describe how a distributed amplifier circuit functions.

2·11 In the low-frequency compensation circuit shown, find the following if the lowest desirable frequency is 10 cps: (*a*) C_F; (*b*) R_F min.

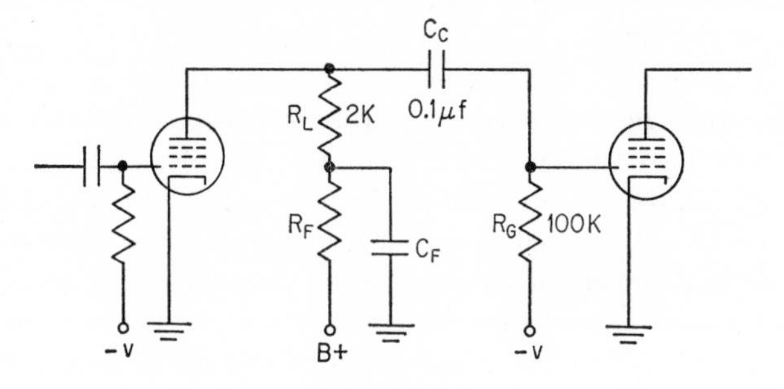

Problem 2·11

CHAPTER 3

LINEAR WAVE SHAPING

3·1 General

Pulses, square waves, and other nonsinusoidal waves can have their shapes changed by a resistor-capacitor combination or by an inductor-resistor combination. In this chapter, the effect of linear elements such as capacitors, resistors, and inductors on these nonsinusoidal waves is discussed using a simple, almost nonmathematical approach. Later, in Chap. 11 covering transient analysis of these linear elements used in a multivibrator circuit, a more rigorous discussion is given using some slightly more advanced mathematics. A thorough understanding of this more elementary chapter should be gained before the more advanced discussion of Chap. 11 is attempted.

3·2 R-C Time Constant

An R-C *time constant* is simply the product of the value of the resistor, in ohms, and the capacitor, in farads, resulting in a time in seconds. The R-C time constant is defined as the time required for the capacitor to charge up to approximately 63 per cent of the applied voltage.

In Fig. 3·1, the universal time-constant chart shows the percentage of the applied voltage that appears across the capacitor and across the resistor for various time constants. As an example, at instant t_1, when switch S_1 in Fig. 3·2*a* is closed and S_2 is open, 100 volts d-c is applied to C and R in series. As shown by curve A in Fig. 3·1, the capacitor charges up 63 per cent of the applied voltage in the first time constant (first TC), 86 per cent in the second time constant, about 95 per cent in the third period (third TC), etc., becoming practically fully charged in the fifth period (TC). If C is 1 μf, and R is 1 megohm, then the R-C time constant is $1 \times 10^6(1 \times 10^{-6}) = 1$ sec. This means that C in Fig. 3·2*a* charges up to about 63 volts (63 per cent of 100 volts) in 1 sec, 86 volts in the next second, etc., reaching about 100 volts in 5 sec (the fifth time constant). In Fig. 3·2*b*, the capacitor voltage, E_C, reaches 100 volts at instant X. The time from instant t_1 to instant X is five R-C time constants, in this case 5 sec.

In the universal time-constant chart of Fig. 3·1, curve B represents, among other things, the voltage across the resistor, E_R. It can be seen that the sum of the voltages, E_C and E_R (curves A and B respectively), adds up to the maximum applied voltage. For example, at time constant

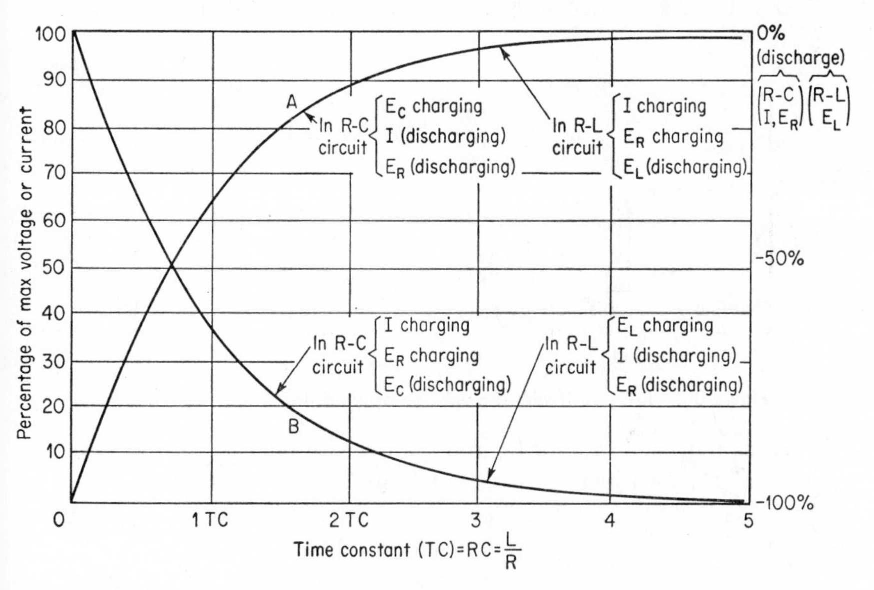

Figure 3·1 Universal time-constant chart.

zero, E_C is at zero volts (curve A, 0 per cent), and E_R is at 100 volts (curve B, 100 per cent), assuming 100 volts applied. At the first R-C time constant, E_C is at 63 volts (curve A, 63 per cent), while E_R is down at 37 volts (curve B, 37 per cent). This 63 volts E_C, and 37 volts E_R, add up to the 100 volts applied, agreeing, of course, with Kirchhoff's voltage law. At the fifth time constant, E_C is 100 volts (curve A, 100 per cent), and E_R is at zero volts (curve B, 0 per cent), again adding up to the total applied voltage.

In Fig. 3·2*b*, E_R at instant t_1 is +100 volts (the full applied voltage), while E_C is still at zero. Charging current, as shown in Fig. 3·2*a*, flows *up* through the resistor and is maximum at first, as shown by curve B of Fig. 3·1. As the capacitor charges up, current decreases, and E_R falls, as shown by curve B in Fig. 3·1 and also in Fig. 3·2*b*. E_C reaches the full applied 100 volts at instant X (Fig. 3·2b), and E_R decreases to zero at this instant. E_C remains fully charged, and E_R remains zero the rest of the time that the applied voltage is at 100 volts.

At instant t_2 (Fig. 3·2b), the switches S_1 and S_2 (Fig. 3·2a) are reversed, opening S_1 and closing S_2. The applied voltage is removed from the circuit, effectively becoming zero at instant t_2, as shown in Fig. 3·2b. The capacitor now starts discharging, and current flows *down* through resistor R (Fig. 3·2a). This is a reversal of current direction, producing a negative voltage across R. At instant t_2, E_C is still 100 volts, and this acts like an applied voltage as far as the resistor is concerned. E_R at instant t_2 becomes -100 volts. As E_C decreases (curve B, Fig. 3·1), the discharge current

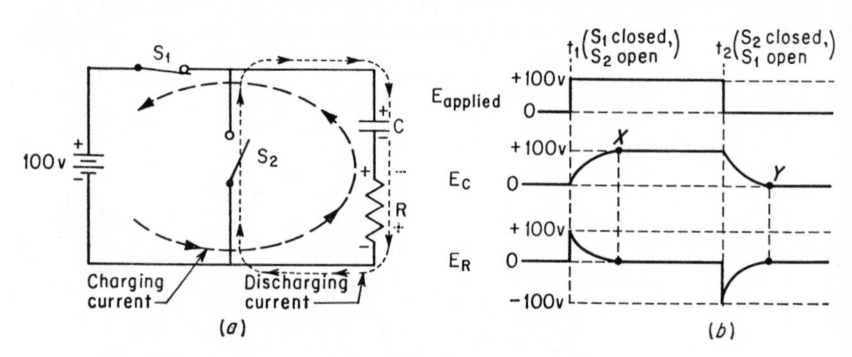

Figure 3·2 R-C circuit with d-c voltage applied. (a) Circuit; (b) voltages.

decreases (curve A, Fig. 3·1), and E_R likewise falls (curve A, Fig. 3·1). This is also shown just after instant t_2 in Fig. 3·2b. Instant Y (Fig. 3·2b) represents five time constants after instant t_2. At instant Y, E_C has fallen to zero, and E_R has therefore also decreased to zero.

In the previous discussion, it was assumed that C was 1 μf, and R was 1 megohm, producing an R-C product or time constant of 1 sec. As another example, assume that C is 0.05 μf and that R is 10 kilohms (10,000 ohms). The R-C time constant here is

$$1 \times 10^4(0.05 \times 10^{-6}) = 0.05 \times 10^{-2} \text{ sec} = 0.0005 \text{ sec}$$

or 500 μsec. This means that this capacitor will charge up 63 per cent of the applied voltage in 500 μsec (first TC) and requires 2,500 μsec (fifth TC) to charge to almost 100 per cent.

Example 3·1 If 30 volts d-c is applied to a 0.2-μf capacitor in series with a 100-kilohm resistor, find (a) R-C time constant (first TC), (b) current when voltage is first applied, (c) E_C at first TC, (d) E_R at first TC, (e) current at first TC, (f) E_C at fifth TC, (g) E_R at fifth TC, (h) current at fifth TC

Solution. (*a*) *R-C* time constant is

$$\begin{aligned}100 \text{ kilohms } (0.2\ \mu\text{f}) &= 1 \times 10^5(0.2 \times 10^{-6})\\ &= 0.2 \times 10^{-1} \text{ sec}\\ &= 0.02 \text{ sec}\end{aligned}$$

(*b*) Current when voltage is first applied. E_C is still zero, and the entire 30 volts applied is across the resistor R. Therefore,

$$\begin{aligned}I &= \frac{E_R}{R}\\ &= \frac{30}{100 \text{ kilohms}}\\ &= \frac{30}{1 \times 10^5}\\ &= 30 \times 10^{-5} \text{ amp}\\ &= 0.0003 \text{ amp}\end{aligned}$$

and this is the *maximum* current.

(*c*) E_C at first TC. From the universal time-constant chart (Fig. 3·1), it can be seen that E_C (curve A) rises to 63 per cent of the applied voltage at first time constant. Therefore,

$$\begin{aligned}E_C &= 63 \text{ per cent } (E \text{ applied})\\ &= 0.63(30)\\ &= 18.9 \text{ volts}\end{aligned}$$

(*d*) E_R at first TC. From the time-constant chart of Fig. 3·1, E_R (curve B) at the first time constant decreases to 37 per cent of the maximum voltage. Therefore,

$$\begin{aligned}E_R &= 37 \text{ per cent } (E \text{ applied})\\ &= 0.37(30)\\ &= 11.1 \text{ volts}\end{aligned}$$

Another solution for E_R, after E_C [part (*c*) of this example] is known, is to use Kirchhoff's voltage law:

$$\begin{aligned}E \text{ applied} &= E_C + E_R\\ 30 &= 18.9 + E_R\\ 30 - 18.9 &= E_R\\ 11.1 \text{ volts} &= E_R\end{aligned}$$

(*e*) Current at first TC:

$$
\begin{aligned}
I &= \frac{E_R}{R} \\
&= \frac{11.1}{100 \text{ kilohms}} \\
&= \frac{11.1}{1 \times 10^5} \\
&= 11.1 \times 10^{-5} \\
&= 0.000111 \text{ amp}
\end{aligned}
$$

Another solution is to refer to the time-constant chart of Fig. 3·1. From curve *B*, it can be seen that the current at the first TC decreases to 37 per cent of its maximum value [found in part (*b*)]. Therefore,

$$
\begin{aligned}
I \text{ at first TC} &= 37 \text{ per cent } I \text{ max} \\
&= 0.37\ (0.0003) \\
&= 0.000111 \text{ amp}
\end{aligned}
$$

(*f*) E_C at fifth TC. From the time-constant chart of Fig. 3·1, it is seen that E_C (curve *A*) rises to about 100 per cent of the applied voltage in 5 time constants. Therefore,

$$
\begin{aligned}
E_C &= 100 \text{ per cent } E \text{ applied} \\
&= 1(30) \\
&= 30 \text{ volts}
\end{aligned}
$$

(*g*) E_R at fifth TC. From the time-constant chart of Fig. 3·1, it is seen that E_R (curve *B*) decreases to about 0 per cent of the applied voltage at the fifth time constant. Therefore,

$$
\begin{aligned}
E_R &= 0 \text{ per cent } (E \text{ applied}) \\
&= 0(30) \\
&= 0 \text{ volts}
\end{aligned}
$$

Another solution for E_R, after E_C [part (*f*) of this example] is known, is to use Kirchhoff's voltage law:

$$
\begin{aligned}
E \text{ applied} &= E_C + E_R \\
30 &= 30 + E_R \\
30 - 30 &= E_R \\
0 \text{ volts} &= E_R
\end{aligned}
$$

(*h*) Current at fifth TC. From the time-constant chart of Fig. 3·1, it can be seen that the current (curve *B*) decreases to about 0 per cent of maximum current [found in part (*b*)], at the fifth time constant. Therefore,

$$\begin{aligned} I &= 0 \text{ per cent } I \text{ max} \\ &= 0(0.0003) \\ &= 0 \text{ amp} \end{aligned}$$

Another solution is evident from the fact that at the fifth time constant, the capacitor has just about charged completely. As a result, no more current flows.

3·3 *L-R* Time Constant

A circuit containing an inductor L and resistor R, similar to the resistor-capacitor combination, constitutes a time constant. The *L-R* time constant is the inductor L, in henrys, *divided* by the resistor R, in

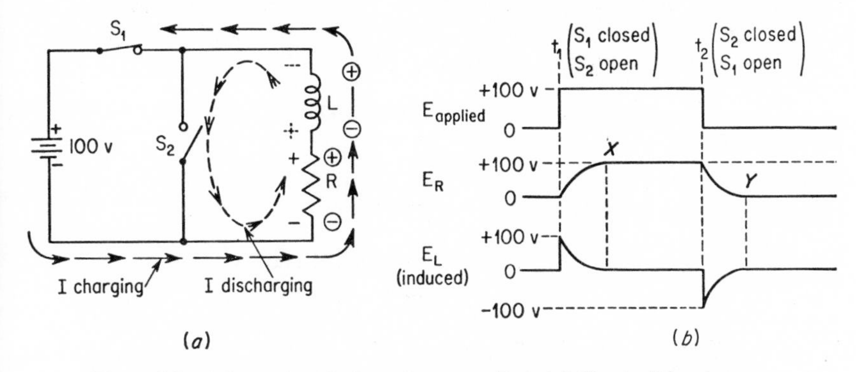

Figure 3·3 L-R circuit with d-c voltage applied. (a) Circuit; (b) voltages.

ohms, or TC $= L/R$. An *L-R* time constant is the time required for the *current* to become 63 per cent of its maximum value and also for the voltage across the resistor, E_R, to become 63 per cent of maximum. This is shown by curve A of the universal time-constant chart of Fig. 3·1. Curve B shows the induced voltage across the inductance, E_L, which is maximum at first, decreasing to 37 per cent of its maximum value in one time constant. The term *charging* is used on curve A as *I charging* and E_R *charging* and also on curve B as E_L *charging* to describe the action during the time that voltage is being applied. The rising current (curve A) causes a magnetic field to build up, and energy is stored in this field; hence the term charging. The term *discharging,* used on curve A as E_L *discharging* and on curve B as *I discharging* and E_R *discharging,* describes the action when the applied voltage is decreased to zero, allowing the magnetic field to collapse; hence the term discharging.

A d-c voltage is applied to an *L-R* series circuit in Fig. 3·3*a* with the voltages shown in (*b*). Switch S_1 is closed at instant t_1, and S_2 is open.

This applies the 100 volts to the coil and resistor. The coil here is assumed to be a pure inductance with an added series resistor. The property of an inductance is that it opposes a *change* of current. With the 100 volts d-c applied, current starts to flow, and a magnetic field builds up around the coil. This field cuts the coil, inducing a voltage which opposes the original current, and delays this current from reaching its maximum value. At the first instant, t_1, therefore, the full applied 100 volts appears across L (E_L in Fig. 3·3*b* and curve B, Fig. 3·1) with zero volts across the resistor (E_R in Fig. 3·3*b*, and curve A, Fig. 3·1).

Current, delayed by the inductance, rises exponentially as shown by curve A in Fig. 3·1, and voltage across the resistor, E_R, similarly increases. In one time constant, I and E_R will rise to 63 per cent of their maximum values, and E_L (curve B) decreases to 37 per cent of its maximum value. If L is 10 mh, and R is 1 kilohm, then an L-R time constant is

$$\begin{aligned}\text{Time constant} &= \frac{L}{R}\\ &= \frac{10\text{ mh}}{1{,}000}\\ &= \frac{0.01}{1{,}000}\\ &= 0.00001\text{ sec}\\ &= 10\ \mu\text{sec}\end{aligned}$$

I and E_R therefore reach 63 per cent of maximum in 10 μsec, and E_L decreases from its maximum to 37 per cent of this value in this time. In five time constants, I and E_R (curve A, Fig. 3·1) reach approximately 100 per cent of their maximum value, while E_L decreases to zero (curve B). In Fig. 3·3*b*, instant X represents five time constants after instant t_1. As shown, E_R rises from zero (at instant t_1) to +100 volts (at instant X). E_L, as shown, falls from +100 volts (at instant t_1) to zero volts (at instant X). This condition (after instant X) continues for the remainder of the time that the 100 volts is applied. Current, called *charging* I, flows *up* through R and L, as shown in Fig. 3·3*a*, producing the polarities across R and L as indicated by the solid-line, encircled signs.

At instant t_2, Fig. 3·3, switch S_1 is opened, and S_2 is closed. This, in effect, brings the applied voltage down to zero. With no applied voltage, current tries to stop. The magnetic field around the coil now starts collapsing, inducing a voltage in the coil with the polarity indicated by the dotted-line signs. The *induced* voltage in the coil is now the cause of a current flow. The top of L is now negative, because of this induced voltage, causing a *discharge* current to flow as indicated by the dotted-line arrows. Note that this current flows from the top of coil L (the negative-voltage end) through switch S_2 *up* through resistor R and back to the lower end of L (the positive voltage end). The *direction* of this *discharge* current is the same as that of the original *charging* current, which was also *up* through resistor R. This is in keeping with the characteristic of an inductance which opposes a *change* of current. The original charging current due to the applied voltage tried to stop at instant t_2 (Fig. 3·3*b*) when the applied voltage became zero. Because of the induced voltage in the coil when the magnetic field collapsed, a current continued to flow in the same direction. This discharge current decreases, becoming zero in about five time constants, as shown by curve B in Fig. 3·1. In Fig. 3·3*b*, instant Y is five time constants after instant t_2. Voltage across the resistor, E_R, decreases from its maximum value of +100 volts (at instant t_2) to zero volts (at instant Y), five time constants later. E_R does *not* reverse polarity, as shown in Fig. 3·3*b*, but simply increases during the time from instant t_1 to instant X and remains constant from instant X until instant t_2. E_R then decreases in the time from instant t_2 to instant Y.

However, voltage across the coil, E_L, does reverse polarity, as shown in Fig. 3·3*b*. During the flow of *charging* current (time from t_1 to t_2), the polarity of voltage across L, as shown in Fig. 3·3*a*, is indicated by the solid-line encircled signs (*positive* at the top of L), while during the *discharge* current (time from t_2 to Y), polarity of voltage across L is indicated by the dotted-line signs (*negative* at the top of L).

Example 3·2 A 200-volt d-c voltage is applied to a 500-μh coil in series with a 100-ohm resistor. Find (*a*) L-R time constant (1 TC), (*b*) E_R at 1 TC, (*c*) E_L at 1 TC, (*d*) E_R at 5 TC, (*e*) E_L at 5 TC, (*f*) I at 5 TC, (*g*) I at 2 TC.

Solution. (*a*) L/R time constant is

$$\frac{L}{R} = \frac{500\ \mu\text{h}}{100\ \text{ohms}} = \frac{500 \times 10^{-6}}{100} = 5 \times 10^{-6}\ \text{sec} = 5\ \mu\text{sec}$$

(*b*) E_R at 1 TC is: from curve A, universal time-constant chart, Fig. 3·1, E_R at 1 TC rises to 63 per cent of the maximum value (applied voltage).

$$E_R \text{ at 1 TC} = 63 \text{ per cent } (E \text{ applied}) = 0.63\ (200) = 126 \text{ volts}$$

(*c*) E_L at 1 TC is as follows: From curve *B*, universal time-constant chart, Fig. 3·1, E_L at 1 TC decreases to 37 per cent of the max value (applied voltage).

$$
\begin{aligned}
E_L \text{ at } 1 \text{ TC} &= 37 \text{ per cent } E \text{ applied} \\
&= 0.37\ (200) \\
&= 74 \text{ volts}
\end{aligned}
$$

Note that $E_R + E_L$, from Kirchhoff's voltage law, must equal the applied voltage. Therefore, an alternate solution for E_L, knowing E_R [part (*a*)], is

$$
\begin{aligned}
E \text{ applied} &= E_R + E_L \\
200 &= 126 + E_L \\
200 - 126 &= E_L \\
74 &= E_L
\end{aligned}
$$

(*d*) E_R at 5 TC is as follows: From curve *A*, Fig. 3·1, E_R at 5 TC rises to 100 per cent of max (applied voltage).

$$
\begin{aligned}
E_R \text{ at } 5 \text{ TC} &= 100 \text{ per cent } (E \text{ applied}) \\
&= 100 \text{ per cent } (200) \\
&= 200 \text{ volts}
\end{aligned}
$$

Five TC is, incidentally, 25 μsec (5×5 μsec).

(*e*) E_L at 5 TC is as follows: From curve *B*, Fig. 3·1, E_L decreases to 0 per cent of maximum (applied voltage) at five time constants

$$
\begin{aligned}
E_L &= 0 \text{ per cent } E \text{ applied} \\
&= 0\ (200) \\
&= 0 \text{ volts}
\end{aligned}
$$

This is also obvious, since at 5 TC, E_R is equal to the full applied 200 volts [part (*d*)], leaving zero volts across the coil.

(*f*) I at 5 TC is as follows: From curve *B*, Fig. 3·1, the current rises to its maximum value at the fifth time constant. Also, at 5 TC, E_R is 200 volts [part (*d*)]. Therefore,

$$
\begin{aligned}
I &= \frac{E_R}{R} \\
&= \frac{200}{100} \\
&= 2 \text{ amp}
\end{aligned}
$$

(*g*) I at 2 TC is as follows: From curve A, Fig. 3·1, the current at two time constants rises to about 86 per cent of its maximum value [2 amp, from part (*f*)]. Therefore,

$$
\begin{aligned}
I \text{ at 2 TC} &= 86 \text{ per cent } (I \text{ max}) \\
&= 0.86\ (2) \\
&= 1.72 \text{ amp}
\end{aligned}
$$

If the maximum value of current is not known, I at the second time constant could be found by using the voltage across the resistor at this instant, as follows:

$$
\begin{aligned}
E_R \text{ at 2 TC} &= 86 \text{ per cent } E \text{ applied} \\
&= 0.86\ (200) \\
&= 172 \text{ volts}
\end{aligned}
$$

then

$$
\begin{aligned}
I \text{ at 2 TC} &= \frac{E_R}{R} \\
&= \frac{172}{100} \\
&= 1.72 \text{ amp}
\end{aligned}
$$

3·4 *R-C* Coupling Circuit

The waveshape of a signal that is coupled from one amplifier to the next ordinarily should not be altered by the coupling components. A sine wave has the unusual property that its *shape* is not subject to change by an *R-C* coupling circuit. Complex waves, on the other hand, are easily altered by this circuit.

The diagram of Fig. 3·4*a* shows the coupling circuit components (R_B-C_C) used between the collector of amplifier transistor T_1 and the base of T_2. As shown in Fig. 3·4*b*, the signal at the collector of transistor T_1 is a rectangular-shaped pulse. This signal is a fluctuating d-c voltage, having a d-c component of +10 volts, and an a-c component of 3 volts peak to peak (+9 volts to +12 volts).

To pass this signal from T_1 collector to T_2 base with a minimum amount of waveshape change, the coupling circuit, C_C-R_B, must be a large or long *R-C* time constant. The term *long R-C* is relative. That is, the *R-C* time constant is much larger than the time duration of the signal pulse. In Fig. 3·4*b*, this time duration is from the *leading* edge of the pulse (instant t_1) to the *lagging* edge (instant t_2). Since the *R-C* time constant is much larger than the period t_1 to t_2, C_C does not have sufficient time in which to charge up to the pulse amplitude of +12 volts. The *R-C* time constant is also made larger than the time period between one pulse and the next, time from instant t_2 to instant t_3. As a result, because of the large time constant, C_C again is not given sufficient time and therefore cannot discharge down to the +9 volts level between pulses.

The step-by-step explanation of Fig. 3·4*a* and *b* is as follows:

1. Before the signal is applied, C_C charges up to the +10 volts d-c which is present at the collector of transistor T_1.

2. With +10 volts d-c applied to the R_B-C_C circuit and with 10 volts charge on C_C, there is zero volts across R_B just before instant t_1.

3. At instant t_1 the voltage at the collector of transistor T_1 (E input) rises to +12 volts. C_C starts charging up, and charging current flows *up* through R_B and R_L, producing the polarity voltage across R_B shown by the solid-line encircled signs, with the top of R_B positive.

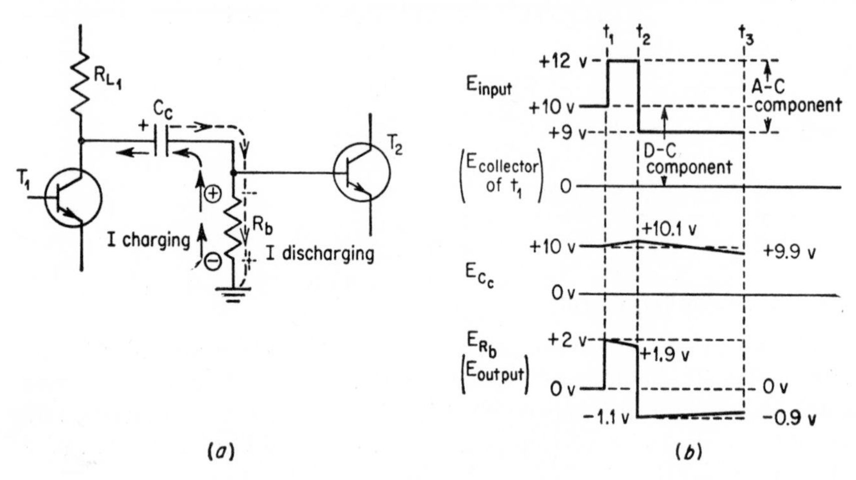

Figure 3·4 R-C coupling. (a) NPN transistors; (b) signal waveshapes.

4. At instant t_1, the applied voltage, E input, is 12 volts, E_{C_C} is still 10 volts, and the voltage across the resistor, E_{R_B}, now becomes 2 volts.

5. During the period from instant t_1 to just before instant t_2, the voltage at T_1 collector remains at +12 volts. C_C charges up only slightly due to the large R-C time constant, and E_{C_C} rises only slightly above its previous charge of 10 volts to a new value of, say, 10.1 volts.

6. With +12 volts, E input, applied, and 10.1 volts now across C_C, the voltage across R_B, E_{R_B}, now becomes +1.9 volts

$$E_{\text{applied}} = E_{C_C} + E_{R_B}$$

7. At instant t_2, E input drops to +9 volts, and C_C starts discharging. Current now flows *down* through R_B, producing the polarity across R_B as shown by the dotted-line signs, negative at the top of the resistor.

8. At instant t_2, with +9 volts applied and with E_{C_C} still at its previous charge of 10.1 volts, E_{R_B} now becomes −1.1 volts.

$$E_{\text{applied}} = E_{C_C} + E_{R_B}$$
$$9 = 10.1 + E_{R_B}$$
$$9 - 10.1 = E_{R_B}$$
$$-1.1 = E_{R_B}$$

9. During the time period from instant t_2 to just before instant t_3, C_C, because of the large R-C time constant, cannot discharge very much. E_{C_C} will change only slightly from its previous value of 10.1 volts to, say, 9.9 volts.

10. Just before instant t_3, with the voltage at the collector of T_1, E input, still at +9 volts and E_{C_C} at 9.9 volts, E_{R_B} now becomes −0.9 volts.

$$E_{\text{applied}} = E_{C_C} + E_{R_B}$$
$$9 = 9.9 + E_{R_B}$$
$$9 - 9.9 = E_{R_B}$$
$$-0.9 = E_{R_B}$$

From the foregoing discussion, and from the waveshapes of Fig. 3·4*b*, it can be seen that the output signal, E_{R_B}, from the R-C coupling circuit is a pure a-c voltage having essentially the same rectangular shape as the input signal to the coupling circuit. Note also that, as expected, C_C blocks the +10 volts d-c component of the input signal so that it does not appear at the output at the base of transistor T_2. For this reason, of course, C_C is often referred to as a *blocking* capacitor. The output signal, E_{R_B}, is almost a replica of the input pulse except for a slight slope at the top and at the bottom. By using a still larger R_B-C_C time constant combination, the downward slope can be further reduced. This slope denotes a leading phase shift and is more fully discussed in Chap. 2.

In the foregoing description of Fig. 3·4, it was assumed for simplicity that the capacitor C_C charged up through R_B and R_{L_1} only. Actually if transistor T_1 is not cut off at instant t_1, the collector resistance R_C (not shown) becomes part of the consideration. Similarly, at instant t_1 the positive-going signal further forward biases the base to emitter of transistor T_2. Input resistance of T_2 parallels R_B, decreasing the time constant.

3·5 *R-C* Differentiating Circuit

When a square or flat-topped wave, such as that shown as *E input* in Fig. 3·4*b* and Fig. 3·5*b* and *c*, is applied to an R-C circuit which consists of a *short* R-C time constant, the output voltage across the resistor bears little or no resemblance to the original waveshape. As shown in Fig. 3·5*b* and c, the output signal, E_{R_1}, consists of very narrow positive-going and negative-going spikes of voltage.

A short R-C with output taken across the resistor is called a *differentiator* and has a time constant which is very much smaller than the time duration of the applied pulse or of the alternation (half cycle) of the square wave. As an example, if the frequency of the square waves in Fig. 3·5b is 1 kc, then the time or period of 1 cycle (from instant t_1 to t_3) is 0.001 sec or 1,000 μsec. The time of an alternation (from t_1 to t_2) is half of 0.001 sec, or 0.0005 sec, or 500 μsec. A short R-C time constant should simply be much less than this 500 μsec. If R_1 in Fig. 3·5a is 50 kilohms,

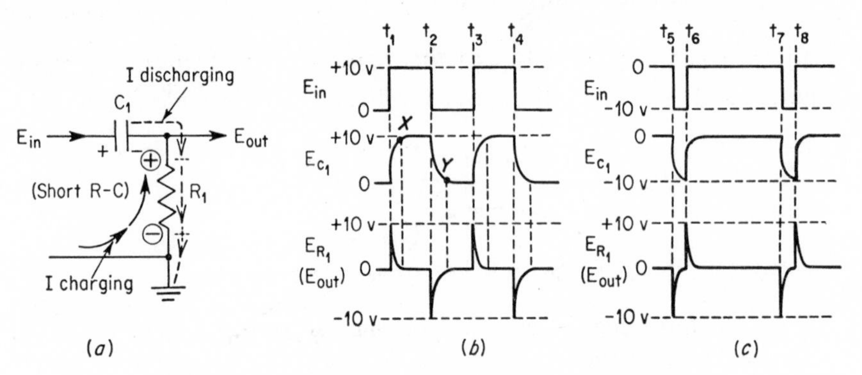

Figure 3·5 **R-C differentiating circuit and waveshapes.**

and C_1 is 500 picofarads or micromicrofarads (pf), then the R_1-C_1 time constant is

$$
\begin{aligned}
R_1C_1 &= 50 \text{ kilohms } (500 \text{ pf}) \\
&= 5 \times 10^4 \, (500 \times 10^{-12}) \\
&= 2{,}500 \times 10^{-8} \text{ sec} \\
&= 25 \times 10^{-6} \text{ sec} \\
&= 25 \ \mu\text{sec}
\end{aligned}
$$

Since this 25-μsec time constant is much smaller than the 500-μsec time of the square-wave alternation (from instant t_1 to t_2 and from t_2 to t_3), then the R_1-C_1 is called a *short* or *small* time constant.

At instant t_1, in Fig. 3·5b, the applied signal, E_{in}, rises abruptly from zero volts to +10 volts. Capacitor C_1, in Fig. 3·5a, starts charging up, and current flows *up* through R_1, producing the voltage polarity shown by the encircled signs—positive at the top of R_1. At this first instant, t_1, capacitor voltage, E_{C_1}, is still zero. As a result, the +10 volts applied all appear across the resistor, and as shown in Fig. 3·5b, E_{R_1} is +10 volts at t_1.

Since R_1-C_1 is a small time constant compared to the time t_1 to t_2, then during this relatively long period, C_1 has sufficient time to charge completely. At instant X, five time constants after instant t_1, E_{C_1} has become $+10$ volts, and E_{R_1} has decreased to zero as shown in Fig. 3·5*b*. For the remainder of the period t_1 to t_2, C_1 remains fully charged, and E_{R_1} stays at zero.

At instant t_2, E_{in} drops abruptly from $+10$ volts to zero, and C_1 starts discharging. Current now flows *down* through R_1, as shown by the dotted-line arrows in Fig. 3·5*a*, producing the polarity across R_1 as shown by the dotted-line signs—negative at the top of R_1. Since the applied voltage, E_{in}, is zero at instant t_2, the 10 volts E_{C_1} is effectively an applied voltage as far as R_1 is concerned. As a result, at instant t_2, when E_{C_1} is 10 volts, E_{R_1} is -10 volts. The short R_1-C_1 time constant permits C_1 to quickly discharge during the relatively long time period, t_2 to t_3. At instant Y, five time constants after instant t_2, E_{C_1} has become zero, discharge current ceases, and E_{R_1} decreases to zero. For the rest of the period up to instant t_3, E_{in} remains at zero, and both E_{C_1} and E_{R_1} stay at zero. As shown in Fig. 3·5*b*, the output, E_{R_1}, consists of 10 volt spikes, *positive* when the input signal goes positive (at instants t_1 and t_3) and *negative* when the input becomes *less* positive (*negative-going* at instants t_2 and t_4).

In Fig. 3·5*c*, the input signal to the differentiator is a narrow -10-volt amplitude pulse, having a time duration from instant t_5 to t_6 and a much longer period between pulses from t_6 to t_7. The R_1-C_1 time constant, operating as a differentiator, is still a smaller value than the pulse duration time, t_5 to t_6. As a result, at instant t_5, when the input goes abruptly from zero to -10 volts, C_1 starts charging to this voltage. Charging current is in the *opposite* direction from that shown in Fig. 3·5*a* and flows *down* through R_1, making the top of R_1 negative. At the first instant, T_5, E_{C_1} is still zero, and E_{R_1} has the full applied -10 volts across it. During the t_5 to t_6 pulse duration, C_1 has sufficient time to charge fully, and E_{C_1} becomes -10 volts (the polarity across C_1 is opposite to that shown in Fig. 3·5*a*). When E_{C_1} is at full 10 volts, E_{R_1} has decreased to zero.

At instant t_6, in Fig. 3·5*c*, E_{in} has abruptly changed from -10 volts to zero. C_1 now quickly discharges, and current flows *up* through R_1 in the opposite direction from that shown in Fig. 3·5*a*. This produces a *positive* voltage at the top of R_1. In the relatively long time interval (t_6 to t_7) between pulses (long compared to the R_1-C_1 time constant) C_1 has ample time in which to discharge, and E_{C_1} decreases to zero, causing E_{R_1} to likewise fall to zero. As shown in Fig. 3·5*c*, the output signal, E_{R_1}, consists of negative and positive spikes.

Note that the output signals of Fig. 3·5*b* and *c*, E_{R_1}, have peak-to-peak values of 20 volts (from $+10$ volts to -10 volts), despite the fact that the input signals are each only 10 volts peak to peak. This is characteristic of a differentiating *R-C* circuit with an input that rises or falls very abruptly. It is not true when the input changes more slowly such as sawtooth or ramp voltages do.

R-C **differentiator effect on ramp voltage.** When the voltage applied to a short *R-C* time constant circuit is not steady as with a flat-topped wave, the voltages across the capacitor and resistor, as depicted in the universal time constant chart of Fig. 3·1, must be revised. A *ramp* voltage such as a sawtooth or triangular-shaped signal does not remain constant but rises and falls relatively slowly. Figure 3·6 shows an applied voltage rising *linearly* (at a constant rate or slope) and then decreasing in the same manner. Current I, resistor voltage E_R, and capacitor voltage E_C are also shown during the time of the rising applied voltage and then during the time that E_{applied} decreases.

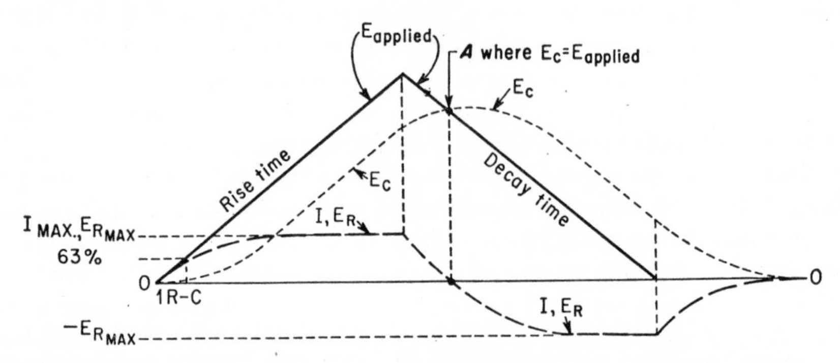

Figure 3·6 **R-C differentiating-circuit effect on applied ramp or triangular voltage.**

When the voltage is first applied, it is very small, and there is, at first, no charge on the capacitor. A small current starts flowing, starting to charge up the capacitor. As E_{applied} continues to increase, the capacitor tries to charge to this larger voltage. Since the voltage now across the capacitor opposes the applied voltage, current increases but at a *slower rate of increase*. The current, I, rises exponentially (fast at first, and then more slowly) as shown by the dashed-line graph of Fig. 3·6. At the end of 1 *R-C* time constant, I has reached approximately 63 per cent of its maximum value. At the fifth time constant, I has become maximum, as shown, remaining at this value for the duration of time that E_{applied} keeps rising at the same linear rate. Since the current flows through the resistor, and voltage across the resistor, E_R, is simply the product I times R, then the E_R graph is the same as the dashed-line I curve of Fig. 3·6. The capacitor voltage, E_C, as shown by the dotted-line curve of Fig. 3·6, rises slowly at first and then more rapidly as the current approaches its maximum. When the current is at maximum and is steady, E_C rises linearly (at a constant rate of increase), following the rising E_{applied}.

In a short R-C, where the current reaches a maximum (at the fifth time constant), the value of this maximum current and the maximum value of the resistor voltage E_R is directly proportional to the slope of E_{applied}.

When the applied voltage reaches its maximum value along its rise and starts decreasing, E_C keeps increasing but at a slower rate until it is equal to E_{applied}. This can be seen in Fig. 3·6 where the dotted-line curve E_C increases linearly with the rise of E_{applied}. When E_{applied} starts decreasing, E_C is shown continuing to rise but more and more gradually. At instant A (indicated in Fig. 3·6), E_C has reached its maximum and is equal to E_{applied}. Since E_C and E_{applied} are equal and opposing at this instant, no current flows, and E_R (the dashed-line curve) is shown at zero. As E_{applied} continues to decrease during the decay time, E_C starts decreasing, and current flows in the reverse direction. A reverse polarity voltage, E_R, is produced across the resistor, reaching its maximum in about five R-C time constants. E_C continues to decrease, following the linear decay of E_{applied}. Finally, as shown, when the applied voltage has dropped to zero, E_C decreases exponentially, reaching zero five time constants later. I and E_R similarly decrease to zero exponentially.

As a capacitor charges, the current depends upon the *rate of increase* of capacitor voltage. This is expressed as

$$I = \frac{dE_C}{dT}\,C \tag{3·1}$$

where dE_C is the *change* of capacitor voltage, dT is the *change* (or period) of time in seconds, and C is the value of the capacitor in farads. The expression dE_C/dT is the rate of change of capacitor voltage. From Eq. (3·1) it can be seen that the current, I, will be *constant* if dE_C/dT is constant. This is shown in the following example.

Example 3·3 (*a*) A 0.01-μf capacitor charges linearly from zero volts to +10 volts in 100 μsec, as shown in Fig. 3·7. Find the value of current during this period.

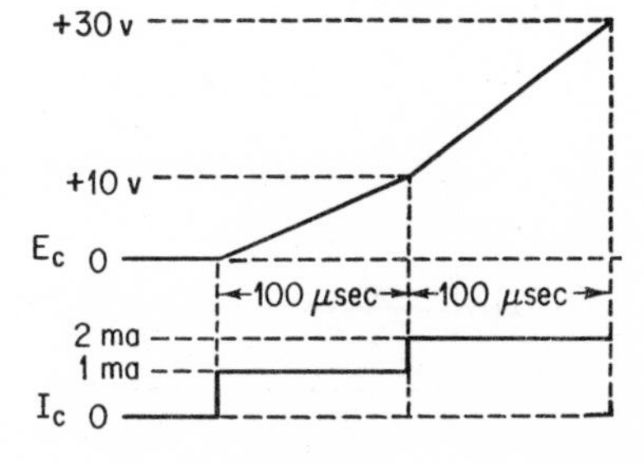

Figure 3·7 Linear voltages across a capacitor, and associated current waveshapes.

(*b*) If this capacitor in the next 100 μsec charges linearly from +10 volts to +30 volts, find the current during this next period.

Solution. (*a*)

$$I = \frac{dE_C}{dT} C \tag{3·1}$$

$$I = \left(\frac{10}{100\ \mu\text{sec}}\right) 0.01\ \mu\text{f}$$

$$= \left(\frac{10}{100 \times 10^{-6}}\right) 0.01 \times 10^{-6}$$

$$= 0.001 \text{ amp}$$

$$= 1 \text{ ma}$$

The current and voltage are shown in Fig. 3·7.

Solution. (*b*)

$$I = \frac{dE_C}{dT} C$$

$$I = \left(\frac{\text{from } +10 \text{ to } +30 \text{ volts}}{100\ \mu\text{sec}}\right) 0.01\ \mu\text{f}$$

$$= \left(\frac{20}{100 \times 10^{-6}}\right) 0.01 \times 10^{-6}$$

$$= 0.002 \text{ amp}$$

$$= 2 \text{ ma}$$

The previous example and the voltage and current curves of Fig. 3·7 show that a constant current flowing to a capacitor is accompanied by a linear or constant *rate of change* of voltage across the capacitor. Also, as in part (*b*) of this example, when the *rate of change* of capacitor voltage (dE_C/dT) doubles, the current doubles.

Returning to the curves of Fig. 3·6, the maximum value of I, which is reached only after five R-C time constants, can be found using Eq. (3·1). As an example, assume that an input (applied) voltage is ramp-shaped,

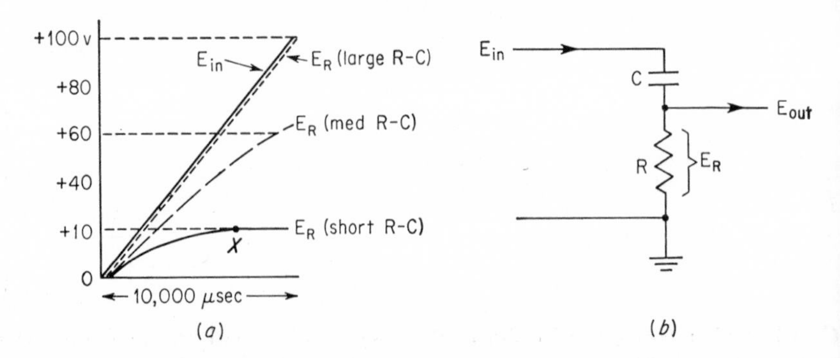

Figure 3·8 **Effect of various *R-C* circuits on applied ramp voltage.**

rising linearly from zero to +100 volts in 10,000 μsec. The maximum value of the current that could flow would depend on the values of resistor and capacitor (*R-C* time constant), since this would determine whether or not the current had sufficient time to reach maximum. If the resistor were zero, then the capacitor would be able to follow the rising input voltage exactly, and E_C would be equal to E_{in}. If C were 0.01 μf and the input (applied) voltage went from zero to +100 volts in 10,000 μsec, as shown in Fig. 3·8*a*, the $I_{\max}$ would be

$$
\begin{aligned}
I_{\max} &= \frac{dE_{\text{in}}}{dT}\,C \qquad\qquad (3\cdot1)\\
&= \left(\frac{100}{10{,}000\ \mu\text{sec}}\right) 0.01\ \mu\text{f}\\
&= \left(\frac{100}{0.01}\right) 0.01 \times 10^{-6}\\
&= 100 \times 10^{-6}\ \text{amp}\\
&= 0.0001\ \text{amp}
\end{aligned}
$$

If a resistor of 100 kilohms were used in series with the 0.01-μf capacitor, the *R-C* time constant would be

$$
\begin{aligned}
R\text{-}C &= 100\ \text{kilohms}\ (0.01\ \mu\text{f})\\
&= 1 \times 10^{5}\ (0.01 \times 10^{-6})\\
&= 0.001\ \text{sec, or,}\\
&= 1{,}000\ \mu\text{sec}
\end{aligned}
$$

Since E_{in} of Fig. 3·8*a* is rising for a period of 10,000 μsec, this is equal to ten *R-C* time constants, sufficient time for I to reach maximum (I rises to 63 per cent of maximum in 1 *R-C*, reaching maximum in about five *R-C* time constants). The output voltage, E_{out} or E_R, is across the resistor of Fig. 3·8*b* and is simply I times R. I (from above) is 0.0001 amp. Therefore, E_R is

$$
\begin{aligned}
E_R &= IR\\
&= 0.0001(100{,}000)\\
&= 10\ \text{volts}
\end{aligned}
$$

This voltage curve is shown in Fig. 3·8*a* by the solid-line graph marked E_R (short *R-C*). Point X on this curve indicates the five *R-C* point where I and E_R reach maximum.

Example 3·4 For the input-ramp voltage of Fig. 3·8*a* and where $C = 0.1$ μf and $R = 100$ kilohms, find (*a*) $I_{\max}$ without R, (*b*) I with R, and (*c*) E_{out} or E_R.

Solution. (*a*) I_{max} without any resistance is the value that I would become if C were in the circuit alone.

$$I_{max} = \frac{dE_{in}}{dT} C \tag{3·1}$$
$$= \left(\frac{100}{10{,}000\ \mu sec}\right) 0.1\ \mu f$$
$$= \left(\frac{100}{0.01}\right) 0.1 \times 10^{-6}$$
$$= 0.001 \text{ amp}$$

(*b*) I with R depends on the R-C time.
The R-C time constant is

$$R\text{-}C = 100 \text{ kilohms } (0.1\ \mu f)$$
$$= 100 \times 10^{3}\ (0.1 \times 10^{-6})$$
$$= 10 \times 10^{-3}$$
$$= 0.01 \text{ sec}$$
$$= 10{,}000\ \mu sec$$

Since E_{in}, from Fig. 3·8*a*, rises in 10,000 μsec, this is equal to one R-C, and in one R-C the current rises exponentially to 63 per cent of its maximum value [found in part (*a*)]. Therefore, I with R is

$$I = 63 \text{ per cent } (I_{max})$$
$$= 0.63\ (0.001)$$
$$= 0.00063 \text{ amp}$$

(*c*) E_{out} or E_R is

$$E_R = IR$$
$$= 0.00063(100{,}000)$$
$$= 63 \text{ volts}$$

Example 3·4 uses a medium-value R-C. The output voltage, E_R, is shown in Fig. 3·8*a* as the dashed-line curve marked E_R (medium R-C).

Example 3·5 For the input ramp voltage of Fig. 3·8*a* and where $C = 1\ \mu f$ and $R = 100$ kilohms, find (*a*) I_{max} without R, (*b*) I with R, and (*c*) E_{out} or E_R.

Solution. (*a*) I_{max} without any resistance is the value that I would become if C were in the circuit alone.

$$I_{\max} = \frac{dE_{\text{in}}}{dT} C \tag{3·1}$$

$$= \left(\frac{100}{10{,}000\ \mu\text{sec}}\right) 1\ \mu\text{f}$$

$$= \left(\frac{100}{0.01}\right) 1. \times 10^{-6}$$

$$= 0.01 \text{ amp}$$

(*b*) I with R depends on the R-C time. The R-C time constant is

$$R\text{-}C = 1\ \mu\text{f}\ (100{,}000)$$

$$= 1 \times 10^{-6}\ (100 \times 10^3)$$

$$= 100 \times 10^{-3} \text{ sec}$$

$$= 0.1 \text{ sec, or,}$$

$$= 100{,}000\ \mu\text{sec}$$

Since E_{in}, from Fig. 3·8*a*, rises in 10,000 μsec, this is only 1⁄10 of the R-C time constant. In 1⁄10 of an R-C, I rises exponentially to about 10 per cent of its maximum value [found in part (*a*)]. Therefore, I with R is

$$I = 10 \text{ per cent } (I_{\max})$$

$$= 10 \text{ per cent } (0.01)$$

$$= 0.001 \text{ amp}$$

(*c*) E_{out} or E_R is

$$E_R = IR$$

$$= 0.001(100{,}000)$$

$$= 100 \text{ volts}$$

Example 3·5 uses a large value R-C. The output voltage, E_R, is shown in Fig. 3·8*a* as the dotted-line curve marked E_R (large R-C). From the curves of Fig. 3·8*a*, the following should be noted: A ramp-shaped input voltage applied to an R-C circuit produces, *across the resistor*, (*a*) a small, flat-topped voltage if the R-C is short; (*b*) a larger, exponentially rising voltage if the R-C is medium; and (*c*) a linearly rising voltage, duplicating the input, if the R-C is large.

***R-C* differentiated trapezoidal pulses.** In Fig. 3·5*b* and *c*, square waves and rectangular pulses were shown applied to a short R-C time constant, resulting in sharp positive- and negative-going spikes. The rise and decay times of these waveforms were assumed to be instantaneous or zero time. In reality, these voltages have finite rise and fall times. In Fig. 3·9*a*, input flat-topped signals are shown having rise times shown as times T_1 and T_5, with decay times shown as T_3 and T_7. These rise and fall times are of much shorter duration than the flat portions of the waves shown as times T_2, T_4, T_6, and T_8.

In Fig. 3·9*b*, voltage across the resistor, E_R, is shown for an *R-C* circuit where the time constant is *smaller* than the times T_2, T_4, and T_6 ($R\text{-}C < T_2$, T_4, and T_6) yet *larger* than the short periods of time T_1, T_3, T_5, and T_7 ($R\text{-}C > T_1$, T_3, T_5, and T_7).

During rise time T_1, the relatively large *R-C* (large compared to T_1) permits E_R to follow the linear rise of E_{in}, as shown in Fig. 3·9*b* and also in Fig. 3·8*a*. At the end of time T_1, E_{in} remains constant at +10 volts for the long period T_2. The short *R-C* (short, compared to T_2) permits the

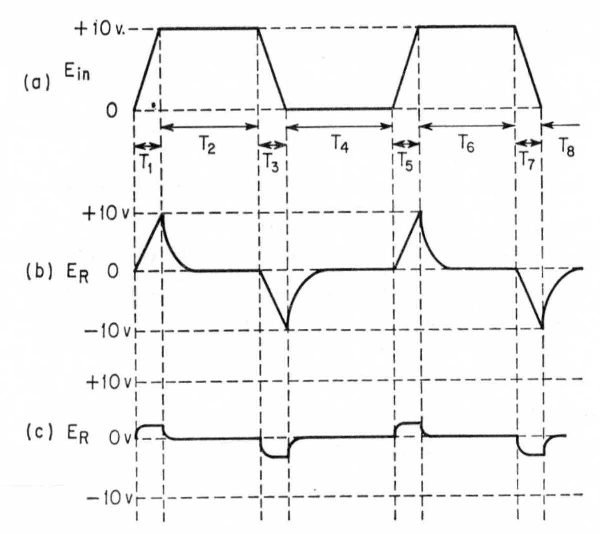

Figure 3·9 **R-C differentiated trapezoidal pulses. (b) where $R\text{-}C > T_1$, T_3, T_5, and T_7, and $R\text{-}C < T_2$, T_4, and T_6; (c) where $R\text{-}C < T_1$, T_3, T_5, and T_7, and $R\text{-}C < T_2$, T_4, and T_6.**

capacitor sufficient time to become fully charged, and E_R decreases to zero exponentially, as shown in Fig. 3·9*b*.

At the end of time T_2, E_{in} decreases linearly to zero in short time T_3. Again, the relatively large *R-C* permits E_R to follow the linearly changing E_{in}, producing −10 volts E_R. At the end of period T_3, E_{in} remains at zero for a long period, T_4. This gives the relatively short *R-C* sufficient time for the capacitor to completely discharge, producing an exponentially decreasing E_R, which reaches zero in about five time constants. The process is repeated during the next cycle, where time T_5 duplicates T_1, T_6 duplicates T_2, T_7 repeats T_3, and T_8 repeats T_4. The signal E_R of Fig. 3·9*b* is similar to that of the spikes of voltage of Fig. 3·5*b* and *c*, except that the finite rise and fall times of E_{in} are duplicated in E_R and are not instantaneous.

In Fig. 3·9*c*, E_R is shown where the *R-C* time constant is *short* for both the fast rise and decay times of E_{in} (T_1, T_3, T_5, and T_7) and also, of course, *short* compared to the longer periods T_2, T_4, T_6, and T_8. As shown in Fig. 3·8*a*, E_R, in a short *R-C* circuit with a ramp-shaped input voltage, is a small, flat-topped voltage. For rising voltage periods T_1 and T_5 of E_{in}, E_R is a small, positive-going, flat-topped wave. For falling voltage periods T_3 and T_7 of E_{in}, E_R is a small, negative-going, flat-topped wave. For the very long periods T_2 and T_6 of E_{in}, the capacitor in the short *R-C* circuit has sufficient time to completely charge. Similarly, during T_4 and T_8 of E_{in}, the capacitor has time to completely discharge. E_R, as shown in Fig. 3·9*c*, consists of exponentially decreasing voltages, going to zero, during these long periods.

3·6 *R-C* Integrating Circuit

A large *R-C* time constant, with the output developed across the capacitor, is called an *integrator*. The circuit is shown in Fig. 3·10*a*. This is just the reverse of an *R-C differentiator*, which uses a small *R-C*, developing its output across the resistor. The term *large time constant* means that the product of *R* in ohms and *C* in farads is larger than some period of time to which the *R-C* time is being compared. This period of time is T_1 in Fig. 3·10*b*, *d*, and *e* and also the periods T_1, T_2, T_3, etc., of Fig. 3·10*c*.

As shown on the universal time-constant chart, Fig. 3·1, a capacitor will charge up to the full applied voltage in about five *R-C* time constants (curve *A*). In the first time constant, *C* charges to about 63 per cent of the applied voltage. If less time is given, *C* will charge to a lesser amount. For example, curve *A* in Fig. 3·1 shows that for 0.5*R-C*, *C* charges to only about 37 per cent; for 0.25*R-C*, *C* charges to about 22 per cent; and for 0.1*R-C*, *C* charges to about 10 per cent of the applied voltage. This means that, as shown in Fig. 3·10*b*, if an input applied voltage of +10 volts were applied for a short time period T_1, the capacitor would only charge up slightly. If the *R-C* time constant is ten times the period of time T_1, then during T_1, *C* only has 0.1*R-C* in which to charge up. As a result, *C* charges to about 10 per cent (curve *A*, Fig. 3·1) of the +10 volts applied, and E_C rises during time T_1 from zero to about 1 volt, as shown in Fig. 3·10*b*. Curve $B(E_R)$ of Fig. 3·1 is equal to the full applied voltage at the first instant, before *C* has had a chance to charge, and E_C is still at zero. Also, from Fig. 3·1, it can be seen that $E_C + E_R = E_{applied}$. Therefore, when E_C (curve *A*) is about 10 per cent, at 0.1*R-C*, E_R (curve *B*) is about 90 per cent of $E_{applied}$. With +10 volts applied, then, as shown in Fig. 3·10*b*, E_R becomes +10 volts at the first instant, with E_C at zero. During the period T_1, E_C rises to +1 volt, and E_R decreases to +9 volts. The rise of E_C, in a large *R-C* circuit, is quite linear.

An applied square-wave voltage to an R-C integrator results in a triangular voltage waveshape across the capacitor, as shown in Fig. 3·10c. The square wave, E_{in}, is varying from zero to +10 volts, having a d-c component of +5 volts. Assuming that the R-C time constant is ten times the period T_1 and that $T_1 = T_2 = T_3 =$ etc., then the capacitor

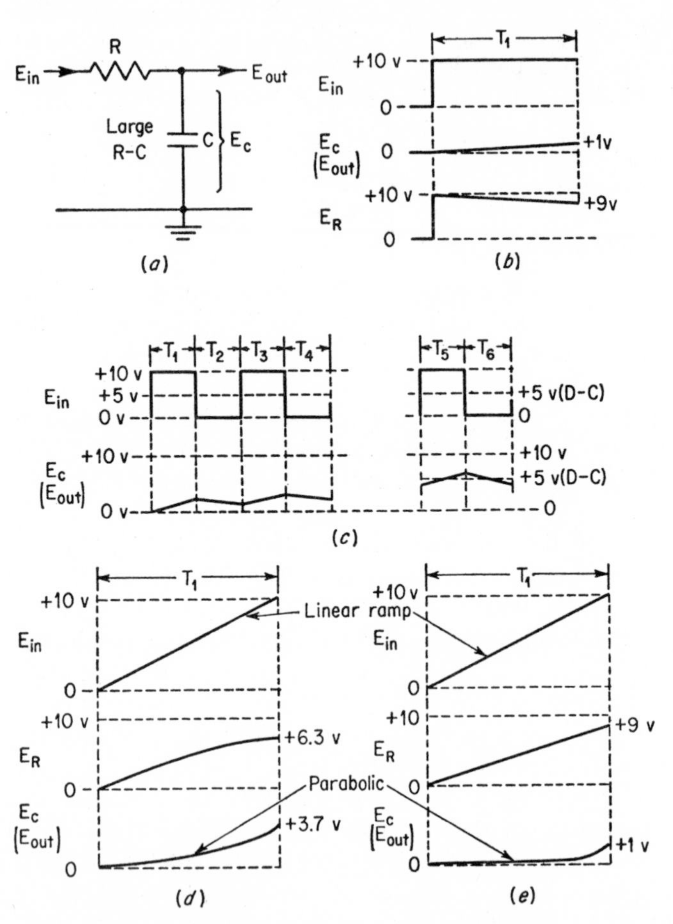

Figure 3·10 R-C integrator circuit and waveshapes. (a) Circuit; (b) $RC = 10T_1$; (c) $RC = 10T_1 = 10T_2 = 10T_3$, etc.; (d) $RC = T_1$; (e) $RC = 10T_1$.

does not have sufficient time to fully charge up during time T_1 and similarly does not have enough time to fully discharge during time T_2. During T_1, E_C rises to 10 per cent of the applied voltage (+10 volts), becoming +1 volt. During T_2, the capacitor discharges only about 10 per cent of its voltage, decreasing only about 0.1 volt, and E_C becomes about

+0.9 volts. During T_3, the applied +10 volts and the +0.9-volt charge on the capacitor produces a net difference of 9.1 volts. E_C gains another 10 per cent of this 9.1 volts during T_3, rising to about 1.81 volts (10 per cent of 9.1 volts is 0.91 volt; adding this 0.91 volt to the previous 0.9 volt on C gives 1.81 volts E_C). During each positive-going alternation of E_{in}, periods T_1, T_3, T_5, etc., E_C rises higher and higher, whereas during each negative-going alternation of E_{in}, periods T_2, T_4, T_6, etc., E_C decreases slightly but never goes lower than it had decreased previously. After several cycles of E_{in}, E_C reaches the d-c voltage component of E_{in} (+5 volts), varying about this value as its axis as C charges up slightly above +5 volts (time T_5) and then discharges slightly below +5 volts (time T_6). This linear rise and fall of E_C is a triangular waveshape, being produced by integrating the square wave.

R-C integrating circuit and ramp voltage. When the applied voltage to a large R-C circuit is a linearly changing voltage, such as shown in Fig. 3·10*d* and *e*, the result on E_R and E_C is as shown. These can be explained by referring to Figs. 3·6 and 3·8, illustrating the effect of an input ramp-shaped voltage on an R-C circuit. In Fig. 3·6, the current, I, and resistor voltage, E_R, rise in the first R-C time constant to 63 per cent of some constant value called $E_{R_{max}}$. This $E_{R_{max}}$ is reached only in a *short* R-C circuit, requiring about five time constants. In such a circuit, E_C rises slowly at first, then rises more rapidly, and after the fifth R-C time constant, E_C rises still more steeply, rising linearly with the increasing $E_{applied}$. The sum of E_R and E_C is always equal to $E_{applied}$.

In a large R-C, where the time of the applied input ramp voltage is much less than an R-C, the operation would be a small portion of the curves at the extreme left side, or at the beginning of Fig. 3·6, or between zero and the first R-C. This would produce an E_R which rises almost linearly and almost at the same rate as $E_{applied}$. E_C, as shown, rises very slowly at first and then more rapidly. In Fig. 3·8 (E_R, medium R-C) and also in Fig. 3·10*d*, where the time duration of the applied ramp is equal to the R-C, E_R is shown rising to 63 per cent of $E_{applied}$. E_C, in Fig. 3·10*d*, is shown slowly increasing and then rising more rapidly to 37 per cent of the +10 volts E_{in}, or E_C becomes +3.7 volts. The E_C waveshape is called *parabolic*. In Fig. 3·8 (E_R, large R-C), and also in Fig. 3·10*e*, the effect of a ramp-input voltage in a large R-C circuit is illustrated. E_R rises linearly along with the ramp-shaped E_{in}, practically reaching the full applied voltage. In Fig. 3·10*e*, E_R rises to +9 volts, with E_{in} being +10 volts. E_C describes a parabolic curve, rising only to about +1 volt. A ramp-shaped voltage such as a triangular signal or a sawtooth signal is changed to a parabolic wave by an R-C integrator.

3·7 *L-R* Differentiating Circuit

In Sec. 3·3, a discussion of *L-R time constants* is presented. An *L-R* time constant consists of the value of the inductance, L, in henrys, divided by the resistor, R, value in ohms or *L-R* time constant $= L/R$. In one *L-R* time constant, the current, I, and the resistor voltage, E_R, rise to 63 per cent of their maximum values. This is shown by curve A of the universal time-constant chart, Fig. 3·1. Voltage across the inductance, E_L, rises to its maximum value at the first instant, decreasing to 37 per cent of maximum at the first time constant. This is shown in curve B, Fig. 3·1.

In Fig. 3·11*a*, an *L-R* short time-constant diagram is shown, acting as a differentiating circuit. A differentiator changes a square-wave voltage

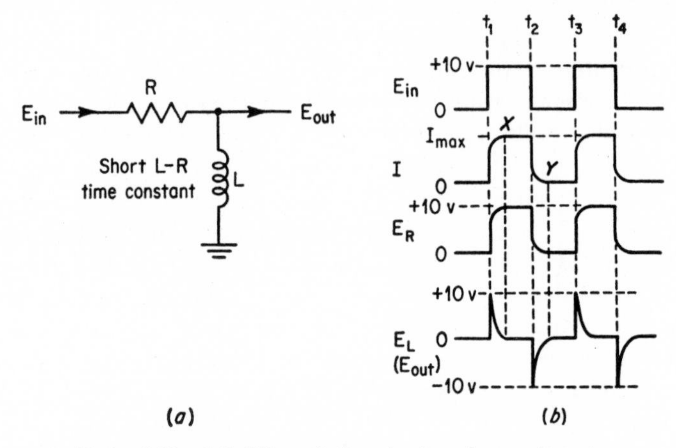

Figure 3·11 **L-R differentiating circuit and waveshapes.**

input into a peaked or spiked output waveshape, as shown in Fig. 3·11*b*. Note that the peaked-output voltage is developed across L. This waveshape is identical to that of the *R-C* differentiator of Fig. 3·5*b* where the output signal is developed across R.

In Fig. 3·11*a* and *b*, a +10-volt square wave is applied to the short *L-R* time constant where the *L-R* time constant is much less than the duration time (t_1 to t_2) of the applied signal or the time between (t_2 to t_3). At instant t_1, E_{in} becomes +10 volts. A current attempts to flow, and a magnetic field is built up around the inductance, inducing a voltage in the coil. At instant t_1, E_L is maximum and opposes the flow of current. I slowly increases, reaching its maximum in about five *L-R* time constants (instant X), and E_R likewise increases from zero (at instant t_1) to maximum (at instant X). As E_R increases, E_L decreases since the sum of the two must equal E_{in}. During the remainder of the period (instant X to instant t_2), I and E_R stay at their maximum values, while E_L remains at zero. During this period (X to t_2), when I is constant, the magnetic field is present but not building up. Therefore, the stationary field no longer induces voltage in the coil.

At instant t_2 (Fig. 3·11*b*), E_{in} drops to zero. I attempts to decrease, and the magnetic field starts collapsing. The moving field once again cuts the coil, this time in the opposite direction, inducing an opposite polarity voltage in the coil. E_L, at instant t_2, is a maximum but in the opposite polarity from what it had been at instant t_1. This is shown in Fig. 3·11*b*. It is this voltage E_L which keeps the current flowing in the original direction even when E_{in} has dropped to zero, at instant t_2. As the field collapses, E_L decreases, I decreases, causing E_R to become smaller. Five time constants after t_2 (instant Y) I, E_R, and E_L all decrease to zero, remaining at zero for the rest of the time that E_{in} is at zero (from Y to t_3). The output is across L and is a peaked-voltage waveshape.

L-R differentiating circuit and ramp voltage. A linearly rising voltage such as a ramp (Fig. 3·12), when applied to an L-R differentiating circuit,

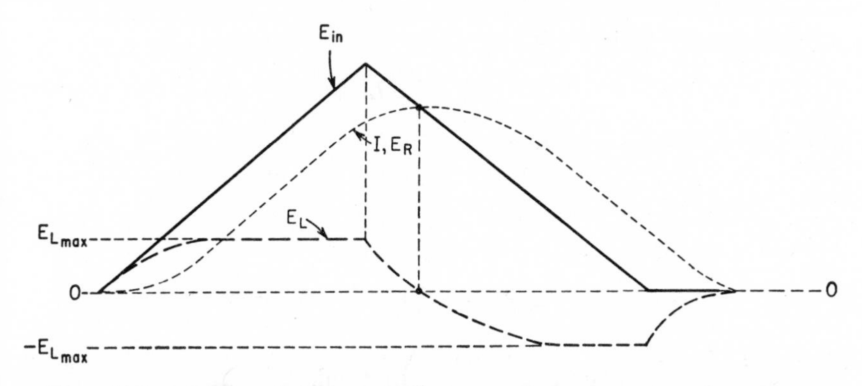

Figure **3·12** ***L-R* differentiating-circuit effect on applied ramp or triangular voltage.**

produces exactly the same waveshapes as those in the R-C circuit as shown in Fig. 3·6. The dashed-line curve (I and E_R in the R-C graphs of Fig. 3·6) is now E_L (in the L-R graphs of Fig. 3·12). The dotted-line curve (E_C in the R-C graphs of Fig. 3·6) is now I and E_R (in the L-R graphs of Fig. 3·12). The following example is similar to that involving a ramp-shaped input to an R-C circuit given in Sec. 3·6 but uses an L-R circuit instead.

Example 3·6 An input ramp-shaped voltage rises from zero to +100 volts in 0.01 sec (10,000 μsec) and is applied to a 100-mh coil with a resistor of 100 ohms in series with it. Find (*a*) L-R time constant, (*b*) I max, (*c*) E_L max.

Solution. (*a*) *L-R* time constant is

$$\text{Time constant} = \frac{L}{R}$$
$$= \frac{100 \text{ mh}}{100 \text{ ohms}}$$
$$= \frac{100 \times 10^{-3}}{100}$$
$$= 1 \times 10^{-3} \text{ sec, or,}$$
$$= 1 \text{ msec, or}$$
$$= 1{,}000\ \mu\text{sec}$$

(*b*) I max is

$$I = \frac{E}{R}$$
$$= \frac{100}{100}$$
$$= 1 \text{ amp}$$

(*c*) E_L maximum is

$$E_L = \frac{dI}{dT} L \tag{3·2}$$

where dI = change of current
dT = change of time
L = value of the inductance, henrys

I rises from zero to 1 amp in 10,000 μsec.

$$E_L = \frac{dI}{dT} L$$
$$= \left(\frac{1 \text{ amp}}{10{,}000\ \mu\text{sec}}\right) 100 \text{ mh}$$
$$= \left(\frac{1}{10{,}000 \times 10^{-6}}\right) 100 \times 10^{-3}$$
$$= 10 \text{ volts}$$

The figures in this example are exactly the same as in the *R-C* differentiator example with the ramp-input voltage given previously (between the numbered Examples 3·3 and 3·4). The output voltage in the *R-C* problem was also 10 volts, except that it was across R instead of being across L. The curves of Fig. 3·12 are typical of the voltages and current such as in the foregoing *L-R* example. These curves are where the *L-R* time constant is short compared to the time of the applied input voltage. As shown in Fig. 3·12, E_L rises exponentially, reaching its maximum in about five *L-R* time constants. I and E_R rise slowly, at first, and after five constants, rise linearly with E_{in}. If the L-R time constant were large, then I and E_R would only have time to rise slightly, whereas the output of the L-R differentiator circuit, E_L, would rise linearly with E_{in}. The voltages shown for the corresponding R-C circuit, shown in Fig. 3·8a and b, would be exactly the same for the L-R circuit, except that the voltages marked E_R (large R-C), etc., would be called E_L (large L-R), etc.

Similarly, the output of an L-R differentiator for a trapezoidal input can be compared to that shown for the R-C circuit in Fig. 3·9. The voltages of (b) and (c) marked E_R would be called E_L in the L-R circuit and would be exactly the same. Also, the terms RC would be replaced by L/R.

3·8 L-R Integrator

A large L-R time constant reacts to an input signal exactly the same as a large R-C circuit. As shown in Fig. 3·13a, the output from the L-R

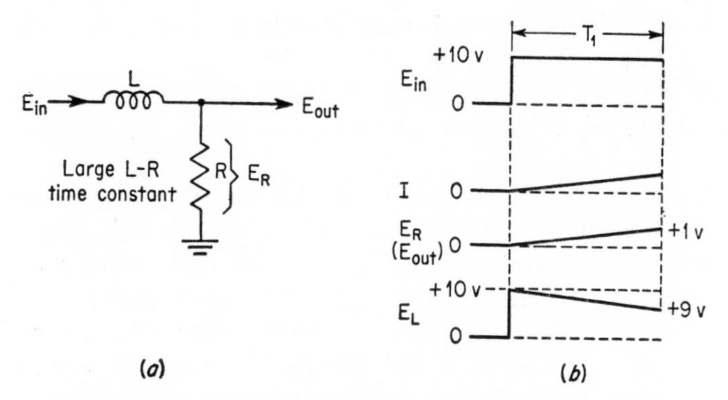

Figure 3·13 **L-R integrator and waveshapes. (b) $L/R = 10T_1$**

circuit is developed across R, whereas in the R-C integrator, it is produced across C (Fig. 3·10a). The +10-volt square wave or flat-topped applied voltage, E_{in}, of Fig. 3·13b, produces a linearly rising current I, a similar rise of E_R, and an immediate rise of E_L, which then decreases slightly. The large L-R time constant, much larger than the period of time T_1, does not allow sufficient time for the current to reach maximum. It would take about five L-R time constants for I to become maximum, as shown in the L-R *differentiating* circuit waveshapes of Fig. 3·11b. In Fig. 3·13b, since I has very little time to increase, it rises only slightly. If, as shown, the L-R time constant is $10T_1$, the current only has $\frac{1}{10}$ of a time constant in which to rise. From the universal time-constant chart of Fig. 3·1, I (in an L-R circuit) rises to only about 10 per cent of its *possible* maximum value of $I = E/R$. As a result, E_R also rises to only about 10 per cent of its maximum value of +10 volts (the applied voltage, E_{in}) or to about

+1 volt, as shown. At the first instant, with +10 volts applied, E_R is zero, and E_L must therefore be +10 volts. At the end of period T_1, with +10 volts still applied, and with E_R at +1 volt, E_L will have decreased to +9 volts, as shown in Fig. 3·13*b*. The output voltage, as indicated, is E_R.

A series of input square waves to an *L-R* integrator, such as those fed to an *R-C* circuit shown in Fig. 3·10*c*, produces an output across the *resistor* of the *L-R* circuit which duplicates E_C of the *R-C* integrator. Similarly, ramp-shaped input signals, such as those shown in Fig. 3·10*d* and *e*, produce output parabolic waveshapes across the resistor of an *L-R* integrator, exactly duplicating the E_C of the *R-C* circuit. The E_R waves of Fig. 3·10*d* and *e*, of the *R-C* circuit, become the E_L waveshapes of the *L-R* integrator.

PROBLEMS

3·1 A 300-volt d-c source is connected to a .02-μf capacitor in series with a 25-kilohm resistor. (*a*) What is the time of one *R-C* time constant? (*b*) What will the voltage be across the capacitor at the end of the first time constant? (*c*) What will the voltage be across the resistor at this time?

3·2 A 150-mh coil in series with a 1-kilohm resistor is connected to 100 volts d-c. (*a*) What is the time of one *L-R* time constant? (*b*) How long will it take, after the voltage is first applied, for E_L to become 100 volts? (*c*) How long will it require for the E_R to become 86 volts?

3·3 A square-wave a-c voltage is applied to a capacitor and resistor in series. (*a*) Draw E_C and E_R if the *R-C* time constant is very short. (*b*) Draw E_C and E_R if the *R-C* time constant is very large.

3·4 A square-wave a-c voltage is applied to a coil and resistor in series. (*a*) Draw E_R and E_L if the *L-R* time constant is very short. (*b*) Draw E_R and E_L if the *L-R* time constant is very large.

MORE CHALLENGING PROBLEMS

3·5 A 10-μf capacitor charges linearly, as shown in the accompanying diagram, from +25 to +50 volts in time T_1 of 50 μsec. In time T_2 of 200 μsec, E_C rises to +90 volts. Find the value of current (*a*) during time T_1, (*b*) during time T_2.

3·6 A linearly rising voltage is applied for a period T to a coil in series with a resistor. If the maximum value of this applied ramp voltage is 100, draw with values (*a*) E_L during time T if the *L-R* time constant is equal to T, (*b*) E_R during time T if the *L-R* time constant is equal to T, (*c*) E_L during time T if the *L-R* time constant is much larger than T, (*d*) E_R during time T if the *L-R* time constant is much larger than T.

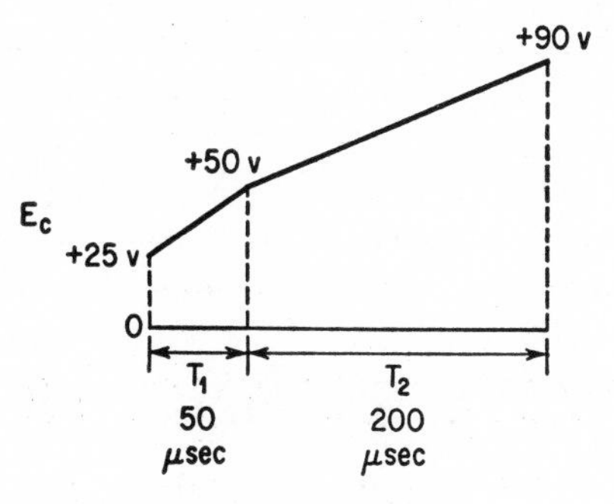

Problem 3·5

3·7 A linearly rising voltage goes from zero to +100 volts in a time period of 1,000 μsec and is applied to a resistor and capacitor in series. If R is 100 kilohms and C is 5,000 pf, find (*a*) I maximum at the end of the time period, (*b*) E_R at the end of the time period.

3·8 A voltage waveshape, such as shown in the accompanying diagram, is applied to a series L-R circuit. Draw the following voltages with

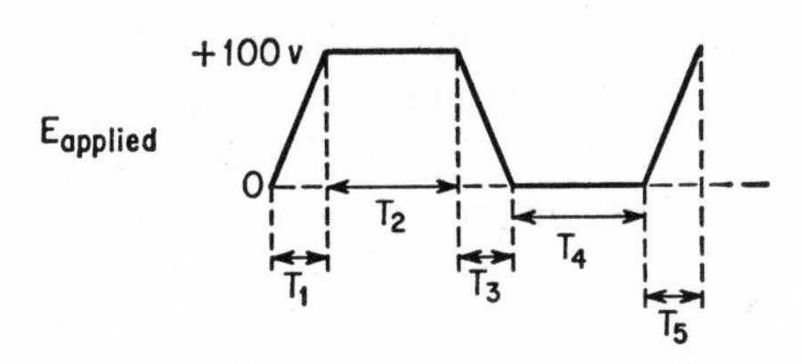

Problem 3·8

values for each of the time periods T_1, T_2, etc: (*a*) E_L, where the L-R time constant is less than T_2 and T_4, but greater than T_1, T_3, and T_5, (*b*) E_R for the same conditions.

3·9 Using the input voltage shown in the Prob. 3·8 circuit, draw the following voltages for each of the time periods T_1, T_2, etc.: (*a*) E_L, where the L-R time constant is less than each of the time periods, (*b*) E_R for the same conditions.

CHAPTER 4

WAVE SHAPING, NONLINEAR

4·1 Clippers or Limiters, General

A clipper, also called a limiter, is a circuit which produces an output signal that falls above or below some voltage level. Either the positive, or the negative, or both extremities are removed. A sine-wave input signal could then produce an output signal with the positive portion flat, or with the negative portion flat, or with both sections flat as in a square wave. Similarly, positive- and negative-going spiked or peaked signals such as differentiated waves (explained in Chap. 3) can be clipped so that the output contains only one type, either positive-going spikes or negative-going ones. Clippers are often used when it is desired to change the shape of a waveform. Diodes, triodes, or pentodes are used as clippers.

4·2 Diode-series Clippers, Unbiased

Figure 4·1 shows a diode-series clipper with input and output signal voltages. Since the load is connected from the output voltage terminal to ground, it is in series with the diode and hence the name "series clipper." Figure 4·1*a* shows a diode vacuum tube, while (*b*) shows a crystal diode. Note that the electron flow in the crystal diode is, as usual, in the opposite direction from the arrow-head symbol of the crystal diode.

Resistor R_1 must be very large compared with the forward resistance (resistance when conducting) of the diode crystal and small compared to the backward resistance of the crystal. A crystal diode used as a clipper should therefore have as high a back-to-forward ratio as possible.

When the input signal goes positive, as shown in Fig. 4·1*c* and *d*, it drives the diode plate positive, causing current flow. This current through

R_1 makes the top of R_1 positive with respect to ground, as shown in Fig. 4·1*a* and *b*. If the input signal is +100 volts peak amplitude, causing current flow, the 100 volts divide up, from Kirchhoff's voltage law, part of them across the small resistance of the conducting diode and part across the high resistance of R_1. Since R_1 is very large, the voltage across it is practically the full +100 volts, with a negligible voltage across the diode. The output, across R_1, is therefore +100 volts, as shown in Fig. 4·1*c* and *d*.

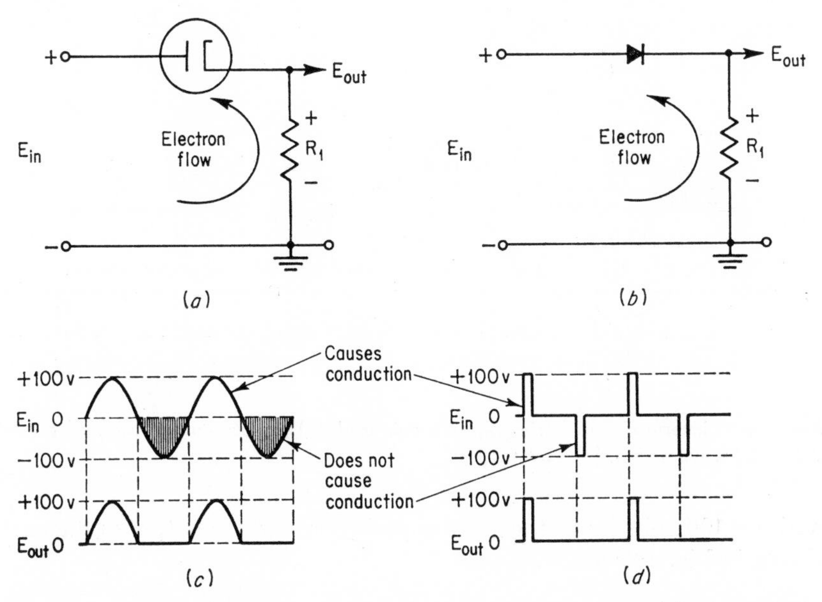

Figure **4·1 Diode series clippers, unbiased. (a) Vacuum-tube diode; (b) crystal diode; (c) waveshapes; (d) waveshapes.**

When the input signal reverses polarity and goes negative, it drives the plate negative, and no current flows. This occurs during the shaded portion of Fig. 4·1*c*. With no current flow through R_1, there is zero volts across it, and the output is zero.

By inverting the diode, as shown in Fig. 4·2, an opposite effect is produced. Now, when the input signal goes positive, as shown by the shaded portion of Fig. 4·2*c*, it drives the cathode positive, preventing current flow. Zero voltage is produced across R_1, and the output is zero. When the input goes negative, it drives the cathode negative, producing a current which flows, as shown in Fig. 4·2*a* and *b*, down through R_1. This makes the upper end of R_1, as shown, negative with respect to ground. The output now becomes a negative voltage. If the input signal has a peak amplitude of 100 volts, then on its negative peak of −100 volts, the low resistance of the conducting diode has practically zero volts across it. The

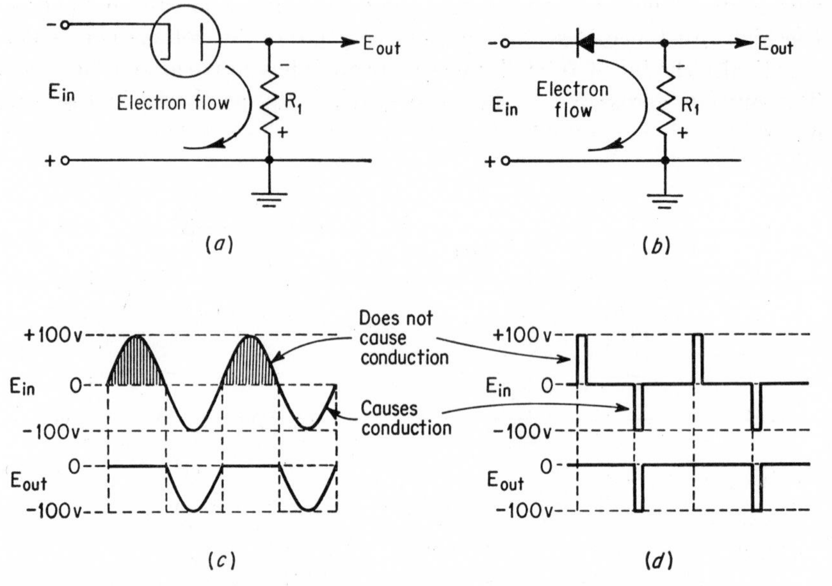

Figure 4·2 Inverted diode series clippers, unbiased. (a) Vacuum-tube diode; (b) crystal diode; (c) waveshapes; (d) waveshapes.

entire −100 volts input therefore appears across R_1 as output voltage, as shown in Fig. 4·2*c* and *d*.

4·3 Diode Series Clippers, Biased

By adding a d-c voltage to the diode clipper, the output signal can be clipped at other levels than zero volts. The added d-c voltage is called a bias voltage, somewhat similar to the fixed bias used between grid and cathode of an amplifier. In Fig. 4·3 the diodes are reverse biased, as shown. This means that the bias voltage is of such polarity that the diode cannot conduct without a signal input. In Fig. 4·3*a* the 30 volts d-c makes the cathode +30 volts, while the plate (with no input signal) is at zero volts. As a result no current flows, and there is zero volts *across* R_1. This means that the upper end of R_1 is at the same potential as the lower end. Since the lower end is +30 volts with respect to ground, the upper end of R_1, which is the output voltage, is also +30 volts. As shown in the voltage waveshapes of Fig. 4·3*a*, when the input is zero, the output is +30 volts.

Current cannot flow, because of the reverse-bias voltage, until the input signal drives the diode plate more positive than the cathode. This occurs when the input exceeds +30 volts. When the input becomes +40 volts, the diode conducts. Current flows, as shown, up through R_1, producing a voltage across R_1 with the polarity shown. The output voltage is now more positive than the +30 volts of the bias source. If the input signal is +40 volts, it is bucking the +30 volts of the bias voltage, leaving a net applied voltage of 10 volts. With the resistance of the conducting diode very small compared to R_1, practically the entire 10 volts is across R_1, making the output voltage E_{R_1} (10 volts) + E bias (30 volts) or +40 volts.

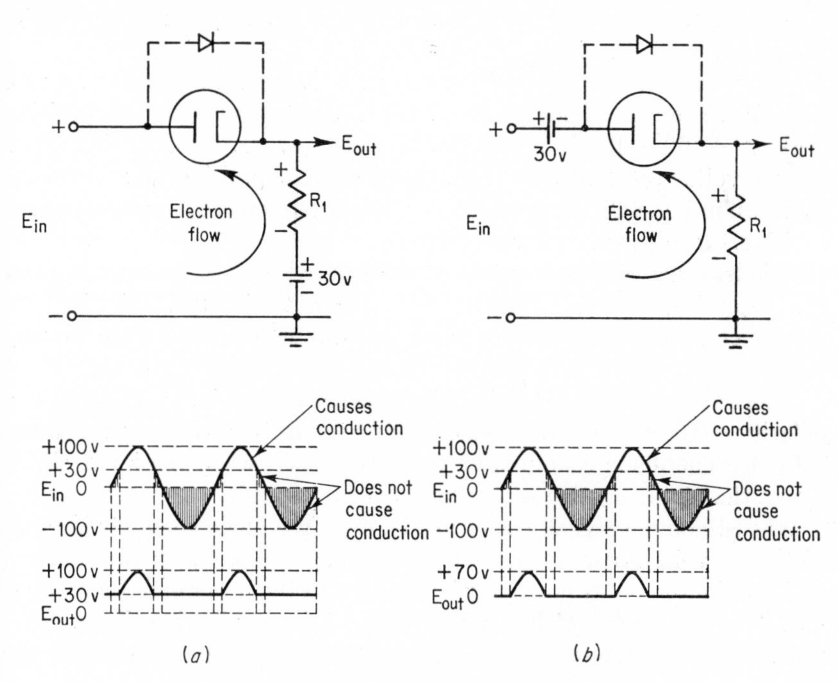

Figure 4·3 **Diode series clippers, reverse biased.**

When the input signal reaches its maximum of +100 volts, it bucks the +30 volts bias, giving a net applied voltage of 70 volts. E_{R_1} is now 70 volts, and the output voltage is now +70 volts (E_{R_1}) added to +30 volts (E bias) or +100 volts, as shown in Fig. 4·3*a*.

When the input signal decreases to +30 volts or less, the diode cuts off. No current flows, and the output is now again +30 volts. The diode remains cut off, and the output stays at +30 volts when the input signal reverses polarity and goes negative, as shown in the waveshapes of Fig. 4·3*a*. The shaded portions of the input signal when the input is less positive than the bias voltage therefore do not produce diode conduction. Note that the output signal is clipped off at +30 volts, extending up to

+100 volts, and has a 70-volt peak-to-peak amplitude. The crystal diode is shown with dotted-line connections. It could be used in place of the diode vacuum tube.

A similar clipper, also producing a 70-volt peak-to-peak clipped output signal but at the zero-volt level, is shown in Fig. 4·3*b*. Here, the reverse bias voltage has been moved to the diode plate, instead of the cathode. This diode, like that in Fig. 4·3*a*, cannot conduct without an input signal since its cathode is zero volts while its plate is −30 volts. When the input is positive but 30 volts or less, the diode remains cut off due to the reverse bias voltage. No current flows, and E_{R_1} and the output are zero volts. This is shown in the waveshapes of Fig. 4·3*b*.

Only when the input is positive and exceeds the voltage of the reverse bias source can the diode conduct. When the input reaches its positive peak of +100 volts, it bucks the 30 volts reverse bias, leaving a net applied voltage of +70 volts, with the diode plate positive. Current flows, as shown in Fig. 4·3*b*, up through R_1, making the top of R_1 and the output positive. Since R_1 is very large compared to the resistance of the conducting diode, practically the entire net voltage of 70 volts is across R_1, and the output is +70 volts. The shaded portions of the input signal, Fig. 4·3*b*, do not cause diode conduction, and the output remains at zero during this time.

A forward-biased diode-series clipper is shown in Fig. 4·4*a* and *b*. The polarity of the bias voltage makes the plate positive with respect to the cathode. The diode conducts all the time except when the input signal drives the plate negative with respect to the cathode.

When the input signal is zero in Fig. 4·4*a*, the diode conducts because of the 30 volts forward bias. Current flows, as shown, up through R_1, making the top of R_1 positive with respect to the lower end. Since R_1 is very large compared with the resistance of the conducting diode, practically the full 30 volts will be across R_1. The output voltage which is the sum of E_{R_1} and the bias voltage will be zero since the two voltages are equal but of opposing polarities, as shown in Fig. 4·4*a*.

When the input signal is +100 volts, it makes the diode plate more positive with respect to its cathode. Current, as shown, again flows up through R_1, and the top of R_1 is positive with respect to the lower end. The total applied voltage is now the +100 volts of the input signal added to the bias voltage for a total of 130 volts. Practically all of this 130 volts is dropped across the large R_1. E_{R_1} is now +130 volts, while the bias voltage of 30 volts is of the opposite polarity. Output voltage, the sum of these two, is now +100 volts, as shown in the waveshapes of Fig. 4·4*a*.

Output voltage becomes −30 volts when current flow through R_1 ceases. This occurs during the shaded areas of the input signal of Fig. 4·4*a* when the input becomes −30 volts or larger. When the input signal is −30 volts, the diode plate is then at the same potential as the cathode, and no current flows. The voltage across R_1 is now zero which means that the upper end of R_1 (output voltage point) is at the same potential as the lower end (the −30 volts of the bias voltage). Output voltage is therefore −30 volts and remains at this level during the time that the diode is inoperative.

Another forward-biased diode-series clipper is shown in Fig. 4·4*b*. Here, the bias voltage has been placed in the plate instead of the cathode as in Fig. 4·4*a*. Again, the bias-voltage polarity makes the diode plate positive

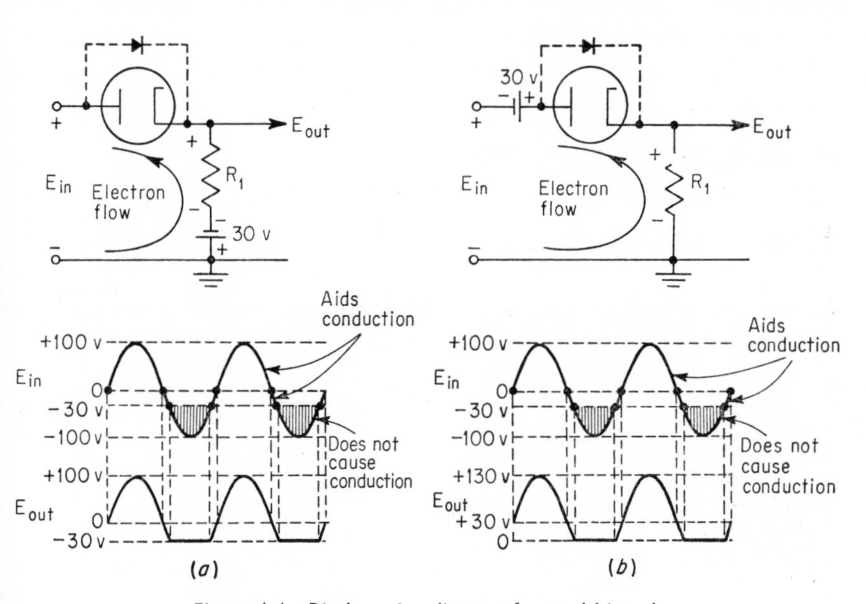

Figure 4·4 **Diode series clippers, forward biased.**

with respect to the cathode, and current always flows until the input signal drives the plate negative.

When input signal is zero, the 30-volt forward bias voltage causes current flow up through R_1. Practically the full 30 volts appears across the large R_1, and output voltage, as shown in the waveshapes of Fig. 4·4*b*, is +30 volts. As the input-signal voltage becomes more positive, driving the plate more positive, the diode conducts more heavily. With more current flowing through R_1, the voltage across it, which is the output voltage, increases. When the input signal reaches its maximum of +100 volts, it adds to the +30 volts bias on the plate, making a net total applied to the plate of +130 volts. Practically the entire 130 volts is dropped across the large R_1, and the output becomes +130 volts, as shown in Fig. 4·4*b*.

When the input voltage reverses polarity and becomes -30 volts and larger, the diode cuts off. With no current flowing through R_1, there is zero volts across it, and the output voltage becomes zero. At -30 volts input, this input voltage is of opposite polarity from the 30 volts bias, producing zero volts on the diode plate. The diode cuts off, and no current flows, producing zero volts at the output. The output voltage remains at this zero level during the shaded portion of the input signal of Fig. 4·4*b*. During this time, the plate is either the same as or negative in respect to the cathode. As a result, the diode remains cut off. Note that the output signals of Fig. 4·4*a* and *b* are similar, both having a peak-to-peak amplitude of 130 volts (*a* going from $+100$ volts to -30 volts and *b* going from $+130$ volts to zero). Both outputs are almost a reproduction of the input, except for the clipping of the negative portion.

4·4 Diode Shunt Clippers, Unbiased

Figure 4·5 shows an unbiased diode shunt clipper. Either the diode vacuum tube or the crystal diode (shown connected with dashed lines)

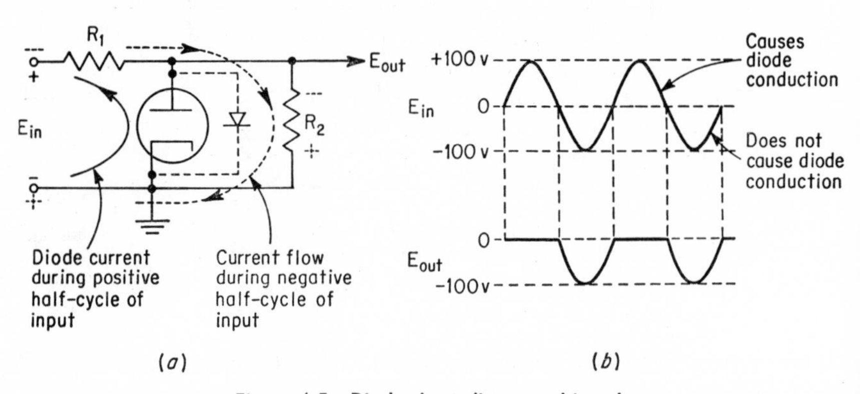

Figure 4·5 **Diode shunt clipper, unbiased.**

could be used. The load is connected between the output voltage point and ground. The diode is therefore in parallel with the load; hence, the circuit name of *shunt clipper*.

R_1 must be much larger than the resistance of the conducting diode (R_P), and R_2 must be much larger than R_1. As shown in Fig. 4·5*a*, current flows through the diode on the positive half cycle of the input signal. Since R_P is so much smaller than R_2, practically no current flows through R_2 now. On the negative half cycle of the input signal, the diode cuts off, and current flows, as shown by the dotted-line arrow of Fig. 4·5*a*, down through R_2.

When the input signal is zero, no current flows, and the output voltage is zero. When the input signal goes positive, it drives the diode plate positive, and the diode conducts. This current, as shown by the solid-line arrow of Fig. 4·5*a*, flows through the diode and through R_1. Since R_1 is much larger than the resistance (R_P) of the conducting diode, practically the full input voltage is across R_1, leaving nothing across R_P. The output, as shown in Fig. 4·5*b*, is therefore zero during the positive half cycle of the input signal.

When the input reverses polarity and becomes negative, it drives the diode plate negative, cutting off the diode. Now, as indicated by the dotted voltage polarities and the dotted-line arrows of Fig. 4·5*a*, current flows through R_1 and down through R_2. This makes the top of R_2 negative with respect to the lower end, and the output is now a negative voltage.

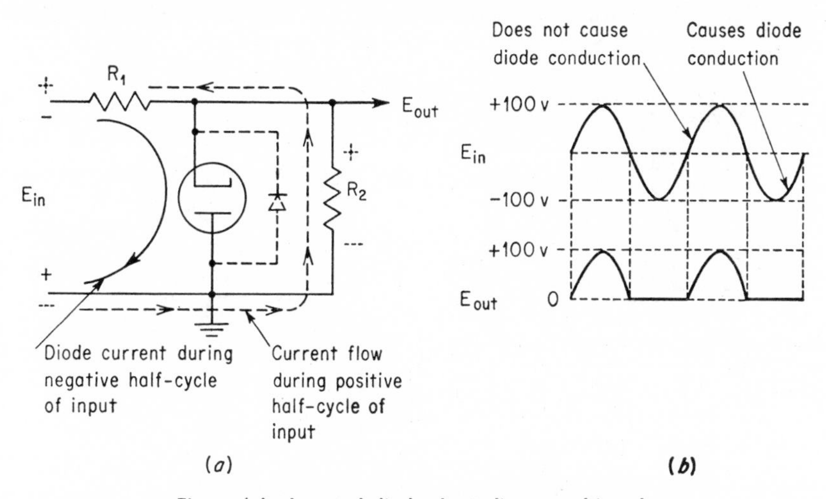

Figure 4·6 **Inverted diode shunt clipper, unbiased.**

If the input signal is −100 volts, then practically this entire voltage is across the very large R_2, and very little across R_1. The output, across R_2, is therefore −100 volts, as shown in Fig. 4·5*b*.

An inverted diode shunt clipper is shown in Fig. 4·6*a*. The operation is the opposite to that described for the previous circuit of Fig. 4·5. The inverted diode conducts during the *negative* half cycle of the input signal. This current flows, as shown by the solid-line arrow of Fig. 4·6*a*, through R_1 and the diode. Since R_1 is much larger than the R_P of the conducting diode, practically the full voltage of the input negative half cycle appears across R_1, leaving zero volts across the diode. Output voltage is therefore at zero during the negative half cycle of the input signal, as shown in Fig. 4·6*b*.

When the input signal is on its positive half cycle, it drives the diode cathode positive, cutting off the diode. Current now flows, as indicated by the dotted-line arrows of Fig. 4·6*a*, up through R_2 and through R_1.

This current makes the top of R_2 positive with respect to the lower end. Since R_2 is much larger than R_1, practically the full voltage of the input positive half cycle appears across R_2. Output voltage, across R_2, is therefore positive, as shown in Fig. 4·6*b*.

4·5 Diode Shunt Clippers, Biased

By adding a d-c voltage to the shunt clippers discussed previously, the output signal can be clipped at some other level than zero volts. The d-c voltage used is called the bias voltage. If the polarity of this voltage prevents diode current, then it is called "reverse bias." An example of this is shown in Fig. 4·7. If the bias-voltage polarity enables the diode to conduct, it is called "forward bias," as shown in Fig. 4·8. Either the diode

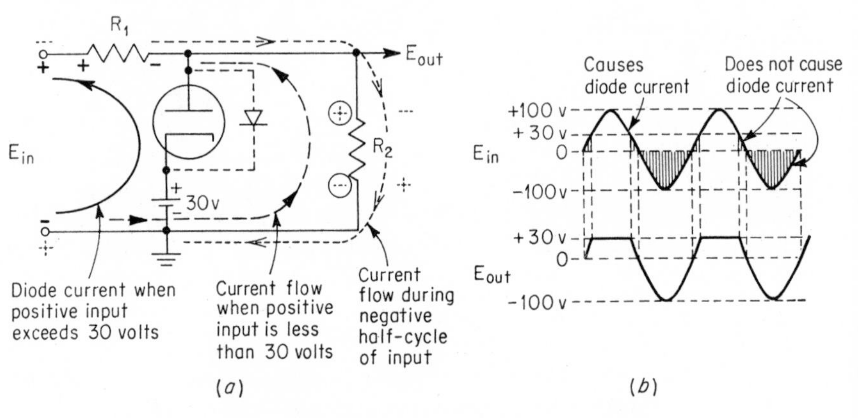

Figure 4·7 **Diode shunt clipper, reverse biased.**

vacuum tube or the crystal diode (shown connected with dashed lines) may be used in these circuits.

The discussion of the reverse-biased shunt clipper of Fig. 4·7 will be given first. It can be seen from this diagram that the 30 volts bias makes the diode cathode +30 volts. This means that the diode cannot conduct until an input signal drives the plate more positive than the +30 volts on the cathode.

When the input signal is zero, no current flows anywhere, and the output voltage across R_2 is zero, as shown in the waveshapes of Fig. 4·7*b*. As the input signal becomes positive, but less than 30 volts, current flows *up* through R_2 (as indicated by the *dashed*-line arrows of Fig. 4·7*a*) and through R_1. Since R_2 is very much larger than R_1, practically the entire applied input voltage now appears across R_2, and the output voltage is this positive value, as shown by the dashed-line, encircled polarities across R_2 of Fig. 4·7*a*.

As the input signal increases in the positive direction, exceeding +30 volts, it makes the diode plate more positive than the cathode, and diode current flows. If the input signal is +100 volts, diode current flows through R_1, as indicated by the solid-line arrow of Fig. 4·7*a*. Note from the diagram that the polarities of the positive input signal and the bias voltage are opposing, resulting in a net voltage applied of 70 volts (the difference between the 100 volts and the 30 volts). Since R_1 is much larger than the resistance R_P of the conducting diode, practically the entire 70 volts will be across R_1, with the polarity shown in the diagram. This leaves almost zero volts across the diode. The output voltage is therefore +30 volts as shown in Fig. 4·7*b* and can be computed in either of the following two ways:

1. With just about zero volts *across* the diode, the plate is almost the same as the +30 volts cathode, making the output (which is connected to the plate) just about +30 volts.

2. The 70 volts across R_1 with the polarity shown (negative at the right side) is opposite to the polarity of the +100 volts input. Therefore, from the output connection to ground, going through R_1 and the input, gives a −70 volts and a +100 volts, for a result of +30 volts.

When the input signal reverses polarity and goes negative, as shown by the dotted-line polarity at the input terminals (Fig. 4·7*a*), it drives the diode plate negative. The diode is inoperative, actually becoming cut off when the input dropped to +30 volts. Current now flows through R_1 and *down* through R_2, as shown by the dotted-line arrows of Fig. 4·7*a*. If the input signal is −100 volts, practically the entire voltage appears across the very large R_2, and the output is −100 volts. This is shown by the dotted-line polarities across R_2 (Fig. 4·7*a*) and the waveshapes of Fig. 4·7*b*. Note that the output signal, across R_2, has the positive portion clipped at +30 volts, while the negative half cycle is complete, going down to −100 volts. The positive output voltage is produced either when the diode conducts or when current flows *up* through R_2. Negative output voltage is produced when current flows *down* through R_2.

Figure 4·8 shows the diagram and waveshapes for a forward-biased shunt clipper. The 30 volts of the bias voltage makes the diode cathode −30 volts. Without an input signal, the plate appears positive with respect to the negative cathode, and the diode conducts. Diode current flows, as shown by the solid-line arrow of Fig. 4·8*a*, through R_1. Since R_1 is much larger than R_P of the conducting diode, almost the entire 30 volts appears across R_1, leaving practically zero *across* the diode. With almost zero voltage across the diode, the plate is just about the same as the −30-volt cathode. Output voltage, connected to the diode plate, is therefore also −30 volts.

When the input signal goes positive, driving the plate move positive with respect to the cathode, the diode conducts more heavily. If the input is +100 volts, it is in series aiding with the 30-volt bias voltage, as shown in Fig. 4·8*a*. The total net voltage applied is then 130 volts, practically all of which is dropped across R_1. Output voltage remains at −30 volts since there is still almost zero volts across the diode, making the diode plate just about the same as the −30-volt cathode. The −30 volts output could also have been determined by the fact that the 130 volts across R_1, with the right end negative (Fig. 4·8*a*), is opposite in polarity to the +100 volts of the input. Going from the output voltage terminal to ground through R_1 and the input terminals produces a −130 volts (E_{R_1}) and then a +100 volts (E input) for an output voltage of −30 volts. As long as the diode conducts, the output is just about −30 volts. The diode

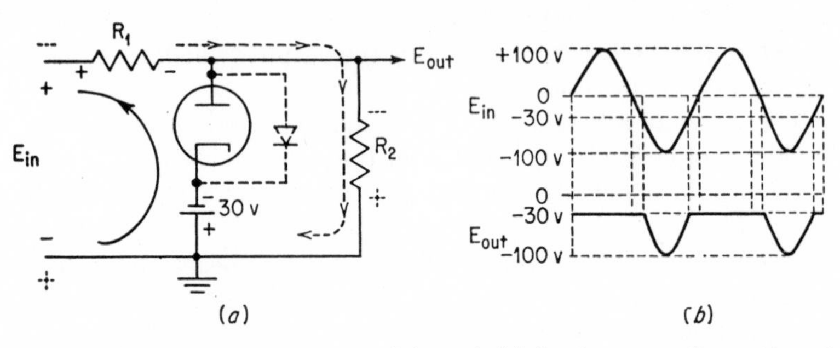

Figure 4·8 Diode shunt clipper, forward biased. (a) Diode current flow without input signal and on positive half cycle of input and when negative input does not exceed 30 volts, is shown by solid-line arrow. Current flow when negative input exceeds 30 volts is shown by dotted-line arrow.

will conduct when the signal is zero, also during the positive half cycle and during that portion of the negative input that does not exceed 30 volts.

When the input signal is larger than −30 volts, it drives the diode plate negative with respect to the cathode, and the diode cuts off. If the input is −100 volts, current flows left to right through R_1 and down through R_2, as shown by the dotted-line arrows of Fig. 4·8*a*. Since R_2 is much larger than R_1, practically all of the −100 volts of the applied input is dropped across R_2, making the output −100 volts.

Double clipping is produced by the two diodes of Fig. 4·9*a*. The sine-wave input signal produces a square-wave output, as shown in Fig. 4·9*b*. The circuit is simply one diode shunt clipper, reverse biased (V_1), and a second diode inverted shunt clipper, also reverse biased (V_2).

No current flows anywhere when the input signal is zero, and zero volts output is produced then.

When the input signal goes positive but does not exceed 30 volts, V_1 and V_2 are still inoperative due to their reverse-bias voltages. Current, however, flows *up* through R_2, as shown by the dashed-line arrows of Fig. 4·9*a* and from right to left through R_1. Polarity of voltage across R_2, which is also the output voltage, is, as shown by the *encircled* dashed-line polarities, positive at the upper end and negative at the lower. Since R_2 is very much larger than R_1, then practically all the positive input appears as positive output voltage across R_2. As the input increases and becomes more positive (but not exceeding 30 volts), the output likewise becomes more positive. This is shown in the E_{out} waveshape of Fig. 4·9*b*. When the input becomes +30 volts, the output becomes just about +30 volts also, less the tiny amount dropped across R_1.

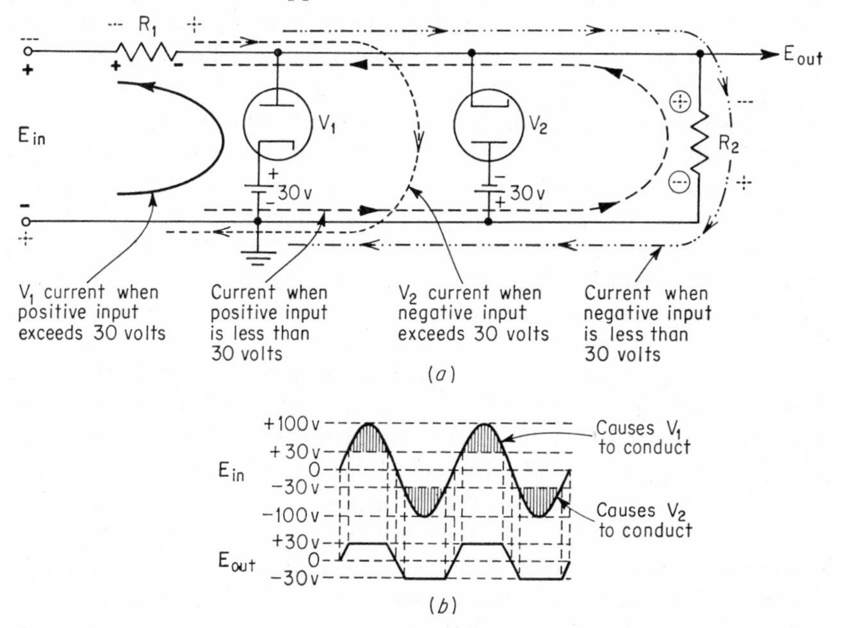

Figure 4·9 **Double-diode clipper.**

When the positive input exceeds 30 volts, it drives the plate of V_1 more positive than its cathode, and V_1 now conducts as shown by the solid-line arrow plate current of Fig. 4·9*a*. This plate current flows from right to left through R_1, producing the voltage polarity shown, the negative end at the right. If the positive input is 100, its polarity is bucking (series opposing) the 30 volts V_1 bias, giving a net applied voltage to the diode V_1 of 70 volts. Since R_1 is much larger than R_P of V_1, almost the full 70 volts is across R_1, leaving almost zero across V_1. The output voltage is therefore +30 volts. This can be seen from either of the following:

1. With practically zero volts *across* V_1, its plate (the output voltage terminal) is just about the same as its cathode (+30 volts). Therefore, the output is +30 volts.

2. Going from the plate of V_1 (the output voltage point) to ground, around to the left in Fig. 4·9*a*, E_{R_1} (70 volts) is bucking the polarity of the positive input voltage (100 volts). The right end of R_1, being the negative end, is 70 volts negative with respect to the +100 volts input, making this right end of R_1 (the output voltage point) +30 volts with respect to ground.

The output remains clipped at +30 volts as long as V_1 conducts. This period is the positive shaded area of E_{in} of Fig. 4·9*b*.

When the input reverses polarity, as shown by the dotted-line polarities of E_{in} of Fig. 4·9*a*, but does not exceed 30 volts, V_1 and V_2 are inoperative. Current flows *down* through R_2, as indicated by the *dashed- and dotted*-line arrows. This produces the voltage across R_2 shown by the dotted-line polarities at the right side of R_2 in Fig. 4·9*a*. Since R_2 is very large, practically the entire negative input voltage (when 30 volts or less) appears across R_2, and the output is this negative voltage. An increase in the negative input voltage (not exceeding 30 volts) produces a larger negative output. This can be seen in the E_{out} waveshape of Fig. 4·9*b*.

When the negative input exceeds 30 volts, it drives the cathode of V_2 more negative than its plate, and V_2 now conducts. V_2 plate current is shown by the dotted-line arrows flowing from left to right through R_1 and then through V_2. This current produces a voltage across R_1 as shown by the dotted-line polarities, with the positive end at the right. If the negative input voltage is −100 volts, V_2 conducts. From the diagram of Fig. 4·9*a*, it can be seen that the dotted-line polarity of E input is bucking (series opposing) the 30 volts V_2 bias, giving a net applied voltage to the V_2 diode of 70 volts. Most of this voltage appears across R_1, with just about zero volts across V_2. The output voltage is therefore −30 volts. This can be explained in either of the following:

1. With practically zero volts across V_2, the cathode is about the same potential as the plate (−30 volts). The output terminal is connected to the V_2 cathode and is therefore −30 volts.

2. Going from the cathode of V_2 to ground, to the left in the diagram, E_{R_1} (70 volts) is of opposite polarity from the negative E_{in} (100 volts), using the dotted-line polarities. The right end of E_{R_1} (which is the output voltage) is therefore +70 volts with respect to the −100 volts E_{in} and is therefore −30 volts.

The output remains clipped at −30 volts as long as V_2 conducts. This period is the negative shaded area of E_{in} of Fig. 4·9*b*.

4·6 Triode and Pentode Clippers

Waveshape clipping can be achieved by using triodes and pentodes as well as the diode circuits explained previously. The amplifier clippers have the advantage of gain which diodes, of course, have not. Three types of clipping are possible: grid clipping, cutoff clipping, saturation clipping, and combinations of these.

Figure 4·10 shows the circuit and waveshapes for grid clipping. The amplifier, which could be a triode or a pentode, is unbiased in Fig. 4·10 but biased at −10 volts in Fig. 4·11. The 15 volts peak-value input signal in Fig. 4·10 drives the grid circuit +15 volts and −15 volts. When the input signal drives the grid above the cathode potential of zero volts, the grid draws electrons from the cathode, much the same as a diode plate acts. Grid current flows through R_1 as shown, producing a voltage across it with the polarity shown. R_1 is very much larger than the cathode to grid resistance when the grid draws current. As a result, the entire positive

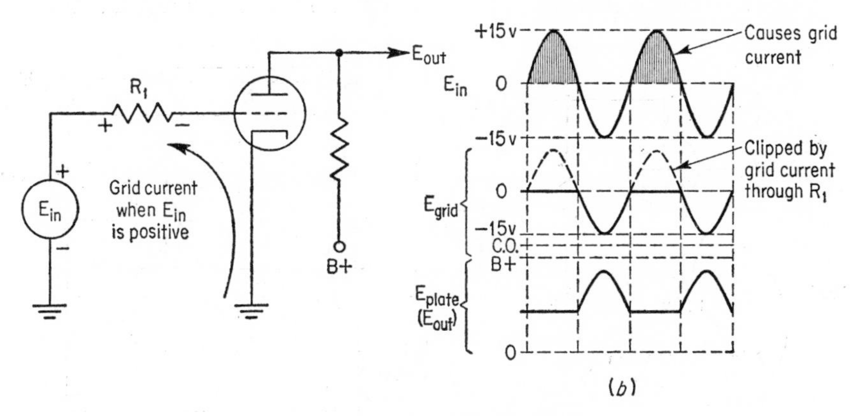

Figure 4·10 Grid clipping, zero bias.

half cycle of the input signal appears across the large R_1, leaving practically zero volts on the grid. This is similar to the unbiased diode shunt clipper explained previously. As shown in Fig. 4·10*b*, the grid voltage waveshape has the positive half cycle clipped off at zero. When the input signal goes negative, it drives the grid to the peak value of −15 volts. Since no grid current flows during this negative half cycle, the cathode-to-grid resistance is now much larger than R_1. As a result, practically the full negative half cycle appears on the grid as shown. It is assumed here that the negative input signal does not drive the grid to the cutoff value. Plate voltage, or output signal, is the amplified and inverted reproduction of the clipped grid signal.

In Fig. 4·11, another example of grid clipping is shown, except that a 10-volt bias is used. As usual, the input signal varies about the bias as its axis or average value. This means that the +15 volts peak input drives the grid from its −10 volts bias (grid appears −10 volts with respect to the cathode) up to a positive peak of +5 volts. When the grid rises above the cathode, grid current flows. This occurs during the shaded area of

Fig. 4·11*b*. Grid-current flow through R_1 produces the clipped signal on the grid, as shown. The negative peak of -15 volts of the input signal drives the grid (with respect to the cathode) from its bias of -10 volts down to -25 volts. Again, it is assumed here that the signal is of insufficient amplitude to drive the grid to cut off. As a result, plate voltage output signal is the inverted and amplified version of the clipped grid signal, as shown in Fig. 4·11*b*. The difference between the output signals of Figs. 4·10 and 4·11 is the same as that between the unbiased and reverse-biased diode shunt clippers (Figs. 4·5 and 4·7). The biased circuit clips off less of the signal than does the unbiased stage.

Figure 4·12 is a combination of grid clipping and cutoff clipping, resulting in an output signal that has the top and bottom squared off. In this

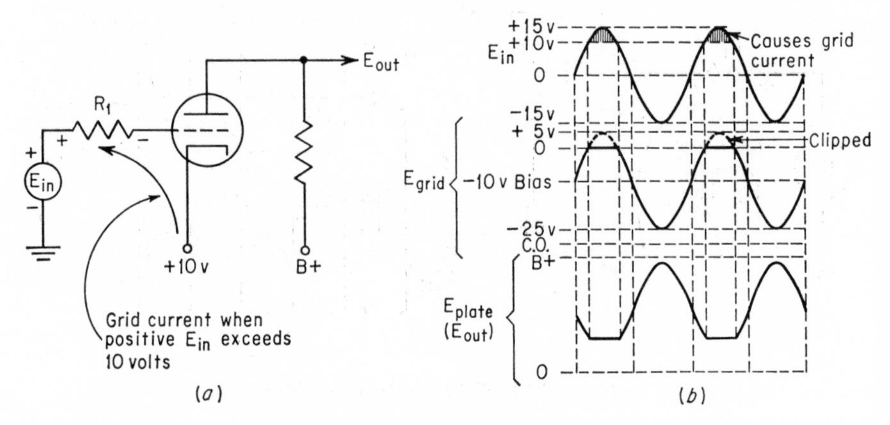

Figure 4·11 Grid clipping, biased.

diagram, it is assumed that the input signal, on its negative portion, drives the grid more negative than cutoff, while on its positive excursion, it drives the grid positive with respect to the cathode. As shown in the E_{grid} waveshape of Fig. 4·12*b*, the positive portion of input signal on the grid is clipped off at zero, while the negative portion is not clipped on the *grid*. Clipping is produced at the plate where the most negative part of the grid waveshape cuts off the tube. With no plate current flowing through R_L, there is zero voltage drop across R_L, and plate voltage becomes full B+. Output signal then, as shown in Fig. 4·12*b*, is double clipped, at B+ due to plate current cutoff and also clipped at the other extremity due to grid-current clipping.

Double clipping is also produced by the circuit of Fig. 4·13. Here, saturation clipping and cutoff clipping is used. As shown, the most positive part of the input signal drives the grid more positive than the cathode. Although grid current flows, no clipping of grid-voltage waveshape occurs since R_1 is not used (as it was in Figs. 4·10 to 4·12). However, the positive-going grid is being operated in the saturation region. In this region, a positive-going increase of grid voltage does not produce much of an increase of plate current. As a result, as the grid voltage goes positive into the saturation region, plate current does not rise but flattens out.

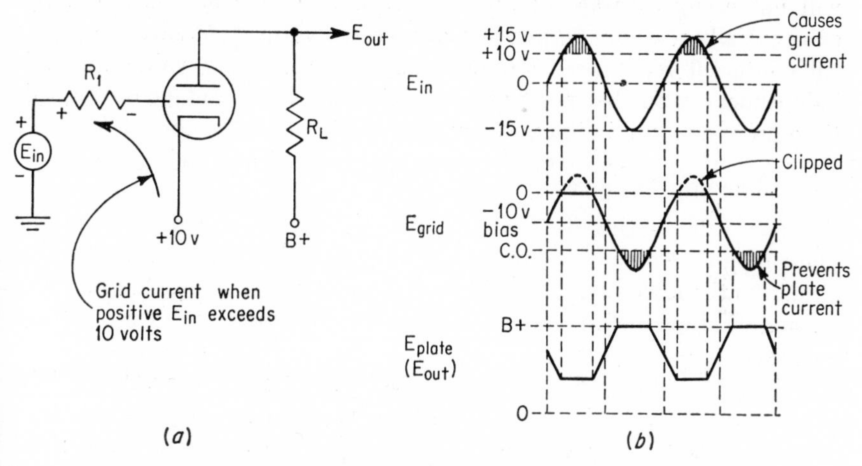

Figure 4·12 Combination of grid clipping and cutoff clipping.

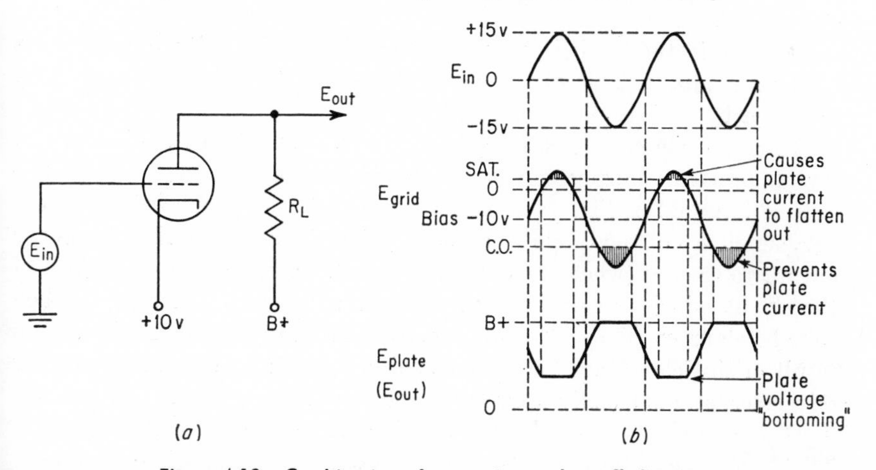

Figure 4·13 Combination of saturation and cutoff clipping.

The fairly constant plate current produces a constant voltage across R_L. Plate voltage, as shown at the lowest point of Fig. 4·13*b*, is therefore steady or clipped during the time that the grid signal is in the saturation region.

Since the most negative portion of the grid voltage drives the tube beyond cutoff, then plate voltage rises to full B+ during the time that the tube is inoperative. As shown in Fig. 4·13*b*, E_{out} is clipped off at the top and bottom extremities.

In Fig. 4·13, it was assumed that the negative peaks of the output signal voltage at the plate were clipped because of plate current saturation. Actually however, a tube employing an indirectly heated cathode will not easily saturate unless the filament voltage has been reduced substantially to about 70 per cent of its rated voltage. At saturation, the tube internal resistance, R_B, becomes a constant low value which is independent of further positive grid excitation. Plate voltage during this time drops to a minimum value and is said to have "bottomed."

4·7 Clamping, General

When a fluctuating d-c voltage, such as that which is present at the plate of an amplifier, is applied through a resistor-capacitor coupling circuit to the next stage, the coupling capacitor (also called the blocking capacitor) stops the d-c voltage component, allowing only the a-c voltage component to pass through. This can be seen in Fig. 4·14*a*, where the signal input to the *R-C* circuit is a positive fluctuating d-c voltage. The average of 200 volts is the d-c component, while the a-c component is 100 volts peak to peak, from +250 volts to +150 volts. The output voltage of Fig. 4·14*a* contains only pure a-c (+50 volts and −50 volts) and no d-c.

In many instances, it is necessary to restore a d-c voltage component to the a-c output. This can be accomplished by the use of a diode (Fig. 4·14*b*) or by an inverted diode (Fig. 4·14*c*).

When a diode conducts, its cathode-to-plate resistance R_P is low and for all practical purposes may be regarded as almost zero ohms. The output terminal of Fig. 4·14*b* and *c* is connected to ground through the very small R_P of the diode when it conducts. This practically shorts the output to ground (when the diode conducts), clamping that peak part of the signal, which causes diode conduction, to ground. The diode is then called a clamper. Since a d-c voltage is produced in the output signal as a result of this clamping, the circuit is also known as a d-c restorer or d-c reinserter, and is usually abbreviated to DCR.

4·8 Clamping Circuits

The diode circuit of Fig. 4·14*b* produces the output signal shown by the accompanying waveshape. Note that the output waveshape shown has its most positive peak at zero volts. This occurs when the input signal, shown in Fig. 4·14*a*, is going positive, becoming +250 volts. The positive-going input voltage causes C_1 to charge up through the low R_P of the conducting diode vacuum tube, or through the crystal diode (shown in the dashed-line connections) low forward resistance. The charging of C_1

in Fig. 4·14*b* is readily accomplished since C_1 and R_P make up a short or small *R-C*. When the input voltage decreases (negative-going), C_1 starts discharging through R_1. This combination, R_1 and C_1, makes up a long or large *R-C*, and C_1 is unable to discharge.

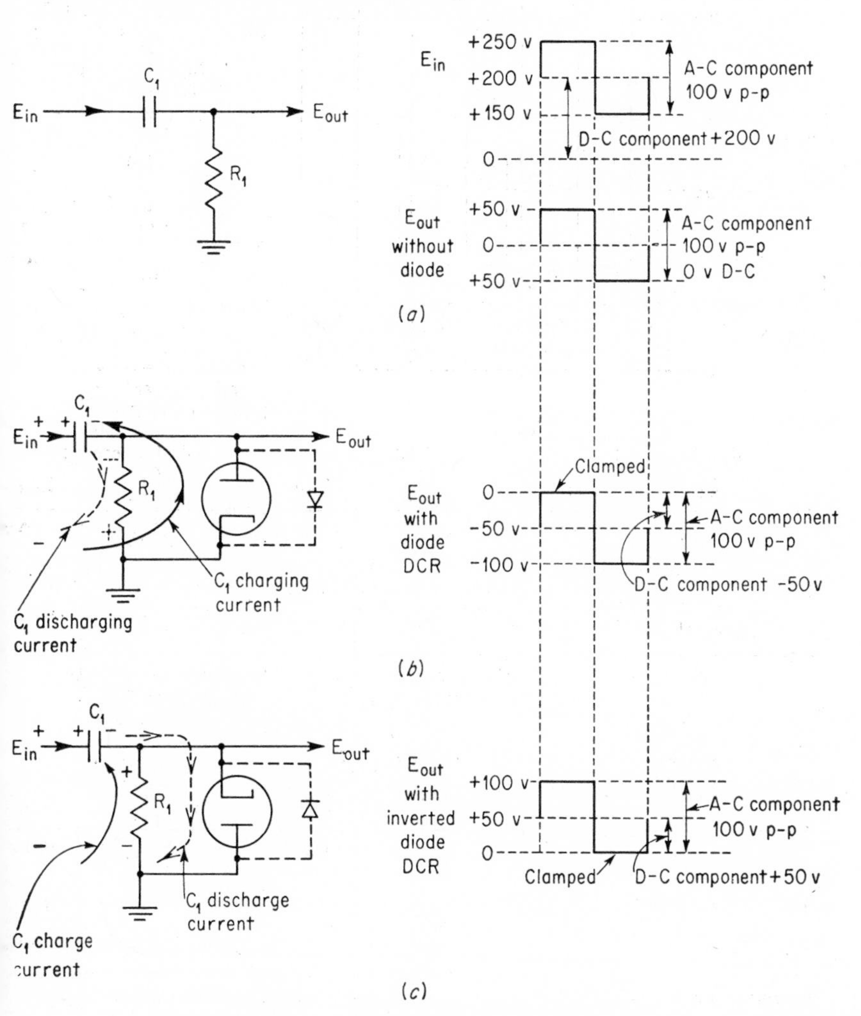

Figure 4·14 **Diode clamping or d-c restoration.**

The detailed theory of the clamping or d-c restoration action of the circuit of Fig. 4·14*b* follows. First, reference to the graph of Fig. 5·3 should be made. This graph shows the exponential rise and fall of a capacitor as it charges and discharges. Voltage across the resistor, shown by the dotted-line graph, is immediately (instant t_1 of Fig. 5·3) the full

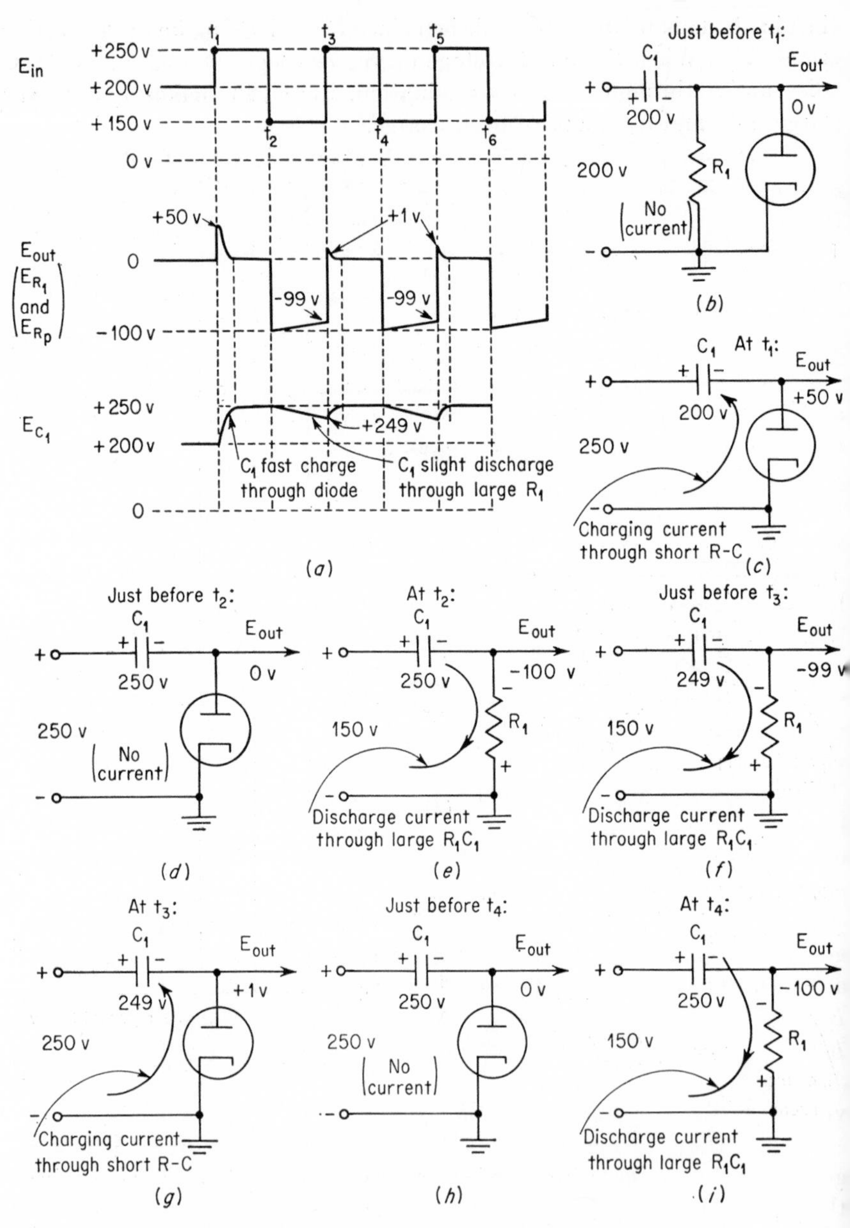

Figure 4·15 **Details of diode negative DCR.**

applied voltage, decreasing down to zero as the capacitor charges. Similarly, the instant the capacitor starts discharging (instant t_2 of Fig. 5·3), resistor voltage becomes immediately the full negative voltage. Resistor voltage decreases as the capacitor discharges.

The waveshapes and circuits of Fig. 4·15*a* through (*i*) should now be referred to in the following detailed explanation of the clamper circuit of Fig. 4·14*b*.

1. Assume first that the input voltage is +200 volts for a *long* time before instant t_1. As a result, C_1, as shown in Fig. 4·15*b*, has fully charged to the +200 volts. No current flows in the circuit of *b* after C_1 has charged, and E_{out} is zero.

2. At instant t_1, E_{in} has risen 50 volts to +250 volts. In Fig. 4·15*c*, C_1 starts charging to the input +250 volts, having to charge up 50 volts more. Charging current, as shown, flows through the diode. The small *R-C* time constant allows C_1 to quickly charge up to the full +250 volts. At instant t_1, when the input increased 50 volts, E_{out} across the diode R_P goes positive 50 volts, as shown in Fig. 4·15*a*. When C_1 has charged to +250 volts, E_{out} across R_P decreases back to zero. This is seen in (*a*) and in (*d*) just before t_2.

3. At instant t_2, E_{in} has decreased 100 volts down to +150 volts, as shown in (*a*) and (*e*). C_1 now tries to discharge its +250 volts down to +150 volts. Discharge current now flows down through the large R_1, making E_{out} immediately −100 volts. The large R_1-C_1 time constant prevents C_1 from discharging very much.

4. Just before instant t_3, as shown in Fig. 4·15*a* and *f*, C_1, due to the large *R-C*, has only lost 1 volt, going from its full charge of +250 volts down to +249 volts. E_{out} across R_1 has now decreased 1 volt, going from −100 volts to −99 volts.

5. At instant t_3, in Fig. 4·15*a* and *g*, the input has risen 100 volts again, going from +150 volts to +250 volts. The charge on C_1 is +249 volts, and C_1 starts charging up through the diode R_P to the higher +250 volts. E_{out} across R_P is therefore +1 volt at instant t_3. Because of the small *R-C*, C_1 quickly charges up, and E_{out} again decreases to zero.

6. Just before instant t_4, as shown in Fig. 4·15*a* and *h*, C_1 is at +250 volts charge, no current flows, and E_{out} is zero.

7. At instant t_4, Fig. 4·15*a* and *i*, the input has decreased 100 volts, going from +250 volts down to +150 volts. C_1 starts discharging down through the large R_1, attempting to discharge its +250 volts down to +150 volts. E_{out} across R_1 now becomes −100 volts. Because of the large R_1C_1 time constant, C_1 cannot discharge very much, and E_{out} only decreases slightly, as before, to −99 volts.

8. The remaining instants, t_5, t_6, etc., are repeats of the previous ones, and E_{out} as shown in Fig. 4·15*a*, is essentially the same shape as E_{in}.

Note, however, that E_{out}, except for the first positive-going spike at t_1, is a fluctuating *negative* d-c voltage, varying from zero volts down to -100 volts. The diode clamp of Fig. 4·14*b* is therefore called a *negative* d-c restorer.

The diode clamp circuit of Fig. 4·14*c* is an inverted diode. This diode conducts on the negative-going portion of E_{in}, clamping the corresponding part of E_{out} at zero, as shown in Fig. 4·14*c*. This produces an E_{out} which is

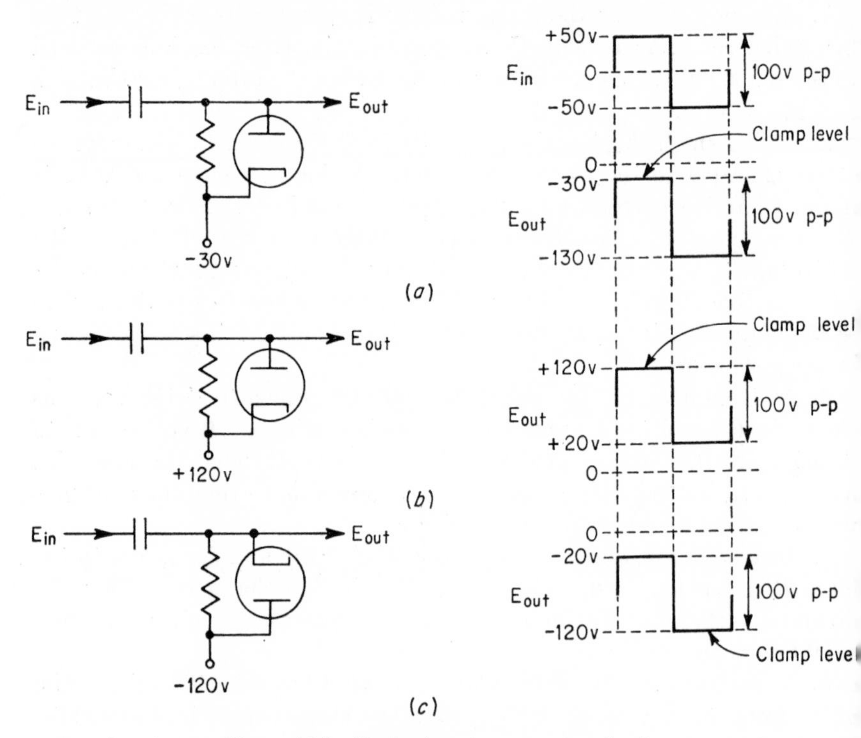

Figure 4·16 **Diode clamping at various levels.**

a *positive* fluctuating d-c voltage. The inverted diode clamp of Fig. 4·14*c* is therefore called a *positive* d-c restorer.

In the circuits of Fig. 4·14*b* and *c*, clamping took place at the zero level since the diode was connected to ground. The clamping level will be at whatever d-c voltage is connected to the diode.

In Fig. 4·16, E_{in} is a 100-volt peak-to-peak signal to all the circuits of (*a*), (*b*), and (*c*). In (*a*), a *negative* DCR is used. The diode cathode is connected to -30 volts, and that becomes the clamp level. E_{out} in (*a*) is clamped at -30 volts, with the signal extending negatively 100 volts down to -130 volts.

In Fig. 4·16*b* another *negative* DCR is shown; this time the diode cathode is connected to +120 volts. E_{out}, as shown, is clamped at +120 volts, extending negatively downward 100 volts to +20 volts.

Figure 4·16*c* is a positive DCR with the inverted diode plate connected to −120 volts. E_{out} is therefore clamped at this −120 volts and extends positively upwards 100 volts to −20 volts.

A triode clamping circuit is illustrated in Fig. 4·17*a*. The grid, before input a-c signal is applied, is zero biased. That is, grid and cathode are at the same potential. The +200 volts d-c component has charged up C_1 to this value. When the input rises, it drives the grid positive, and grid current flows. The grid acts like a diode plate, and resistance between

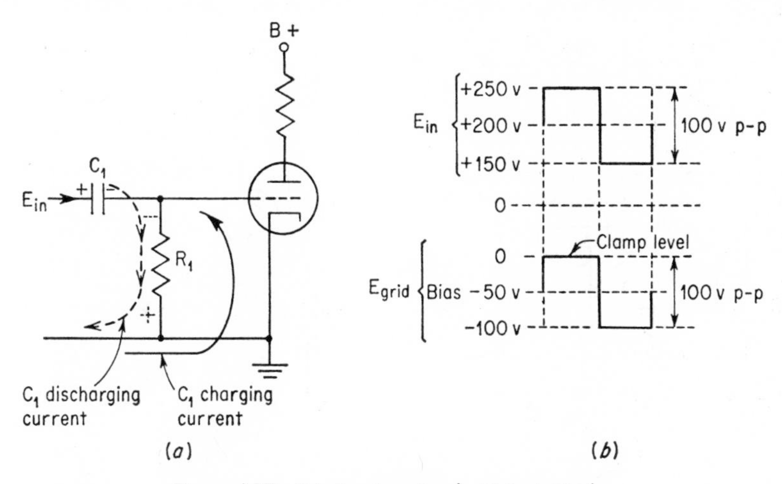

Figure 4·17 Triode clamping (grid-leak bias).

cathode and grid is very small, allowing C_1 to quickly charge to the +250 volts of the applied signal. As explained previously for the diode negative DCR, voltage across the small resistance falls to zero when C_1 has quickly charged. In Fig. 4·17*b*, the grid is at zero volts when input signal is at +250 volts. When E_{in} decreases 100 volts from +250 volts down to +150 volts, C_1 starts discharging down through the very large R_1. Voltage across R_1 now becomes −100 volts. Because of the large R_1-C_1 time constant, C_1 cannot discharge. The −100 volts across R_1 therefore remains fairly constant during the time that E_{in} is down at +150 volts, as shown in Fig. 4·17*b*.

Grid voltage, as shown, is a fluctuating negative d-c, varying from zero volts and down to −100 volts, with a d-c component of −50 volts. The −50 volts is the bias on the grid and is only present when input signal is being applied. This method of bias is called grid-leak bias.

4·9 Schmitt Trigger

The Schmitt trigger or *cathode-coupled bistable multivibrator* is shown in Fig. 4·18. The discussion of this circuit is given here rather than in Chap. 5 covering multivibrators because this circuit is more commonly employed as a *squaring* circuit. A sine wave, sawtooth, or other input signals result in an excellent square or rectangular output-voltage waveshape.

In the diagram of Fig. 4·18*a*, V_2 conducts heavily due to its positive grid. Plate current of V_2, flowing through R_K, makes the top of R_K and both cathodes positive. The grid of V_2 draws current from its cathode, and cathode and grid clamp at some positive voltage. Since the grid of V_1 is connected through R_g to ground, this grid is at zero volts and

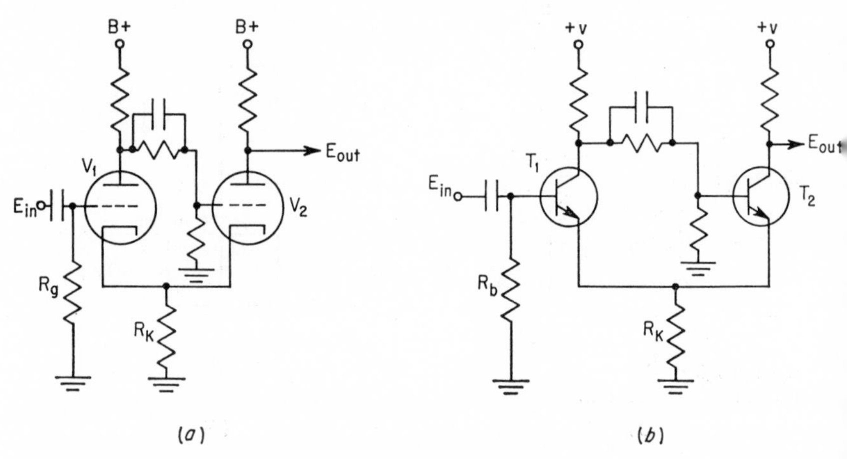

Figure **4·18 Schmitt trigger or squaring circuit (cathode-coupled bistable multivibrator). (a) Vacuum-tube circuit, V_2 normally ON; (b) NPN-transistor circuit, T_2 normally ON.**

appears sufficiently negative with respect to its positive cathode as to keep V_1 cut off. With V_1 inoperative, its plate voltage is practically the full B+, and the grid of V_2 is very positive, keeping V_2 conducting heavily.

With a sine-wave input, as shown in Fig. 4·19*a*, applied to the grid of V_1, V_1 conducts when the input signal drives V_1 grid sufficiently positive. Assume that -5 volts grid bias is the cutoff for V_1. Also assume that when V_2 is heavily conducting, E_{R_K} is 30 volts, making both cathodes +30 volts. At instant t_1, the input sine wave is at +25.1 volts. With V_1 cathode at +30 volts and its grid at +25.1 volts, the grid bias is now at -4.9 volts. V_1 now starts conducting, and its plate voltage decreases. This lowers the grid voltage of V_2, and V_2 conducts less. E_{R_K} decreases, becoming less positive, permitting V_1 to conduct more heavily. V_1 plate voltage falls further, and V_2 now cuts off. E_{R_K}, with only V_1 conducting, is less positive than when V_2 heavy plate current flowed through R_K. Assume that E_{R_K}, as shown in Fig. 4·19*a*, is now +25 volts.

V_1 stays ON (conducts), and V_2 remains OFF (nonconducting) until the input signal cuts off V_1 again. The output signal at V_2 plate, as shown in Fig. 4·19*a*, jumped from a low steady value, when V_2 conducted heavily (before instant t_1), to full B+, when V_2 became inoperative (at instant t_1).

At instant t_2, the input signal has returned to its +25.1-volt value (as it was at t_1). However, E_{R_K} is now +25 volts, and V_1 will not cut off at this instant. It was assumed that cutoff for V_1 is −5 volts bias, therefore V_1 will cut off when the input signal drops to a lower value than that which originally triggered V_1 into conduction. When the input signal

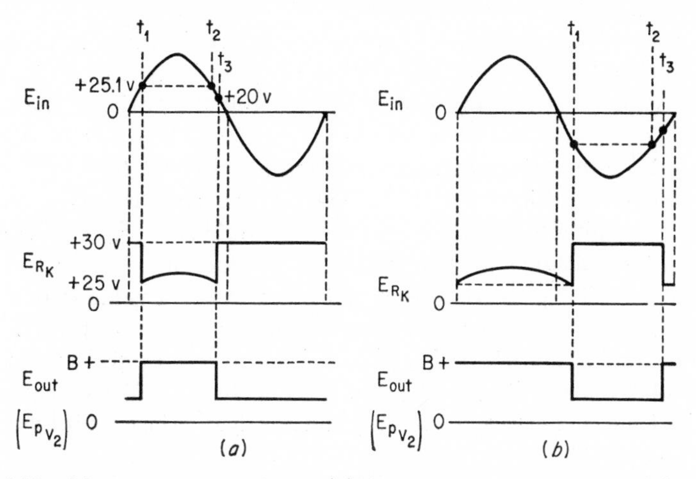

Figure 4·19 **Schmitt trigger waveshapes. (a) With V_2 the normally ON tube; (b) if V_1 were made the normally ON tube.**

drops to about +20 volts at instant t_3 and E_{R_K} is +25 volts, the bias becomes −5 volts. Now V_1 becomes inoperative, and its plate voltage rises sharply to almost B+, making V_2 grid very positive again. V_2 again conducts heavily at instant t_3, and its plate voltage falls abruptly down from B+. Note that in the E_{in} waveshape of Fig. 4·19*a*, V_1 is triggered into conduction when E_{in} rises to the +25.1-volt level at instant t_1, but is not cut off when E_{in} decreases to this level at instant t_2. A short time later at instant t_3, when E_{in} has decreased further down to +20 volts, V_1 becomes cut off, and V_2 conducts again. This time lag or delay between instant t_2 and instant t_3 is called *hysteresis*.

The NPN transistor Schmitt trigger circuit shown in Fig. 4·18*b* is identical in operation to the vacuum-tube counterpart. Transistor T_2 is normally ON since its P-type base is at a positive voltage and is therefore *forward*-biased. T_2 current flows up through R_K, making the N-type emitter of T_1 *positive,* while its base is at ground potential. Transistor T_1 is therefore *reverse* biased and nonconducting. As in the discussion for the vacuum-tube circuit of Fig. 4·18*a*, when the input signal becomes sufficiently positive, it *forward* biases transistor T_1, triggering this stage into the ON condition.

Another possible operation of the circuits of Fig. 4·18 is to have V_1 (and T_1) normally conducting. This could be brought about by connecting the lower end of R_g, Fig. 4·18*a*, and R_B, Fig. 4·18*b*, to a positive voltage. This would make V_1 grid (and T_1 base) positive, and these stages would conduct. The conducting V_1 makes its cathode clamp to about this same positive voltage. Plate voltage of conducting V_1 is now low, which makes V_2 grid low. V_2 is therefore inoperative.

When the input signal goes sufficiently negative, it drives V_1 grid down from its normal positive voltage, cutting off V_1 at instant t_1 of Fig. 4·19*b*. V_1 plate voltage rises sharply to practically full B+, driving the grid of V_2 positive. V_2 now conducts heavily, making E_{R_K} larger than when V_1 conducted. Plate voltage of V_2, E_{out}, drops abruptly at t_1.

At instant t_2 of Fig. 4·19*b*, E_{in} is back at the same negative value as that which originally cut off V_1 (at instant t_1). However, because of the more positive cathode (due to heavy V_2 conduction) V_1 will not start conducting until E_{in} has risen to a much less negative value than that at instants t_1 and t_2. This occurs a little later at t_3. At this instant, V_1 goes back into conduction. Its falling plate voltage lowers V_2 grid voltage, and V_2 goes OFF. Plate voltage of V_2 at instant t_3 (Fig. 4·19*b*) rises sharply to B+. The time lag between instants t_2 and t_3 is due to the circuit hysteresis.

PROBLEMS

4·1 From the diagram shown, draw the output signal with values at (*a*), (*b*), (*c*), and (*d*).

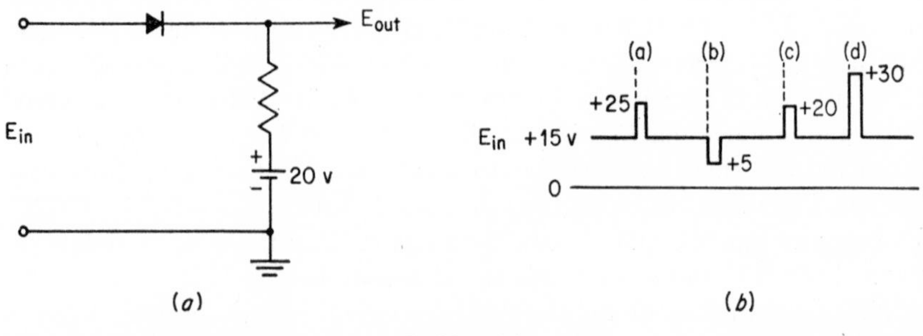

Problem 4·1

4·2 From the circuit accompanying this problem and using the input shown in the diagram of Prob. 4·1, draw the output signal with values at (*a*), (*b*), (*c*), and (*d*).

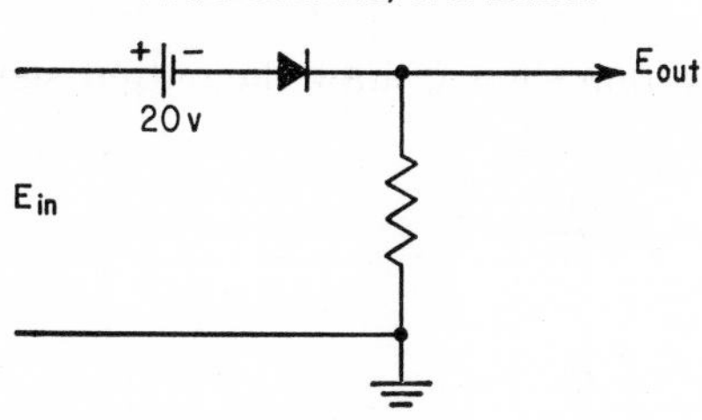

Problem 4·2

4·3 From the diagram shown, draw the output signal with values at (*a*), (*b*), (*c*), and (*d*).

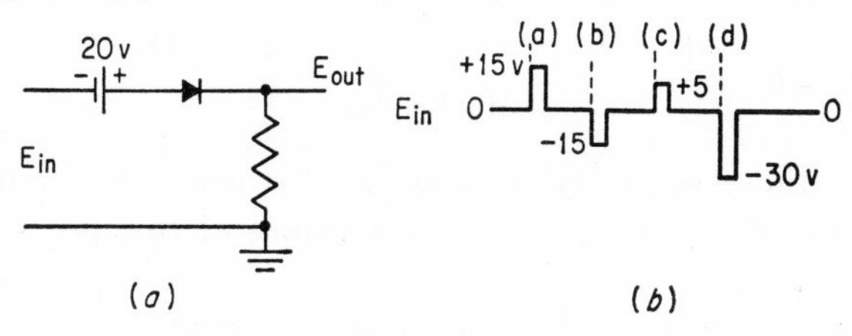

Problem 4·3

MORE CHALLENGING PROBLEMS

4·4 From the circuit accompanying this problem and using the input signal shown in Prob. 4·3, draw the output signal with values at (*a*), (*b*), (*c*), and (*d*).

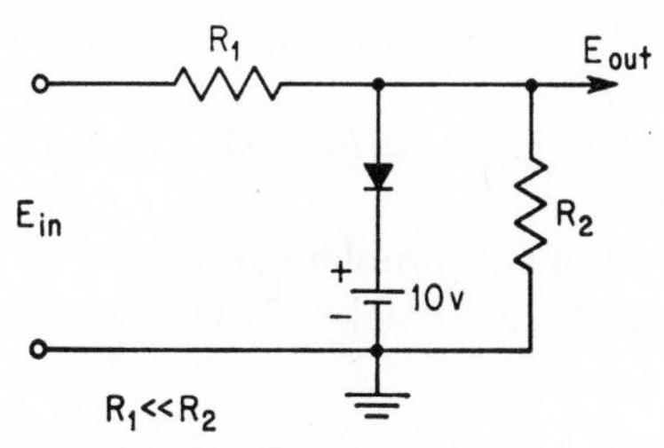

Problem 4·4

4·5 From the circuit for this problem and using the input signal shown in Prob. 4·3, draw the output signal at (*a*), (*b*), (*c*), and (*d*).

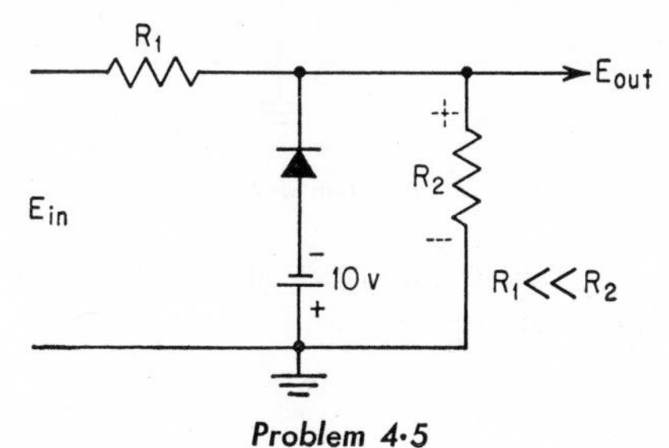

Problem 4·5

4·6 From the diagram shown, draw the output signal for both input cycles.

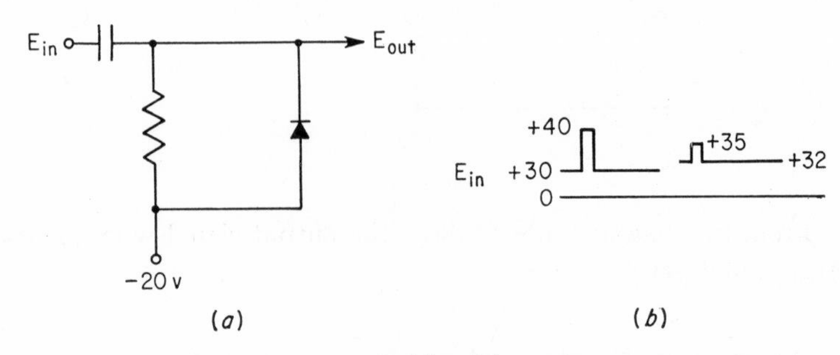

Problem 4·6

4·7 From the accompanying diagram and using the input signal of Prob. 4·3, draw the output signal with values at (*a*), (*b*), (*c*), and (*d*).

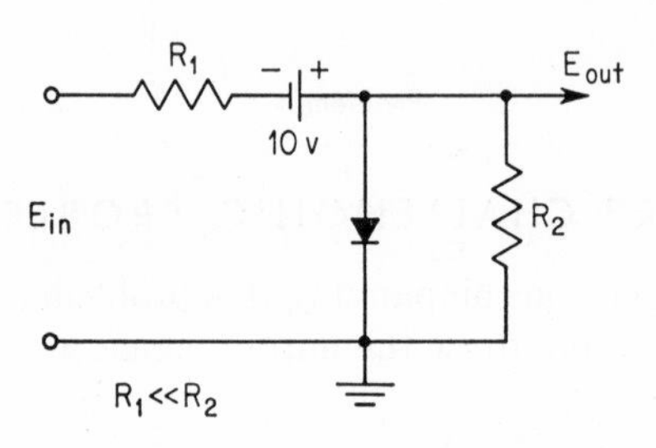

Problem 4·7

4·8 Describe how clipping is performed at the grid and at the plate of a triode amplifier tube.

4·9 From the circuit of this problem and using the input signal shown in Prob. 4·3, draw the output signal with values at (*a*), (*b*), (*c*), and (*d*).

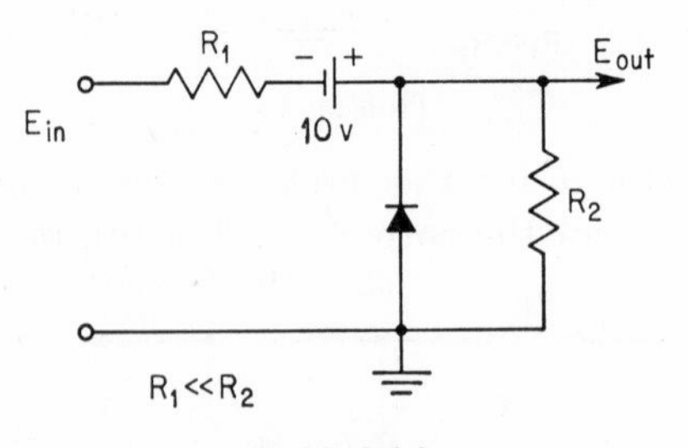

Problem 4·9

4·10 Draw the schematic diagram of a Schmitt trigger using PNP transistors, and describe its operation.

CHAPTER 5

MULTIVIBRATORS

5·1 Multivibrator, General

A multivibrator consists of two triodes, either vacuum tubes or transistors. Only one triode conducts at a time. They may be designed to operate alternately by themselves in which case the multivibrator acts as an oscillator. This type is called "free-running" or "astable" and uses input signals only to synchronize or stabilize the operation. A type which requires an input signal in order to make it operate is called a "driven" multivibrator. This type can be subdivided further into two classes: (*a*) monostable and (*b*) bistable.

A *monostable multivibrator* is one which has only one stable state of operation and requires input signals in order to function. Of the two triodes, one always conducts, while the other is always cut off. The circuit remains in this stable state until an input signal is applied. When the input signal is applied, the process *temporarily* reverses, and the conducting triode becomes inoperative, while the previously cutoff stage now conducts. After a period of time, determined by a resistor and capacitor (*R-C* time), the circuit reverts back to its first state with the original triode again conducting, and the other becoming and remaining cut off. The circuit remains in this stable state until the next input signal is applied. A monostable multivibrator is also called a "one-shot" multivibrator.

A *bistable multivibrator* is one which has two stable states of operation and, like the monostable, requires input signals to operate. In the bistable type, one triode always conducts while the other remains cut off. Then, when an input signal is applied, the process reverses but not temporarily as in the monostable. The conducting triode now cuts off, and the other triode now conducts. The circuit now stays in this mode of operation permanently and does not revert back to its original state by itself. It requires a second input signal to reverse the process, putting the original triode back into operation and cutting off the other. The circuit remains in this condition, changing only when one input arrives and reverting back again with the next input signal. Another name for the bistable multivibrator is the "flip-flop" or *Eccles-Jordan* (E-J) *circuit*, although the term flip-flop sometimes refers to the monostable multivibrator. The original operation of this circuit is often referred to as its "flop" state,

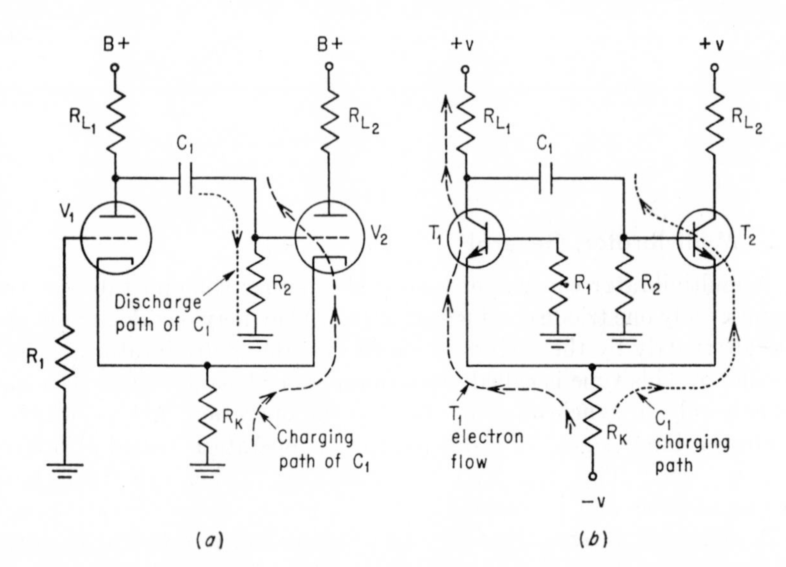

Figure 5·1 Cathode-coupled astable multivibrators using direct coupling between cathodes or emitters. (a) Tube type; (b) NPN-transistor type.

remaining in this condition until an input signal arrives. When the first input signal reverses the operation, the circuit goes into its "flip" state, remaining in this condition until the second input signal "flops" it again.

5·2 Cathode-coupled Astable Multivibrator (Direct Coupling)

The cathode-coupled multivibrator shown in Fig. 5·1 is directly coupled between cathodes and operates as a free-running or astable type. This multivibrator is an oscillator. Voltage waveforms or waveshapes are shown in Figs. 5·2 and 5·3. Refer to these three figures during the following steps of the theory discussion.

1. When B+ is first applied, triodes V_1 and V_2 start conducting.

2. Capacitor C_1 starts charging by pulling electrons from the grid of V_2, making that grid immediately positive (instant t_1 of Figs. 5·2 and 5·3).

3. V_2 grid current flows, producing a small resistance between cathode and grid of V_2, called R_{GK}.

4. C_1 can quickly charge up through this small R_{GK} of V_2 in series with R_K and R_{L_1}.

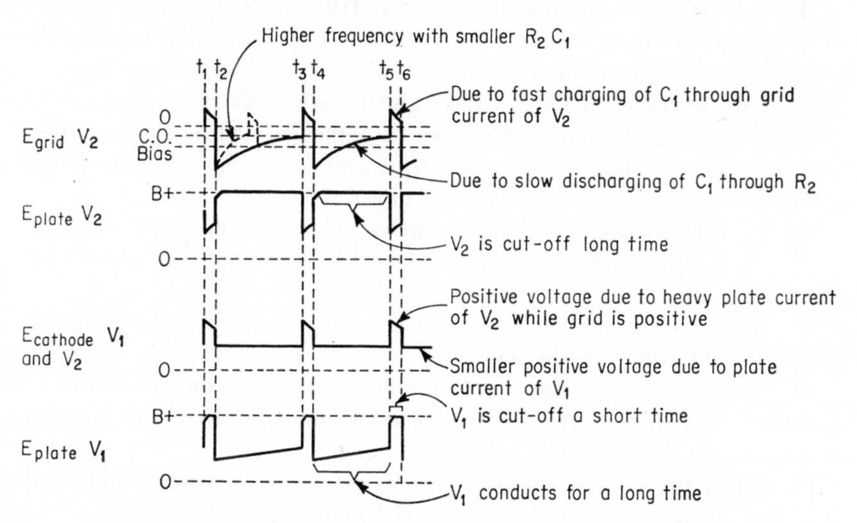

Figure 5·2 Waveshapes of cathode-coupled multivibrator of Fig. 5·1.

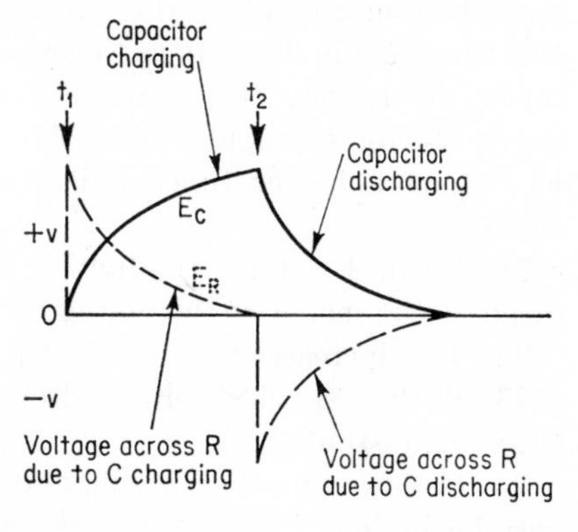

Figure 5·3 Voltages across a capacitor and a resistor during the charge and discharge periods.

5. With a positive grid, V_2 will conduct heavily.

6. Plate current of V_2, flowing through R_K, makes the top of R_K positive with respect to ground.

7. With the correct value of R_K, the heavy plate current of V_2 drives the cathode of V_1 sufficiently positive (making the grid of V_1 appear negative) to bias V_1 beyond cutoff.

8. With V_1 cut off, its plate voltage rises toward B+, and C_1 charges to this value.

9. As C_1 charges, the positive voltage on the grid of V_2 (actually across R_{GK}) decreases from its maximum value (between instant t_1 and t_2 of Fig. 5·2 and voltage across R, of Fig. 5·3, due to C charging).

10. V_2 now conducts less current, and the voltage across R_K decreases.

11. This decreases the bias on V_1, permitting V_1 to conduct.

12. Plate current of V_1 produces a voltage drop across R_{L_1}, causing the plate voltage of V_1 to fall (instant t_2, E plate V_1 of Fig. 5·2).

13. This decreasing voltage causes C_1 to start discharging.

14. A discharge current flows down through R_2, making the top of R_2 and the grid of V_2 negative (instant t_2, E grid V_2 of Fig. 5·2 and voltage across R, of Fig. 5·3, due to C discharging).

15. V_2, with its grid at a large negative value, now cuts off.

16. Plate voltage of V_2 now rises toward B+ (instant t_2, E plate V_2 of Fig. 5·2).

17. With V_2 cut off, V_1 is now conducting alone, its plate current limited by its own cathode bias.

18. As C_1 discharges, the negative voltage across R_2 on the grid of V_2 decreases (Fig. 5·2, E grid V_2, between instant t_2 and t_3, and Fig. 5·3, voltage across R due to C discharging).

19. When the negative voltage on V_2 grid has decreased sufficiently (Fig. 5·2, instant t_3), it no longer keeps V_2 cut off.

20. The complete discharge path of C_1 is down through R_2 to chassis ground, from ground up through R_K, and through R_P of V_1. This total R-C time mainly determines the length of time that V_2 is held cut off and hence determines the frequency of the multivibrator operation. A larger R_2, for example, would slow down the discharge of C_1, keeping V_2 cut off a longer time. This would result in a lower frequency of operation.

21. When V_2 starts conducting again, its plate current, flowing up through R_K, increases the voltage across R_K.

22. This increases the bias on V_1, and it conducts less.

23. With a smaller voltage drop across R_{L_1} plate voltage of V_1 rises (Fig. 5·2, instant t_3).

24. C_1 now stops discharging and again starts charging as it did at the very beginning.

25. Charging current of C_1 now drives the grid of V_2 positive (Fig. 5·2, instant t_3).

26. V_2, with a positive grid, now conducts very heavily.

27. This increases the voltage across R_K, increasing the bias on V_1, cutting this tube off.

28. The cycle is now complete since V_1 is again cut off, and V_2 is conducting alone.

V_2 grid voltage waveshape shown in Fig. 5·2 is mainly a negative fluctuating voltage. Its average value is the negative d-c voltage which can be measured across R_2, indicating that the circuit is oscillating. This average negative d-c voltage is the grid-leak bias voltage. Total bias of V_2 is this negative voltage across R_2 added to the average positive d-c voltage across R_K. For example, if the d-c voltage across R_2 is negative 40 volts and the d-c voltage across R_K is positive 5 volts, then the total bias on V_2 is negative 45 volts (grid with respect to cathode).

The frequency of this astable multivibrator is determined chiefly by the R-C time constant in the grid circuit (R_2-C_1 in Fig. 5·1). If either or both are decreased, the multivibrator frequency goes higher. This is illustrated in Fig. 5·2, V_2-grid voltage wave.

An *emitter-coupled astable* multivibrator (direct coupling) using NPN transistors is shown in Fig. 5·1*b*. To conduct, it will be recalled that a transistor must be forward biased between emitter and base and reverse biased between collector and base. Forward bias simply means that the polarity of the applied d-c voltage is negative at the N-type emitter and positive at the P-type base, as shown in Fig. 5·1*b*. Reverse bias at the collector consists of a positive voltage applied to the N-type collector. With forward bias between emitter and base, a small base current flows, producing a small resistance between emitter and base. At the same time, the reverse bias at the collector permits a current flow between emitter and collector. The direction of base and collector currents is backwards to the emitter arrow symbol. In the NPN-type transistor, electron flow is from emitter to base, and also from emitter to collector, as shown by the dashed-line and dotted-line arrows of Fig. 5·1*b*. Collector current is directly proportional to the amount of forward bias between emitter and base.

The circuit theory of the emitter-coupled astable multivibrator of Fig. 5·1*b* is very similar to that of its cathode-coupled vacuum-tube counterpart. A brief description follows.

1. Applying the forward bias and reverse bias voltages, as shown in Fig. 5·1*b*, causes both transistors, T_1 and T_2, to conduct. Collector and base currents flow in both.

2. Capacitor C_1 starts charging up through R_2 and even more so through the small parallel resistor path of T_2 emitter to base.

3. The charging of C_1 drives T_2 base positive, increasing the forward bias on the P-type base and increasing base current.

4. T_2 collector current increases, making the top of R_K sufficiently positive to *reverse* bias the N-type emitter of T_1.

5. T_1 now cuts off, and T_2 conducts alone.

6. C_1 will quickly charge up through the low resistance of T_2 emitter to base. After C_1 has charged, T_2 base current decreases, decreasing T_2 collector current.

7. With less T_2 collector current flowing, voltage across R_K decreases, and the reverse bias at the top of R_K, on T_1 emitter, disappears.

8. T_1 now starts conducting, and electron flow is from emitter to collector, as shown by the dashed-line arrows in Fig. 5·1*b*.

9. This current flow through R_{L_1} produces a voltage drop across R_{L_1}, and the positive voltage at the lower end of R_{L_1} falls.

10. This falling voltage causes C_1 to start discharging, and discharge current flows down through R_2, making the top of R_2 and the base of T_2 negative.

11. This reverse biases the P-type base of T_2 and cuts off this transistor.

12. T_1 now conducts alone, while T_2 remains inoperative.

13. T_2 is kept cut off while C_1 is discharging down through R_2. As C_1 discharges, the negative voltage across R_2 decreases.

14. When the negative voltage at the top of R_2, which acts as reverse bias on T_2 base, decreases sufficiently, T_2 again conducts.

15. T_2 collector current, flowing up through R_K, makes the top of R_K positive going.

16. This reduces the forward bias of T_1, and T_1 conducts less.

17. With less current flowing through R_{L_1}, voltage at the collector of T_1 goes positive, causing C_1 to start charging.

18. C_1 charges mainly through the emitter-to-base current of T_2, as shown by the dotted-line arrows of Fig. 5·1*b*.

19. This greatly increases T_2 collector current, and the next cycle of operation begins.

The explanation of this multivibrator using PNP transistors is exactly the same as that given previously for NPN transistors except that the polarities of the applied d-c voltages are reversed, and electron flow through the transistors is from *base* to *emitter* and from *collector* to emitter.

5·3 Synchronization

The cathode-coupled multivibrator of Fig. 5·1 is a free-running oscillator generating the voltage waveshapes shown in Fig. 5·2 and requires no input signal to operate. Input signals, however, are usually applied to prevent the multivibrator frequency from drifting. If the frequency of input signals is slightly higher than that of the free-running multivibrator, the multivibrator operation becomes speeded up to the higher frequency of the input signals. This is called *synchronization* and is illustrated in Fig. 5·4.

The multivibrator free-running signal voltage at the grid of V_2, shown as the dashed-line wave, is unaffected by input pulses number 1 and 2. This is true of any input pulses which occur when the grid is much more negative than cutoff. However, as shown, pulse number 3 occurs when the multivibrator grid signal is only slightly below cutoff. This positive-going input pulse drives the grid above cutoff, and V_2 conducts slightly sooner than it would have by itself. Once having been triggered into conduction by an input pulse, each successive pulse continues to trigger the tube. This is true when, as was mentioned previously, the input-pulse frequency is slightly higher than the free-running frequency. As a typical example, a vertical multivibrator oscillator in a television receiver may be free-running at about 55 cps. The vertical synchronizing pulses, having a frequency of 60 cps, speed up and lock the vertical multivibrator to this higher frequency. Similarly, the horizontal oscillator, having a free-running frequency of about 15 kc, is sped up and locked to the higher frequency of 15,750 cps of the horizontal synchronizing pulses.

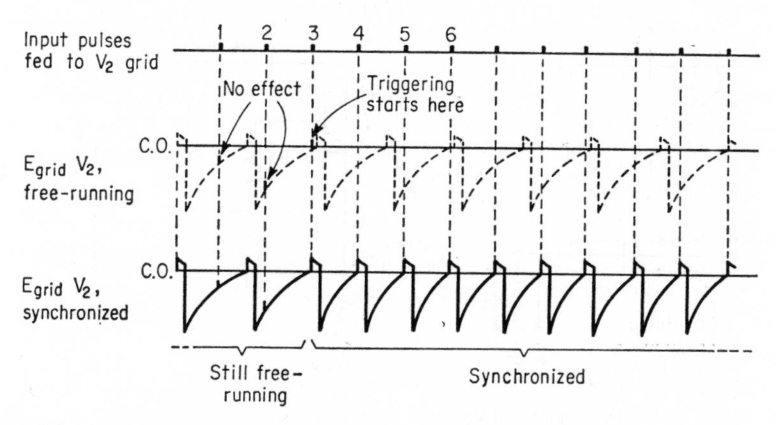

Figure 5·4 **Astable multivibrator grid voltage waveshapes, free-running and synchronized.**

5·4 Cathode-coupled Astable Multivibrator (Capacitance Coupling)

The cathode-coupled multivibrator shown in Fig. 5·5 uses capacitance coupling between cathodes and operates as a free-running or astable type. This multivibrator is an oscillator. Figure 5·6 shows the waveshapes for this circuit.

Refer to Figs. 5·5, 5·6, and 5·3 during the following steps of the theory discussion.

1. Applying B+ will cause both V_1 and V_2 to start conducting at first.

2. C_1 will start charging up through R_2 and also through the cathode-to-grid resistance of V_2 in series with R_3.

3. This makes the top of R_2 and the grid of V_2 positive.

4. The positive grid draws electrons from the cathode, and C_1 charges mainly through the small resistance between the cathode and grid (R_{GK}) of V_2.

5. V_2, with a positive grid, conducts heavily, making the upper end of R_3 positive.

6. C_2 starts charging up to the positive voltage across R_3.

7. Charging current of C_2 flows up through R_1 to the left side of C_2, continuing through the internal plate to cathode resistance (R_P) of V_2 and through R_{L_2}.

8. C_2 charging drives the top of R_1 and the cathode of V_1 positive.

9. This cuts off V_1, since driving its cathode positive has the same effect as driving its grid negative.

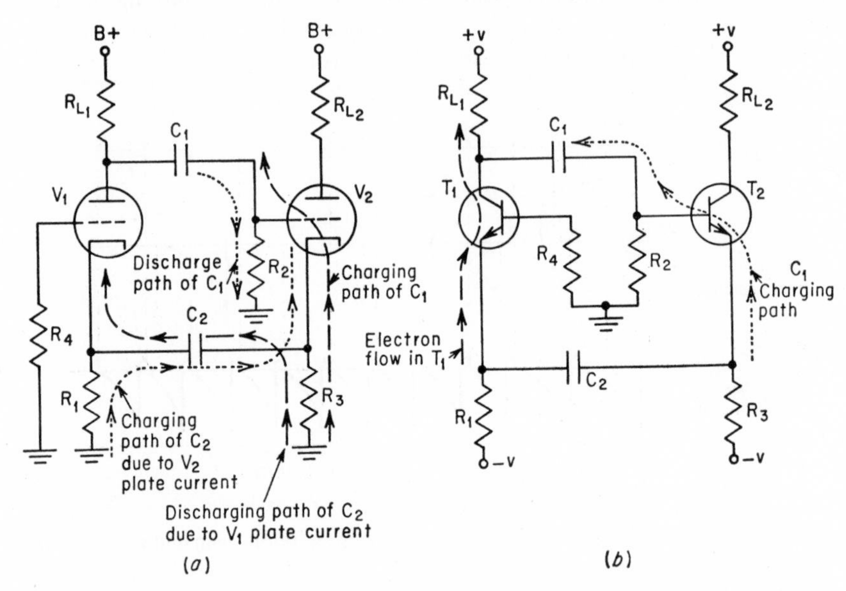

Figure 5·5 **Cathode-coupled astable multivibrators using capacitance coupling between cathodes or emitters. (a) Tube type; (b) NPN-transistor type.**

10. As C_2 charges, voltage across R_1 decreases. This is illustrated in Fig. 5·3 in the period between t_1 and t_2 and also in Fig. 5·6 for the same period.

11. When the cathode voltage of V_1 (which is the voltage across R_1) decreases sufficiently, V_1 will start conducting.

12. Plate current of V_1 produces a voltage drop across R_{L_1}, and V_1 plate voltage decreases.

13. C_1 now starts discharging down through R_2.

14. Discharge current makes the top of R_2 and the grid of V_2 negative, cutting off V_2 at instant t_2.

15. V_2 cathode voltage falls abruptly, and this sharp voltage decrease is coupled to the V_1 cathode.

16. Plate current of V_1 drives the top of R_1 more positive again. As V_1 conducts steadily, its cathode voltage levels off (Fig. 5·6, period t_2 to t_3).

17. C_2 starts discharging and attempts to charge up to the opposite polarity, to the voltage across R_1.

18. C_2 discharging current now flows up through R_3 to the right side of C_2, continuing through the internal plate to cathode resistance (R_P) of V_1 and through R_{L_1}.

19. As C_2 discharges (or attempts to charge up to the reverse polarity), voltage across R_3 decreases and becomes less positive. This is shown in Fig. 5·6, V_2 cathode voltage during the t_2 to t_3 period.

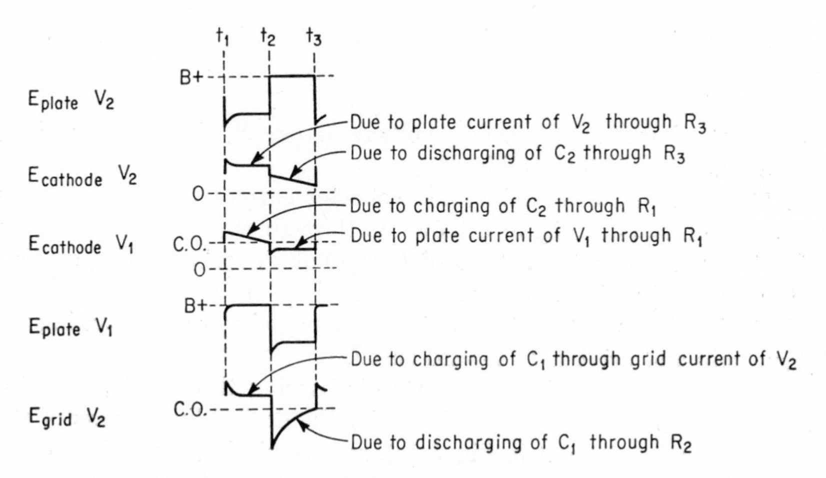

Figure 5·6 **Waveshapes of cathode-coupled multivibrator of Fig. 5·5.**

20. When the combination voltage on V_2 (negative on the grid and positive on the cathode) decreases sufficiently, V_2 again starts conducting, and the next cycle of oscillation begins.

The frequency of oscillation of this free-running multivibrator can be synchronized by an input pulse signal as explained in Sec. 5·3 and illustrated in Fig. 5·4. R_2-C_1, in the grid of V_2 of Fig. 5·5, mainly determines how long V_2 is held cutoff, and R_1-C_2, in the cathode of V_1, mainly determines how long V_1 is cut off. The multivibrator frequency therefore depends on these R-C time constants.

An *emitter-coupled astable* multivibrator (capacitance coupling) using NPN transistors is shown in Fig. 5·5*b*. Circuit operation is very similar to the discussion presented previously for the cathode-coupled circuit of Fig. 5·5*a*. This previous discussion should be read first, and then the following briefer theory of the transistorized version should be studied.

The theory of the emitter-coupled astable multivibrator of Fig. 5·5*b* follows.

1. Applying the negative d-c voltage to the N-type emitters forward biases them, while the positive voltage applied to the N-type collectors reverse biases these, and transistors T_1 and T_2 both start conducting.

2. Capacitor C_1 starts charging through R_2 and even more so through the parallel, small emitter-to-base resistance of T_2.

3. The charging of C_1 drives T_2 base positive, increasing the forward bias on the P-type base and increasing base current.

4. T_2 collector current increases, making the top of R_3 positive-going.

5. Capacitor C_2 starts charging up through R_1 (as shown by the dotted-line arrows of the vacuum tube circuit Fig. 5·5*a*) to the voltage at the top of R_3.

6. C_2 charging current, flowing up through R_1, drives the top of R_1 and the emitter of T_1 positive, reverse biasing T_1.

7. T_1, with its emitter reverse biased, cuts off.

8. As C_2 charges up, positive voltage across R_1 decreases until T_1 emitter is no longer reverse biased.

9. T_1 now starts conducting, and collector current (as shown by the dashed-line arrows of Fig. 5·5*b*) flows up through R_{L_1}.

10. The voltage at the bottom of R_{L_1} and at T_1 collector falls, causing C_1 to start discharging down through R_2.

11. This drives the top of R_2 and T_2 base negative, reverse biasing the P-type base and driving T_2 toward cutoff.

12. With T_1 conducting, C_2 starts discharging through T_1 collector current and up through R_3 (as shown by the dashed-line arrows of the vacuum-tube counterpart in Fig. 5·5*a*), reverse biasing the emitter of T_2 and helping to cut off T_2.

13. As C_1 discharges down through R_2, the negative voltage across R_2 at the base of T_2 decreases, allowing T_2 to again conduct, and the next cycle of oscillation begins.

If PNP transistors were used in this multivibrator instead of the NPN types discussed here, the circuit theory would be the same, with the main differences being the opposite polarities of *all* applied d-c voltages and the electron flow going from *base* to *emitter* and from *collector* to *emitter*.

5·5 Plate-coupled Astable Multivibrator

The plate-coupled multivibrator shown in Fig. 5·7 is an astable or free-running type. Waveshapes are shown in Fig. 5·8. Reference should be made to both these figures and also to Fig. 5·3 during the following theory discussion.

1. Applying B+ will start V_1 and V_2 conducting, and C_1 and C_2 will charge up to the plate voltages of V_1 and V_2 respectively.

2. V_1 and V_2, even though they may be matched tubes, will not *exactly* duplicate each other's characteristics. For example, as the B+

supply voltage ripples, due to imperfect but normal filtering, V_1 and V_2 will conduct slightly more heavily as B+ increases and slightly less as B+ decreases. However, the instant that B+ increases, one of the two tubes will start conducting more heavily a bit sooner than the other. Which of the two does this makes no real difference. As long as one triode "beats" the other to the "punch," multivibrator oscillation begins. Other factors that could start the "ball rolling," besides B+ ripple, are as follows: The a-c voltage usually used for the heaters produces slight variations of cathode temperature and causes changes of cathode emission; slight differences in tube construction such as the allowable tolerance of tube element spacing, grid-wire spacing etc.

3. Assume that with both tubes operating, V_1 is the tube that suddenly conducts slightly more heavily.

4. Plate voltage of V_1 falls slightly because of the increased voltage drop across R_{L_1} (see instant t_1, Fig. 5·8).

5. C_1 starts discharging to the lower plate voltage on V_1.

6. Discharge current flows down through R_1, making the top end of R_1 and the grid of V_2 slightly negative.

7. V_2, with its grid voltage becoming a small negative value, conducts less.

8. Because of the gain of V_2, its plate voltage now rises substantially.

9. C_2 charges to the higher voltage on the plate of V_2.

10. The charging of C_2, at first up through R_2, makes the top of R_2 and the grid of V_1 positive.

11. The positive grid draws current from the cathode, and C_2 charges quickly through the low resistance between V_1 cathode and grid (R_{GK}).

12. V_1, with its grid positive, conducts very heavily, and plate voltage falls substantially to a low value (period t_1 to t_2, Fig. 5·8).

13. C_1 attempts to discharge to the low voltage on the V_1 plate.

14. This drives the grid of V_2 very negative, cutting off V_2 (Fig. 5·8, instant t_1).

15. Plate voltage of V_2 rises to full B+, slightly delayed in its rise due to C_2 not being able to charge instantaneously. This is shown by the rounded, rather than squared, leading edge of V_2 plate voltage as it rises toward B+ in Fig. 5·8 immediately after instant t_1.

16. The complete charging path for C_2 is from the grounded (B−) cathode of V_1 to the grid, on to the left side of C_2, then up through R_{L_2} to B+ power supply.

17. V_2 will be held cut off by the discharge of C_1 down through R_1. As C_1 discharges, the negative voltage across R_1 on V_2 grid decreases (period t_1 to t_2, Fig. 5·8).

18. When this negative voltage decreases sufficiently, it can no longer hold V_2 cut off, and V_2 now starts conducting (instant t_2, Fig. 5·8).

19. V_2 plate current produces a voltage drop across R_{L_2}, and V_2 plate voltage falls down from its B+ value (Fig. 5·8, instant t_2).

20. C_2 now starts discharging down through R_2.

21. This produces a negative voltage at the top of R_2 and at the grid of V_1.

22. V_1 now conducts less, and its plate voltage rises due to the small drop across R_{L_1}.

23. C_1 stops discharging and starts charging up to the rising plate voltage of V_1.

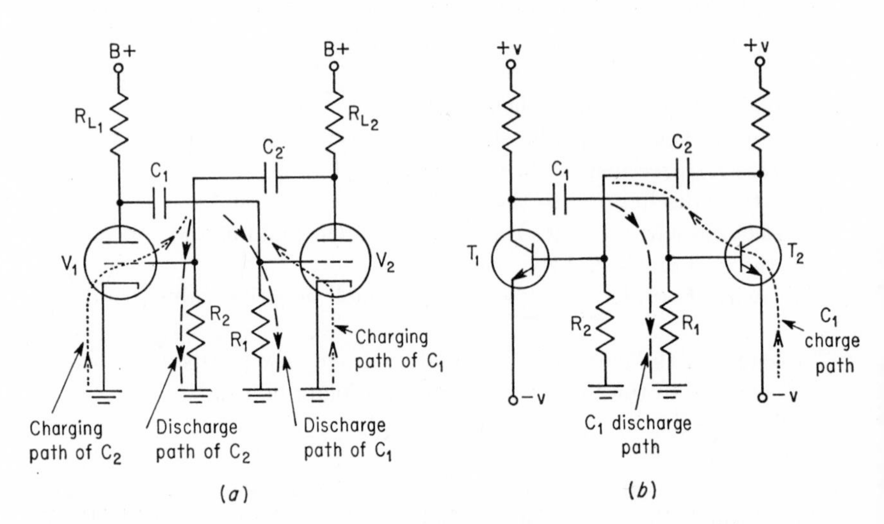

Figure 5·7 **Plate-coupled (collector-coupled) astable multivibrators. (a) Tube type; (b) NPN-transistor type.**

24. The charging C_1 drives the grid of V_2 positive (instant t_2, Fig. 5·8).

25. V_2 now conducts heavily, and its plate voltage falls substantially to a low value.

26. C_2 starts discharging to this low plate voltage of V_2.

27. C_2 discharge current flows down through R_2, making the grid of V_1 very negative (Fig. 5·8, instant t_2), and V_1 now cuts off.

28. Plate voltage of V_1 rises to B+, delayed slightly by the required charging of C_1.

29. V_1 is held cut off by the discharge of C_2 through R_2.

30. As C_2 discharges, the negative voltage on V_1 grid decreases (period t_2 to t_3, Fig. 5·8).

31. When this negative voltage on V_1 grid decreases sufficiently, it no longer holds V_1 cut off (instant t_3, Fig. 5·8).

32. V_1 now starts conducting, and its plate voltage falls.

33. C_1 starts discharging down through R_1, driving the grid of V_2 negative (Fig. 5·8, instant t_3).

34. V_2 conducts less, and its plate voltage rises.

35. C_2 starts charging to this rising plate voltage, driving V_1 grid positive.

36. The heavy conduction of V_1 causes its plate voltage to decrease substantially.

37. C_1 now discharges heavily down through R_1, driving V_2 grid very negative and cutting off V_2 (Fig. 5·8, instant t_3).

38. The cycle is now complete and about to repeat itself.

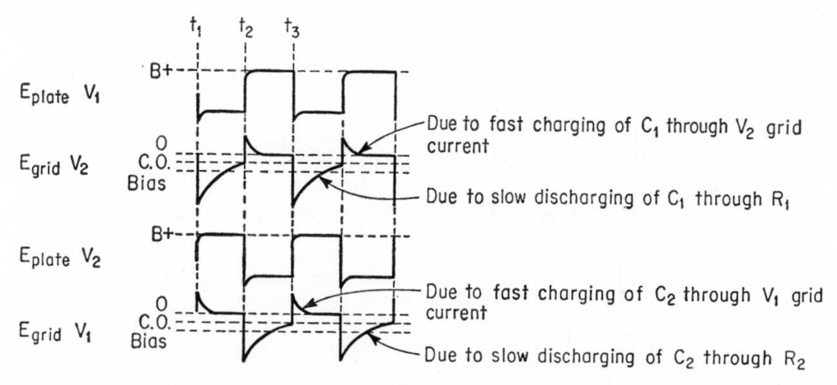

Figure 5·8 Waveshapes of plate-coupled multivibrator of Fig. 5·7.

The plate-coupled astable multivibrator shown in Fig. 5·7 is an oscillator, the frequency of which depends mainly on the *R-C* time constants in *both* grids (R_1-C_1 and R_2-C_2). If R_1-C_1 is equal to R_2-C_2, and R_{L_1} is equal to R_{L_2}, the circuit is said to be balanced or symmetrical. V_1 and V_2 will be cut off for *equal* periods of time but not during the *same* periods. V_2 is cut off during the period t_1 to t_2 of Fig. 5·8, and V_1 is inoperative during the t_2 to t_3 period. In a balanced circuit, period t_1 to t_2 is equal to period t_2 to t_3. This is also shown in Fig. 5·9*a*.

If both grid resistors R_1 and R_2 were changed so that one were increased and the other decreased in the same proportion, the multivibrator frequency would *not* be changed. However, the circuit would now be unbalanced or asymmetrical. If R_1 were increased, V_2 would be kept cut off a longer time. Decreasing R_2 would keep V_1 cut off less time than originally. The balanced square waves at the plates of V_1 and V_2, as shown in Fig. 5·9*a*, would be changed to unbalanced square waves or more rectangular as shown in Fig. 5·9*b*. Period t_1 to t_2 is now longer than t_2 to t_3, but the total time, t_1 to t_3, is unchanged. This means that the frequency remains the same in (*b*) as in (*a*). In Fig. 5·9*c*, *both* resistors have again been changed. This time R_2 has been increased and R_1 decreased. V_1 is now cut off (period t_2 to t_3) longer than V_2 (period t_1 to t_2). Again, however, the frequency is unchanged.

To change the frequency of the plate-coupled astable multivibrator of Fig. 5·7, only *one* resistor should be changed. As shown in Fig. 5·9*d*, R_2 has been decreased, and R_1 is unchanged. The smaller R_2 will now allow C_2 (Fig. 5·7) to discharge faster, thus holding V_1 cut off less time than originally. V_2 will still be cut off the same time as it had been originally.

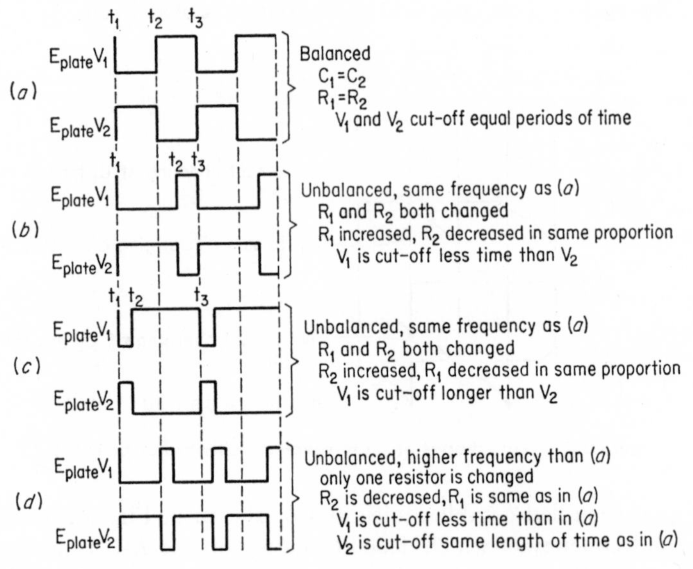

Figure 5·9 **Waveshapes of plate-coupled multivibrator of Fig. 5·7 with variations of R_1 and R_2.**

The result is a higher signal frequency from the multivibrator, as illustrated in Fig. 5·9*d* where almost three cycles are shown in the same space as only two cycles occupied before in (*a*), (*b*), and (*c*).

The grid-voltage waveshapes shown in Fig. 5·8 are mainly negative fluctuating d-c voltages. The average value on each grid is the *bias* for that stage, the bias being produced by the signal itself causing grid current (grid-leak bias). Measurement of this negative d-c voltage across either R_1 or R_2 would indicate whether the multivibrator is oscillating or not.

Since an astable multivibrator such as Fig. 5·7 is quite unstable in frequency, synchronizing pulses are usually applied to the grid in order to lock the multivibrator frequency to that of some stable signal. The discussion of synchronization is given in Sec. 5·3.

A *collector-coupled astable* multivibrator using NPN transistors is shown in Fig. 5·7*b*. The circuit theory is very similar to that described previously for the vacuum-tube plate-coupled astable multivibrator. A better understanding of the transistor circuit can be gained if the vacuum tube detailed circuit theory explanation is read first. A briefer theory explanation of the transistor circuit of Fig. 5·7*b* follows.

1. The applied d-c voltages provide forward biasing between emitter and base (negative voltage applied to the N-type emitter) and reverse bias on the collector (positive voltage applied to the N-type collector).

2. Both transistors T_1 and T_2 conduct, and C_1 and C_2 charge up to the voltages at their respective collectors.

3. As explained in the previous vacuum-tube multivibrator discussion, a slightly greater current will suddenly flow in one circuit, say T_1, than in the other. This is due to some slight difference in transistor construction, difference in the values of components, and slight changes or ripple of the applied d-c voltages.

4. Assuming that T_1 suddenly conducts slightly more, a larger voltage drop occurs across R_{L_1}, and the collector voltage of T_1 falls.

5. C_1 discharges down through R_1, making the top of R_1 and the base of T_2 negative and reverse biasing the P-type base of T_2.

6. T_2 cuts off, and its collector voltage goes more positive, causing C_2 to charge up to this higher voltage.

7. C_2 charges up through T_1 emitter to base current, causing T_1 collector current to increase to saturation.

8. T_2 is kept cut off by the discharge of C_1 through R_1. As C_1 discharges, voltage across R_1 becomes less negative.

9. When the negative voltage across R_1 decreases sufficiently, the base of T_2 is no longer reverse biased, and T_2 starts conducting.

10. T_2 collector current flows up through R_{L_2}, and the collector voltage falls, causing C_2 to start discharging down through R_2.

11. This makes the P-type base of T_1 negative, reverse biasing T_1 to cutoff.

12. T_1 collector voltage rises, causing C_1 to charge up through T_2 base current, driving T_2 collector current to saturation.

13. T_1 is held cut off by the discharge of C_2 down through R_2. As C_2 discharges, voltage across R_2 becomes less negative.

14. When the negative voltage across R_2 decreases sufficiently, the base of T_1 is no longer reverse biased, and T_1 starts conducting, and the next cycle starts.

If PNP transistors were used in this circuit instead of the NPN types discussed here, the circuit operation would be the same except that the applied d-c voltages would have opposite polarities, and electron flow occurs from *base* to *emitter* and from *collector* to *emitter*.

5·6 Cathode-controlled Astable Multivibrator

The cathode-*controlled* astable multivibrator of Fig. 5·10 is a free-running oscillator which, by itself, produces excellent symmetrical square waves without the spiked overshoots common to other multivibrators. (See plate-voltage waveshapes of Figs. 5·2, 5·6, and 5·8 as compared to those of Fig. 5·11.) This overshoot of plate voltage as it falls is due to the grid being driven positive with respect to cathode, the usual occurrence in multivibrators but not true of this circuit. The positive spike on the

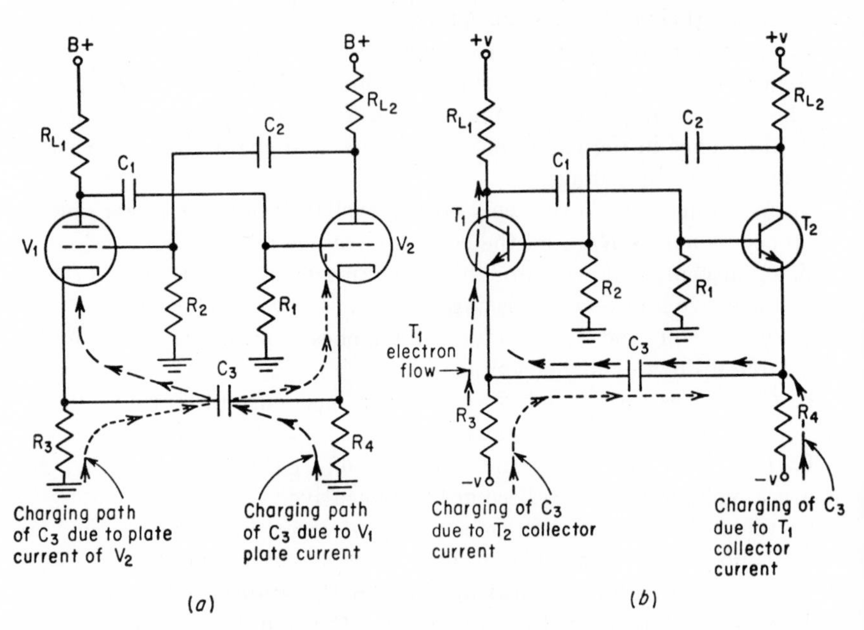

Figure 5·10 Cathode-controlled (emitter-controlled) astable multivibrators. (a) Tube type; (b) NPN-transistor type.

grid (as shown at t_1 on the grid of V_1 in Fig. 5·8) results in excessive plate current and minimum plate voltage. As the grid voltage decreases to zero and remains there, plate current decreases to a steady value, and plate voltage rises slightly and levels off.

The cathode-controlled astable multivibrator resembles the plate-coupled type (Fig. 5·7) and cathode-*coupled* type (Fig. 5·5) but should not be confused with either.

In the following theory discussion, refer to the schematic of Fig. 5·10, the waveshapes of Fig. 5·11, and the basic capacitor and resistor graphs of Fig. 5·3.

1. When B+ is applied to V_1 and V_2, both tubes start conducting.

2. The circuit is symmetrical or balanced. That is, all components on one side are equal to the associated components on the other side; R_{L_1} is equal to R_{L_2}, $C_1 = C_2$, $R_2 = R_1$, and $R_3 = R_4$; and V_1 and V_2 are the same tube type, being usually halves of a duo-triode.

3. Despite this balanced circuit, V_1 and V_2 will not *always* conduct equally. One tube, and it could be either one, will momentarily conduct more heavily than the other. This could be caused by any one of several conditions such as B+ ripple, periodic temperature changes of filament due to the a-c voltage source, slight differences of tube construction between V_1 and V_2, etc.

4. Assume that V_2 happens to be the tube that suddenly conducts slightly more than V_1 (Fig. 5·11, instant t_1).

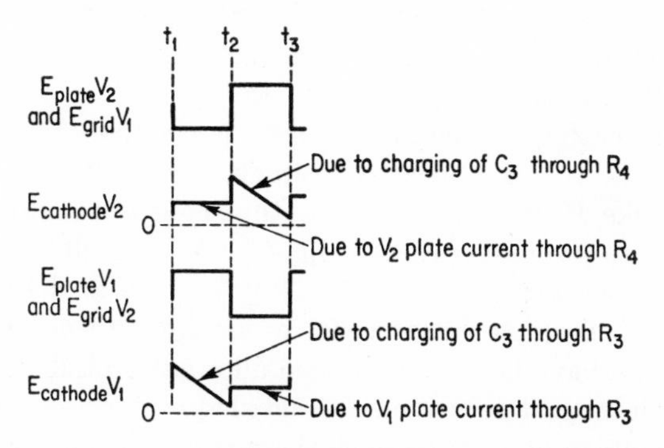

Figure 5·11 Waveshapes of cathode-controlled multivibrator of Fig. 5·10.

5. This slight increase of V_2 plate current produces a larger voltage drop across R_{L_2}, causing V_2 plate voltage to fall.

6. At the same time, the increased V_2 plate current produces a larger voltage across R_4, and the cathode of V_2 goes positive.

7. The decreasing plate voltage is coupled through C_2 to the grid of V_1 as a negative voltage.

8. C_2 and R_2 make up a very large R-C time constant so that C_2 cannot discharge, except slightly. Since the voltage across C_2 cannot change very much, the falling or negative-going voltage at V_2 plate all appears across R_2, at V_1 grid, with no decrease.

9. With a negative grid, V_1 conducts less.

10. To aid V_1 in conducting less, its cathode is driven positive by the positive-going voltage at V_2 cathode when V_2 first started conducting more heavily. This rise of V_2 cathode voltage is coupled through C_3 to the V_1 cathode.

11. As V_1 conducts less, its plate voltage increases towards B+ because of the smaller voltage drop across R_{L_1}.

12. The positive-going voltage at V_1 plate is coupled through C_1 to V_2 grid, driving V_2 into heavier conduction.

13. V_2 plate voltage now decreases further, and its cathode voltage rises higher (still instant t_1, Fig. 5·11).

14. V_1 grid is driven sufficiently negative while its cathode goes positive so that V_1 now cuts off.

15. C_3 charges up to the positive voltage at the V_2 cathode, its charging path being up through R_3 to the left side of C_3, up through the R_P of V_2, then up through R_{L_2}.

16. As C_3 charges, the positive voltage across R_3, at V_1 cathode, decreases. (See Fig. 5·11, period t_1 to t_2. See also Fig. 5·3, period t_1 to t_2.)

17. How long V_1 is kept cut off is determined by the positive voltage at its cathode. This depends mainly on the R_3-C_3 time constant.

18. When the voltage across R_3 at V_1 cathode has decreased sufficiently, V_1 is no longer held cut off (Fig. 5·11, instant t_2).

19. As V_1 starts conducting, its plate voltage falls due to the voltage drop across R_L (instant t_2).

20. The decreasing voltage at V_1 plate is coupled through C_1 to the grid of V_2, across R_1, as a negative voltage.

21. R_1-C_1, like R_2-C_2, is a very large time constant and therefore does not permit C_1 to discharge, except slightly. As a result, since C_1 cannot change its voltage very much, the falling voltage at V_1 plate appears across R_1, at V_2 grid, as a flat negative voltage with no decreasing tilt, such as it would have been had C_1 been allowed to discharge.

22. V_2, with its grid negative, conducts less.

23. Plate voltage of V_2 rises since there is now a smaller voltage drop across R_{L_2}.

24. This rise of voltage is coupled through C_2 and appears across R_2 at V_1 grid as a positive voltage.

25. V_1 plate current flows up through R_3, producing a more positive voltage across R_3 at the cathode of V_1 (instant t_2, Fig. 5·11).

26. This positive-going sharp rise of voltage at V_1 cathode is coupled to V_2 cathode through C_3.

27. C_3 now starts discharging and attempts to charge up to the opposite polarity. With V_1 conducting, the discharge path (and the new charging path) is up through R_4 to the right side of C_3, then up through R_P of V_1, and up through R_{L_1}.

28. V_2, with its grid at a negative voltage and with its cathode positive, cuts off.

29. V_2 is held cut off by the discharging (and then, charging) of C_3 up through R_4, making V_2 cathode positive.

30. As C_3 loses charge and then starts charging to the opposite polarity, voltage across R_4 decreases (Fig. 5·11, period t_2 to t_3).

31. When the voltage across R_4 has decreased sufficiently, V_2 is no longer held cut off (Fig. 5·11, instant t_3).

32. As V_2 again starts conducting, the next cycle of oscillation begins.

The square-wave voltages at each plate, as shown in Fig. 5·11, are also at the opposite grids. Since R_1-C_1 and R_2-C_2 are very long time constants, each square-wave plate signal is coupled to the other grid with no tilt or decrease such as would occur if the R-C were small. These large time constants have no effect on the frequency of a cathode-controlled multivibrator.

R_{L_1} and R_{L_2} are made much smaller than the cathode resistors R_3 and R_4. As a result, each stage has a gain of less than one from grid to plate. With no appreciable gain, the circuit could not oscillate if C_3 were opened, even though the circuit would look like a plate-coupled multivibrator. It is C_3 discharging and charging alternately through R_3 and R_4 that is responsible for and determines the frequency of this multivibrator. Varying C_3 alone would change the frequency while still maintaining balanced or symmetrical square waves at each plate. Higher frequency of oscillation would be produced by a smaller-value C_3.

An *emitter-controlled astable* multivibrator using NPN transistors is shown in Fig. 5·10*b*. The circuit operation is practically identical to that of the vacuum-tube circuit of Fig. 5·10*a* which has just been described in this section. A brief description of the transistor circuit follows, but it is recommended that the more detailed vacuum tube circuit discussion be studied first.

1. The applied d-c voltages, negative to the N-type emitter for forward bias and positive to the N-type collector for reverse bias, permit both transistors T_1 and T_2 to start conducting and capacitors C_1 and C_2 to charge to their respective collector voltages.

2. Owing to some slight differences of transistor construction, component values, and a variation or ripple of the applied d-c voltages, one of the transistors will suddenly conduct more heavily than the other.

3. Assume that T_2 suddenly conducts more heavily, producing a larger voltage drop across R_{L_2}, causing T_2 collector voltage to decrease.

4. C_2 starts discharging down through R_2, driving the base of T_1 negative.

5. Because of T_2 heavier conduction, voltage across R_4 increases, and the top of R_4 is positive-going, causing C_3 to start charging up to this voltage through R_3 (as shown by the dotted-line arrows of Fig. 5·10*b*).

6. This makes the top of R_3 and the emitter of T_1 positive-going. With this N-type emitter going positive, while the P-type base is negative-going (due to C_2 discharging through R_2), T_1 is reverse biased and becomes inoperative, and C_1 charges fully through T_2 emitter-to-base current, driving T_2 to saturation collector current.

7. T_1 is kept cut off by the charging of C_3 up through R_3. As C_3 charges, voltage across R_3 decreases.

8. When the voltage across R_3, at the emitter of T_1, becomes sufficiently small, the reverse bias is no longer present, and T_1 starts conducting.

9. T_1 collector current, flowing up through R_{L_1}, causes its collector voltage to fall, and C_1 starts discharging down through R_1, driving the P-type base of T_2 negative.

10. With T_1 conducting, voltage across R_3 is positive-going, causing C_3 to start discharging and charging up to the reverse polarity. This new charging path of C_3 is up through R_4 (as shown by the dashed-line arrows of Fig. 5·10*b*).

11. This current flow up through R_4 makes the N-type emitter of T_2 positive-going.

12. T_2 now cuts off due to its N-type emitter going positive, while its P-type base goes negative (due to discharging C_1), and collector voltage of T_2 rises to the full positive voltage.

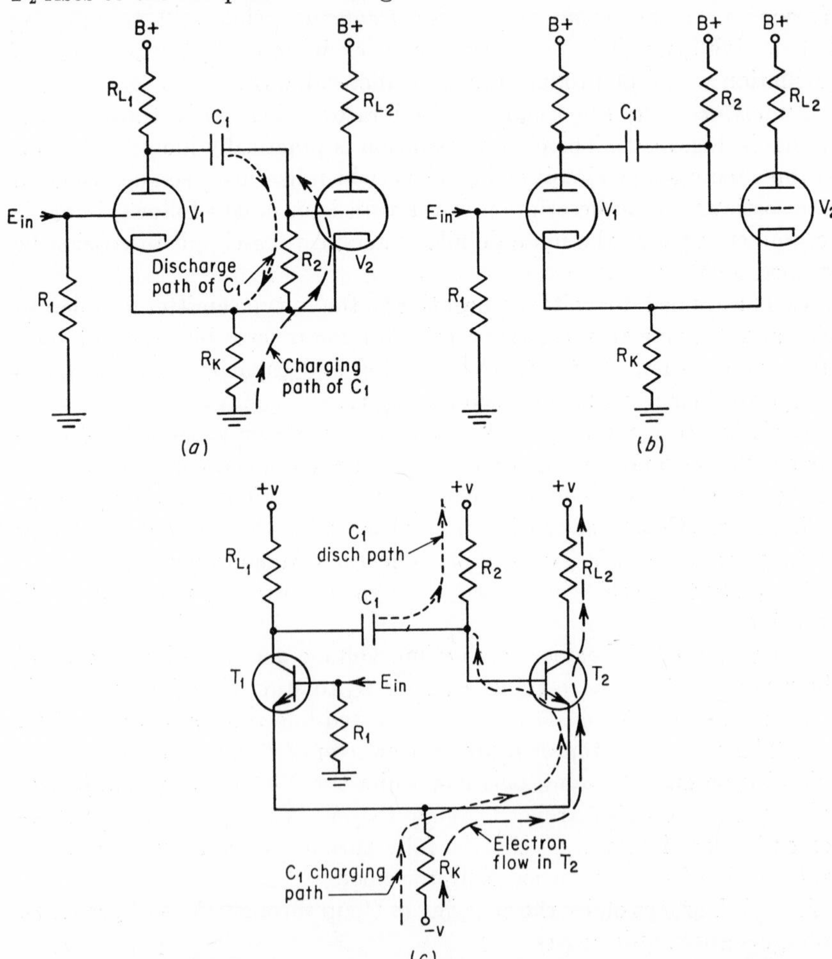

Figure 5·12 Cathode-coupled (emitter-coupled) monostable multivibrators. (*a*) V_2 zero biased; (*b*) V_2 positive biased; (*c*) NPN-transistor type.

13. With T_2 cut off, C_2 charges up through the emitter-to-base current of T_1, driving T_1 collector current to saturation.

14. T_2 is kept cut off by the charging of C_3 up through R_4. As C_3 charges, voltage across R_4 decreases.

15. When this voltage has decreased sufficiently, the reverse bias on the emitter of T_2 is no longer present, and T_2 again starts conducting, and the next cycle of oscillation begins.

Circuit operation would be the same if PNP transistors were used instead of the NPN types shown in Fig. 5·10*b*. The polarities of *all* applied d-c voltages would, of course, be opposite to that shown. Electron flow through the transistors would be from *base* to *emitter* and from *collector* to emitter.

5·7 Cathode-coupled Monostable Multivibrator

The circuit of Fig. 5·12 is a cathode-coupled monostable multivibrator. This multivibrator is *not* able to function by itself but requires input signals in order to operate, producing one cycle of operation for each input signal voltage. The monostable multivibrator has only one stable state. One tube always conducts, and the other is always cut off. An input signal *temporarily* reverses things, and then the circuit reverts back to its permanent stable state by itself until the next input triggering signal. Another name for this circuit is a one-shot multivibrator. The temporary reversal of operation due to an input signal is called its "quasi-stable" state. This means that it resembles, but is not actually, a stable state. The multivibrator remains in this quasi-stable condition for only a very brief period.

Refer to the schematic diagram Fig. 5·12, the voltage waveshapes of Fig. 5·13, and the basic resistor and capacitor voltages of Fig. 5·3 during the following theory discussion.

1. Applying B+ voltage causes V_1 and V_2 to start conducting.

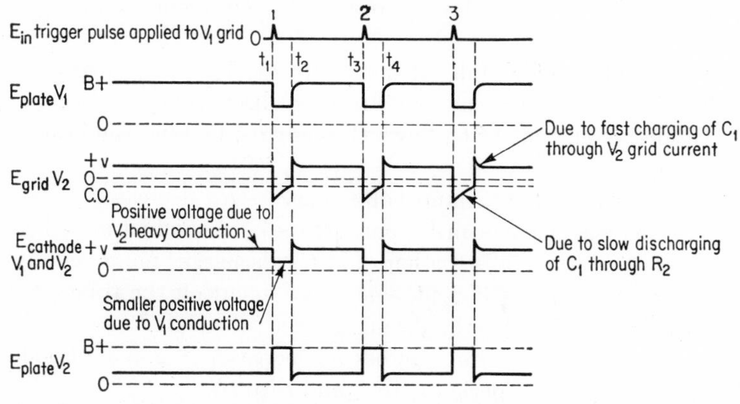

Figure 5·13 Waveshapes of cathode-coupled monostable multivibrator of Fig. 5·12.

2. In Fig. 5·12*a* V_2 is zero biased at first since no current flows through R_2. This means that with zero voltage drop across R_2, the grid is the same potential as the cathode, or, zero voltage from grid to cathode. V_2 will conduct very heavily with zero bias.

3. Heavy plate current of V_2 flows up through R_K. With proper value of R_K, the voltage produced across it biases the cathode of V_1 beyond cutoff. Actually, both cathodes become the same positive voltage, but since V_2 grid is at this same value, V_2 is zero biased and conducts heavily. V_1, with its grid at ground potential (zero volts) and its cathode positive, becomes cut off. If R_K is too small, the voltage produced across it by the heavy zero-bias V_2 plate current will not be sufficient to cut off V_1. The circuit then would operate as a free-running or astable cathode-coupled multivibrator, as explained in Sec. 5·2.

4. C_1 will charge up to the positive voltage at V_1 plate. C_1 charging path causes current to flow up through R_2, making the top of R_2 and the grid of V_2 positive with respect to the lower end of R_2 which is the cathode. The positive grid, with respect to cathode, draws current from the cathode, producing a small resistance, R_{GK}. As a result C_1 charges rapidly through this small R_{GK}. V_2 grid thus becomes more positive than its cathode for a short time until C_1 charges.

5. In Fig. 5·12*b* the grid is connected through R_2 to B+, keeping the grid positive with respect to its cathode. Grid current flows, and C_1 charges up in this circuit through grid current. Here too, heavy V_2 plate current flows up through R_K, biasing V_1 beyond cutoff.

6. V_2, in both the (*a*) and (*b*) diagrams, conducts heavily all the time, holding V_1 cut off permanently.

7. This condition continues until a positive-going input pulse is applied to the V_1 grid.

8. At instant t_1 of Fig. 5·13, pulse number 1 is applied.

9. This drives V_1 into conduction for the first time.

10. Plate current of V_1 flows through R_{L_1}, producing a voltage drop across it.

11. Plate voltage of V_1 falls at instant t_1.

12. This falling voltage causes C_1 to start discharging.

13. Discharge current of C_1 flows through R_2, making the grid end of R_2 negative.

14. V_2 now conducts less, and voltage across R_K decreases.

15. This reduces the bias on V_1, permitting V_1 to conduct more heavily.

16. V_1 plate voltage falls further, and C_1 discharges more heavily.

17. V_2 grid is driven sufficiently negative to cut off the tube.

18. V_2 is held cut off by the discharge of C_1 through R_2.

19. As C_1 discharges, voltage across R_2 decreases (Fig. 5·3, period after instant t_2 and Fig. 5·13, period t_1 to t_2 and t_3 to t_4).

20. When the negative voltage at V_2 grid becomes small enough, it no longer can hold V_2 cut off (Fig. 5·13, instant t_2).

21. V_2 now again starts conducting, and voltage across R_K is increased, with the top end going more positive.

22. This increases the bias on V_1, and this tube conducts less.

23. Less voltage drop is now produced across R_{L_1}, and V_1 plate voltage rises.

24. This rising voltage causes C_1 to stop discharging and to begin charging up again.

25. V_2 grid is now driven positive by C_1 charging current, and V_2 conducts heavily.

26. Voltage across R_K rises, biasing V_1 again beyond cutoff.

27. V_2 now continues to conduct heavily, even after C_1 has rapidly charged, because of the zero-bias condition of Fig. 5·12*a* or the positive-bias condition of Fig. 5·12*b*.

28. This stable condition of V_2 conducting and V_1 cutoff continues forever or until the next trigger pulse (number 2 in Fig. 5·13) is applied.

29. Then the next cycle is produced, starting at instant t_3 of Fig. 5·13.

30. The period of time (t_1 to t_2 and t_3 to t_4, Fig. 5·13) which is the quasi-stable condition is determined mainly by the R_2-C_1 time constant.

An *emitter-coupled monostable* multivibrator using NPN transistors is shown in Fig. 5·12*c*. The theory of this transistor circuit is very similar to that explained previously for the vacuum-tube counterparts of Fig. 5·12*a* and *b*. The previous explanation should be read first, and then the following briefer discussion of the emitter-coupled circuit should be studied.

1. Transistor T_2 will conduct heavily because of the forward bias applied between emitter and base; the negative d-c voltage applied through R_K to the N-type emitter, and the positive d-c voltage applied through R_2 to the P-type base.

2. The large base current of T_2, from emitter to base, keeps T_2 heavily conducting and also charges capacitor C_1.

3. T_2 emitter saturation current flows up through R_K, making the top of R_K sufficiently positive as to reverse bias T_1 emitter, keeping T_1 cut off.

4. This condition (T_2 heavily conducting and T_1 inoperative) is permanent until an input pulse is applied to the circuit of Fig. 5·12*c*.

5. A positive-going pulse fed to the base of T_1 forward biases it, causing T_1 to conduct, and T_1 collector current flows up through R_{L_1}.

6. The collector voltage of T_1 falls. This phase inversion, of course, is typical of all common emitter (CE) circuits, where a positive pulse applied to the base results in a negative-going pulse at the collector.

7. The drop of T_1 collector voltage starts C_1 discharging through R_2, as shown in Fig. 5·12*c*, driving the lower end of R_2 and T_2 base negative-going, reverse-biasing it, and cutting off T_2.

8. Now T_1 conducts and T_2 is inoperative, but this is only a temporary condition called the quasi-stable state.

9. As C_1 discharges through R_2, the voltage drop across R_2 decreases. When it decreases sufficiently, the applied positive d-c voltage at the upper end of R_2 again forward biases T_2 base, and T_2 again conducts.

10. Emitter current of T_2, flowing up through R_K, starts reverse biasing the emitter of T_1, and T_1 conducts less.

11. T_1 collector voltage rises, causing C_1 to start charging up again through T_2 base current.

12. T_2 now conducts very heavily, reverse biasing T_1 emitter via R_K, and T_1 cuts off.

13. T_2 continues to conduct heavily *permanently* because of the positive voltage applied to its base, causing large base current.

14. T_1 remains cut off, and the circuit is back in its stable state again, temporarily changing at the next input pulse.

If PNP transistors were used in the circuit of Fig. 5·12*c*, *all* applied d-c voltages would be reversed in polarity, and electron flow through the transistors would take place from *base* to *emitter* and from *collector* to emitter.

5·8 Plate-coupled Monostable Multivibrator

The plate-coupled monostable multivibrator of Fig. 5·14 is not an oscillator. A monostable multivibrator has only one stable state. One tube always conducts, and the other is always cut off. This condition can never

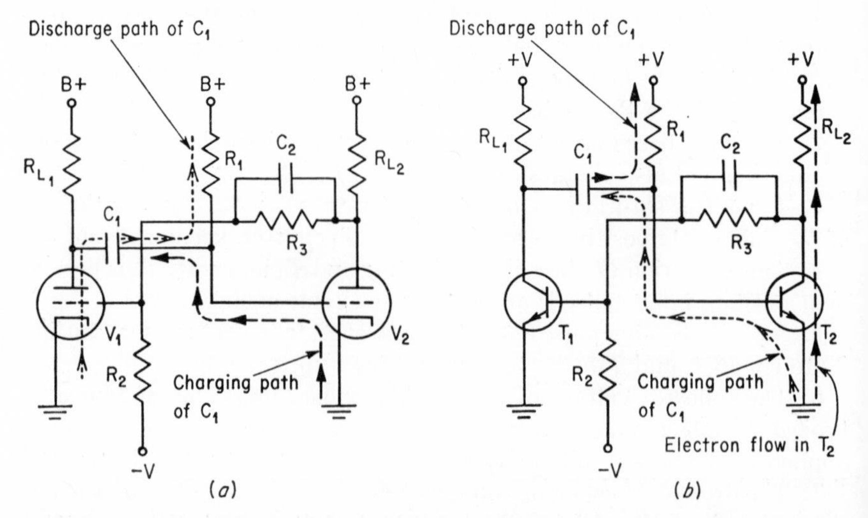

Figure 5·14 **Plate-coupled (collector-coupled) monostable multivibrator. (*a*) Tube type; (*b*) NPN-transistor type.**

change by itself. It requires an input trigger signal to reverse the operation, and then it is only a temporary reversal. By itself, the circuit reverts back to its original stable state, remaining like this until the next input trigger signal. The temporary reversed condition, due to the input signal, is called the quasi-stable state.

In the following theory discussion, reference should be made to the circuit diagram Fig. 5·14, the voltage waveshapes Fig. 5·15, and the basic capacitor- and resistor-voltage graphs of Fig. 5·3.

1. Applying B+ to the circuit will cause only V_2 to conduct since its grid is connected to B+ through R_1.

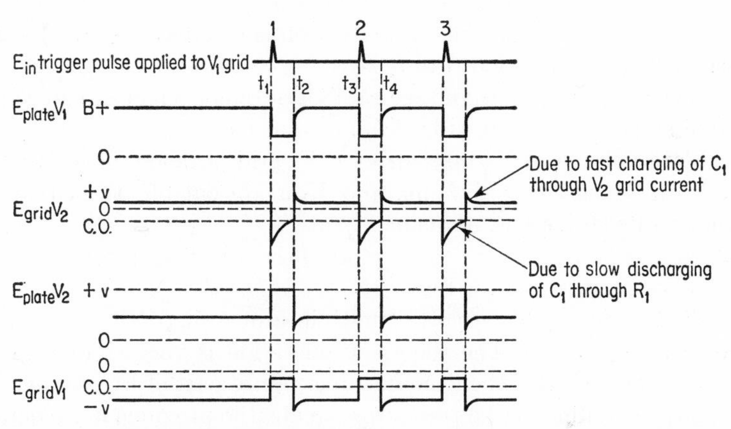

Figure 5·15 **Waveshapes of plate-coupled monostable multivibrator of Fig. 5·14.**

2. The grid of V_1 is connected to a resistor voltage divider, R_2 and R_3, between the negative bias voltage and the positive voltage at V_2 plate.

3. Values of R_2 and R_3 are such that when V_2 conducts, producing a low voltage at the plate, the grid of V_1 is sufficiently negative to cut off V_1.

4. This stable-state condition of V_2 heavily conducting because of its positive grid and V_1 cut off because of its negative grid would continue indefinitely if input trigger signal were not applied.

5. The application of the positive-going trigger input pulse to the grid of the cutoff V_1 drives V_1 into conduction.

6. V_1 plate current flows through R_{L_1}, producing a voltage drop across it, and V_1 plate voltage decreases (Fig. 5·15, instant t_1).

7. This falling voltage starts C_1 discharging through R_1, making the grid of V_2 negative.

8. V_2 conducts less, and its plate voltage rises because of the smaller voltage drop across R_{L_2}.

9. The positive-going voltage at V_2 plate is coupled to V_1 grid through resistors R_3 and R_2, driving the grid positive (actually less negative).

10. This aids the original positive-going input pulse and drives V_1 into heavier conduction.

11. C_2, in shunt with R_3, is called a "commutating" or "speedup" capacitor. It improves the high-frequency response of the multivibrator by compensating for shunt capacitance at V_1 grid. It allows the multivibrator to respond to short-duration input trigger pulses, which might otherwise have had no effect.

12. With V_1 conducting more heavily, its plate voltage decreases further, driving V_2 beyond cutoff.

13. V_2 is held cut off by the discharge of C_1 through R_1. This condition is only temporary and is the quasi-stable state.

14. As C_1 discharges, the negative voltage across R_1, at V_2 grid, decreases (Fig. 5·15, period t_1 to t_2).

15. When this negative voltage at V_2 grid has decreased sufficiently, V_2 again starts conducting (Fig. 5·15, instant t_2).

16. Plate voltage of V_2 falls, driving V_1 grid negative beyond cutoff.

17. With V_2 again conducting, and V_1 again cut off, the circuit has returned to its stable and permanent state.

18. This stable condition continues until the next trigger pulse (number 2) is applied at instant t_3 of Fig. 5·15.

A *collector-coupled monostable* multivibrator using NPN transistors is shown in Fig. 5·14*b*. The theory of operation is very similar to that described previously for the vacuum-tube counterpart of Fig. 5·14*a*. For a better understanding of the transistor circuit, the previous more detailed discussion given for the vacuum-tube circuit should be studied first. Then, the following briefer transistor-circuit theory should be read.

1. The d-c voltages shown in Fig. 5·14*b* forward-bias T_2 base to emitter (positive voltage to the P-type base) and reverse-bias T_2 collector (positive voltage to N-type collector), permitting T_2 to conduct.

2. T_1 is prevented from conducting because of the reverse bias between base and emitter (negative voltage to P-type base).

3. The heavy emitter-to-collector current of T_2, flowing up through R_{L_2}, produces a voltage drop across R_{L_2} which brings the collector voltage of T_2 down to practically zero.

4. The positive base of T_2 draws current from its emitter, and C_1 charges up through this base current.

5. R_2 and R_3 form a voltage divider between the collector voltage of T_2 and the negative applied d-c voltage. With T_2 heavily conducting and its collector voltage at practically zero, the P-type base of T_1 at the junction of R_2 and R_3 is at a negative voltage, keeping T_1 permanently cut off.

6. This is the *stable* state of this monostable multivibrator and is only changed temporarily by an input pulse.

7. If a positive-going pulse is applied to the P-type base of T_1, it overcomes the negative voltage there and now forward-biases this base, causing T_1 to conduct.

8. T_1 emitter-to-collector current flows up through R_{L_1}, producing a voltage drop across R_{L_1}, and the collector voltage of T_1 falls. (This, of course, is the typical phase inversion of a common emitter circuit where a positive-going pulse applied to the base, produces a negative-going pulse at the collector.)

9. The falling collector voltage of T_1 causes C_1 to start discharging through R_1, driving the base of T_2 negative and reverse-biasing T_2 to cutoff.

10. With T_2 inoperative, its collector voltage rises to the full applied positive voltage, aiding the positive-going input signal by making the base of T_1 positive, keeping T_1 in conduction even after the end of the input pulse.

11. T_2 is kept cut off by the discharging C_1 through R_1. As C_1 discharges, the voltage across R_1 decreases, until it no longer can hold T_2 inoperative.

12. Conduction of T_2 causes its collector voltage to fall, driving the base of T_1 negative, again cutting off T_1, and the circuit is back in its permanent and stable state.

If PNP transistors were used in this circuit, the explanation would be the same as that given above for the NPN circuit of Fig. 5·14*b* except that electron flow through the transistors is from *base* to *emitter* and from *collector* to emitter. The polarities of *all* applied d-c voltages are reversed from those shown in Fig. 5·14*b*.

5·9 Bistable Multivibrator

Figure 5·16 illustrates the circuit of the bistable multivibrator, also known as the Eccles-Jordan and popularly called the "flip-flop."

The bistable multivibrator has two permanent or stable methods of operation. Either of the two tubes could conduct alone *forever*. The circuit is not an oscillator but requires an input trigger to produce an output. The first input signal reverses the circuit operation so that the original conducting tube cuts off, and the other tube now conducts. The circuit has been "rolled" into its flip state and remains in this condition permanently or until the next trigger pulse again rolls the operation back to its original flop state. Unlike a monostable multivibrator, the bistable circuit does not revert back to its original condition *by itself*.

In the following theory discussion, refer to the schematic diagram Fig. 5·16 and to the voltage waveshapes Fig. 5·17.

1. When B+ is applied to the circuit of Fig. 5·16*a*, V_1 or V_2, or both, may start conducting. Whether they do or do not depends on their grid voltages. Each grid is connected to a voltage divider between the negative

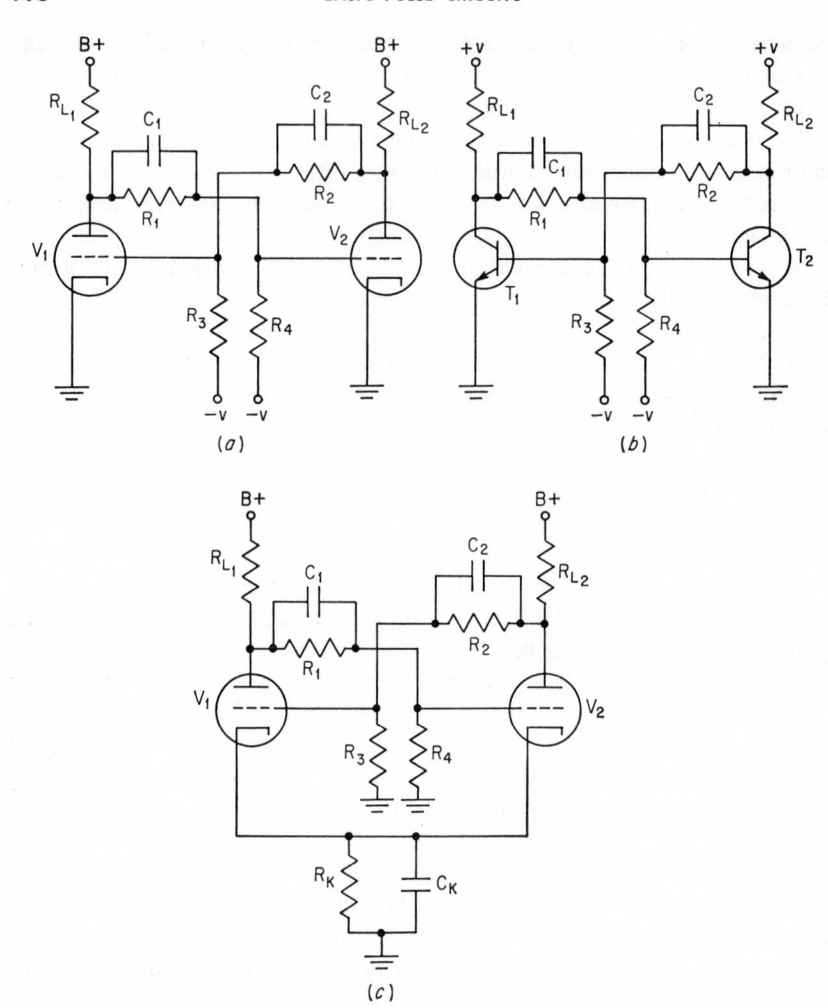

Figure 5·16 Bistable multivibrators. (*a*) Grid biased; (*b*) NPN-transistor type; (*c*) cathode biased.

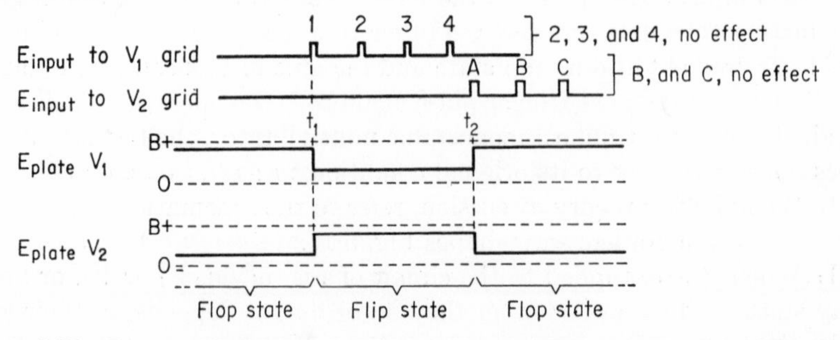

Figure 5·17 Unsymmetrical triggering waveshapes, bistable multivibrator of Fig. 5·16.

voltage and the positive voltage at the plate of the opposite tube. For example, V_1 grid connects to the resistor-voltage divider, R_3 and R_2, while V_2 grid uses R_4 and R_1.

2. If both V_1 and V_2 did conduct at first, this condition would quickly change to one tube conducting and the other cut off. This would be brought about by *either* one of the tubes suddenly conducting slightly more, possibly because of ripple of B+, or slight filament temperature changes caused by the a-c filament supply, or differences in the construction between V_1 and V_2.

3. Either tube could unpredictably be the one that suddenly conducts slightly more heavily, or one tube, say V_2, could be forced into this heavier conduction. For example, a manual switch (not shown in the diagram) could momentarily open R_4.

4. This opens the R_4-R_1 voltage divider and disconnects the grid of V_2 from the negative voltage.

5. V_2 grid is now positive, and the tube conducts heavily.

6. Plate current of V_2 flows through R_{L_2}, producing a large voltage drop across it, and V_2-plate voltage drops to a small positive value.

7. This low voltage at V_2 plate brings the grid of V_1, due to the R_3-R_2 divider, to a negative voltage which cuts off V_1.

8. This condition of V_2 conducting and V_1 cut off remains permanently and is one of the stable states. It is called the flop state.

9. V_1 remains cut off, and V_2 keeps conducting until a proper polarity trigger pulse rolls the circuit into its flip state.

10. In Fig. 5·16, input trigger pulses could be applied separately to the grids of V_1 and V_2. To have an effect, the trigger pulse must either be positive-going to the grid of the cutoff tube or negative-going to the grid of the conducting tube. When pulses are applied separately to *each* grid, it is called unsymmetrical triggering. A later discussion will cover symmetrical triggering which applies the same pulse simultaneously to both grids.

11. In the period before instant t_1 of Fig. 5·17, V_1 is in a cutoff condition and its plate voltage is high, while V_2 is conducting and its plate voltage is low.

12. Trigger pulse number 1 applied to the grid of V_1 drives V_1 into conduction at instant t_1.

13. V_1 plate voltage falls due to the voltage drop across R_{L_1}.

14. The lower voltage at the plate of V_1 brings the voltage at V_2 grid down from its former positive value to a negative voltage.

15. V_2 conducts less, and its plate voltage rises due to the smaller voltage drop across R_{L_2}. This V_2 plate-voltage rise occurs at instant t_1 of Fig. 5·17.

16. The rise of voltage at V_2 plate drives the grid of V_1 more positive, aiding the effect of positive-going trigger pulse number 1, a regenerative action.

17. V_1 conducts more heavily, driving its plate voltage still lower.

18. This sends the grid of V_2 further negative, cutting off V_2.

19. With V_1 conducting and V_2 cut off, the circuit is in its flip state, the second stable state in which it could continue indefinitely.

20. Any further positive-going pulses applied to the grid of V_1 while it is conducting have no effect. This is shown in Fig. 5·17 by pulses 2, 3, and 4.

21. The circuit continues in this flip state until a positive-going pulse is applied to the grid of the cutoff V_2. This is shown as pulse A in Fig. 5·17.

22. At this instant, t_2, pulse A drives V_2 back into conduction.

23. Plate current of V_2, flowing through R_{L_2}, produces a voltage drop across it, and the plate of V_2 falls.

24. This falling voltage drives the grid of V_1 down from its positive value to a negative voltage.

25. V_1 now conducts less, and its plate voltage rises at instant t_2.

26. This drives the grid of V_2 more positive, aiding the effect of trigger pulse A (regeneration), and V_2 conducts very heavily.

27. With V_2 in heavy conduction again, because of trigger pulse A and the regenerative feedback, plate voltage of V_2 is very low.

28. This brings the grid of V_1 down to a negative voltage sufficient to cut off V_1.

29. At instant t_2, with V_2 again conducting and V_1 cut off, the circuit is back in its flop state again.

30. Pulses B, C, and D have no effect since they are being applied to the grid of V_2 while it is conducting.

31. The period t_1 to t_2 of Fig. 5·17, or the flip state, depends on the time duration between pulse 1 (which flipped V_1 into conduction) and pulse A (which flopped V_2 back into conduction).

Figure 5·16*c* differs from (*a*) only in the method of bias. In (*a*) the grids are connected to voltage dividers, the ends of which are at a negative voltage. In (*c*) the voltage dividers connect to ground. This makes each grid somewhat positive. However, plate current flowing up through R_K makes the cathodes positive, providing the necessary negative bias (grid with respect to cathode) to cut off one of the tubes.

A *transistor bistable multivibrator* using NPN types is shown in Fig. 5·16*b*. The theory of its operation is very similar to the discussion covered previously in this section for the vacuum-tube circuit of Fig. 5·16*a*. For a better understanding of the transistor version, the previous more detailed discussion of the vacuum-tube circuit should be read first. A brief explanation of the transistor bistable multivibrator of Fig. 5·16*b* follows.

1. The applied d-c voltages forward bias the bases to emitters of transistors T_1 and T_2 so that both transistors are permitted to conduct. Both base voltages would be positive if T_1 and T_2 current were momentarily ignored. Resistors R_3, R_2, and R_{L_2} form a voltage divider between the negative d-c and the positive d-c applied voltages. Resistor values are such that the top of R_3 (the P-type base of T_1) is positive with respect to the grounded N-type emitter of T_1, producing forward bias. An identical voltage divider, R_4, R_1, and R_{L_1}, produces the required positive voltage on the base of T_2 for its forward bias.

2. Although both transistors are free to conduct, conduction of one will permanently cut off the other. When the d-c voltages are first applied, one of the transistors, because of some slight difference between T_1 and T_2, will start conducting before the other. It could be either T_1 or T_2 that starts operating first.

3. Assume that T_2 is the one which goes into conduction first. Collector current of T_2 flows up through R_{L_2}, producing a voltage drop across it, and T_2 collector voltage falls to practically zero.

4. With almost zero volts at T_2 collector, R_3 and R_2 act as a voltage divider, producing a *negative* voltage on T_1 P-type base, reverse biasing T_1 to cutoff.

5. With T_1 inoperative and T_2 heavily conducting, the bistable multivibrator is in one of its stable states and will continue like this indefinitely.

6. If a positive-going pulse were now applied to the base of T_1, overcoming the reverse bias there, T_1 will start conducting, and its collecto- voltage now falls to almost zero.

7. This brings the P-type base of T_2 down to a negative voltage, via the R_4-R_1 voltage divider, reverse biasing T_2 to cutoff, and the bistable multivibrator is now in its second stable state.

8. The circuit remains in this condition with T_1 conducting and T_2 cut off permanently until a positive-going trigger pulse is applied to the base of T_2.

9. A positive-going pulse applied to T_2 base, while T_2 is cut off, overcomes the reverse bias, and T_2 conducts.

10. Collector voltage of T_2 falls, and the base of T_1 is driven negative again, reverse biasing T_1 to cutoff, and the circuit is back in its original stable state again.

If PNP transistors were used in the circuit of Fig. 5·16*b*, the circuit explanation would be the same except that electron flow through each transistor takes place from collector to emitter. Also, polarities of *all* voltages shown would be reversed.

It was stated previously that to force one of the tubes into operation to start the flop state, a manual switch could be inserted in Fig. 5·16. The switch could be momentarily opened to open R_4. This opens the R_4-R_1 voltage divider, making the grid of V_2 positive and driving V_2 into heavy

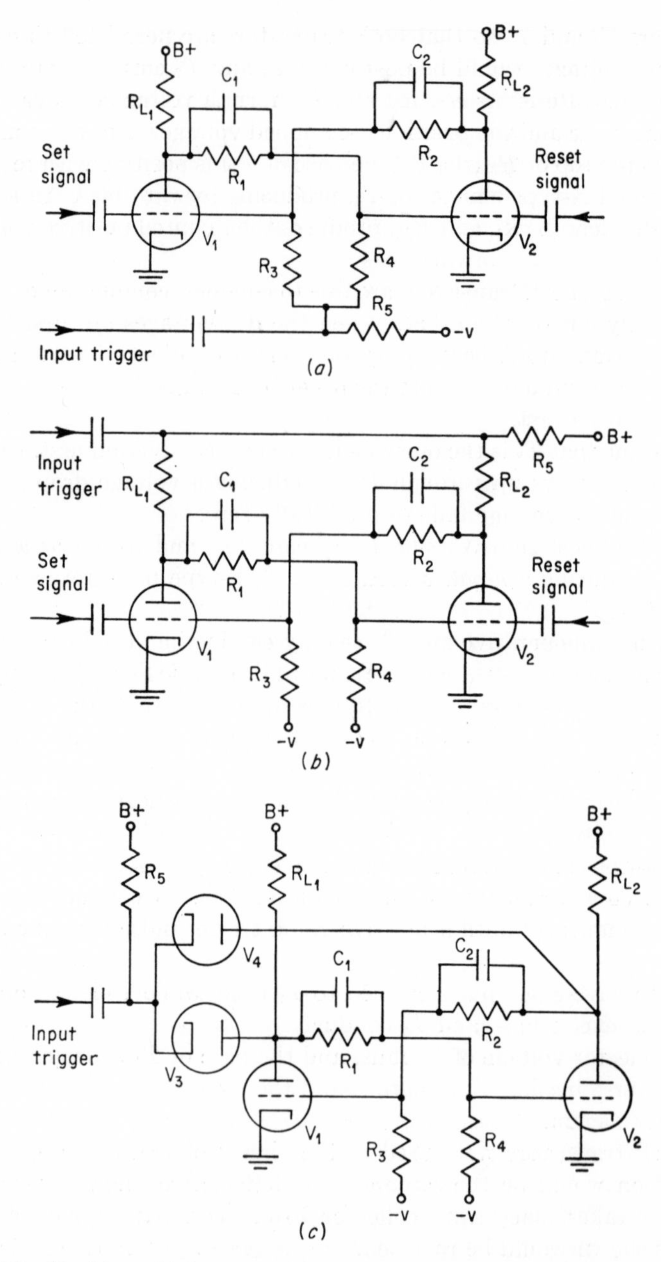

Figure 5·18 Bistable multivibrators using symmetrical triggering. (*a*) Symmetrical grid triggering; (*b*) symmetrical plate triggering; (*c*) symmetrical triggering using steering diodes.

conduction. This flop state is often referred to as "reset" and the switch as the reset switch. Another method of reset is to apply a positive-going pulse (*not* the trigger pulse) to the V_2 grid. This signal is called the "reset signal." Similarly, another signal called the *set signal* could, instead, be applied to the grid of V_1 to turn V_1 ON, placing the circuit in its flip or "set" state. These set or reset signals are not the trigger pulses which must now be applied in a different manner than that explained previously.

Figure 5·18 shows bistable multivibrators where set and reset signals could be applied as well as the input trigger pulses. Since input trigger pulses are applied from one line to both grids in (*a*) or to both plates in (*b*), it is called single-line or symmetrical triggering.

The circuits of Fig. 5·18 could be placed in the flop state or reset, which means that it is desired to have V_2 conducting, by applying a positive pulse to the grid of V_2 *alone*. If V_2 happens to be conducting, then the reset pulse does not change anything (such as pulses B, C, and D in Fig. 5·17). If V_2 happens to be cut off, the reset pulse drives the tube into conduction (as did pulse A in Fig. 5·17). The circuits of Fig. 5·18 could be placed in the flip state or set, which means that V_1 is desired to operate by applying a positive pulse to V_1 grid alone. This signal will drive V_1 into conduction if it happens to be cut off (pulse 1, Fig. 5·17) or will not change V_1 if it is already conducting (pulses 2, 3, and 4, Fig. 5·17).

The symmetrical single-line triggering of Fig. 5·18 is applied simultaneously to both grids in (*a*), or both plates in (*b*), or to the *proper* plate in (*c*) as determined by the steering diodes. These diodes will be discussed later in this section.

These single-line triggered circuits of Fig. 5·18 are much more susceptible to negative pulses than to positive-going ones. One reason for this is the characteristics of a tube. If a tube is biased much more negative than cutoff, it requires a large positive-going pulse to make that tube start conducting, and even then only the most positive part of the pulse has any effect. This is shown by the shaded areas of the pulses in Fig. 5·19*a*. Normal sized, small positive-going pulses will not have any triggering effect.

A small negative-going pulse, on the other hand, will immediately make the conducting tube conduct less. Its rising plate voltage, larger than the input pulse because of the tube gain, drives the cutoff tube grid positive, and it starts conducting. This rolls the circuit.

Positive-going pulses have almost no triggering effect in Fig. 5·18*a* or *b* and cannot get through the diodes V_3 or V_4 in (*c*). If a positive-going pulse were applied to both grids in (*a*), it would make the conducting tube, say V_2, conduct more heavily. Plate voltage of V_2 would fall, driving V_1 grid more negative. This feedback from the conducting tube, V_2 plate to the tube already cut off, V_1, keeps V_1 cut off and also is out of phase with the positive-going pulse being applied. Because of the gain of V_2, its negative-going plate signal is larger than the positive-going input applied to V_1 and V_2 grids, and effectively obliterates the original input of V_1 grid. Thus, a normal positive input pulse would have no rolling effect.

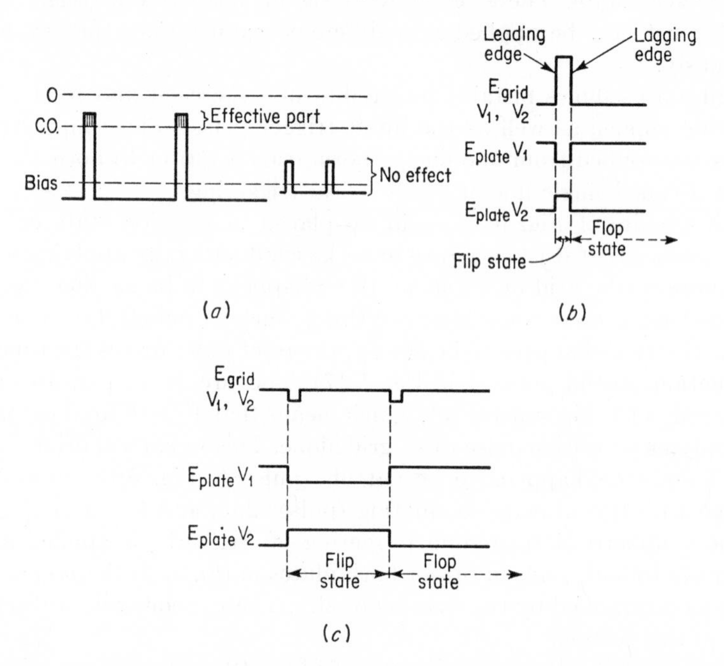

Figure **5·19 Symmetrical triggering. (*a*) Positive-going trigger pulses; (*b*) effect of large "brute force" positive-going trigger pulse; (*c*) effect of normal-sized negative-going trigger pulse.**

A very large positive-going trigger input could "brute-force" V_1 into conduction and roll the circuit. However, as shown in Fig. 5·19*b*, a positive-going pulse really has a positive-going leading edge, followed closely by a negative-going trailing or lagging edge. If the positive-going input pulse is sufficiently large, it will "brute-force" V_1 into conduction, and roll the circuit into its flip state (Fig. 5·19*b*). However, a moment later, the negative-going lagging edge of the pulse again rolls the circuit back into its flop state. This, of course, is undesirable since the circuit should flip at the first input pulse and should only flop at the second input pulse (Fig. 5·19*c*) and not flip *and* flop because of *one* input pulse (Fig. 5·19*b*).

Normal-sized trigger pulses are small in amplitude as shown in (*c*). If this negative-going pulse were simultaneously applied to the grids of V_1 and V_2 in Fig. 5·18*a*, it would have no *immediate* effect on the tube which is already cut off, say V_1. However, if V_2 is the conducting tube, the negative input pulse causes V_2 to conduct less. The rise of V_2-plate voltage is larger, because of V_2 gain, than the negative input grid pulse. This large rise of V_2 plate voltage drives the grid of V_1 positive, and V_1 starts conducting. V_1 plate voltage decreases, driving V_2 grid more negative, cutting off V_2. With V_1 now conducting and V_2 cut off, the circuit has rolled to its flip state (Fig. 5·19*c*). The positive-going lagging edge of this small negative-going pulse is too small to affect the circuit.

The circuit remains in this stable state until the next negative-going trigger pulse to both grids reverses the process and again rolls the circuit into its flop state, as shown in Fig. 5·19*c*.

The capacitors C_1 and C_2 in Fig. 5·18*a* are not vitally necessary. These capacitors are called commutating or speed-up capacitors. Without them, the multivibrator might not roll when the input trigger pulse is of very short duration. This would be due to the plate and grid capacitance-to-ground of each tube not allowing the short-duration pulse to affect the circuit. C_1 and C_2 act as frequency-compensating devices to improve the high-frequency response of the circuit.

In Fig. 5·18*b*, C_1 and C_2 are more than only high-frequency compensating capacitors and are necessary for the operation. In (*b*), the input signals are applied to the plates of V_1 and V_2 and through C_1 and C_2 to both grids. Without these capacitors, the pulses would be greatly attenuated at the grids due to the voltage dividers, R_1-R_4 and R_2-R_3.

In Fig. 5·18*b*, these capacitors are charged to different voltages. If V_2 is the conducting or ON tube, its plate voltage is low. This keeps the grid of V_1 negative, and V_1 is the OFF tube. Plate voltage of V_1 is high, driving the grid of V_2 positive and keeping V_2 conducting. The grid of V_2 will draw current from its cathode, and V_2 grid voltage is clamped to almost zero (the cathode potential). Resistance between grid and cathode of V_2 (R_{G-K}) is small and parallels R_4, practically shorting R_4. Charge on C_2 is the voltage across R_2 which is ($R_2/R_2 + R_3$) times the voltage difference between the low V_2 plate voltage and the negative bias voltage. E_{C_2} is quite small. Charge on C_1 is the voltage across R_1 which is approximately ($R_1/R_1 + R_{L_1} + R_{G-K}$) times the full applied B+ voltage. E_{C_1} is quite large, with V_1 inoperative.

With V_2 the conducting tube, and V_1 the cutoff stage, a negative-going trigger pulse should reverse this condition. The trigger pulse must be of short duration compared to R_1-C_1 and R_2-C_2 of Fig. 5·18*b*. When the negative trigger pulse is applied, it drives both plates and both grids negative. V_2 now becomes cut off along with V_1. While the trigger is present then, both stages are inoperative. When the pulse has passed, the charges on C_1 and C_2 act as if they alone "remembered" which stage was originally the ON and which was the OFF stage. With V_1 and V_2 inoperative, C_1 and C_2 start recharging to their respective high plate voltages. Since C_1 is already quite fully charged, it will not require much additional charging. However, C_2, the capacitor at the formerly conducting V_2 plate, is at a low charge and now starts charging up, driving the grid of V_1 (formerly the OFF stage) positive. V_1 now starts conducting. Thus, the operation of the circuit has reversed, with V_1 now the ON and V_2 now the OFF stages.

In Fig. 5·18*c*, the input trigger pulses are applied to the cathodes of diodes V_3 and V_4. These are called "steering" diodes since they steer the pulses to the V_1 or V_2 tube which requires pulses for rolling. The steering is accomplished as follows: If V_1 is the cutoff tube and V_2 is conducting, the plate voltage of triode V_1 is high, while triode V_2 plate voltage is low. This makes diode V_3 plate voltage high (connected to V_1 plate) and diode V_4 plate voltage low (connected to V_2 plate). Both diode cathodes are at B+ voltage, since no current is flowing through R_5. Neither diode will conduct until an input trigger pulse is applied. Positive-going pulses will drive the diode cathodes positive, keeping them cut off. As a result, positive pulses are prevented from getting into the multivibrator.

A negative-going input pulse drives both diode cathodes negative-going (actually less positive). Since the plate of diode V_4 is at a low positive voltage, V_4 cannot conduct. However, diode V_3 plate voltage is high. When V_3 cathode voltage goes down because of the negative-going input pulse, V_3 conducts. This makes the triode V_1 plate voltage fall, driving the grid of V_2 negative. Thus, steering diode V_3 permits the negative-going input trigger pulse to reach the grid of the conducting tube, resulting in a rolling of the multivibrator into its flip state.

When V_1 is the conducting tube, its low plate voltage keeps diode V_3 cut off. The negative-going input pulse causes V_4 to conduct, allowing the trigger pulse to get to the plate of V_2 and through C_2 to the grid of the conducting V_1, rolling the circuit back to flop state.

5·10 Multivibrator Improved Modifications

Improvement in multivibrator frequency stability and output voltage waveshape can be achieved by modifying the basic astable plate-coupled multivibrator of Fig. 5·20*a* to that of diagram (*b*).

In Fig. 5·20*b* the grid resistors R_1 and R_2 are connected to B+ instead of to ground as it is in the basic circuit of (*a*). Also in (*b*), limiting resistors R_3 and R_4 are added in the grid circuits for waveshape improvement.

By connecting the grid resistors R_1 and R_2 to B+, a more linear grid-voltage waveshape is produced, with greater frequency stability. This is illustrated in Fig. 5·21*a*. Capacitor C_1 in Fig. 5·20*a* tries to discharge toward zero as the plate voltage of V_1 falls. This produces the curved or nonlinear negative grid voltage on V_2 as shown in the solid-line waveshape of Fig. 5·21*a* since a large portion of the nonlinear capacitor-discharge curve is used. When the positive voltage is connected to the grids through R_1 and R_2 (Fig. 5·20*b*), C_1 not only discharges toward zero but starts

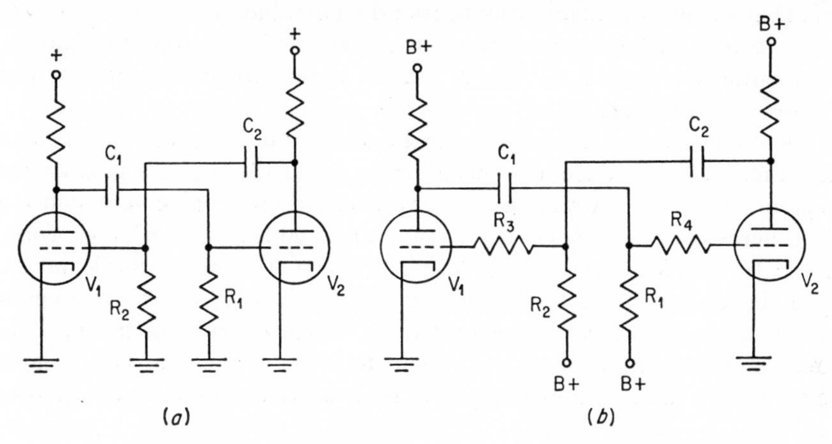

Figure 5·20 **Astable plate-coupled multivibrators. (a) Basic circuit; (b) modified circuit.**

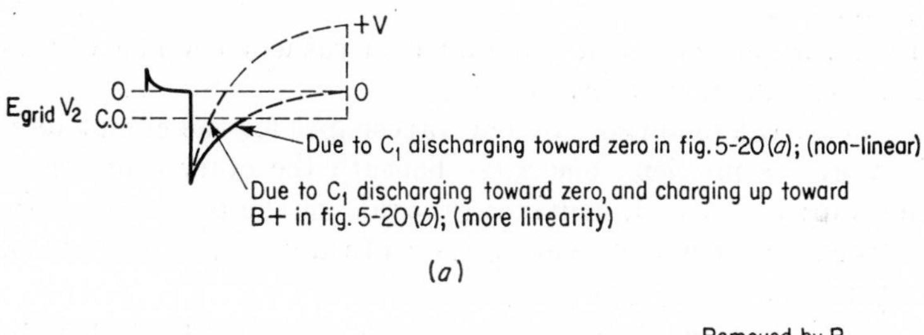

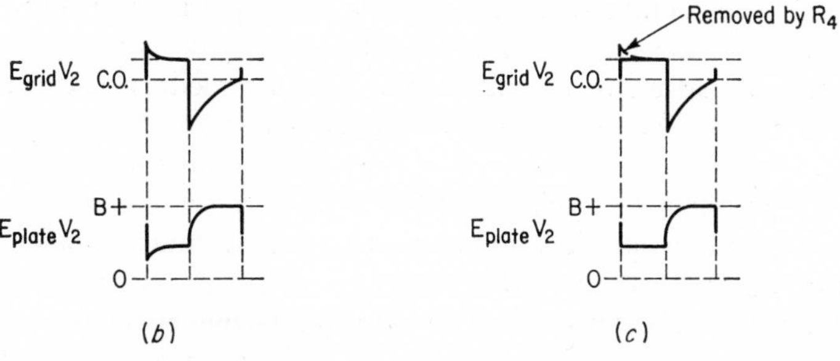

Figure 5·21 **Multivibrator improved waveshapes. (a) Effect of positive grid return; (b) without R_4; (c) with R_4.**

charging up to this positive voltage. The result is a steeper, more linear negative grid voltage on V_2 as shown by the dashed-line portion of V_2 grid wave (Fig. 5·21*a*). Here, a smaller portion of the capacitor discharge curve is used. Variations of B+ and changes of tube characteristics have less effect on multivibrator frequency when the negative grid approaches cutoff more sharply than when the grid rises toward cutoff gradually. The positive voltage on the grids increases the frequency since cutoff is reached sooner. However, by using larger *R-C* circuits, the frequency can be brought back to its original value.

Plate-voltage output signal produced by the multivibrator of Fig. 5·20*a* is essentially a square wave (Fig. 5·21*b*), except for the negative-going overshoot. This is produced by the positive-going spike on the grid, due to the fast charge of C_1 through V_2 grid current. Limiting resistors R_3 and R_4 (Fig. 5·20*b*) produce a limiting action when the positive grid draws current. This grid current of V_2 flows through R_4, producing a voltage drop across it. Since R_4 is very large, about 100 kilohms or larger, compared to the small resistance existing between the positive grid and its cathode, then most of the positive voltage due to C_1 charging appears across R_4 and almost none across the cathode-to-grid resistance. This removes the positive-going spike from the grid-voltage waveshape and results in a squarer plate-voltage waveshape as shown in Fig. 5·21*c*.

PROBLEMS

5·1 Draw the schematic diagram of a cathode-coupled (direct coupling) astable multivibrator.

5·2 Draw the following voltage waveshapes for the circuit requested in the previous problem, one wave beneath the other in correct phase relationship: (*a*) cathode voltage; (*b*) grid voltage of the tube which is cut off the longer time; (*c*) plate voltage of this tube; (*d*) plate voltage of the other tube.

5·3 Draw an emitter-coupled astable multivibrator, using PNP transistors.

5·4 Draw a cathode-coupled (capacitance-coupled) astable multivibrator.

5·5 An astable multivibrator is free-running at approximately 950 cps. What is the frequency of this stage when each of the following (small amplitude) input signals is applied to the proper multivibrator grid: (*a*) 1 kc; (*b*) 2 kc; (*c*) 3 kc; (*d*) 700 cps?

5·6 Draw the schematic diagram of a plate-coupled astable multivibrator.

5·7 Draw the schematic diagram of a collector-coupled astable multivibrator using PNP transistors.

5·8 Draw the following voltage waveshapes for an unbalanced (asymmetrical) plate-coupled astable multivibrator, showing the waves one below the other in correct phase relationship: (*a*) grid of tube cut off a longer time; (*b*) plate of this tube; (*c*) grid of the other tube; (*d*) plate of this tube.

5·9 Draw the schematic of a cathode-controlled astable multivibrator.

5·10 Draw the following voltage waveshapes of the cathode-controlled astable multivibrator, showing one wave beneath the other in its proper phase relationship: (*a*) cathode of one tube; (*b*) plate of this tube; (*c*) grid of the second tube; (*d*) cathode of the second tube; (*e*) plate of the second tube; (*f*) grid of the first tube.

5·11 Explain why a change of capacitance (C_3) between cathodes of Fig. 5·10 will change the frequency but not the flip-flop ratio.

5·12 Draw the schematic of a cathode-coupled monostable multivibrator.

5·13 Draw the following voltage waveshapes of the cathode-coupled monostable multivibrator, showing the waves one under the other in proper phase relationship: (*a*) single narrow positive trigger pulse applied to the grid of the cutoff tube; (*b*) plate of this tube; (*c*) grid of the other tube; (*d*) cathode; (*e*) plate of this stage.

5·14 Draw a bistable multivibrator using fixed bias on the grids and using symmetrical (single-line) grid triggering.

5·15 What polarity input pulse would "roll" the circuit of the previous problem? Also explain why an opposite polarity input pulse would not roll the stage.

5·16 Draw a bistable multivibrator using PNP transistors.

5·17 Explain why a monostable multivibrator acts as a 1:1 (input to output) ratio circuit, while a bistable acts as a divide-by-2 circuit.

CHAPTER 6

TIME BASE OSCILLATORS AND GENERATORS

6·1 General

The horizontal deflection or sweep of the electron beam in a cathode ray tube is produced by a sawtooth or modified sawtooth voltage which is generated by the *deflection oscillator* or *time-base oscillator*, also called the *sweep oscillator*. In this chapter several types of these circuits are discussed in detail, starting with the simple neon gas diode.

6·2 Neon-tube Relaxation Oscillator

A simple circuit which produces a sawtooth output voltage is the *neon-tube relaxation oscillator* shown in Fig. 6·1*a*. The 100 volts d-c (E_T) is applied across the series circuit R_1 and C_1. The neon tube and resistor R_2 are ignored for the time being. Capacitor C_1 starts charging up exponentially as has been discussed in Chap. 3, *Linear Wave Shaping, R-C Circuits.* Voltage across C_1 rises exponentially as shown in the first curve of Fig. 6·1*b*. Without the gas diode, C_1 would charge fully to the 100 volts applied (E_T) in approximately five or six R_1-C_1 time constants.

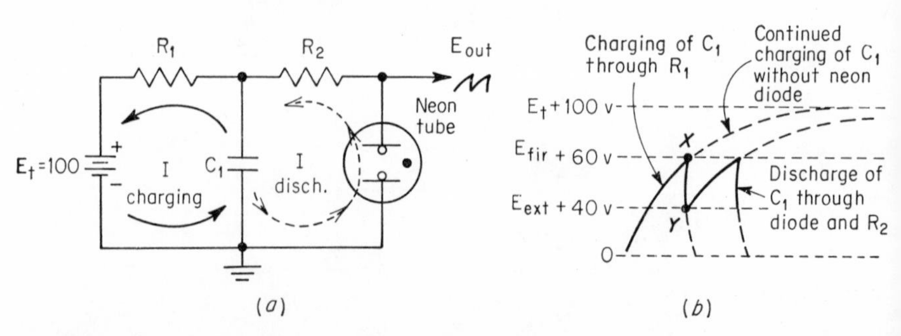

Figure 6·1 Neon-tube relaxation oscillator. (*a*) Circuit; (*b*) E_{out} (E_{C_1}) waveshape.

A neon diode conducts when the voltage across it is sufficiently large to ionize the gas. This voltage is called the *ionization* potential or the *firing* voltage and is shown in Fig. 6·1*b* as E_{fir}. *Ionization* is the changing of gas atoms into ions due to the electrons leaving the atoms and going to the positive electrode of the diode. The atoms, having lost electrons, are now positively charged ions. These ions drift to the negative electrode, taking electrons from this electrode and becoming neutral atoms again, and the process keeps repeating. In the process, the diode is said to be conducting since electrons go from the negative electrode to the positive one. The internal resistance of the ionizing tube is very small, providing a low resistance path for C_1 to discharge. Once ionization takes place, it continues even though the voltage across the tube drops below the firing point. However, when the voltage falls to a sufficiently small value (the de-ionization or extinguishing voltage), ionization ceases, and the tube becomes cut off.

When the 100 volts E_T is first applied, C_1 starts charging up through R_1. The neon diode does not conduct until the voltage across it, E_{C_1}, reaches the firing potential (+60 volts in this example). When E_{C_1} reaches point X, Fig. 6·1*b*, the gas ionizes, and the conducting diode permits C_1 to discharge through the tube and R_2. This resistor simply limits the discharge current through the tube. C_1 discharges quickly through this low resistance path, and E_{C_1} falls toward zero.

When E_{C_1} reaches point Y, Fig. 6·1*b*, the extinguishing point (E_{ext} of +40 volts), the neon tube cuts off, and C_1 no longer can continue to discharge. It now starts charging again through R_1, and E_{C_1} again rises toward the applied E_T, and the next sawtooth waveshape starts. The sawtooth will be quite linear if the peak-to-peak amplitude is small compared to the total d-c voltage applied (E_T).

The frequency of this relatively slowly rising and rapidly falling voltage, or sawtooth, depends on many factors. These are the applied d-c voltage E_T, the firing and extinguishing voltages of the gas diode, and the values of R_1 and C_1. These factors appear in the equation for the sawtooth frequency derived in the following.

Referring to the exponential charging curve for a capacitor, Fig. 6·2, it can be seen that

$$E_T = E_1 + E_{\text{ext}} \tag{6·1}$$

Solving for E_1 gives

$$E_1 = E_T - E_{\text{ext}} \tag{6·2}$$

and from Fig. 6·2

$$E_T = E_2 + E_{\text{fir}} \tag{6·3}$$

Solving for E_2 gives

$$E_2 = E_T - E_{\text{fir}} \tag{6·4}$$

A basic equation for the charging of a capacitor is

$$\frac{E_2}{E_1} = \epsilon^{-\frac{T}{RC}} \qquad (6\cdot5)$$

where ϵ is the mathematical symbol called *epsilon* and has a value of 2.718. Equation (6·5) is also given in Chap. 11, Transient Analysis, where

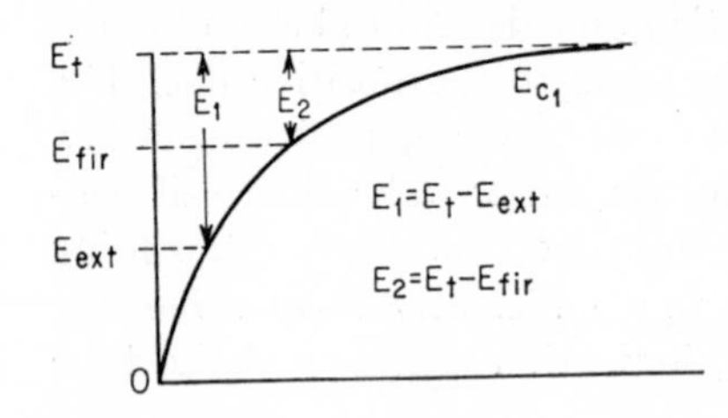

Figure 6·2 Capacitor-charging curve.

a detailed discussion is given. From Eq. (6·5), the following equation is derived in detail in Chap. 11 [see Eqs. (11·1) to (11·11)]:

$$T = 2.3RC \log \frac{E_1}{E_2} \qquad (6\cdot6) \text{ (and also 11·11)}$$

A review of logarithms is also given in Chap. 11. Substituting for E_1 [from Eq. (6·2)] and for E_2 [from Eq. (6·4)] in Eq. (6·6) gives

$$T = 2.3RC \log \frac{E_T - E_{\text{ext}}}{E_T - E_{\text{fir}}} \qquad (6\cdot7)$$

where T is the time in seconds for one cycle of sawtooth, and the discharge time is negligible. Since frequency is the reciprocal of the time for one cycle, then

$$F = \frac{1}{T} \qquad (6\cdot8)$$

where F is the frequency in cycles per second.

Example 6·1 Find the frequency of a neon-tube relaxation oscillator circuit similar to that of Fig. 6·1 but where $E_T = 135$ volts, $R_1 = 400$ kilohms, $C_1 = 0.04$ μf, and where the resulting linear sawtooth is 20 volts peak to peak with a +75 volts d-c component.

Solution. Since the 20 volts peak-to-peak sawtooth is linear, the average value or d-c component of +75 volts is also the center of the wave, with the sawtooth going 10 volts *below* the average, or +65 volts, and 10 volts *above* the average, or +85 volts. The +65 volts at the bottom of the sawtooth is then the deionization or extinguishing voltage (E_{ext}), and the +85 volts at the top of the sawtooth is the ionization or firing voltage (E_{fir}). The time (T) of one sawtooth cycle is then

$$T = 2.3RC \log \frac{E_T - E_{ext}}{E_T - E_{fir}} \qquad (6\cdot7)$$

$$= 2.3(400 \times 10^3)(.04 \times 10^{-6}) \left(\log \frac{135 - 65}{135 - 85}\right)$$

$$= 36.8 \times 10^{-3} \left(\log \frac{70}{50}\right)$$

$$= 36.8 \times 10^{-3}(\log 1.4)$$

$$= 36.8 \times 10^{-3}(0.146)$$

$$= 5.37 \times 10^{-3} \text{ sec,}$$

$$= 0.00537 \text{ sec}$$

The frequency is then

$$F = \frac{1}{T} \qquad (6\cdot8)$$

$$= \frac{1}{5.37 \times 10^{-3}}$$

$$= 0.186 \times 10^{+3}$$

$$= 186 \text{ cps}$$

A neon-tube sawtooth oscillator has several disadvantages. Among these are the following: (*a*) The required ionization and de-ionization voltages vary, resulting in changes in amplitude and frequency, and (*b*) the generated sawtooth cannot be synchronized to the frequency of another signal.

6·3 Thyratron Relaxation Oscillator

By using a gas triode tube such as the *thyratron* in a relaxation circuit, Fig. 6·3, the chief disadvantages of the neon-tube circuit are not present here. The operation is fundamentally the same as the neon-tube oscillator in that C_1 attempts to charge up to the applied B+ voltage through R_1, with the negative grid keeping the tube inoperative. When E_{C_1} becomes sufficiently large, the gas ionizes and triode-plate current flows, and the grid no longer has control. C_1 quickly discharges through the conducting tube. When the voltage across the tube, or E_{C_1}, falls low enough, the gas de-ionizes, and the tube cuts off with the grid regaining control. C_1 now starts charging up again, and the next cycle begins.

The ionization voltage required depends on the grid bias. As the negative voltage on the grid is decreased, the voltage required to produce ionization is less, and the tube "fires" at a smaller value of E_{C_1}. This enables the firing of the tube to be controlled. A signal is applied to the grid, varying the bias. Each time the bias decreases sufficiently, the tube is capable of conducting. It will conduct if E_{C_1} is then also at a sufficient voltage. This synchronization by an input signal results in the sawtooth oscillator frequency becoming locked to that of the synchronizing signal or at some submultiple of the input-signal frequency. In an oscilloscope, the sync signal is usually the signal which is applied to the vertical input terminals for visual display. One stationary, complete cycle of this signal is then observed if the sawtooth oscillator is locked to the same frequency as the input signal. Three cycles of this signal may be observed if the sawtooth oscillator is locked to one-third of the signal frequency. The frequency of the sawtooth oscillator is changed manually by varying the

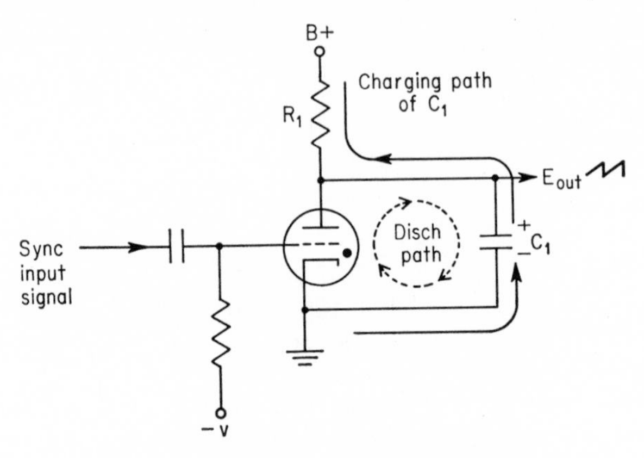

Figure 6·3 **Thyratron relaxation oscillator.**

R_1-C_1 time constant. Usually a switch selects different values of capacitors for C_1, while R_1 is a potentiometer. The free-running sawtooth oscillator frequency is adjusted by means of R_1 and C_1 so that the input sync signal frequency is slightly higher, or a multiple of this. The sync signal then causes the sawtooth to speed up and become locked to the frequency of the input signal or a submultiple of it.

The calculation of sawtooth frequency, using the sawtooth waveshapes, is exactly the same as for the neon-tube circuit, employing Eqs. (6·7) and (6·8). When the sawtooth waveshape is produced by a free-running oscillator, it is usually referred to as a *recurrent* sawtooth. If the signal waveshape to be observed on the oscilloscope is of very short duration compared to the time of a complete cycle, then using a recurrent sawtooth for electron beam deflection will only display the signal as a very narrow one occupying a very small part of the horizontal sweep or time-base line. As an example, a pulse signal at a frequency of 1 kc has 1,000 μsec from the start of one pulse to the beginning of the next. If the pulse is 10 μsec in duration, the pulse occupies only $^{10}/_{1,000}$ or $^{1}/_{100}$ of the cycle time. With a sawtooth locked to this same 1-kc frequency, the time of the time-base line is 1,000 μsec. The 10-μsec signal pulse would be seen as a narrow spike only 1/100 of the width of the time-base line. The details of such a narrow pulse can hardly be observed. In a case such as this, the sawtooth is usually changed from *recurrent* to *driven* or *triggered* operation. This is discussed in the following section.

Thyratron-driven (triggered) sawtooth circuit. The thyratron sawtooth circuit of Fig. 6·4, unlike the two previous circuits of Figs. 6·1 and 6·3, is not an oscillator. The diode clamper prevents the circuit from generating a sawtooth before an input signal is applied. Bias on the

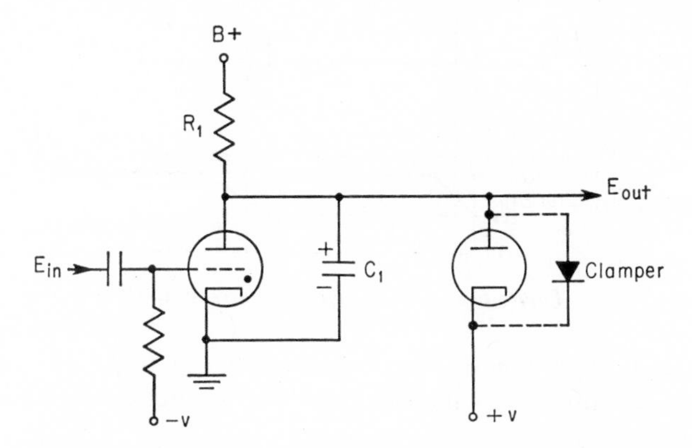

Figure 6·4 Thyratron-driven or triggered sawtooth circuit.

thyratron is such that the ionization voltage required is higher than the positive voltage on the clamper cathode. With only the d-c voltages applied, but without an input signal, the thyratron tube is held cut off by its negative grid. Capacitor C_1 starts charging up towards B+, through R_1. When E_{C_1} exceeds the positive voltage on the diode cathode, the diode conducts. The low resistance of the conducting diode (assume that it is a short) results in this positive cathode voltage also appearing practically at the diode plate and at the output point. C_1 remains charged, with the polarity shown in Fig. 6·4, to this positive voltage. (Clamper circuits are discussed in detail in Chap. 4, Wave-shaping, Nonlinear.)

When the input signal pulse goes positive, it decreases the bias on the thyratron. This now decreases the amount of voltage required for ionization to a smaller value than that which is applied to the diode cathode and to which C_1 is charged. At this instant, instant A in the waveshape diagram of Fig. 6·5, ionization occurs, and C_1 now quickly discharges through the conducting thyratron. Voltage across C_1 falls sharply, and the diode becomes inoperative. When E_{C_1} decreases to the extinguishing or de-ionization voltage of the thyratron (E_{ext}), the tube cuts off. This is instant B in Fig. 6·5. With the thyratron inoperative, C_1 starts charging up again towards B+. When E_{C_1} rises to an amount slightly above the positive voltage on the clamper cathode, the diode again conducts, clamping its anode to just about this voltage and preventing C_1 from charging any higher. This occurs at instant C of Fig. 6·5. The rise of E_{C_1} from instant B to instant C of Fig. 6·5 is the deflection or sweep voltage for the cathode-ray tube electron beam. Note that the entire sweep time is only a little longer than the width of the narrow input-pulse signal. This results in the displayed pulse on the oscilloscope screen being spread out, making the details of the waveshape more discernible. The B-to-C period (Fig. 6·5) or *time* of the sweep can be changed by varying the R_1-C_1 time constant, thus increasing or decreasing the width of the displayed pulse.

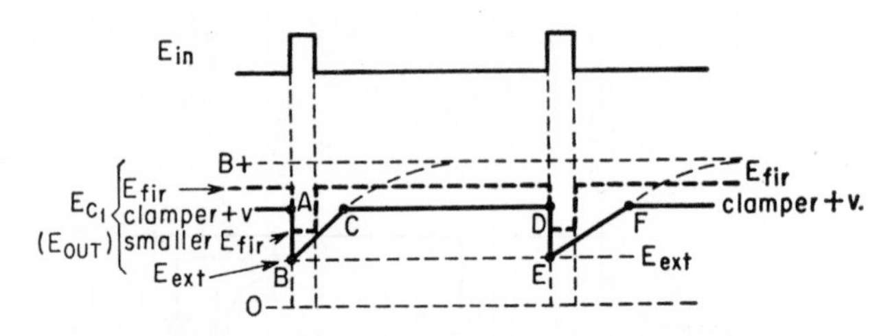

Figure 6·5 **Driven thyratron waveshapes.**

Capacitor C_1 remains charged to the positive voltage of the clamper. Without an input signal, this voltage on C_1 is less than that required for ionization, and the thyratron remains cut off until the next positive-going input signal. At instant D (Fig. 6·5), the leading edge of the input pulse again decreases the ionization voltage requirement, and E_{C_1} is then sufficient to "fire" the tube, and the next sawtooth cycle occurs. The deflection or sweep of the electron beam of the cathode ray tube occurs during the B-to-C period and again during the E-to-F period. Retrace or flyback of the beam occurs during the A-to-B time and again during the D-to-E time. No movement of the beam occurs during the C-to-D time.

Thyratron sawtooth circuits, recurrent or driven, are inadequate at higher sweep frequencies due to the need of ionizing and de-ionizing the gas. Oscilloscopes employing higher frequency sweeps, and radar indicators and other equipment too, therefore use vacuum tubes or transistors as the sawtooth generating device.

6·4 Vacuum Tube (Hard Tube) Sawtooth Circuit

A vacuum tube, often called a *hard* tube to differentiate between it and a gas tube such as a thyratron which is referred to as a *soft* tube, is often used as the sawtooth generator. A sawtooth-producing circuit using a vacuum tube is shown in Fig. 6·6. The tube, without an input signal, is normally conducting heavily since no bias is being used. Resistor R_G is shown connected between grid and ground, and the triode is zero-biased. Alternatively, this grid resistor could have been connected to B+ instead of to ground. This would make the grid positive but not as positive as B+. The resultant cathode-to-grid current produces a small internal resistance between these tube elements, and grid clamping occurs (see Chap. 4), clamping the grid at a small positive voltage.

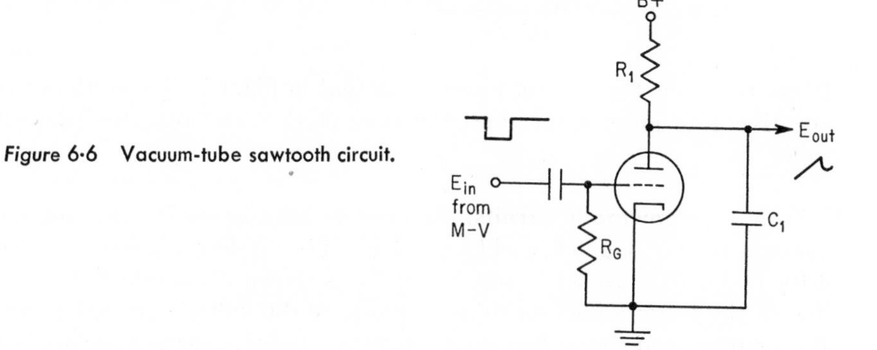

Figure 6·6 Vacuum-tube sawtooth circuit.

In either case, zero biased or positive biased, the tube conducts heavily, and its d-c plate voltage is quite low. Capacitor C_1 charges up, and remains charged, to this d-c plate potential. A sawtooth is produced when an input signal "gates" the tube OFF. This input signal is usually a rectangular waveshape and is fed from either an astable or a monostable multivibrator (see Chap. 5).

As shown in the waveshape diagram of Fig. 6·7, the negative-going part of the input rectangle gates the tube OFF, and plate voltage tries to rise, causing C_1 to start charging. At instant A, C_1 starts charging up through R_1, trying to charge up to the B+ supply. However, the R_1-C_1 time constant is made large compared to the width of the negative-going gate signal. As a result, C_1 cannot charge up very much during the time that the tube is inoperative and by using only a fraction of its exponential charging curve, the charge is linear between instant A and instant B.

When the negative-going gate signal from the multivibrator has finished, instant B of Fig. 6·7, the tube goes back to its heavy conduction, and C_1 quickly discharges through the conducting tube. C_1 discharges to the low positive d-c plate voltage, in the time from instant B to instant C. The sweep or deflection of the cathode-ray tube electron beam occurs during the A-to-B time, with retrace or flyback occurring during the B to C period of Fig. 6·7. The next sweep does not start until the next multivibrator gate signal appears. In an oscilloscope, the multivibrator is usually a monostable type which is triggered or flipped by the signal that is to be displayed on the screen. This produces a *driven* sweep. If the multi is an astable or free-running type, it is simply synchronized to the displayed signal frequency. In a radar indicator, the multivibrator is

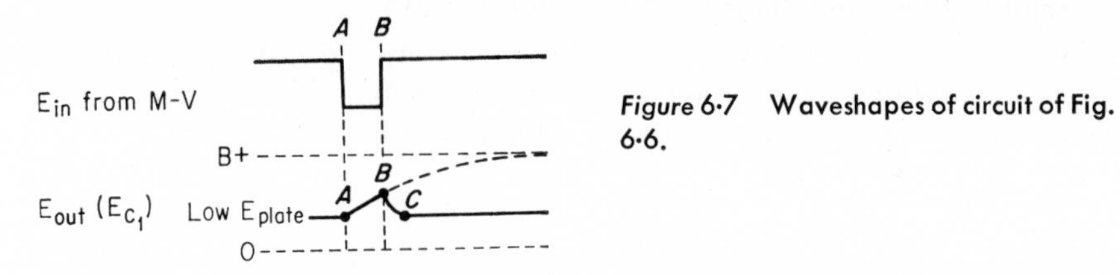

Figure 6·7 Waveshapes of circuit of Fig. 6·6.

triggered by the transmitter pulse. Note that in Fig. 6·7, the width of the negative gate pulse determines the time of the sawtooth rise (A-to-B time).

Transistor sawtooth circuits. Transistors may be substituted for the vacuum tube of Fig. 6·6, and two such circuits are shown in Figs. 6·8 and 6·10. In Fig. 6·8, an NPN transistor, T_1, is shown. T_1 conducts heavily due to the *forward* bias on its base (positive d-c voltage applied to the P-type base) and *reverse* bias on its collector (positive d-c voltage applied

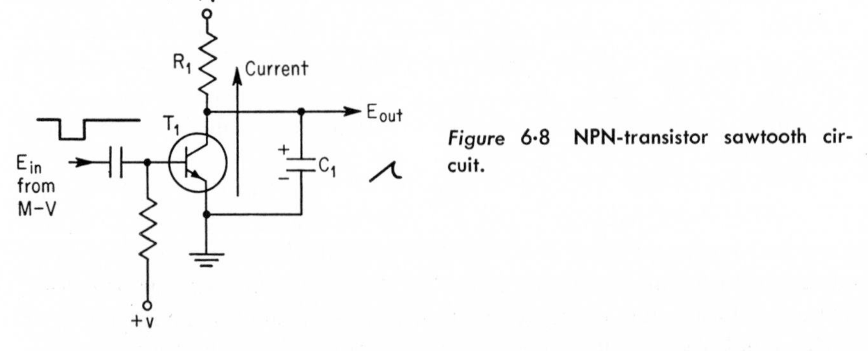

Figure 6·8 NPN-transistor sawtooth circuit.

to its N-type collector). T_1 current is *up* through the transistor and R_1. Collector voltage is only slightly positive due to the low internal resistance compared to resistor R_1, and C_1 is charged to this small voltage.

When the negative gate from a multivibrator is applied to the P-type base, it *reverse* biases the base, and T_1 cuts off. This occurs at instant A of Fig. 6·9, and C_1 starts charging up. Before E_{C_1} can rise very much, the gate is completed, and T_1 goes into conduction again (instant B). C_1 now quickly discharges through T_1 to the low positive voltage on the collector (instant C).

A *PNP transistor sawtooth circuit* is shown in Fig. 6·10. As in the previous circuit, this transistor, T_2, also heavily conducts because of its base being *forward* biased (negative d-c voltage on its N-type base). T_2 current (electron flow) is *down* through R_1 and *down* through the transistor. Collector voltage is practically zero, being slightly negative, and C_1 is charged to this voltage, with the polarity shown.

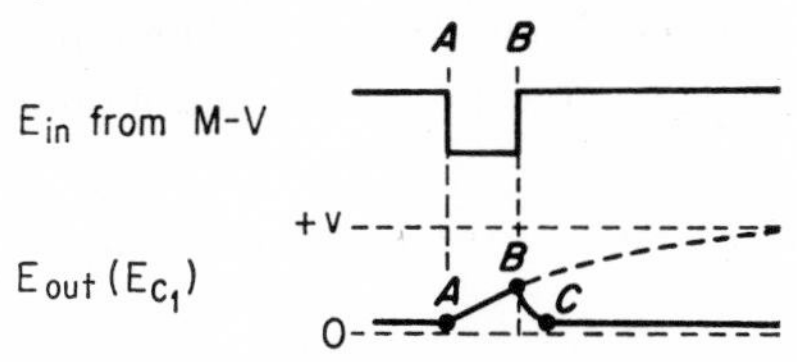

Figure 6·9 Waveshapes of circuit of Fig. 6·8.

A positive-going signal from a multivibrator gates the transistor off, since it *reverse* biases the N-type base. This is at instant A of Fig. 6·11. With T_2 cut off, C_1 now charges up through R_1, trying to charge to the full applied negative d-c voltage. At the end of the positive gate (instant B), T_2 again conducts, and C_1 quickly discharges to the low negative d-c

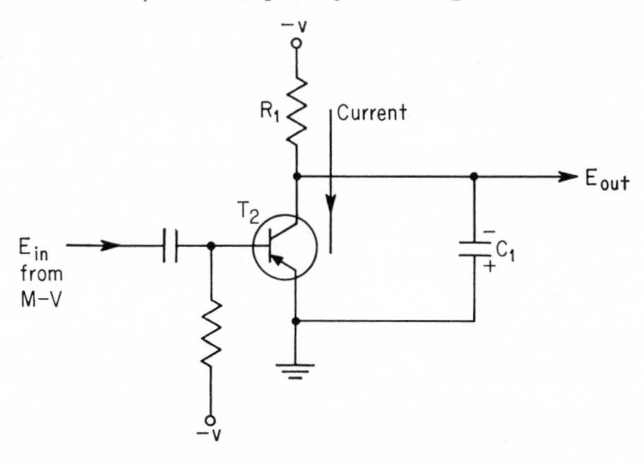

Figure 6·10 PNP-transistor sawtooth circuit.

voltage of the collector (instant C). In Fig. 6·11, the sweep or deflection of the electron beam of the cathode ray tube occurs during the A to B time, with retrace or flyback occurring during the B-to-C period.

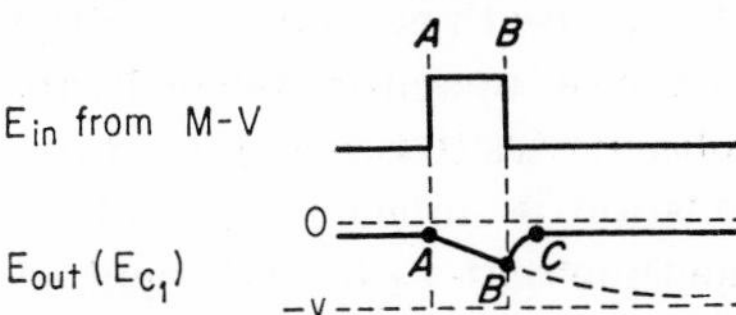

Figure 6·11 Waveshapes of circuit of Fig. 6·10.

Vacuum tubes and transistors are also employed as sawtooth generators where they are often part of a multivibrator oscillator or a blocking oscillator.

6·5 Multivibrator Sawtooth Generator

Multivibrators are discussed in Chap. 5. The astable or free-running type is an oscillator which produces square or rectangular waves (see Fig. 5·2). By adding a capacitor in the plate or collector circuit of a multivibrator, as shown in Fig. 6·12*a* and *b*, the rectangular output signal is changed to a sawtooth. The output signal, E_{out}, without the capacitor C_1 in the circuit, is the rectangular wave shown by the dotted line. The

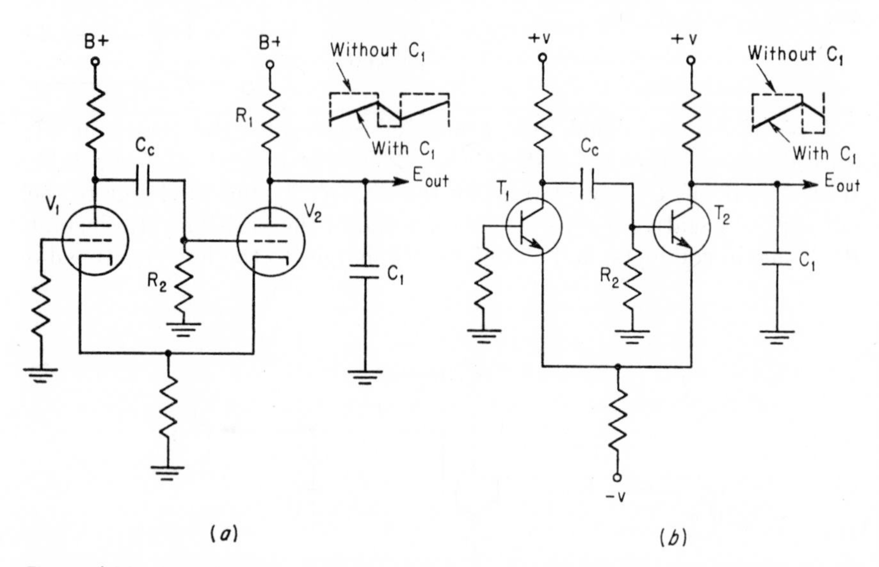

Figure **6·12 Multivibrator sawtooth generators. (*a*) Cathode-coupled astable multivibrator-discharge tube; (*b*) emitter-coupled astable multivibrator-discharge stage (NPN transistors).**

positive-going wide portion of the dotted-line wave occurs when tube V_2, or transistor T_2, is held cut off by the discharge of C_C down through R_2. The negative-going narrow portion occurs when V_2 (or T_2) is heavily in conduction due to C_C charging up through grid current or base current. The sawtooth is produced by the action of C_1, often called the *sawtooth* or *discharge* capacitor. When V_2 (or T_2) is nonconducting, plate or collector voltage tries to rise, and C_1 starts charging up through R_1. The R_1-C_1 time constant is sufficiently large so that C_1 cannot charge very much during the time that V_2 (or T_2) is cut off. As a result the rising E_{C_1} is quite linear. Then, when the tube, or transistor, goes into conduction, its plate (or collector) voltage falls, and C_1 quickly discharges through the low internal resistance of the conducting tube (or transistor). The addition of C_1 to the V_2 (or T_2) circuit changes this triode from simply being half of the multivibrator into also functioning as the discharge tube. This circuit has been used for vertical deflection of the electron beam in many television picture and camera tubes.

This astable multivibrator is an unbalanced or nonsymmetrical multi, which means that one half, V_2, is nonconducting for a longer time than the other half, V_1. The discharge capacitor is always placed in the half that is cut off for the longer time. Synchronization of the multivibrator-discharge tube circuit is achieved by applying pulses of the correct polarity to either grid (or base), as discussed previously in Chap. 5.

The sawtooth output waveshape of Fig. 6·12 can be varied in frequency and in amplitude by making R_2 and R_1 variable. A larger value of the grid (or base) resistor increases the R_2-C_C time constant, keeping the grid of V_2 (or the base of T_2) negative for a longer time. This holds V_2 (or T_2) cut off for a longer time, decreasing the frequency of the multivibrator. A byproduct of this, however, is that it also affects the sawtooth amplitude. A lower frequency, with V_2 cut off for a longer time, gives the discharge capacitor, C_1, more time in which to charge. As a result, C_1 charges to a higher value, producing a larger amplitude sawtooth. The amplitude is generally changed by varying R_1. A larger R_1 increases the R_1-C_1 time constant, causing C_1 to charge more slowly. As a result, in the time that V_2 (or T_2) is inoperative, C_1, now charging more slowly, reaches a lower voltage. The amplitude of the sawtooth is therefore decreased. A side effect of varying R_1 is that the frequency is changed slightly. This is due to the change of d-c plate voltage caused by a different value of R_1. This, in turn, changes the required grid cutoff value, with a subsequent change in frequency.

Another sawtooth generator circuit is basically the *blocking oscillator*, modified in a similar manner as the multivibrator sawtooth circuit by adding a capacitor and a resistor in the plate circuit. The next section deals first with the theory of *blocking oscillators*. Then the *blocking oscillator-discharge* tube circuit is discussed.

6·6 Blocking Oscillators, General

A blocking oscillator is a combination of a feedback, sine-wave oscillator like the Armstrong, and a relaxation oscillator such as the multivibrator described in Chap. 5. Because of excessive feedback, however, the signal that is produced is not a sine wave but a narrow pulse. The large feedback signal results in the grid being driven beyond cutoff and held that way for a comparatively long time while a capacitor slowly discharges.

Similar to multivibrators, the blocking oscillator may be free-running or astable and can be locked or synchronized or may be a one-shot or monostable type, requiring an input signal to function.

6·7 Astable Blocking Oscillator

A free-running or astable blocking oscillator operates by itself, not requiring any input signal to generate its output. Several variations of blocking oscillators are shown in Fig. 6·13, with the resultant waveshapes in Fig. 6·14. In the following theory discussion, constant reference should be made to these diagrams.

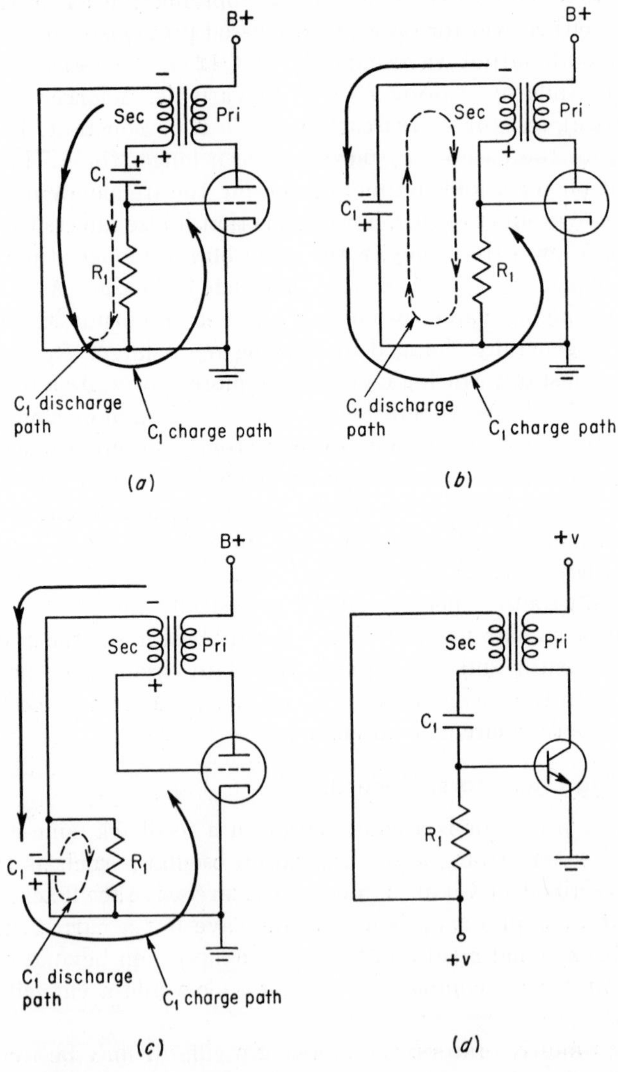

Figure 6·13 Astable blocking oscillators. (*a*) Blocking oscillator; (*b*) blocking oscillator; (*c*) blocking oscillator; (*d*) NPN-transistor type.

1. When B+ is applied, plate (or collector) current starts flowing through the primary, and a magnetic field builds up.

2. This expanding magnetic field cuts the secondary, inducing a voltage in the secondary.

3. The polarity of the induced voltage is, as shown, positive at the grid (or base) end of the secondary. This is due to the phase inversion of the transformer. If the secondary connections were reversed, an incorrect polarity voltage would result, preventing operation of the circuit.

4. The secondary induced voltage, with the polarity shown in Fig. 6·13, causes a current flow shown by the solid-line arrow.

5. This secondary voltage causes C_1 to charge up to the polarity shown. Charging of C_1 draws electrons off the grid (or base) making the grid (or base) positive, resulting in grid (or base) current.

6. C_1 charging path in Fig. 6·13*a* and *d* is from cathode (or emitter) to grid (or base) (very small resistance), onto the lower plate of C_1, off the upper plate of C_1, then through the secondary, and back to the cathode (or emitter).

7. In Fig. 6·13*b* and *c* a similar charging path for C_1 is indicated by the solid-line arrow.

8. The positive grid (or base) drives plate (or collector) current to saturation.

9. At saturation, plate (or collector) current is steady, and the magnetic field around the primary becomes stationary.

10. The stationary field no longer cuts across the secondary and no longer induces voltage in the secondary.

11. With zero volts in the secondary, charging of C_1 ceases, and C_1 now starts discharging.

12. Discharge current of C_1 flows down through R_1, as shown by the dotted-line arrow of Fig. 6·13*a*, *b*, *c*, and *d*.

13. This current, flowing down through R_1, makes the top of R_1 and the grid (or base) negative, cutting off the tube (or transistor).

14. As plate (or collector) current ceases, the magnetic field about the primary starts collapsing, inducing an opposite polarity voltage in the secondary, with the grid (or base) end negative.

15. The rapidly collapsing magnetic-field shock excites the transformer into high-frequency sine-wave oscillations, shown as the dotted waves on the grid of Fig. 6·14. If the transformer has high losses (low Q) or a resistor is shunted across it, the sine-wave oscillations exist for only about a half cycle, as shown in Fig. 6·14. The frequency of these sine waves depends on the L and C of the transformer.

16. The tube (or transistor) is held cut off by the discharge of C_1 through R_1. As a capacitor discharges, the negative voltage across the resistor decreases toward zero (see Fig. 3·1). When the negative voltage across R_1 has decreased sufficiently, it no longer holds the tube (or transistor) cut off. The value of R_1-C_1 determines how long the stage is kept cut off. Varying R_1 therefore changes the frequency.

17. Plate (or collector) current again starts flowing up through the primary, and the magnetic field starts building up again, and the next cycle begins.

18. The grid voltage waveshape of Fig. 6·14, as shown, is mainly a fluctuating negative d-c voltage. The average d-c, read by a voltmeter, is

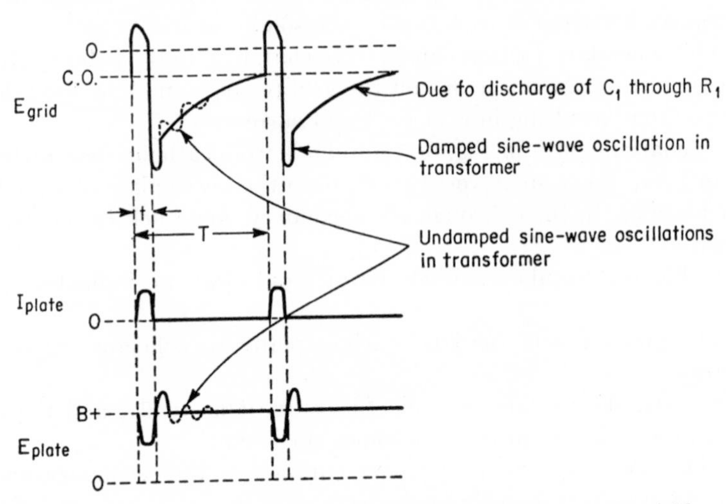

Figure 6·14 **Waveshapes of astable blocking oscillator of Fig. 6·13.**

the negative bias. This bias is present only if signal is being generated and is grid-leak bias. Measuring this bias is an indication as to whether the circuit is functioning or not.

19. The conduction time of the tube, shown as period t in the waveshapes of Fig. 6·14, is determined mainly by the inductance of the transformer and the value of C_1. The period t is directly proportional to the inductance and the size of C_1.

20. The period T, shown in Fig. 6·14, is the time of one cycle of the blocking-oscillator signal. T minus t is the period between pulses and is determined mainly by the R_1-C_1 time constant which determines the frequency.

21. Output signals are available at either the grid to ground or plate (or collector) to ground. As shown in the waveshapes of Fig. 6·14 and also Fig. 6·15*b*, the output signal voltage at the grid is a positive-going pulse with negative overshoot. At the plate a negative-going pulse with positive overshoot is available. A third winding on the transformer, called the "tertiary," shown in Fig. 6·15*a*, also provides a similar output voltage as the plate. Additional output pulses, without any overshoot, could be produced by adding a small resistor R_2 in the plate circuit. The pulse of plate current (Fig. 6·14) would produce a negative-going voltage pulse between the lower end of R_2 and ground (point X of Fig. 6·15*a* and *b*). The pulse of plate current (Fig. 6·14) would produce a positive-going voltage across R_3, a small resistor added in the cathode circuit. This is also shown in Fig. 6·15*a* and *b*.

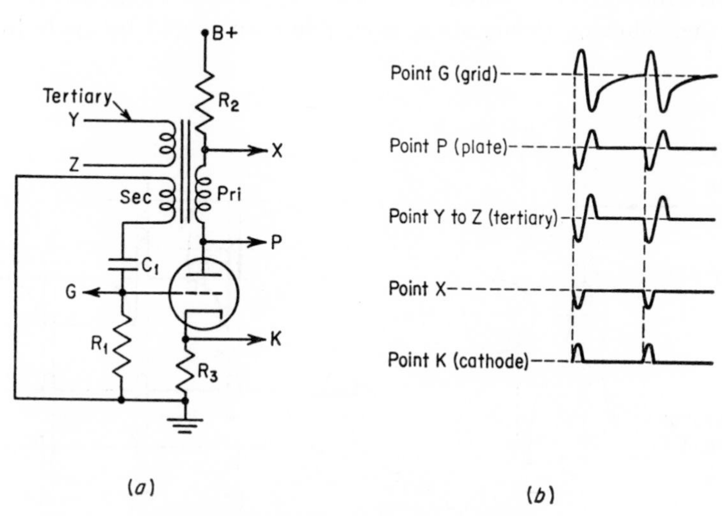

Figure 6·15 **Blocking-oscillator outputs.**

6·8 Synchronization

The astable blocking oscillator of Fig. 6·13 is a free-running device and requires no input signal to function. However, the frequency of operation is unstable, and an input signal is usually applied to lock or synchronize the blocking oscillator to that of some more stable signal. The input synchronizing pulse can be applied to the grid (or base) and must drive the grid (or base) positive. To be effective, the input-pulse frequency must either be slightly higher than the free-running blocking oscillator frequency or a multiple of this higher frequency. That is, if the free-running frequency of the blocking oscillator is 900 cps and a 1-kc pulse is applied, the blocking-oscillator frequency increases to 1 kc. If a 2-kc input signal were applied, only every *second* pulse would trigger the tube (or transistor) into conduction, and the oscillator frequency would become *half* that of the input signal or 1 kc. The result of the free-running and synchronized waveshapes would be similar to those of the astable multivibrator shown in Fig. 5·4.

6·9 Monostable Blocking Oscillator

The monostable blocking oscillator of Fig. 6·16*a* cannot operate without a trigger input pulse. It differs from the astable oscillator of Fig. 6·13 in that a negative d-c voltage is applied to the grid, keeping the tube permanently cut off. Only when a positive-going input trigger is applied to the grid will plate current flow. A negative-going trigger (not shown) could be applied to the plate, and because of the inversion of the transformer, produce the required positive-going pulse at the grid.

In the following theory discussion, reference should be made to Fig. 6·16.

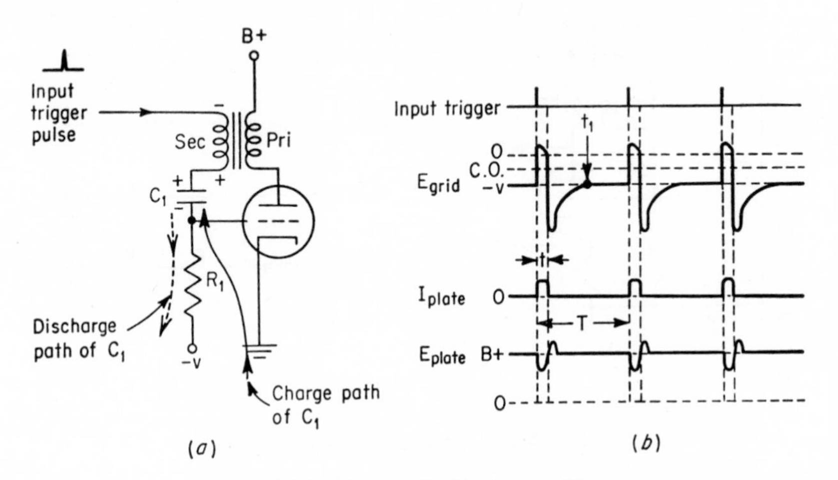

Figure 6·16 Monostable blocking oscillator.

1. The tube is inoperative due to the negative d-c voltage applied at the lower end of R_1. The circuit would remain in this state permanently if no input signal were applied.

2. Applying a positive-going input pulse to the grid causes plate current to flow through the primary, and a magnetic field builds up.

3. The expanding magnetic field cuts across the secondary, inducing a voltage in it.

4. Since the transformer is connected as a phase inverter, the induced secondary voltage has the polarity shown in Fig. 6·16, with the grid end positive.

5. This voltage causes C_1 to start charging up.

6. C_1 charging current takes electrons off the grid, making the grid positive.

7. The positive grid draws current from its cathode, and C_1 quickly charges through the low resistance between the cathode and the positive grid.

8. With a positive grid, plate current rises to saturation, and the magnetic field becomes stationary.

9. The stationary field no longer cuts the secondary, and induced voltage becomes zero.

10. C_1 now starts discharging down through R_1, as shown by the dotted-line arrow in Fig. 6·16, making the top of R_1 and the grid negative. The grid voltage is now the sum of the voltage across R_1 and the negative d-c voltage applied at the lower end of R_1.

11. The tube now cuts off, and the magnetic field collapses.

12. This induces a reverse polarity voltage in the secondary, with the grid end negative, driving the grid more negative.

13. The collapsing field shock excites the transformer into a high-frequency sine-wave oscillation. Because of the high losses in the transformer (low Q), these oscillations damp out after about one half cycle as shown in Fig. 6·16*b*.

14. As C_1 discharges, the negative voltage across R_1 decreases (see Fig. 6·16*b* and Fig. 3·1).

15. When C_1 has discharged completely, there is zero volts across R_1, and the grid voltage now becomes only that of the negative d-c applied at the lower end of R_1. This is shown at instant t_1 on the grid-voltage waveshape of Fig. 6·16*b*.

16. As stated previously, this voltage alone keeps the tube cut off.

17. The tube remains cut off, and the circuit nonoperating, until the next trigger input pulse sets off one more cycle of operation.

18. The width or duration of the plate signal pulse, period t of Fig. 6·16*b*, is independent of the narrow trigger input pulse, depending mainly upon the inductance of the transformer and the size of C_1.

19. The period of time, T, depends only on the frequency of the input trigger pulses.

6·10 Blocking Oscillator Sawtooth Generator

The output-signal waveshapes generated by a blocking oscillator are shown in Fig. 6·14 and in Fig. 6·15*b*. A sawtooth voltage can be produced by adding a capacitor (called the *discharge* or *sawtooth* capacitor), and a resistor in the plate (or collector) circuit of Fig. 6·13, modifying it as shown in Fig. 6·17*a* and *b*. C_2 is the sawtooth capacitor. When the tube (or transistor) is cut off, C_2 charges up through R_2, as shown by the solid-line arrows in Fig. 6·17*a* and *b*. Because of the large R_2-C_2 time constant, C_2 cannot charge up very much in the relatively short time that the stage is inoperative. As a result, E_{C_2} rises fairly linearly. When plate (or collector) current flows, C_2 discharges through the conducting tube (or transistor) as shown by the dotted-line arrows. The effect of the capacitor C_2 is to change the rectangular voltage, shown as the dotted-line E_{out} wave of Fig. 6·17, into the sawtooth shown as the solid-line E_{out} wave.

The sawtooth frequency and amplitude can be varied by making R_1 and R_2 adjustable. The R_1-C_1 time constant in the grid (or base) circuit determines how long the stage is held inoperative and therefore directly affects the *frequency*. A *smaller* value of R_1 allows C_1 to discharge more

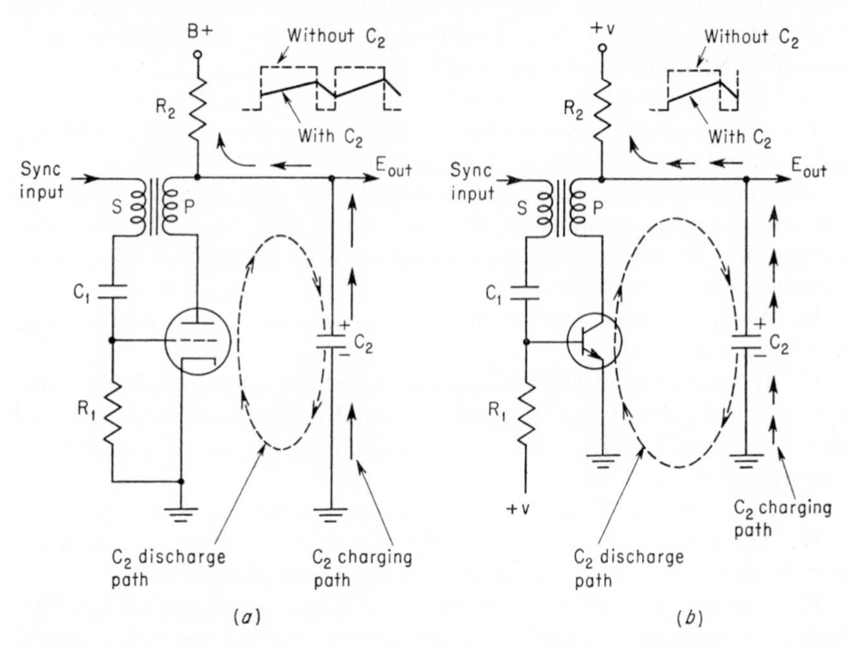

Figure 6·17 **Blocking-oscillator discharge stages. (*a*) Vacuum-tube circuit; (*b*) NPN-transistor circuit.**

quickly, keeping the stage cut off for less time, resulting in a *higher frequency*. A higher frequency, in turn, allows the sawtooth capacitor C_2 less time in which to charge up. Now C_2 can only charge to a smaller voltage. This results in a smaller amplitude sawtooth.

Changing R_2 will similarly affect both sawtooth *amplitude* and frequency. Making R_2 larger, for example, will cause the sawtooth capacitor C_2 to charge up more slowly. Now C_2 can only charge to a smaller value of voltage, resulting in a smaller amplitude wave. At the same time, a larger R_2 decreases plate (or collector) current. A smaller current flows through the transformer primary, inducing less voltage in the secondary. C_1 now charges to a smaller voltage. When C_1 discharges, therefore, a smaller negative voltage is produced across R_1 (see grid wave of Fig.

6·18). As a result, as C_1 discharges, the negative decreasing voltage across R_1 on the grid (or base), starting from a smaller negative value, reaches

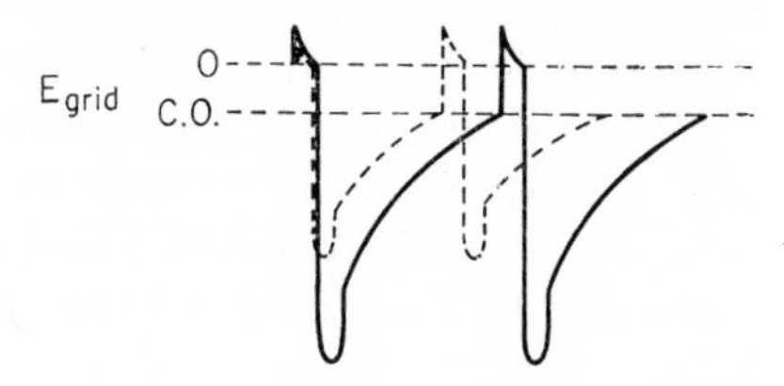

Figure 6·18 **Blocking-oscillator-grid waveshape. Smaller signal due to larger R_2 in Fig. 6·17a resulting in higher frequency is shown by the dotted-line waveshape.**

the cutoff level sooner. The next cycle starts earlier, resulting in an *increased frequency*.

6·11 Miller Integrator

General. A circuit capable of producing an extremely linear slope or sawtooth voltage is called the *Miller integrator*. Before going into the actual circuit shown in Fig. 6·21, a simple theoretical one is presented to show a method of producing a linearly sloping voltage waveshape.

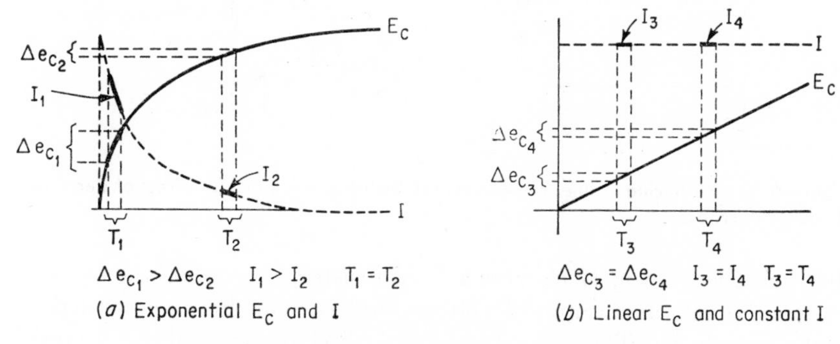

Figure 6·19 **Capacitor voltages and currents.**

When a capacitor normally charges or discharges through a resistor, it does this exponentially; that is, voltage across the capacitor (during the charge) rises rapidly at first, then less rapidly, and finally, very slowly. This is shown in the E_C graph of Fig. 6·19a. Charging current, I in Fig. 6·19a, is maximum at first and becomes progressively less and less. The magnitude of the current is given in the following relationship:

$$I = C\frac{\Delta e_C}{\Delta t} \qquad (6\cdot9)$$

where I = current, amp
C = capacitance, farads
Δe_C = *change* in capacitor voltage from one instant to the next
Δt = time duration between the instants

As shown in Fig. 6·19*a* on the E_C curve, the change of capacitor voltage (Δe_{C_1}) during the time period $T_1(\Delta t)$ is greater than that during the later but equal period T_2. As a result of this ($\Delta e_{C_1} > \Delta e_{C_2}$), the value of current I_1 is greater during T_1 than I_2 during T_2.

If the charging of a capacitor were linear, such as E_C in Fig. 6·19*b*, then $\Delta e_{C_3} = \Delta e_{C_4}$. Now the current, from Eq. (6·9,) would be *equal* during periods T_3 and T_4. As shown in Fig. 6·19*b*, the current I graph is therefore *constant* if E_C rises *linearly*. Conversely, if I is constant, then E_C must be linear.

The diagram of Fig. 6·20*a* shows a theoretical circuit which, because of the variable resistor R_2, could produce a linear sawtooth voltage. To begin with, assume that capacitor C has been previously charged to

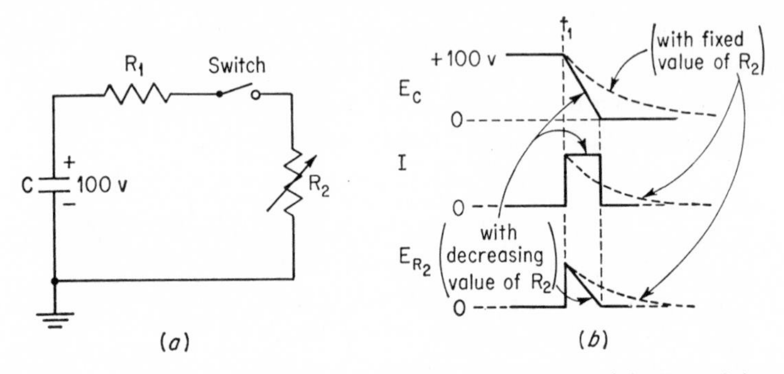

Figure 6·20 Producing a linear sawtooth. (a) Switch closes at t_1; (b) effect of decreasing R_2 as E_C decreases.

100 volts, as shown. If R_2 were a fixed resistor, then when the switch is closed, E_C acts as an applied voltage, and current flows up through R_2 and through R_1. The capacitor is now discharging, and E_C decreases exponentially as shown by the dotted-line curve of Fig. 6·20*b*. With a smaller voltage, current also decreases in the identical fashion (since $I = E/R$), as shown by the I graph dotted line. As a result, voltage across R_2, E_{R_2}, similarly decreases (since $E_R = IR$), as shown by the dotted-line graph in Fig. 6·20*b*. The dotted-line curves are the normal ones for a capacitor and resistor circuit.

If R_2 could be varied correctly, then as the capacitor discharges, a linear E_C decrease rather than an exponential one could be produced. As E_C decreases, if R_2 is decreased at the same rate, then the current I would remain *constant*, since $I = E/R$. Then, as stated previously, and illustrated in Fig. 6·19*b*, a constant I means that E_C is changing at a linear rate (or a constant rate of change). The constant I is shown by the solid-line graph of Fig. 6·20*b*, along with those of the linearly sloped E_C and the linearly sloped E_{R_2}. Voltage across R_2, E_{R_2}, is linear since it was assumed that R_2 is decreasing at the same rate as E_C. E_{R_2} is a negative-going (or decreasing) linear voltage since current I is constant, while R_2 is decreasing linearly (and $E_{R_2} = IR_2$).

Miller-integrator circuits. The diagram of the *Miller integrator* is shown in Figs. 6·21 and 6·23. A vacuum-tube circuit is shown in Fig. 6·21 and an NPN transistorized version in Fig. 6·23. The operation of the two are identical. The vacuum-tube circuit description follows. V_2 is the actual Miller integrator, while V_1 simply acts as a switch. Capacitor C_1 is the sawtooth capacitor, charging and discharging. The internal resistance of

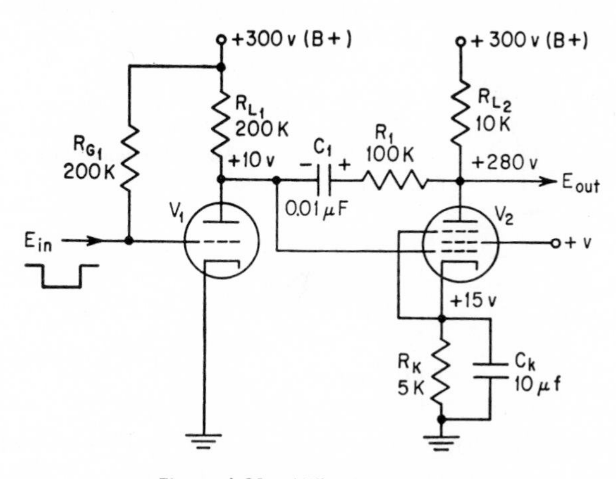

Figure 6·21 Miller integrator.

V_2 acts as the variable resistor R_2 did in the circuit of Fig. 6·20, decreasing in value as the capacitor discharges through it. The lowering of resistance as E_{C_1} decreases keeps the current constant. A constant current flow "through" a capacitor results in a linearly sloped voltage across the capacitor and across the linearly decreasing resistor.

The switch tube V_1 normally conducts heavily since its grid is connected to B+ through a large resistor R_{G_1} (about 200 kilohms). The grid clamps to about zero volts while the plate voltage is quite low, about +10 volts, due to the voltage drop across the large R_{L_1} (about 200 kilohms).

V_2 normally conducts only slightly because of its large cathode-bias circuit (R_K about 5 kilohms and C_K about 10 μf), with the cathode at about +15 volts while the grid is +10 volts. Plate voltage on V_2 is almost the full B+, or about +280 volts. Capacitor C_1, connected between the plate of V_1 (+10 volts) and the plate of V_2 (+280 volts), is charged up therefore to about 270 volts with the polarity shown in Fig. 6·21.

A negative-going gate applied to the grid of V_1 cuts off this tube, and its plate voltage immediately rises, as shown at instant t_1 in Fig. 6.22. This abrupt positive rise appears on the grid of V_2 (directly coupled) and also on the plate of V_2 (coupled through capacitor C_1). Resistor R_1, in series with C_1 between the two plates, will reduce or even eliminate the abrupt positive rise at the plate of V_2, if R_1 is made equal to $1/G_M$. The positive rise is cancelled at the plate of V_2 because of the pentode amplification and inversion between its grid signal and plate signal.

As V_1 is cut off, causing its plate voltage to rise, C_1 starts discharging. Discharge current flows up through R_{L_1} to B+ power supply, up through R_K and C_K, and then up through the pentode tube V_2, which has started

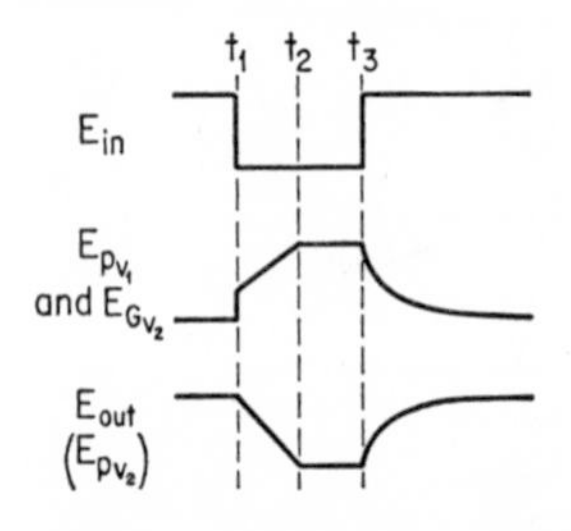

Figure 6·22 Miller integrator waveshapes of circuit of Fig. 6·21.

to conduct because of its positive-going grid. The rising voltage at V_1 plate is headed towards +300 volts and is slowed down by the discharging of C_1. This exponentially rising voltage at V_1 plate and at V_2 grid causes V_2 to conduct more heavily. As a result, the internal resistance of V_2 decreases at the same rate that E_{C_1} is decreasing. As explained previously in Fig. 6·20, this keeps the discharge current constant, making E_{C_1} fall linearly. Plate voltage of V_1 now rises toward +300 volts linearly instead of exponentially and so does the grid of V_2, as shown during the t_1 to t_2 period of Fig. 6·22. Resistance of pentode V_2 therefore decreases at this same linear rate. With a constant current flowing through V_2 and with its internal resistance decreasing linearly, voltage across the tube, E_{out}, is decreasing in the same fashion, as shown in Fig. 6·22. At the end of the negative-going input gate signal at instant t_3, V_1 again returns to its heavy conduction, and its plate voltage falls. C_1 now starts charging up again, and charging current flows up through V_1, through R_1, and up through R_{L_2} to B+. If the negative gate duration is longer than the discharge time of C_1, as shown in Fig. 6·22, then the capacitor completes its linear discharge (t_1 to t_2 time), and voltage waveshapes level off for the remainder of the gate time (t_2 to t_3 time).

Linear sawtooth sweep time is determined mainly by R_{L_1}, C_1, R_1, and V_2 internal resistance (the discharge path of C_1). C_1 is kept small to produce a fast retrace or recovery time (the charging path).

Discharge linearity of C_1 is enhanced because of the Miller effect of V_2 producing a much larger R-C discharge circuit. Disregarding the grid-to-plate interelectrode capacitance of V_2 (very small in a pentode) and also the stray wiring capacitance, which are small compared to C_1, the discharge capacitance is not C_1 alone but is effectively C_1 times the factor $1 + A$ (the Miller effect). The gain A can be substantial, resulting in a circuit discharge capacitor which is many times the value of C_1. A larger R-C results in a more linear discharge.

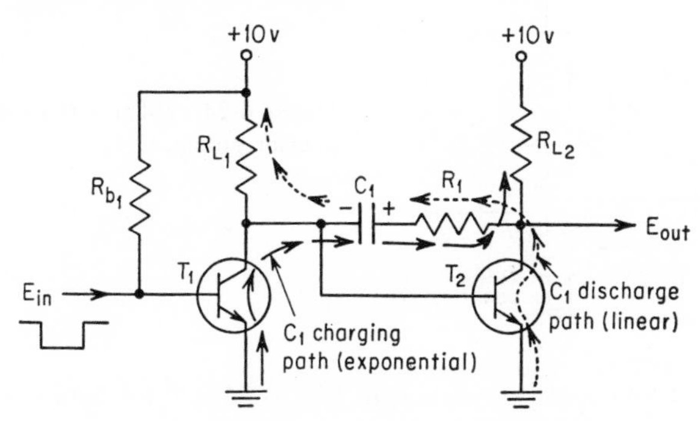

Figure 6·23 NPN-transistor Miller integrator.

The NPN-transistor Miller integrator of Fig. 6·23 is practically identical in operation to its vacuum-tube counterpart of Fig. 6·21. T_1 is the switching device, gated off by a negative-going wide pulse. T_1 normally conducts heavily due to the large forward bias voltage between its emitter and base. Collector voltage of T_1 is low, practically zero, making the base of T_2 this same small positive value. T_2 therefore conducts weakly, and its collector is practically the full +10 volts applied. C_1 charges to this difference of voltage between the collectors. Charging of C_1 is exponential and is, as shown by the solid-line arrows of Fig. 6·23, up through T_1, through R_1, and up through R_{L_2} to the applied +10-volt source.

The negative gate input cuts off T_1, and its collector voltage rises, causing C_1 to start discharging in the dotted-line arrow path of Fig. 6·23. The positive-going T_1 collector voltage drives the base of T_2 positive, and T_2 starts conducting more heavily. As C_1 discharges, the collector of T_1 and also the base of T_2 rise toward the +10 volts. T_2 conducts more heavily, and its internal resistance decreases. This decreasing resistance in the discharge path of C_1, as shown previously in Fig. 6·20, keeps the discharge current constant, and voltage across C_1 drops *linearly*. T_1 collector and T_2 base voltage likewise rise linearly, and T_2 collector voltage decreases at this same linear rate. Waveshapes are the same as shown for the vacuum-tube circuit and are those of Fig. 6·22.

Suppressor grid gated Miller integrator. Another Miller integrator sawtooth sweep generator using a single pentode tube is shown in Fig. 6·24, with the waveshapes shown in Fig. 6·25. The pentode used has a sharp suppressor grid-plate current cutoff characteristic. The suppressor grid is connected, as shown, to a negative d-c voltage sufficient to prevent plate current flow. With no voltage drop across R_{L_1}, plate voltage is at the

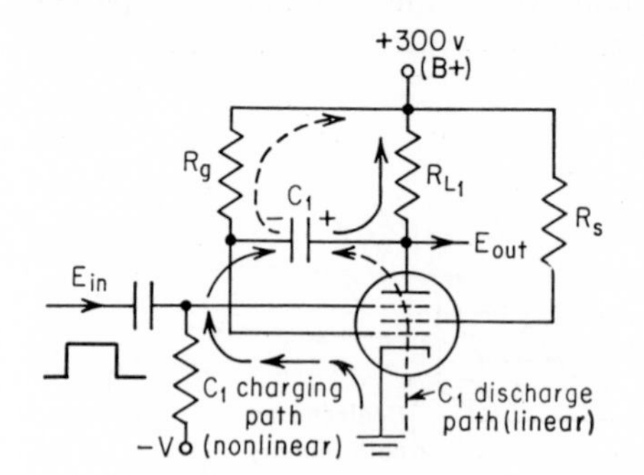

Figure **6·24 Miller integrator using suppressor gating.**

full applied +300 volts, as shown in Fig. 6·25 before instant t_1. The control grid is connected through R_G to B+ supply. Cathode-to-grid current flows, clamping the grid to a very small positive voltage, practically zero, also illustrated in Fig. 6·25 before instant t_1. Screen current flows, and screen voltage (not shown) is low. Capacitor C_1, connected between the grid (small + voltage) and the plate (full +300 volts), as shown in Fig.

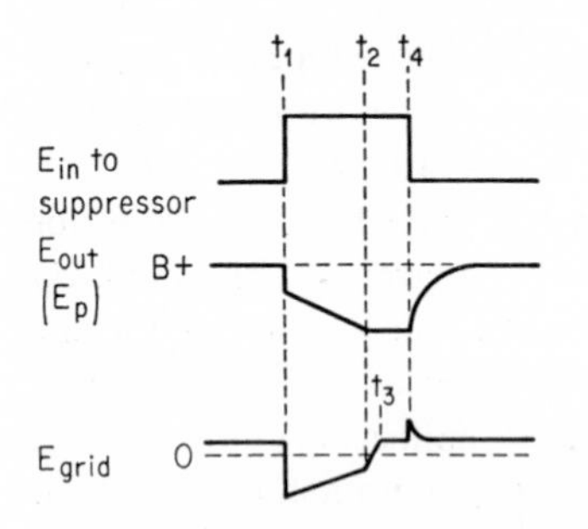

Figure **6·25 Waveshapes of suppressor gated Miller integrator circuit of Fig. 6·24.**

6·24, charges up to the difference of potential between these points. Charging path of C_1 is from cathode to grid, and up through R_{L_1}, as indicated by the solid-line arrows.

When a positive-going gate signal is applied to the suppressor, plate current flows. The drop across R_{L_1} causes plate voltage to fall abruptly, as shown at instant t_1 of Fig. 6·25. This instantaneous voltage decrease is coupled through C_1 to the grid, driving the grid negative. C_1 starts discharging. Discharge current flow, as shown by the dotted-line arrows of

Fig. 6·24, is up through the tube from cathode to plate and up through resistor R_G. As C_1 discharges, voltage across R_G decreases, and the negative grid heads back toward its clamped, slightly positive voltage. As grid voltage goes less negative, after instant t_1 in Fig. 6·25, plate current increases. The internal resistance of the tube, R_B, now decreases. This resistance is in series with C_1 discharge path. As shown previously in Fig. 6·20, a resistor decreasing as the capacitor voltage decreases keeps the discharge current constant. With a steady discharge current flowing, E_{C_1} discharges linearly. Grid voltage therefore rises linearly, as shown in Fig. 6·25, between t_1 and t_2, and tube resistance R_B decreases linearly. Voltage across the pentode, at the plate, with a linearly decreasing resistance and a constant discharge current flowing through it, is a linearly negative-sloping waveshape, as shown in Fig. 6·25, between instant t_1 and t_2.

When C_1 has completed its discharge, plate voltage no longer falls and levels off (instant t_2). Without discharge current flowing through R_G, the grid end of this resistor rises to the grid clamp voltage, and C_1 charges slightly through cathode to grid resistance and R_{L_1}. The grid rises to its clamp level during the t_2 to t_3 time of Fig. 6·25.

At the completion of the positive-gate input signal (instant t_4), to the suppressor grid, plate current again becomes cut off. Plate voltage rises sharply at instant t_4, and this is coupled through C_1 to the grid. C_1 now resumes charging up again, and pentode plate voltage rises exponentially, as shown after instant t_4.

6·12 Phantastron Circuits

A linear sawtooth voltage generator similar to the Miller sweep circuits of Sec. 6·11 is called the *phantastron*. A simple circuit is shown in Fig. 6·26, with the waveshapes for this circuit shown in Fig. 6·27. Besides the sawtooth waveshape being used as a sweep, the phantastron is also often used as a delay circuit, as will be shown. A narrow input pulse is employed to trigger the circuit which then is self-gating for a much longer period than the trigger-pulse duration. The duration of this self-gate signal is the delay time and is also the sweep duration time.

The theory of operation of the phantastron may seem slightly *fantastic*, which is actually the origin of its name. In the following discussion, reference should be made to the schematic and to the waveshapes (Figs. 6·26 and 6·27).

Suppressor-screen-coupled phantastron. The circuit is operated here as a suppressor-screen gated phantastron. Before an input pulse is applied, the control grid is clamped to almost zero (the cathode potential) since the grid is connected to the B+ supply through resistor R_G. The positive grid draws current from the cathode, becoming clamped to slightly above zero. Because of the voltage divider R_1, R_2, and R_3 connected between B+ and the negative voltage (with respect to cathode ground), the suppressor grid is sufficiently negative to prevent plate current flow, and plate voltage is at full B+. Capacitor C_1 charges up as shown by the solid-line arrows. The screen grid is connected to a positive

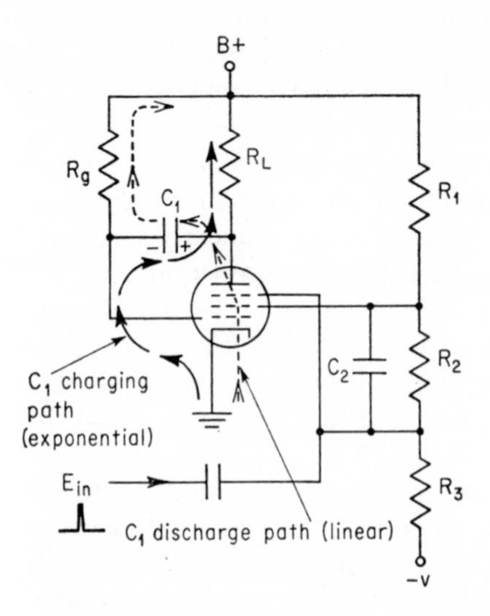

Figure **6·26 Suppressor and screen gated phantastron.**

voltage point on the resistor voltage divider, and screen current therefore flows from cathode to screen. Screen voltage is low.

At instant t_1 of Fig. 6·27, a positive-going pulse is applied to the suppressor grid. This permits plate current to flow, and plate voltage falls sharply, driving the control grid negative. Screen current is decreased,

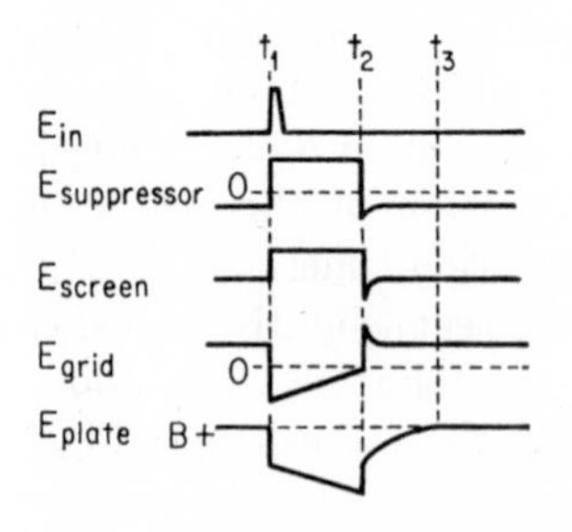

Figure **6·27 Waveshapes of suppressor and screen gated phantastron of Fig. 6·26.**

driving the screen voltage positive. This drives the suppressor positive also since the suppressor is coupled to the screen by C_2 and also via the common resistor voltage divider. The suppressor is now being driven positive even after the trigger pulse has ended. The falling plate voltage starts C_1 discharging. Discharge path is indicated by the dotted-line arrows in Fig. 6·26 and occurs through the tube from cathode to plate, "through" C_1 and up through R_G to the B+ power source. This current through R_G, as C_1 discharges, decreases the negative voltage on the grid, and plate current increases. The internal tube resistance R_B is now decreasing. As shown previously in Fig. 6·20, a decreasing resistance in a capacitor discharge path keeps the current steady even though E_C is decreasing. A constant discharge current means that E_{C_1} is decreasing *linearly*, and voltage across the tube (plate voltage) is similarly running downward at a linear slope. This is a typical Miller sweep operation, and the waveshapes are shown during the t_1 to t_2 period of Fig. 6·27.

When the grid voltage rises to slightly less than zero (just before instant t_2) and C_1 has almost discharged, plate voltage is at its lowest possible value and is said to have "bottomed." Since plate voltage is just about constant now, C_1 stops discharging, driving the grid slightly positive again at instant t_2. Screen current starts increasing, and screen voltage falls abruptly (instant t_2). This causes the suppressor grid voltage to go negative, reducing plate current. Plate voltage starts rising, and C_1 starts charging again, causing the grid to go positive. Screen current increases, and screen voltage falls further, driving the suppressor more negative and cutting off plate current. Plate voltage rises, causing C_1 to charge exponentially (after instant t_2) through the solid-line arrow path of Fig. 6·26. Grid voltage rises above its normal clamp level and then returns to this level after C_1 has charged. The positive overshoot of grid voltage at t_2 causes a momentary rise of screen current with its resultant negative-going undershoot at the screen and also at the suppressor to which it is coupled.

The plate voltage waveshape of Fig. 6·27 between instants t_1 and t_2 is used as the linear sweep output signal. The gate signal on the screen (and the suppressor) may be used as the delay-period output signal. Recovery time for the sawtooth is shown as the time between t_2 and t_3. Sweep time can be varied by changing the value of B+, R_G or C_1.

Cathode-coupled phantastron. Another version of a phantastron circuit is shown in Fig. 6·28, with its waveshapes in Fig. 6·29. In this discussion, reference should be made to these diagrams. Control-grid voltage is clamped to the positive voltage of the cathode. The grid is connected to B+ through R_G, making the grid draw current from the cathode. Cathode voltage, because of the large resistor R_K (about 10 kilohms), is positive with respect to ground. Screen grid is connected through R_S to B+, and heavy screen current flows, making the screen voltage much less than B+. This is shown before instant t_1 in Fig. 6·29. The suppressor grid is connected to the resistor voltage divider, R_1, R_2, R_3, R_4, and R_5, and the suppressor tap is such that its voltage is much less positive than the cathode. The suppressor therefore appears *negative* with respect to the cathode, preventing any plate-current flow. Plate voltage is not at full B+ due to the *plate-catching diode* D_1. The cathode voltage of diode D_1 is E_1, which can be varied by the setting of the arm of R_2 which is part of the voltage divider R_1 through R_5. With no pentode plate current, the plate voltage of the pentode V_1 should become the full applied B+. However, this is greater than voltage E_1 on the diode cathode, and D_1

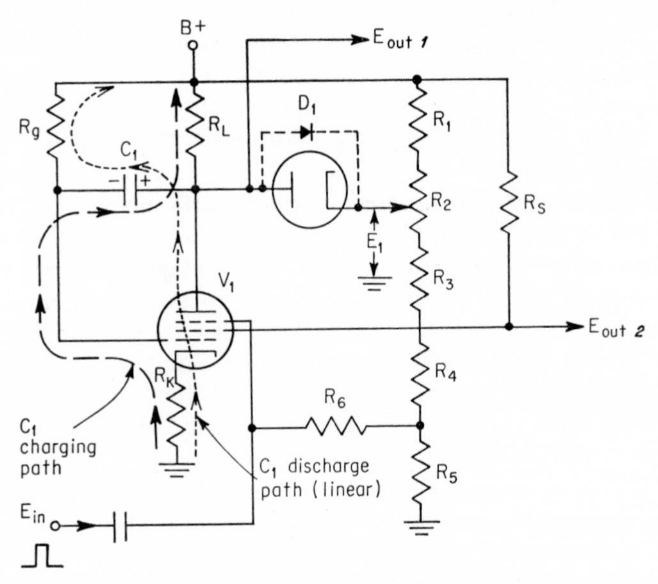

Figure 6·28 **Cathode-coupled phantastron.**

conducts. This clamps the diode plate, and the pentode plate, to just about voltage E_1, preventing the plate from going above this value. Capacitor C_1 charges up through cathode-to-grid current and then up through resistor R_L to B+.

At instant t_1 of Fig. 6·29, a narrow positive-going input pulse is applied to the suppressor grid. Plate current now flows, and plate voltage falls sharply. This abrupt drop is coupled to the grid, driving the grid negative, and cathode current decreases. Cathode voltage likewise decreases. The fact that plate current has just started, despite the negative-going grid and decrease of cathode current, seems irreconcilable. However, this is possible because screen current now decreases. The new plate current therefore results from the plate drawing current that previously went to the screen. The decrease of screen current causes screen voltage to rise sharply.

Plate current continues to flow even after the completion of the narrow positive input pulse at the suppressor since the falling or negative-going cathode now makes the suppressor appear to be more positive. Plate current flow and the sharp drop in plate voltage at instant t_1 start C_1 discharging. Discharge current flows up through R_G to B+, then from B− (or ground) up through R_K, and up through the pentode from cathode to plate. As C_1 discharges, voltage across R_G decreases, and the negative grid voltage decreases as the grid heads back towards a positive value. The positive-going grid causes plate current to rise, decreasing the pentode internal resistance R_B. As shown in Fig. 6·20, a decreasing resistor in

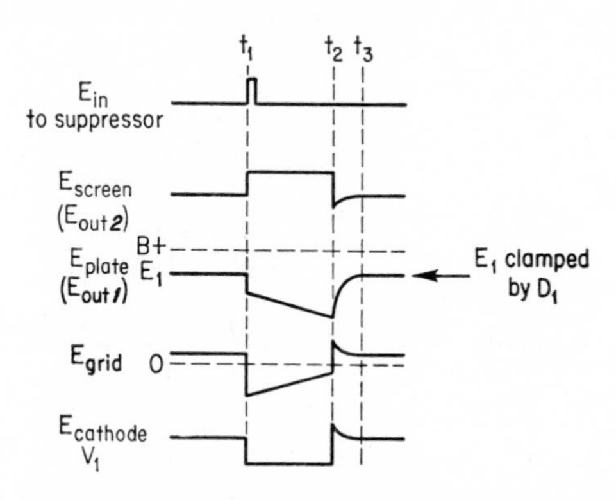

Figure 6·29 **Waveshapes of cathode-coupled phantastron of Fig. 6·28.**

series with a capacitor discharge results in a linear sawtooth voltage across the decreasing resistor and a linear voltage discharge of the capacitor. Pentode plate voltage during the t_1 to t_2 period of Fig. 6·29 is therefore a linear downward-sloping voltage. Just before instant t_2 grid voltage is only slightly negative, and C_1 has almost discharged. Plate voltage has run down to its lowest value and is said to have "bottomed." With a steady plate voltage, C_1 stops discharging, driving the grid positive at instant t_2. The sharp rise of grid voltage is coupled immediately to the plate. The positive grid increases cathode current, and the cathode voltage goes positive because of the larger voltage now appearing across R_K. The positive-going cathode now makes the suppressor appear to be negative-going, cutting off plate current. As a result, screen current increases, and screen voltage falls. Without plate current, plate voltage rises, and C_1 charges up exponentially. The grid is driven positive at instant t_2 above its normal clamped value by the charging of C_1. The positive grid overshoot results in a similar overshoot at the cathode, making the suppressor appear to go even more negative, turning back more electrons to the screen. Screen current rises slightly, causing a small negative-going undershoot of screen voltage just after instant t_2 of Fig. 6·29. The charging of C_1 exponentially raises the plate voltage of the pentode towards B+. When it exceeds the voltage E_1 at the cathode of the plate-catching diode D_1, conduction of the diode occurs, clamping the pentode plate at just slightly above E_1 at instant t_3.

The sweep time is the linear rundown of pentode-plate voltage during the t_1 to t_2 period. Recovery time is the exponential rise of plate voltage during the t_2 to t_3 period. The rectangular voltage at the screen could be

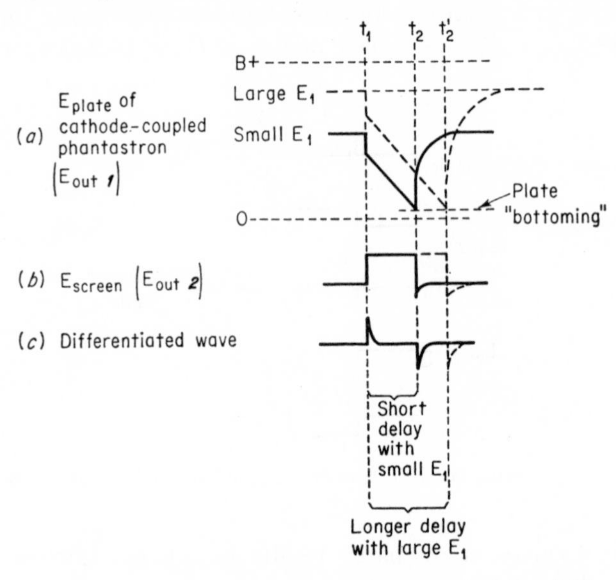

Figure 6·30 **Effect on time delay of varying E_1 of Fig. 6·28.**

used as a delay period between instant t_1 and instant t_2. If the screen rectangular wave were applied to a differentiator *R-C* circuit (short *R-C* time constant), it would result in a sharp positive spike at instant t_1, followed by a sharp negative spike of voltage at t_2, as shown in Fig. 6·30*b* and *c*.

Variable delay time. The delay period between t_1 and t_2 can be varied by means of potentiometer R_2 of Fig. 6·28. R_2 varies the voltage E_1 to which the pentode-plate voltage becomes clamped on its most positive excursion. As shown in Fig. 6·30*a*, a larger value of E_1 clamps the pentode plate at this more positive voltage. The linear plate voltage rundown is due to the discharging of capacitor C_1 through R_G and the decreasing pentode R_B of Fig. 6·28. With a larger value of E_1 due to the setting of R_2, the sloping plate waveshape starts at a higher value, but the *slope* of the run-down remains the same since neither the resistance nor the capacitance has been changed. The value of the plate voltage at its "bottoming" or minimum value remains the same. As a result, if the plate voltage starts at a higher value of E_1, as shown by the dotted-line waveshape of Fig. 6·30*a*, it runs down to the bottom taking a longer time (t_1 to t_2') than the solid-line waveshape did (t_1 to t_2) for the smaller value of E_1. The rectangular screen voltage is shown by the solid-line drawing for the smaller E_1 and by the dotted-line wave for the larger E_1 in Fig. 6·30*b*.

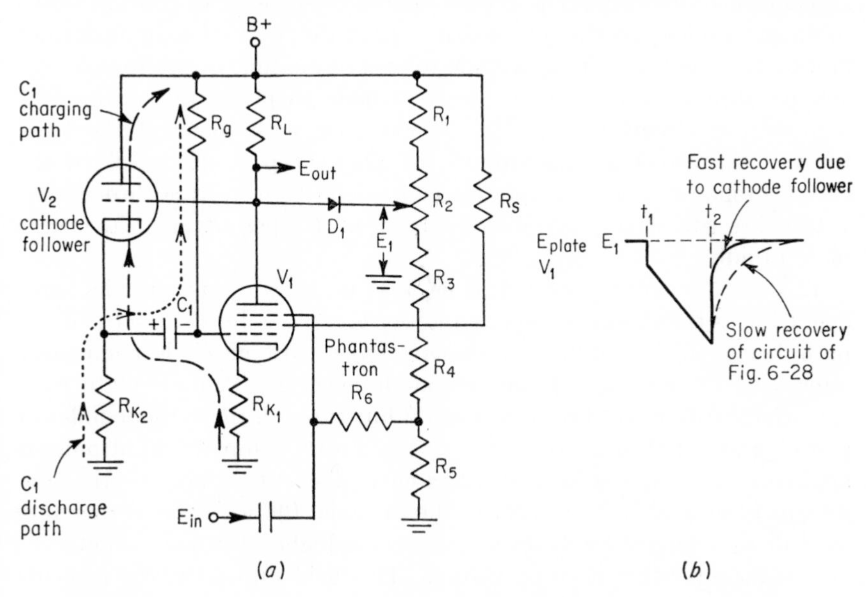

Figure 6·31 Fast-recovery cathode-coupled phantastron with cathode follower.

The differentiated screen voltage is shown as the spikes in Fig. 6·30*c*. As indicated, the time delay between the positive spike and the negative one is longer (t_1 to t_2') for the larger E_1.

Fast recovery. The recovery time of the phantastron circuit of Fig. 6.28 can be decreased by the addition of a *cathode follower* V_2 as shown in Fig. 6·31*a*. The recovery time is shown in the plate-voltage waveshape of Fig. 6·29 between instant t_2 and instant t_3 and is due to the charging of C_1 in Fig. 6·28 as indicated, up through R_K, from cathode to grid, "through" C_1, up through R_L to B+ power supply.

The *fast-recovery phantastron* of Fig. 6·31*a* produces the same waveshapes as the phantastron of Fig. 6·28 in essentially the same manner. The fast recovery time is due to the plate voltage of V_1 in Fig. 6·31 being able to rise (at instant t_2) rapidly without having to charge up C_1 as was required in previous circuits. When plate voltage of pentode V_1 rises, it is free to do so, uninhibited by any capacitor except tube interelectrode and stray wiring capacitance. The fast rise (fast recovery) time is shown after t_2 in Fig. 6·31*b*. The increase of V_1 plate voltage drives the grid of the cathode follower V_2 more positive. V_2 conducts more heavily, and voltage across its cathode resistor R_{K_2} rises, and the cathode becomes more positive. This is typical of a cathode follower in that a positive-going grid produces a positive-going cathode, and a negative-going grid causes the cathode to do likewise. The positive-going cathode of V_2 charges up C_1 in the path shown by the solid-line arrows of Fig. 6·31*a*. Charging current flows up through R_{K_1} from cathode to grid of pentode V_1, "through" C_1, and up through the triode V_2 from cathode to plate. In the previous phantastron circuit (Fig. 6·28), C_1 charging path included the large resistor R_L (about 1 megohm) in the plate of pentode V_1. With the cathode follower V_2, C_1 charges through triode V_2 instead of through R_L. Internal resistance of V_2 is much smaller than R_L, permitting C_1 to charge more rapidly.

The fast recovery phantastron of Fig. 6·31*a* is essentially the same operation as the slower recovery circuit of Fig. 6·28. Besides the C_1 charge path being slightly different, the *discharge* path of C_1 is completely altered by the addition of the cathode follower V_2. Pentode plate V_1 is essentially still coupled back to its grid but through the cathode follower and C_1 and not through C_1 alone. In Fig. 6·31*a*, before an input pulse is applied to the suppressor grid, plate current is cut off because the suppressor is negative with respect to the cathode. Plate voltage is therefore the full E_1, clamped by diode D_1 from going higher. Cathode follower V_2 grid is at this same positive voltage. This grid draws current from its cathode, and cathode and grid of V_2 clamp at this voltage. The left side of C_1 is therefore at the same potential as the plate of pentode V_1. The right side of C_1, of course, is at V_1 grid potential. When V_1 plate voltage decreases at instant t_1, V_2 positive-grid voltage decreases, and V_2 conducts less. Through cathode follower action, V_2 cathode becomes less positive. C_1 starts discharging, and discharge current flows up through R_G to B+ from B− (ground) up through R_{K_2} to C_1, as shown by the dotted-line arrows of Fig. 6·31*a*.

Certain pentode tubes such as the 6AS6, having a sharp suppressor grid-plate current cutoff, and the pentagrid converter tube 6SA7 are ideally suited in phantastron circuits.

PROBLEMS

6·1 Draw the schematic of a thyratron relaxation oscillator.

6·2 (*a*) Using a smaller *R-C* time constant in the circuit of the previous problem has what effect on the sawtooth frequency?

(*b*) Explain how an input signal to the grid can synchronize the sawtooth frequency.

6·3 Draw the schematic of a thyratron driven sawtooth generator.

6·4 Explain the difference between a *synchronized* sweep and a *driven* sweep.

6·5 (*a*) A thyratron relaxation oscillator has a recurrent frequency of approximately 900 cps. What is its frequency if a small input pulse is applied having a frequency of 1 kc?

(*b*) If the pulse frequency is raised to 3 kc, what is the thyratron frequency?

(*c*) If the thyratron sawtooth generator is a driven circuit, what is its frequency when the input pulse is 3 kc?

6·6 Draw the schematic of a driven vacuum-tube sawtooth generator.

6·7 (*a*) Draw the plate-voltage waveshape of the circuit of the previous problem (6·6).

(*b*) If the input to the grid of a *driven* sawtooth generator is a pulse 50 μsec in duration, what is the duration of the time-base line?

6·8 Draw the schematic of a driven sawtooth generator using a PNP transistor.

6·9 Draw the following voltage waveshapes of the circuit of the previous problem (6·8) one beneath the other in correct phase relationship: (*a*) E_{in} to base; (*b*) E_{out} from collector.

6·10 Draw the schematic of an astable blocking oscillator using a PNP transistor.

6·11 Draw the schematic of a monostable blocking oscillator using a PNP transistor.

6·12 Draw the following voltage waveshapes of an astable blocking oscillator-discharge tube, showing one wave beneath the other in proper phase relationship: (*a*) grid voltage; (*b*) output signal across the discharge capacitor.

6·13 Draw the schematic of an astable emitter-coupled multivibrator-discharge stage using PNP transistors.

MORE CHALLENGING PROBLEMS

6·14 Refer to the diagram of Fig. 6·12*b*. If PNP transistors were employed here instead of the NPN types, draw the output voltage waveshape at the collector of T_2 with C_1 in the circuit.

6·15 In the circuit accompanying this problem, $R_1 = R_2$ at the start. At instant (*a*) the switch is closed, and capacitor C_1 starts charging. At instant (*b*), C_1 has charged up to half of the applied voltage. At instant (*c*), C_1 has become completely charged. Draw the waveshapes of E_{C_1}, current I, E_{R_2}, and E_{R_1} with values just before instant (*a*), at instant (*a*), at instant (*b*), and at instant (*c*) if the value of R_2 is being *decreased* exactly in step with the *increasing* E_{C_1}.

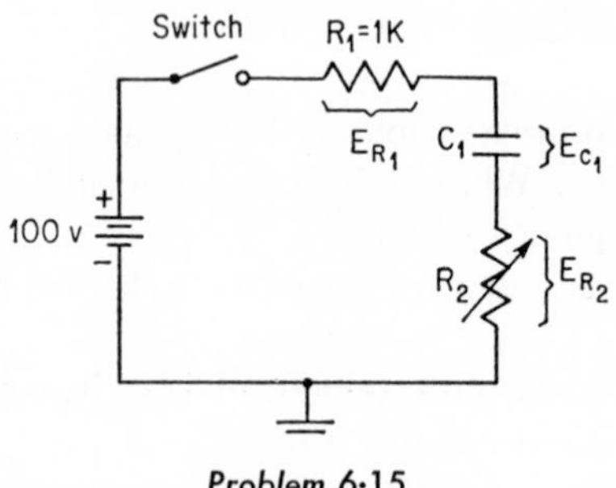

Problem 6·15

6·16 The plate-voltage waveshape of a sawtooth generator circuit is shown in the accompanying diagram. If the sawtooth capacitor is 0.005 μf and charges through a 250-kilohm resistor, find (*a*) the time of one cycle (neglecting the retrace time); (*b*) the frequency of the sawtooth.

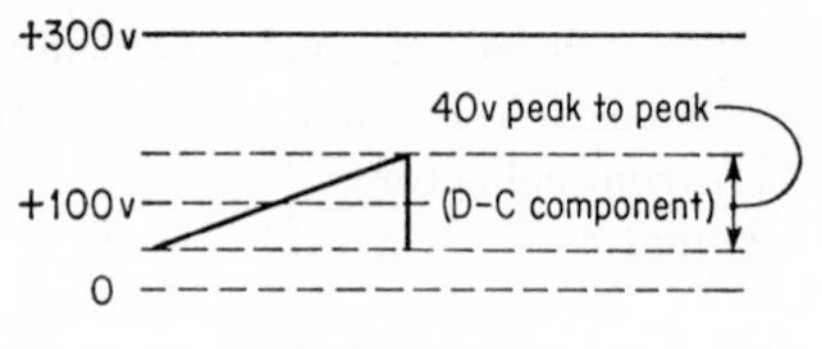

Problem 6·16

6·17 In the diagram of Fig. 6·12*a*, what effect would making R_1 smaller have on the sawtooth output waveshape?

6·18 In the phantastron circuit of Fig. 6·28, what would happen to the sweep duration and to the delay time if diode D_1 should become inoperative?

CHAPTER 7

BINARY AND OCTAL SYSTEMS AND ELECTRONIC COUNTERS

7·1 Introduction

Electronic counters make extensive use of bistable multivibrators, which are described in Chap. 5, Sec. 5·9. This multivibrator circuit produces one complete output cycle for every two input signals and is called a binary circuit. A system of numbers which uses the number 2 as its base is called binary. The number system that we are familiar with uses 10 as its base and is called the decimal system. Most electronic counters use the binary system since it is directly related to the operation of bistable multivibrators.

7·2 Decimal System

The following discussion of decimal system numbers, obvious though it may be, is presented in order to establish the idea of the *base* number. This will then make the binary system much easier to understand.

The number 469. is read as "four hundred and sixty-nine." The first digit at the left, 4, is called "four hundred" because it is in the third position or column to the left of the decimal point. This column is the "hundreds" column, as shown in Fig. 7·1. Since 100 is 10×10 or two 10s multiplied together, 100 could be written as the number 10 raised to the second power (exponent), or 10^2. The digit 4 in this column really means 4×10^2, or 4×100, or 400.

The digit 6 in the example of 469. is read as "sixty" since the 6 is in the second column to the left of the decimal point which is the "tens" column, as shown in Fig. 7·1. Ten, of course, is only one 10 and could be written as 10^1. The digit 6 in this second column then is really 6×10^1, or 6×10, or 60.

The digit 9 in the first column to the left of the decimal point is simply "nine" since it is in the "units" column as shown in Fig. 7·1. This first column, or "units" column or "ones" column, could be written as 10 to the zero power or 10^0 since 10 or *any* number raised to the zero power is always equal to 1. The 9 in this first column is then 9×10^0, or 9×1, or simply 9. The entire number 469. is then $9 + 60 + 400$, or a total of 469.

In the decimal system, as shown in Fig. 7·1, each column has a value of 10 to some power with the first column being 10^0 (or 1), the second column being 10^1 (or 10), the third column 10^2 (or 100), etc. The number 10 is the base number. As you know, in our decimal system we have ten

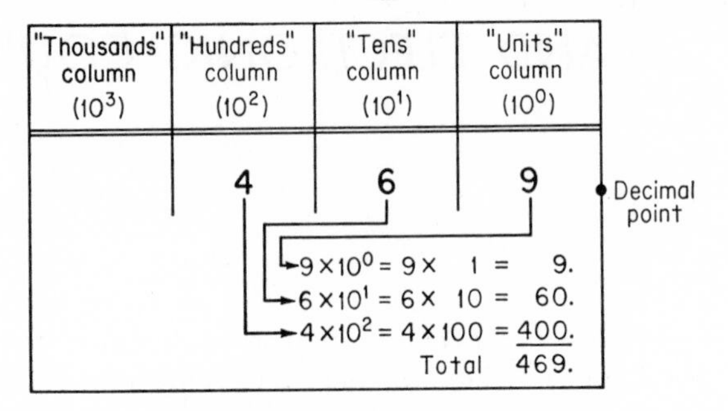

Figure 7·1 Decimal system.

numerical symbols, these being 0, 1, 2, 3, 4, 5, 6, 7, 8, and 9, hence the name "decimal," which means ten.

7·3 Binary System

The binary system uses the number 2 as its base. That is, the value of each column, as shown in Fig. 7·2, is 2 raised to some power. This is similar to the decimal system where 10 is the base number and where each column is 10 raised to some power. In the binary system, only two digits or symbols are possible, either 0 or 1.

The *first* column to the left of the binary point (which corresponds to the decimal point) is the units column, or 2^0, or a column value of 1. The digit 0 in this column, as shown in Fig. 7·2*b*, means that it has a value of zero. That is, $0 \times 2^0 = 0 \times 1 = 0$.

The digit 1 in this first column, as shown in Fig. 7·2*a* and *c*, means that it has a value of 1. That is, $1 \times 2^0 = 1 \times 1 = 1$.

The *second* column to the left of the binary point has a column value of 2, or 2^1. The digit 0 in this column (Fig. 7·2*c*) means that it has a value of 0; that is $0 \times 2^1 = 0 \times 2 = 0$.

A 1 in this second column (Fig. 7·2*a* and *b*) has a value of two; that is, $1 \times 2^1 = 1 \times 2 = 2$.

The *third* column to the left of the binary point has a column value of 4, or 2^2, as shown in Fig. 7·2. A 0 in this third column has a value of 0;

that is, $0 \times 2^2 = 0 \times 4 = 0$. The digit 1 in this third column (Fig. 7·2*c*) has a value of 4; that is, $1 \times 2^2 = 1 \times 4 = 4$.

The fourth column to the left of the binary point has a column value of 8, or 2^3, as shown in Fig. 7·2. (Note that 2^3 is three 2s multiplied together, $2 \times 2 \times 2 = 8$, and *not* 6.) A 0 in this column, as in *any* column, has a value of zero; that is $0 \times 2^3 = 0 \times 8 = 0$. A 1 in this fourth column (Fig. 7·2*b* and *c*) has a value of 8, or $1 \times 2^3 = 1 \times 8 = 8$.

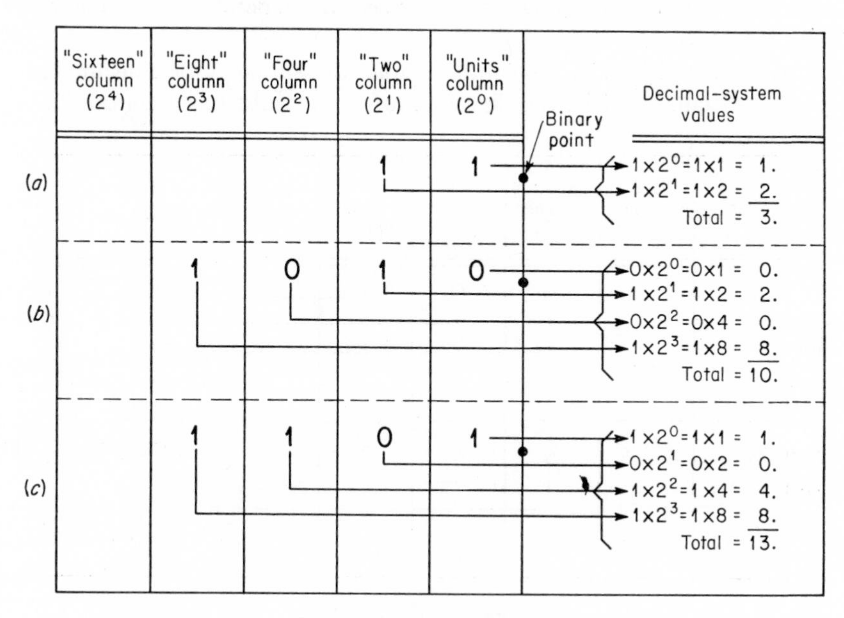

Figure 7·2 Binary system.

Each successive column going to the left *doubles* in value over the previous column. Therefore, the fifth column has a column value of 16 (Fig. 7·2). The sixth column, not shown in the figure, has a value of 32 or 2^5; the seventh column a value of 64 or 2^6, etc.

In the examples shown in Fig. 7·2, the first one (*a*) is 11. This is *not* read as "eleven" as it would be in the decimal system; instead, it should be read as simply "one, one." When writing this number, the base 2 could be added as a subscript to indicate that it is a binary number, such as $11._2$. As shown in Fig. 7·2*a*, this number $11._2$ has a total value of 3. This number 3, is, of course, not a binary symbol. It is the decimal number 3 and could be written to signify the base 10 (decimal system) or $3._{10}$.

In the second example, shown in Fig. 7·2*b* as 1010. (to be read as "one, zero, one, zero"), its value decimally is 8 + 2 or 10. Therefore, the binary number 1010. = the decimal number 10, or $1010._2 = 10._{10}$.

The third example, Fig. 7·2c, 1101. (to be read "one, one, zero, one"), has a decimal value from left to right of $8 + 4 + 0 + 1 = 13$. Therefore, $1101._2 = 13._{10}$.

7·4 Octal System

In the octal system, the number 8 is used as the base instead of the 10 used by the decimal system (Sec. 7·2) and the 2 used by the binary system (Sec. 7·3). Eight numeric symbols are used in the octal system. These are 0, 1, 2, 3, 4, 5, 6, and 7.

As shown in Fig. 7·3, the column values are similar to the decimal and binary systems; that is, the first column at the right has a column value

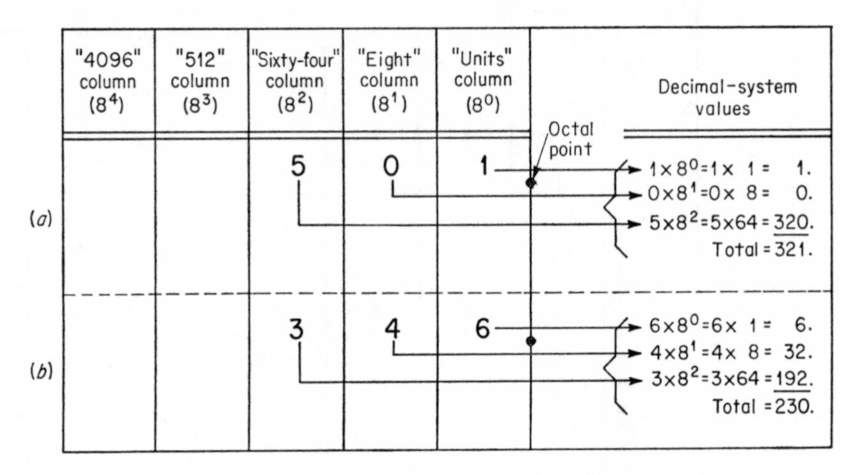

Figure 7·3 Octal system.

of the base number raised to the zero power. In the case of the octal system, the first column value is 8^0 or 1. The second column value is 8^1 or 8. The third column value is 8^2 or 64. Each successive column value is eight times the previous column.

The numbers in the columns may be any of the 8 symbols from 0 through 7. For example, if the number is 6 in the first or right-hand column (Fig. 7·3b), the value is 6 or $6 \times 8^0 = 6 \times 1 = 6$.

A 4 in the second column (Fig. 7·3b) has a value of 32 or

$$4 \times 8^1 = 4 \times 8 = 32.$$

The digit 3 in the third column (Fig. 7·3b) has a value of 192 or

$$3 \times 8^2 = 3 \times 64 = 192.$$

The example shown in Fig. 7·3b, 346 (to be read as "three, four, six"), is the octal number 346 which can be written as $346._8$. This has a left-to-right decimal-system value of $3 \times 8^2 + 4 \times 8^1 + 6 \times 8^0$ or $192 + 32 + 6$, or 230. Therefore, $346._8 = 230._{10}$.

The example shown in Fig. 7·3*a*, 501 (to be read "five, zero, one"), has a left-to-right decimal value of $5 \times 8^2 + 0 \times 8^1 + 1 \times 8^0$, or

$$320 + 0 + 1 = 321.$$

Therefore, $501._8 = 321._{10}$.

7·5 Converting Decimal, Binary, and Octal

A decimal number can be converted into its binary equivalent or into the octal equivalent, and any one of these can be converted from one into the other. The discussions of decimal, binary, and octal of Secs. 7·2 to 7·4 gave examples of each of these numbering systems in relation to the decimal system. In this section several methods of conversion are discussed.

Decimal-to-binary conversion. One method is to select binary column values, the sum of which is equal to the decimal number to be converted. The digit 1 is placed in each column which has been selected, and a zero is placed in each of the other columns. As an example, convert the decimal number 27 to the binary equivalent. As shown in Table 7·1, the sixth

Table 7·1 Column values of decimal, binary, and octal systems

System	7th column	6th column	5th column	4th column	3rd column	2nd column	1st column
Decimal	(10^6) 1,000,000.	(10^5) 100,000.	(10^4) 10,000.	(10^3) 1000.	(10^2) 100.	(10^1) 10.	(10^0) 1.
Binary	(2^6) 64.	(2^5) 32.	(2^4) 16.	(2^3) 8.	(2^2) 4.	(2^1) 2.	(2^0) 1.
Octal	(8^6) 262,144.	(8^5) 32,768.	(8^4) 4096.	(8^3) 512.	(8^2) 64.	(8^1) 8.	(8^0) 1.

column in the binary system has a value of 32. Since this is larger than the decimal number 27, go to the next smaller column (fifth column), which does not exceed the 27. This column value is 16. The first digit of the binary equivalent therefore has a 1 in this fifth column as shown in Fig. 7·4*a*. Subtract the 16 from the 27, leaving 11. Now go to the next lower column value which does not exceed the 11. In this example, it's the fourth column with a value of 8. As shown in Fig. 7·4*a*, the second digit of the binary number also has a 1 in this fourth column. Subtract the

8 (column value) from the 11, leaving 3. Since the next column (third column) has a value of 4 and it is larger than 3, this column cannot be used, and a zero is placed in this position of the binary number (Fig. 7·4*a*). The next column (second column) has a value of 2. Since this is smaller than 3, this column can be used. A 1 is therefore placed in this position of the binary number. The column value of 2 is subtracted from the 3, leaving a difference of 1. The last column (first column) has a value of 1, and this column is used. A 1 is placed in this position of the binary

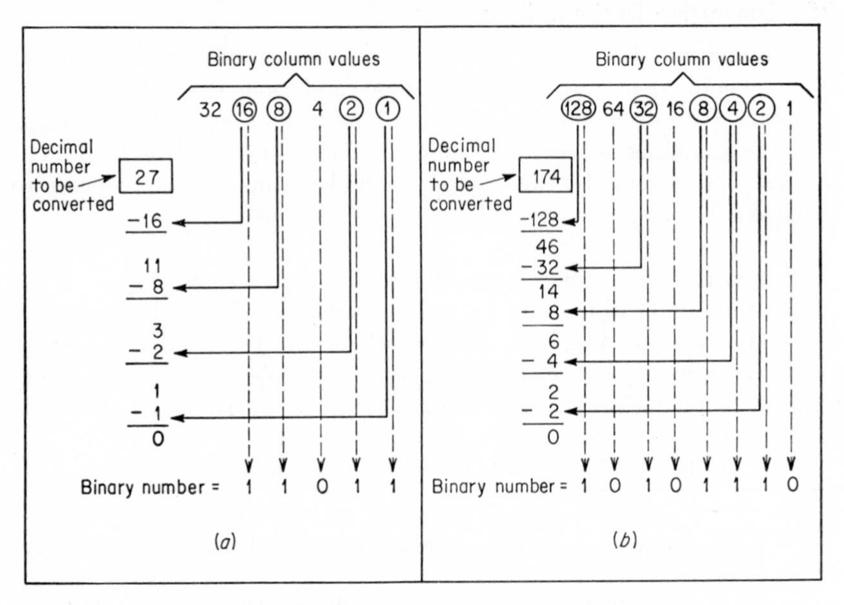

Figure 7·4 **Decimal-to-binary conversion.**

number. The column value of 1 is subtracted from the difference 1, leaving nothing. As shown in Fig. 7·4*a*, the selected binary column values are those which are encircled (16, 8, 2, and 1), and a 1 is placed in each of the *selected* columnar positions. The binary number, as shown, is 11011. Therefore $27._{10}$ is equal to $11011._{2}$.

Another example is shown in Fig. 7·4*b* where the decimal number 174 is to be converted into the equivalent binary. As shown, the selected binary column values are the encircled ones (128, 32, 8, 4, and 2, which add up to 174). The binary number contains the digit 1 in each of the *selected* columnar positions and is 10101110. For ease of reading, these digits may be separated into groups of threes, starting at the right (at the binary point) and working to the left. One or two zeros, to complete a group of three, may be placed in front of the group at the left, if that group only has two or one digits. The binary answer in Fig. 7·4*b*, 10101110., may then be written as 010 101 110. with the first zero at the left having no significance other than to complete a grouping.

Another method of converting from decimal to binary, preferred by some for its simplicity, is called by the colloquial name of *dibble-dabble.* This consists simply of dividing the decimal number by 2 and noting the remainder. (The remainder will always be 1 if the decimal number is odd, and the remainder will always be 0 if the decimal number is even.) The quotient resulting from the division by 2 is then divided by 2 again. Again, note the remainder. This division by 2 is repeated until the

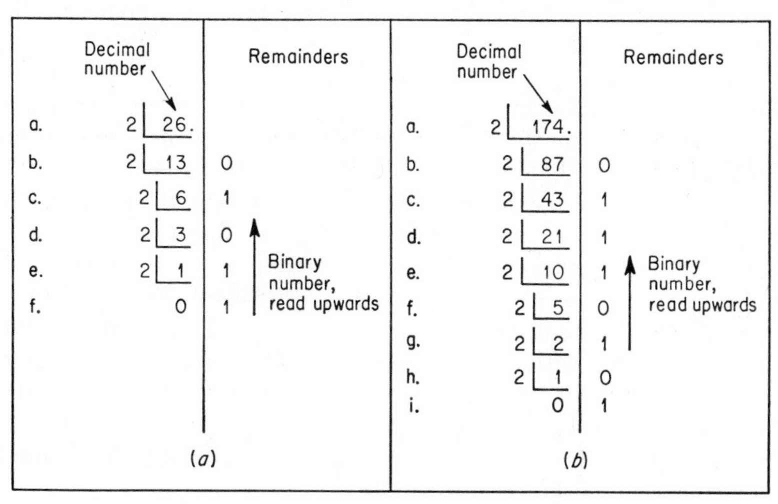

Figure 7·5 **Decimal-to-binary conversion "dibble-dabble" method.**

quotient is zero. The binary number consists of the digits of the *remainders* in the *reverse* order; that is the *last* remainder becomes the *first* digit of the binary.

Figure 7·5 shows two examples of this dibble-dabble method used to convert decimal numbers to binaries. In Fig. 7·5*a*, the decimal number 26 is changed to the binary 11010. in the following steps.

a. Divide 26 by 2, giving a quotient of 13 and a remainder of 0.

b. Divide 13 by 2, giving a quotient of 6 and a remainder of 1.

c. Divide 6 by 2, giving a quotient of 3 and a remainder of 0.

d. Divide 3 by 2, giving a quotient of 1 and a remainder of 1.

e. Divide 1 by 2, giving a quotient of 0 and a remainder of 1.

f. With the quotient 0, no further division is possible. The binary answer is read *upwards* in the *remainder* column of Fig. 7·5*a,* giving an answer of 11010.

In Fig. 7·5*b* the decimal number 174 is used. This is the same as that in the previous example of Fig. 7·4*b*. The dibble-dabble steps of Fig. 7·5*b* are

a. Divide 174 by 2, giving a quotient of 87 and a remainder of 0.
b. Divide 87 by 2, giving a quotient of 43 and a remainder of 1.
c. Divide 43 by 2, giving a quotient of 21 and a remainder of 1.
d. Divide 21 by 2, giving a quotient of 10 and a remainder of 1.
e. Divide 10 by 2, giving a quotient of 5 and a remainder of 0.
f. Divide 5 by 2, giving a quotient of 2 and a remainder of 1.
g. Divide 2 by 2, giving a quotient of 1 and a remainder of 0.
h. Divide 1 by 2, giving a quotient of 0 and a remainder of 1.
i. With a 0 quotient, no further division is possible. The binary answer is read *upwards* in the *remainders* of Fig. 7·5*b*, giving a binary number of 10101110.

Binary-to-decimal conversion. This conversion is just the opposite of that discussed previously. One method is to add up the binary column values. The sum is the decimal equivalent. As an example, refer to Fig. 7·4*a*. The binary number is 11011. Start with the digit at the right, which is the first column; the value of this column is 1 and is added since the binary digit is 1 in this example. The next column (second column from the binary point) has a value of 2 and is also added. The third column with a value of 4 is omitted here since the binary digit is 0 in this example. The fourth column with a value of 8 and the fifth column with a value of 16 are both added since the binary digits in these positions are both 1s. Adding the column value of the binary number 11011. is therefore 1 + 2 + 8 + 16 for a total of 27 which is the decimal-equivalent number. This means that $11011._2$ is equal to $27._{10}$.

In Fig. 7·4*b*, the binary number is 10101110. The values of the columns in which the 1 digits appear are, starting at the right and going to the left, 2, 4, 8, 32, and 128. The sum of these column values is 174, which is the equivalent decimal number.

A second method of binary to decimal conversion is colloquially called "double-dabble." It consists of doubling the first digit (at the left) of the binary number, then adding the next digit to the right and again doubling the sum. This doubling, adding, and doubling is repeated until the last digit (at the right) is reached. This last digit is added, without any further doubling. The result is the decimal number equivalent. As an example, the binary number of Fig. 7·5*a*, 11010. is to be converted to a decimal number. The method is as follows:

a. Double the first digit 1, giving the product 2.
b. Add the next digit 1 to the product 2, giving the sum 3.
c. Double the 3, giving 6.
d. Add the next digit 0, still giving 6.
e. Double the 6, giving 12.
f. Add the next digit 1, giving 13.

g. Double the 13, giving 26.
h. Add the last digit 0, still giving 26.
i. The result, 26, is the decimal number.
Therefore, $11010._2$ is equal to $26._{10}$.

Decimal-to-octal conversion. Converting a decimal number into its octal equivalent can be done by either of the methods described previously for decimal-to-binary conversion, remembering, however, that the octal column values are powers of eight (Fig. 7·3 and Table 7·1) and also that digits from 0 through 7 may be used.

Probably the easier way is the so-called "dibble-dabble" method. In the case of decimal-to-binary conversion, it will be recalled that division

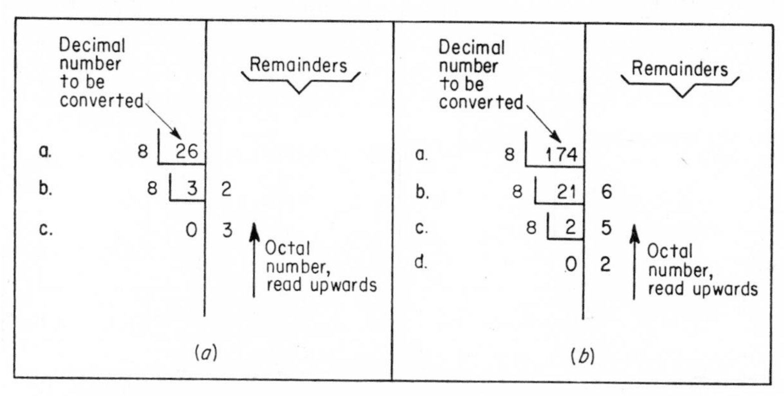

Figure 7·6 Decimal-to-octal conversion "dibble-dabble" method.

by 2's was employed. In decimal-to-octal conversion, division by 8's is used, again noting the remainder left over. The resulting quotient is repeatedly divided by 8, each time noting the remainder, until the quotient of 0 is reached. The octal number is then the remainders read in the reverse order; that is, the remainder from the last division becomes the first digit of the octal number.

In Fig. 7·6 two examples are given, the same decimal numbers as in Fig. 7·5. In Fig. 7·6*a* the decimal number 26 is to be converted to its octal equivalent using the dibble-dabble method. The procedure follows.

a. Divide 26 by 8, giving a quotient of 3 and a remainder of 2.
b. Divide the 3 by 8, giving a quotient of 0 and a remainder of 3.
c. With a quotient of 0, no further division is possible, and the octal number, reading the remainders of Fig. 7·6*a* upwards, is 32.
Therefore, $26._{10}$ is equal to $32._8$.

In Fig. 7·6*b*, the decimal number 174 is changed to the octal number 256 as follows:

a. Divide 174 by 8, giving a quotient of 21 and a remainder of 6.

b. Divide 21 by 8, giving a quotient of 2 and a remainder of 5.

c. Divide 2 by 8, giving a quotient of 0 and a remainder of 2.

d. With a quotient of 0, no further division is possible, and the octal number is 256, reading the remainders of Fig. 7·6*b* upwards. Therefore, $174._{10}$ is equal to $256._{8}$.

A second method of decimal-to-octal conversion is a bit more involved than the dibble-dabble method. Figure 7-7 uses the same decimal numbers, 26 and 174, as the previous examples of Fig. 7·6.

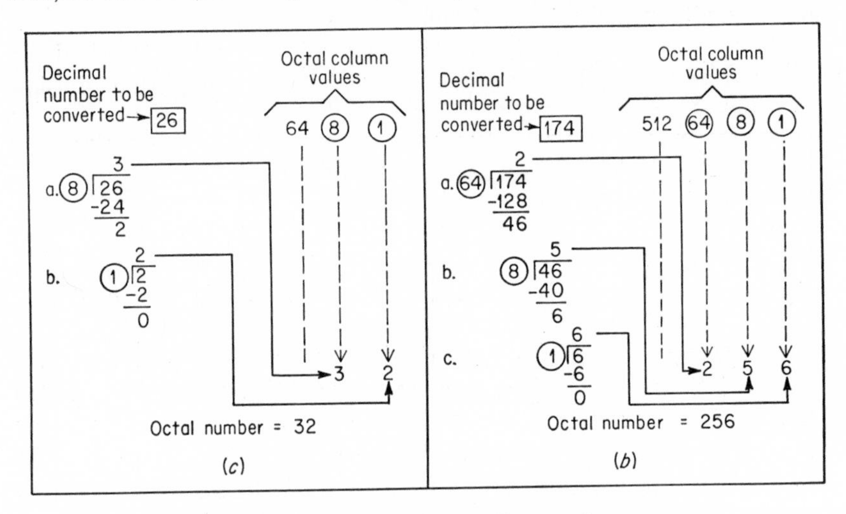

Figure 7·7 Decimal-to-octal conversion.

In Fig. 7·7*a*, the decimal number 26 is converted into the equivalent octal number of 32 by using the column values (shown in Table 7·1) in the following manner:

a. Divide the decimal number 26 by the *highest* octal-column value which does not exceed 26. Since the 64 column is too large, the 8 column must be used. This is shown encircled in Fig. 7·7*a*. Dividing 26 by 8 gives a quotient of 3 and a remainder of 2. The quotient 3 is placed, as shown, as the first octal digit in the 8's column.

b. The remainder of 2 is then divided by the next smaller column value which must not exceed 2. The next smaller column value in this case is 1 and is shown encircled in the figure. Dividing 2 by 1 gives a quotient of 2 with a remainder of 0. The quotient 2 is then placed, as shown, in the 1's column as the second octal digit, giving an octal number of 32. Therefore, $26._{10}$ is equal to $32._{8}$.

In Fig. 7·7*b* the decimal 174 is changed to the octal 256 in the following steps.

a. Divide 174 by the highest octal-column value which must not exceed 174. Since 512 is too large, the next smaller column value is tried. This figure of 64 (shown encircled) is not too large and is then divided into 174, giving a quotient of 2 and a remainder of 46. The quotient 2 is placed, as shown, in the 64's column and is the first digit of the octal number.

b. The remainder 46 is then divided by the next smaller column value which must not exceed 46. This column value is 8 (shown encircled in Fig. 7·7*b*). Dividing 46 by 8 gives a quotient of 5 and a remainder of 6. The quotient 5 is placed, as shown, in the 8's column and becomes the second digit of the octal number.

c. The remainder 6 is then divided by the next smaller column value, which must not exceed 6. This column value is 1 (shown encircled). Dividing 6 by 1 gives a quotient of 6 and a remainder of 0. The quotient 6 is placed, as shown, in the 1's column and becomes the third digit of the octal number. The complete octal number, as shown, is 256.

Therefore, $174._{10}$ is equal to $256._{8}$.

Octal-to-decimal conversion. Octal numbers can be changed to decimal numbers in methods similar to that described previously for binary-to-decimal conversion.

One method is to simply add up the products of the column values and the digits in that column. This is shown in Fig. 7·3 and Table 7·1. In the first example of Fig. 7·3*a*, the octal number is 501. To convert it to the decimal equivalent, the following is done

a. The first digit 5 is in the third column which has a column value of 8^2 or 64. The 5 is multiplied by 64, giving a product of 320.

b. The next digit is 0, which, when multiplied by its column value of 8^1 or 8, results in a product of 0.

c. The third digit 1 is in the column value of 8^0 or 1, giving a product of 1.

d. Adding up these products of 320, 0, and 1 gives 321 which is the decimal equivalent.

Therefore, $501._{8}$ is equal to $321._{10}$.

In Fig. 7·3*b*, a second example is shown. The octal number 346 is changed to the decimal number 230 in the following manner:

a. The first digit 3 is in the column value of 8^2 or 64, and the product of 3 and 64 is 192.

b. The second digit 4 is in the 8^1 or 8-column value, and this product is 32.

c. The third digit 6 is in the 8^0 or 1 column value, and the product of 1 and 6 is 6.

d. Adding up the three products of 192, 32, and 6 gives the decimal equivalent 230.

Therefore, $346._8$ is equal to $230._{10}$.

Another method of octal-to-decimal conversion is similar to the "double-dabble" system of multiplying the first digit by two, adding the next, etc., except that with octals each multiplication is by 8. As an example, take Fig. 7·3*a* where the octal number is 501.

a. Multiply the first digit 5 by 8, giving 40.

b. Add the next digit 0 to the 40, still giving 40.

c. Multiply the 40 by 8, giving 320.

d. Add the last digit 1 to the 320, giving 321.

e. Do *not* multiply any further after adding the *last* digit, and the final result of 321 is the decimal equivalent.

Another example of this octal-to-decimal "double-dabble" conversion is the octal number 346 (Fig. 7·3*b*). The conversion process would be

a. Multiply the first digit 3 by 8, giving a product of 24.

b. Add the next digit 4 to the 24, giving a sum of 28.

c. Multiply the 28 by 8, giving a product of 224.

d. Add the *last* digit 6 to the 224, giving a final result of 230. (Do *not* multiply after adding the *last* digit.)

e. The 230 is the decimal equivalent.

Binary-octal relationship. To convert from a binary number to an octal, or vice versa, it is not necessary to first convert to the decimal number. Table 7·2 shows some decimal numbers, then the binary equivalent, then the binary is repeated but in groups of three digits and finally the octal number.

The last two columns of Table 7·2 show a relationship between the group-of-three binaries and the octals. If each group of digits in the group-of-threes are handled as if they were each the first three columns of a binary number, then the decimal equivalent for *each* group results in the digits of the octal number. For example, the fourth line of Table 7·2, which is the decimal number 3, is the binary group-of-three 011. Consider these three binary digits 011 where the 0 is in the column value of 2^2 or 4, giving a 0×4 or 0 value. The next binary digit 1 is in the 2^1 or 2-column value, giving 1×2 or 2 value. To this 2 must be added the last binary digit 1 which is in the 2^0 or 1-column value, giving a 1×1 or 1 value. The total $2 + 1$ is 3 which is the octal number and also the decimal number. The decimal and octal numbers are the same for numbers up to 7.

Another example of this binary and octal relationship can be seen in Table 7·2 for the decimal number 15. The binary group-of-three is 001 111. Consider the first group 001 as if these digits were the first three binary columns, where the first digit 0 is in the 2^2 or 4 column value, the second digit 0 is in the 2^1 or 2 column, and the third digit 1 is in the 2^0 or 1 column. Therefore, this first group 001 gives a result of $0 + 0 + 1$, or 1.

Consider the next group-of-three 111 as if these digits were in the first three binary columns. The first digit 1 would be in the 2^2 or 4-column value; the second digit 1 in the 2^1 or 2-column value; and the third digit 1 in the 2^0 or 1 column, giving a result of $4 + 2 + 1$, or 7. The binary 001 111 results in the octal number 17, where the first digit 1 results from the first group 001, and the next digit 7 results from the next group 111.

Table 7·2

Decimal number	Binary number	Binary number in groups of three	Octal number
0	0	000	0
1	1	001	1
2	10	010	2
3	11	011	3
4	100	100	4
5	101	101	5
6	110	110	6
7	111	111	7
8	1000	001 000	10
9	1001	001 001	11
10	1010	001 010	12
11	1011	001 011	13
12	1100	001 100	14
13	1101	001 101	15
14	1110	001 110	16
15	1111	001 111	17
16	10000	010 000	20
17	10001	010 001	21
18	10010	010 010	22
19	10011	010 011	23
20	10100	010 100	24
26	11010	011 010	32
27	11011	011 011	33
174	10101110	010 101 110	256

This binary-octal relationship working from the octal to the binary can be seen in the last line of Table 7·2. Here, the octal number is 256. Each of these three digits results in a separate group-of-three, three groups in all. The first digit 2 gives the binary-group 010. The next digit 5 gives the binary-group 101. The last digit 6 gives the group 110. Therefore, the octal 256 gives the binary 010 101 110, as shown in the last line of Table 7·2.

7.6 Unmodified Binary-chain Counter

A simple electronic counter uses bistable multivibrators or flip-flops. The theory of these multivibrators is covered in Sec. 5·9, but a very brief description will be reviewed here. Figure 7·8 shows four of these flip-flops, V_1, V_2, V_3, and V_4, with each tube consisting of two triodes, A and B. A bistable multivibrator has two stable states of operation, either tube A is conducting with B cut off, or B conducts, and A is cut off. To start, one of the tubes, say the B tube, is made to conduct heavily, either by sending in a "reset" pulse signal to that tube grid alone or by momentarily disconnecting that grid from its negative bias voltage. The heavily conducting tube then has a low plate voltage. This is directly coupled to the grid of the other tube, the A tube, making that grid negative, cutting off the A tube. Each grid, as shown in Fig. 7·8, connects to a voltage divider from a negative voltage to the positive voltage at the other plate. The cutoff tube A has a high voltage at its plate, making the grid of the B tube positive, keeping the B tube in heavy conduction. With the B tube conducting or ON, the multivibrator is said to be in its flop or reset state. It remains in this state until a trigger pulse reverses the operation.

A negative trigger input pulse applied, as shown in Fig. 7·8, to the grids of V_{1A} and V_{1B} cuts off the B tube. Its plate voltage rises, driving the A grid positive and putting the A tube into conduction. Plate voltage of tube A is now low, making the B grid negative and keeping tube B cut off. With A conducting or ON, and B OFF, the multivibrator is in its flip or set state. It remains in this state until the next negative-going trigger pulse reverses the operation. One pulse therefore flips the stage, and the next pulse flops the stage.

Figure 7·8 is the circuit diagram of a simple electronic counter with four flip-flop stages, V_1, V_2, V_3, and V_4. Four neon bulbs, N_1, N_2, N_3, and N_4, are shown, one connected in the plate of each B tube. Neon bulbs glow when the voltage applied to them is sufficiently large to ionize the gas. This will happen when a B tube is nonconducting or OFF, since its plate voltage is high then.

The flip-flop stage V_1 is triggered by the negative-going input pulse. V_2 is triggered by a negative-going signal from the plate of V_{1B}. V_3 is similarly triggered by a negative-going signal from the plate of V_{2B}, and the last stage V_4 is triggered by a negative-going signal from V_{3B} plate. A positive-going pulse does not affect these multivibrators which are single-line grid-triggered. This means that stages V_2, V_3, and V_4 are only triggered when the previous B stage *starts* conducting or *comes* ON. Only then does a B-plate voltage fall or become negative-going. When a B stage is *already* conducting, its plate voltage is already low and is not falling, and it does not trigger the next stage. When a B stage becomes

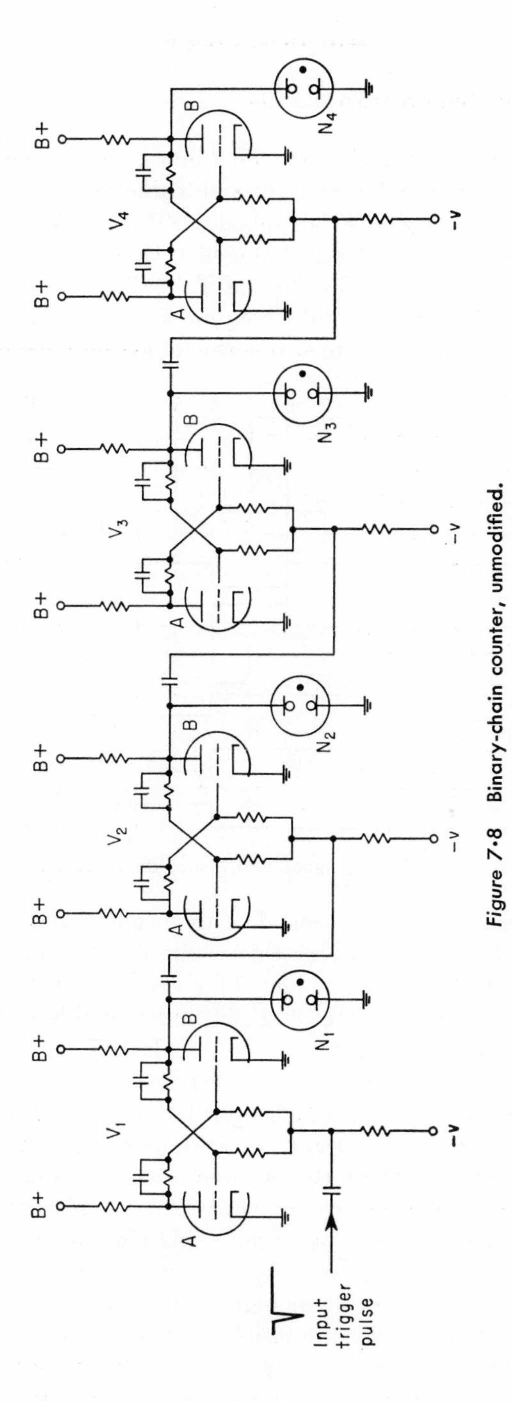

Figure 7·8 **Binary-chain counter, unmodified.**

cut off, its plate voltage rises or is positive-going, and it does not trigger the next stage.

In the following discussion of the binary chain counter, reference should be made to Figs. 7·8 and 7·9 and Table 7·3. Assume that at the start, before any trigger pulse is applied, all B tubes are in operation and all A tubes are cut off, putting all stages in their flop or reset states. To indicate this condition, the first line of Table 7·3, for zero trigger pulse, has a check mark in each B tube column. Check marks, then, indicate that a tube is conducting. Note also in the table that for zero trigger pulse

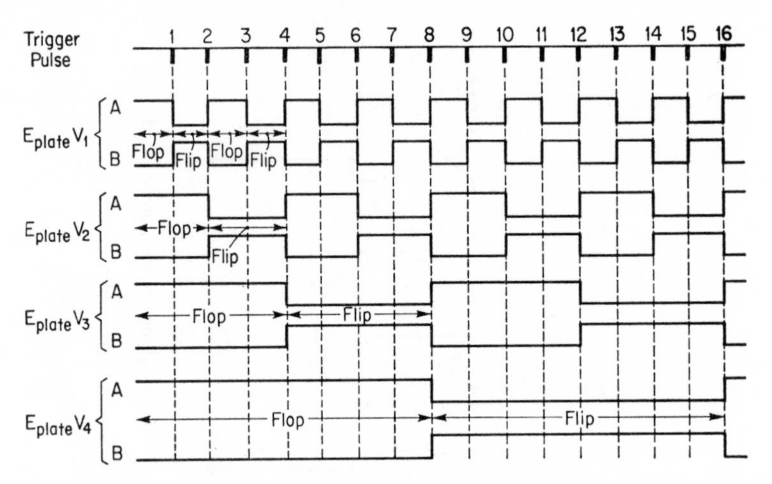

Figure 7·9 **Waveshapes, unmodified binary-chain counter of Fig. 7·8.**

none of the neon bulbs are checked. This means that they are nonconducting or dark. The reason for this is simply that when a B tube is conducting, its plate voltage is low. This low voltage cannot ionize the neon bulb. The waveshapes of Fig. 7·9 indicate this condition, before trigger pulse 1, by showing that all B-tube plate voltages are low and all A-tube plate voltages high.

1. Trigger pulse 1 applied to V_1 rolls V_1 into its flip state, with V_{1B} now cut off. Table 7·3 for trigger pulse 1 now shows a check mark in V_{1A} column, indicating that the A tube is now conducting and the B tube cut off. Waveshapes, Fig. 7·9, show that V_{1A} plate voltage has now fallen and V_{1B} plate voltage has risen. This positive-going voltage does not affect the next stage V_2, and V_2, V_3, and V_4 remain in their previous operation. With V_{1B} plate voltage high, neon bulb N_1 glows, as indicated in Table 7·3. The neon bulbs are visible on the front panel of the counter as "readout" devices which indicate the number of pulses that have been counted. The count or number value indicated therefore when N_1 glows is simply 1.

2. Trigger pulse 2 applied to V_1 again rolls V_1, this time putting V_1 back into its flop state, with V_{1B} coming back into conduction. The falling plate voltage of V_{1B} flips V_2 so that V_{2A} is now conducting. The rise of plate voltage of V_{2B} does not affect V_3 or V_4 which "stay" or "hold" in their previous operation. With V_{1B} conducting, its low plate voltage causes neon bulb N_1 to black out. However, N_2 now glows since plate voltage of cutoff V_{2B} is high (see waveshapes, Fig. 7·9). Since N_2 is now lit

Table 7·3 Table of operations, unmodified binary chain counter

Trigger pulse input	V_1		V_2		V_3		V_4		Binary value of neon bulbs			
	A	*B*	*A*	*B*	*A*	*B*	*A*	*B*	N_1 (1)	N_2 (2)	N_3 (4)	N_4 (8)
0		✓		✓		✓		✓				
1	✓			✓		✓		✓	✓			
2		✓	✓			✓		✓		✓		
3	✓		✓			✓		✓	✓	✓		
4		✓		✓	✓			✓			✓	
5	✓			✓	✓			✓	✓		✓	
6		✓	✓		✓			✓		✓	✓	
7	✓		✓		✓			✓	✓	✓	✓	
8		✓		✓		✓	✓					✓
9	✓			✓		✓	✓		✓			✓
10		✓	✓			✓	✓			✓		✓
11	✓		✓			✓	✓		✓	✓		✓
12		✓		✓	✓		✓				✓	✓
13	✓			✓	✓		✓		✓		✓	✓
14		✓	✓		✓		✓			✓	✓	✓
15	✓		✓		✓		✓		✓	✓	✓	✓
16		✓		✓		✓		✓				

on the second trigger pulse, N_2 has a count value of 2. Note from the waveshapes of Fig. 7·9 that one cycle of square wave was produced at the plates of V_1 for two input pulses. V_1 is therefore called a divide-by-2 stage.

3. Trigger pulse 3 flips V_{1A} into conduction again. V_{1B} plate voltage, as shown in Fig. 7·9, rises again, having no effect on V_2. V_2, V_3, and V_4 hold. With high voltages on the plates of V_{1B} and V_{2B}, neon bulbs N_1 and N_2 both glow. N_1 has a count value of 1, and N_2 has a value of 2. With both lit, the count is $1 + 2$ or 3.

4. Trigger pulse 4 flops V_{1B} into conduction, and its plate voltage falls. This negative-going voltage flops V_2, and V_{2B} comes back into conduction. Plate voltage of V_{2B} falls and flips V_3. V_{3A} now conducts for the first time. Plate voltage of V_{3B} rises and does not affect V_4, which holds in its previous operation. From the waveshapes of Fig. 7·9, it can be seen that low plate voltages exist for pulse 4 at V_{1B}, V_{2B}, and V_{4B}. Only V_{3B}, which is cut off, has a high plate voltage. As a result, only neon bulb N_3 is lit, as shown by the check mark in Table 7·3. Since this is pulse number 4, neon bulb N_3 has a count value of 4. Note that N_1 has a count value of 1; neon bulb N_2, a count value of 2; and N_3, a count value of 4. These are the binary column values, 1, 2, 4, 8, etc., as explained in Sec. 7·3.

5. Trigger pulse 5 flips V_1, and V_{1A} now conducts. The positive-going plate voltage of V_{1B} has no effect on V_2. V_2, V_3, and V_4 hold. Now, as shown in Fig. 7·9, V_{1B} and V_{3B} both have high plate voltages, and N_1 (count value of 1) and N_3 (count value of 4) both glow. This indicates a count of 1 + 4 or 5.

6. Trigger pulse 6 flops V_1, and V_{1B} conducts. This flips V_2, and V_{2A} conducts, not affecting the next stage. V_3 and V_4 hold. High plate voltages now exist at V_{2B} and V_{3B}, lighting N_2 and N_3 for an indicated count value of 2 + 4 or 6.

7. Trigger pulse 7 flips V_1, and V_{1A} conducts, not affecting the next stage. V_2, V_3, and V_4 hold. High plate voltages are now present at V_{1B}, V_{2B}, and V_{3B}, and N_1, N_2, and N_3 glow for an indicated count of 1 + 2 + 4 or 7.

8. Trigger pulse 8 flops V_1, and V_{1B} again conducts. This flops V_2, and V_{2B} conducts. This flops V_3, and V_{3B} conducts. This flips V_4, and V_{4A} conducts for the first time. Only V_{4B} has a high plate voltage, and only N_4 lights, indicating a count of 8.

9. Trigger pulse 9 rolls V_1 into its flip state, and V_{1A} conducts, not affecting V_2. V_2, V_3, and V_4 hold. V_{1B} and V_{4B} now have high plate voltages, lighting N_1 and N_4 for a count value of 1 + 8 or 9.

10. Trigger pulse 10 rolls V_1 into its flop state, and V_{1B} conducts. This rolls V_2 into its flip state, and V_{2A} conducts, not affecting the next stage. V_3 and V_4 hold. V_{2B} and V_{4B} plate voltages are high, lighting N_2 and N_4, indicating a count of 2 + 8 or 10.

11. Trigger pulse 11 rolls V_1 into its flip state, and V_{1A} comes on, not affecting V_2. V_2, V_3, and V_4 hold. High voltages are present at the plates of the cutoff tubes V_{1B}, V_{2B}, and V_{4B}. This lights neons N_1, N_2, and N_4, indicating a count of 1 + 2 + 8 or 11.

12. Trigger pulse 12 flops V_{1B} back into conduction, rolling V_2 also into its flop state, and V_{2B} conducts. This rolls V_3 into its flip state, and V_{3A} conducts, not affecting V_4 which holds. V_{3B} and V_{4B} now have high plate voltages, and N_3 and N_4 glow. This indicates a count of 4 + 8 or 12.

13. Trigger pulse 13 flips V_{1A} into conduction which does not affect V_2. V_2, V_3, and V_4 hold. Now, high voltages are present at the plates of

V_{1B}, V_{3B}, and V_{4B}. Neon bulbs N_1, N_3, and N_4 glow, counting $1 + 4 + 8$ or 13.

14. Trigger pulse 14 flops V_{1B} into conduction, which in turn flips V_{2A} into conduction. V_3 and V_4 hold. V_{2B}, V_{3B}, and V_{4B} all have high plate voltages, lighting neons N_2, N_3, and N_4 for a count value of $2 + 4 + 8$ or 14.

15. Trigger pulse 15 flips V_{1A} into conduction, not affecting V_2. V_2, V_3, and V_4 hold. All *B* plates have high voltages, lighting all the neon bulbs and giving a count reading of $1 + 2 + 4 + 8$ or 15. This is the maximum count for this circuit.

16. Trigger pulse 16 flops V_{1B} into conduction. This flops V_{2B}, which rolls V_3 into its flop state. V_{3B} coming on flops V_{4B} into conduction. All *B* tubes are now in conduction and have low plate voltages. All neon bulbs are dark. This condition is a duplicate of the operation at the beginning, before the first trigger pulse, and the circuit is said to be reset by trigger pulse 16. The neon bulbs indicate a count of zero. From the waveshapes of Fig. 7·9 it can be seen that V_1 produces one cycle for every two trigger pulses and operates as a divide by two stage. V_2, in turn, produces one cycle for every four trigger pulses and operates as a divide by four stage. V_3 produces one cycle for every eight pulses and is called a divide-by-8 stage. Finally, V_4 produces one complete cycle for sixteen trigger pulses and is a divide-by-16 stage. The four flip-flop stages operating as a binary counter are also called a scale of 16.

This counter, as has just been shown, counts in binary fashion. That is, the neon bulbs each have binary-system column values of 1, 2, 4, and 8. By a simple feedback system the circuit can be advanced from a scale of 16 to a scale of 10.

7·7 Modified Binary-chain Decade Counter

A binary-chain counter similar to that of Fig. 7·8 is shown in Fig. 7·10. Four bistable multivibrators or flip-flops, V_1, V_2, V_3, and V_4, are used together with ten neon bulbs. The arrangement of the neon bulbs which give a decimal-system readout instead of the binary system will be discussed later in Sec. 7·8.

As can be seen in the lower right corner of Fig. 7·10, two feedbacks are employed. These are shown by the dashed-line connections marked "Feedback." One feedback connects the plate of V_{4A} to the grid of V_{3B}. A second feedback goes from the plate of V_{3B} to the grid of V_{2A}. The effect of this feedback, as will be shown, is to advance the four-flip-flop binary chain from a divide by 16 (resetting all stages on the sixteenth trigger pulse), to a divide-by-10 circuit (resetting all stages on the tenth pulse). The amount of advance due to feedback can be predicted by noting *to* which stage the feedback is applied. Feedback applied to a flip-flop stage

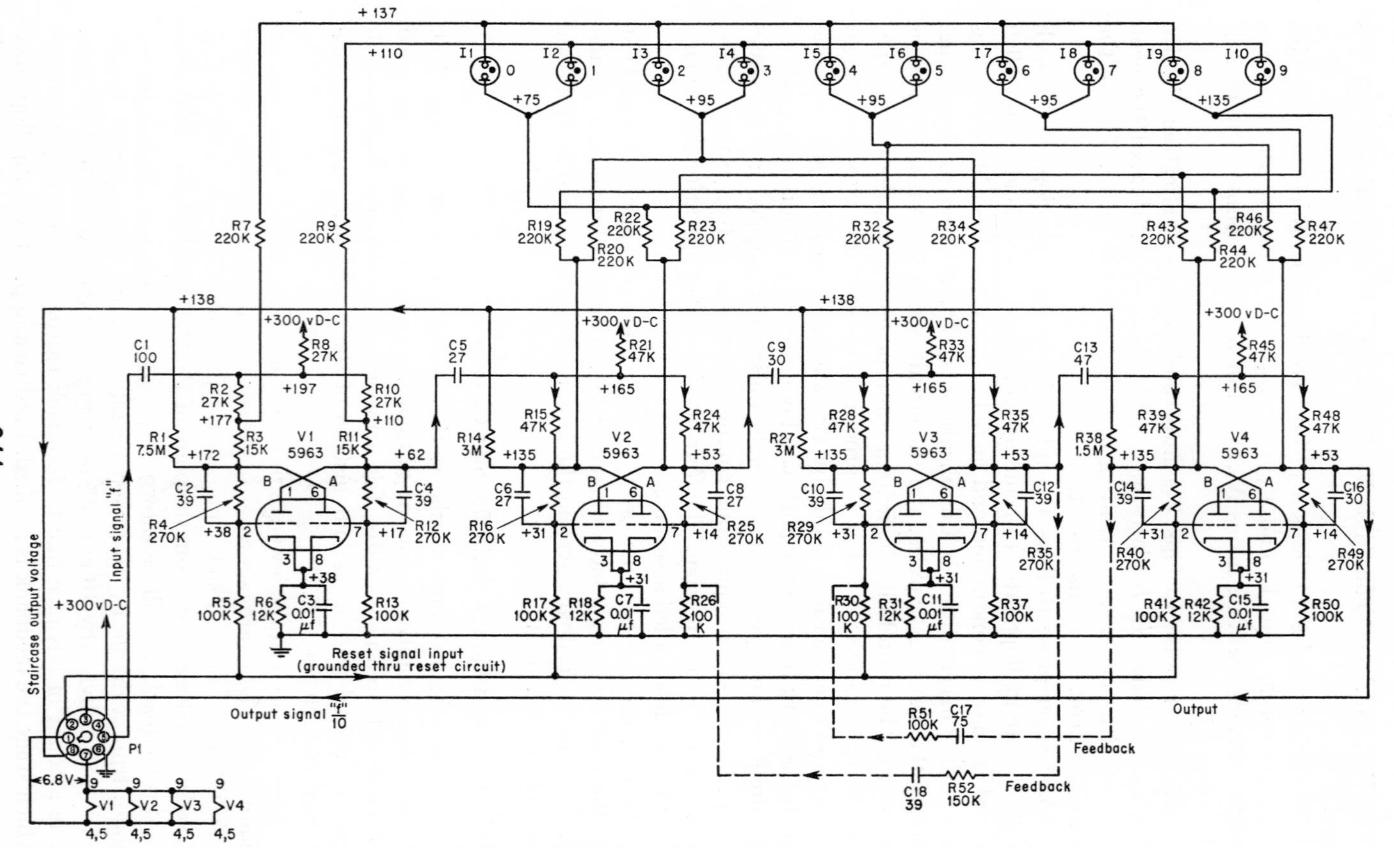

Figure 7.10 Modified binary chain (Hewlett-Packard Co.)

advances the total division of the entire circuit by the binary value of that stage, where these values are 1 for V_1, 2 for V_2, 4 for V_3, and 8 for V_4. Feedback applied to the grid of V_{2A} *alone*, with its count value of 2, would change the entire circuit division by 16 to a new division by 14 (16 − 2). Similarly, feedback to the grid of V_{3B} *alone*, with its count value of 4, would change the original divide by 16 to a new divide by 12 (16 − 4). With *both* feedbacks of Fig. 7·10, to the grid of V_{2A} (count of 2) and to the grid of V_{3B} (count of 4), the original divide by 16 is reduced by 6 (2 + 4) to a divide by 10. For feedback to be effective, the feedback signal (plate voltage either rising or falling) must either be positive-going to the grid of a cutoff tube or negative-going to the grid of a conducting tube.

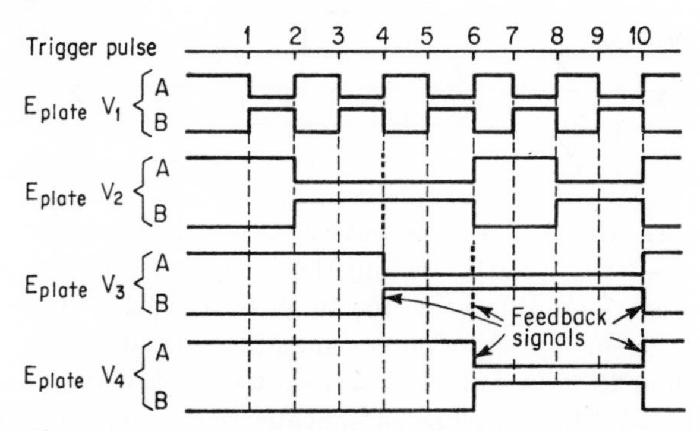

Figure 7·11 Waveshapes of modified binary-chain decade counter of Fig. 7·10.

In the following discussion of the circuit theory of the modified binary-chain decade counter, reference should be made to the schematic diagram of Fig. 7·10, to the waveshapes of Fig. 7·11, and to the operations table 7·4. In going through the circuit of Fig. 7·10 note should be made that each B tube is the *left* half, yet each B-plate circuit is brought over to the *right* side of each stage. Similarly, each A tube is the *right* half, yet its plate circuit is shown at the *left* of the stage.

To start off, each B stage is in conduction (in the flop state) and is made to conduct by the reset switch (not shown) which disconnects all B grids from ground. As explained previously, this opens the resistor voltage divider from B+ to ground, placing a positive voltage on the B grids (+38 volts on V_{1B}, and +31 volts on the others). The heavily conducting B tubes have low plate voltages (+62 volts on V_{1B} and +53 volts on the others) which bring the A grids to low positive voltages (+17 volts on V_{1A} and +14 volts on the others). Cathode bias is used on all stages, making the cathode of V_1 + 38 volts, with +31 volts on all others. With all B tubes conducting, the A tubes are all biased beyond cutoff. V_{1A} has a bias of −21 volts (the difference between the +38 volt cathode and the

Table 7·4 Table of operations, modified binary chain decade counter of Fig. 7·10

Trigger pulse input	V_1		V_2 (Feedback)		V_3	(Feedback)	V_4	
	A	B	Grid A	B	A	Plate Grid B	Plate A	B
0		✓		✓		✓		✓
1	✓			✓		✓		✓
2		✓	✓			✓		✓
3	✓		✓			✓		✓
4		✓	✓	✓ (dotted)	✓			✓
5	✓		✓		✓			✓
6		✓		✓	✓	✓ (dotted)	✓	
7	✓			✓	✓		✓	
8		✓	✓		✓		✓	
9	✓		✓		✓		✓	
10		✓		✓		✓		✓

+17 volt grid). All other A tubes have a bias of -17 volts (the difference between the $+31$ volt cathodes and the $+14$ volt grids). Note that all B tubes are conducting heavily due to their zero bias conditions (all B grids are at the *same* positive potential as the cathodes).

The first line "0 pulse" of Table 7·4 shows check marks in the B-tube columns. The solid-line check mark indicates that that tube is conducting. A dotted-line check mark indicates that a tube starts conducting but is quickly cut off again. This is shown in the table for trigger pulses 4 and 6 and will be explained in the following theory discussion. Feedback signals from the plates of V_{3B} and V_{4A} are present only when those plate voltages either rise or fall and not when the voltage remains steady.

1. Trigger pulse 1 applied to V_1 flips V_{1A} into conduction. Since the plate of V_{1B} is now positive-going (waveshapes, Fig. 7·11), it does not roll V_2. V_2, V_3, and V_4 hold in their original flop stages. No feedback signal is present yet.

2. Trigger pulse 2 flops V_{1B} back into conduction. The negative-going plate voltage of V_{1B} (waveshapes, Fig. 7·11) triggers V_2 into its flip state, and V_{2A} now conducts. Plate voltage of V_{2B} rises (waveshapes, Fig. 7·11), not affecting V_3. V_3 and V_4 hold, and no feedback signal is present yet.

3. Trigger pulse 3 flips V_1, and V_{1A} conducts. The rising plate voltage of V_{1B} does not affect V_2. V_2, V_3, and V_4 hold, and no feedback signal is present yet.

4. Trigger pulse 4 flops V_1, and V_{1B} conducts. The falling plate voltage of V_{1B} flops V_2, and V_{2B} starts conducting. Plate voltage of V_{2B} falls, flipping V_3, and V_{3A} now conducts for the first time. The rising plate voltage of V_{3B} does not affect V_4 which holds. However, the rising plate voltage of V_{3B} is a feedback signal (waveshapes Fig. 7·11) which is applied to the grid of V_{2A}. Since trigger pulse 4 has just resulted in the flop of V_2, V_{2B} is starting to conduct, and V_{2A} has just cut off. The positive-going feedback signal to the grid of V_{2A} now turns V_{2A} back into conduction, again cutting off V_{2B}. In Table 7·4, a dotted-line check mark is shown in V_{2B} column, denoting that the B tube goes into conduction and then almost immediately is cut off again. This is also shown in the waveshapes, Fig. 7·11, as a momentary, dashed-line voltage change. The bent arrow in the V_2 columns for pulses 3 and 4 indicates that for trigger pulse 3, V_{2A} is in conduction; then for trigger pulse 4, conduction is *momentarily* given to V_{2B} and then changed back (due to feedback signal) to V_{2A} again.

5. Trigger pulse 5 flips V_1, and V_{1A} conducts, not affecting V_2. V_2, V_3, and V_4 hold, and no feedback signal is present.

6. Trigger pulse 6 flops V_1, and V_{1B} conducts. The fall of V_{1B} plate voltage flops V_2, and V_{2B} conducts. V_{2B} plate voltage decreases, flopping V_3, and V_{3B} *starts* conducting. The falling plate voltage of V_{3B} flips V_4, and V_{4A} conducts. The negative-going plate voltages of V_{3B} and V_{4A}, as these tubes start conducting, are feedback signals, as indicated in the waveshapes of Fig. 7·11. The negative-going feedback signal from the plate of V_{3B} is applied to the grid of V_{2A}. Since V_{2A} is already cut off, the feedback signal has no effect. The other negative-going feedback signal from the plate of V_{4A} is applied to the grid of V_{3B}. Since V_{3B} has just entered conduction, the negative-going feedback signal applied to its grid cuts off the tube. The dotted-line check mark in Table 7·4 in V_{3B} column denotes that the B tube goes into conduction for an instant, then being cut off again by the feedback signal. (Also see waveshapes, Fig. 7·11, for dashed-line momentary voltage change.) The bent arrow in the V_3 columns for pulses 5 and 6 indicates that for pulse 5, V_{3A} is in conduction; then for pulse 6, conduction *momentarily* switches to V_{3B} and then changes back (because of feedback signal) to V_{3A} again.

7. Trigger pulse 7 flips V_1, and V_{1A} conducts. No effect is had on V_2, and V_2, V_3, and V_4 hold. No feedback signal is produced.

8. Trigger pulse 8 flops V_1, and V_{1B} conducts. Its falling plate voltage rolls V_2 into its flip state, and V_{2A} conducts. No effect is had on V_3. V_3 and V_4 hold, and no feedback is produced.

9. Trigger pulse 9 flips V_1, and V_{1A} conducts, not affecting V_2. V_2, V_3, and V_4 hold, and no feedback is produced.

10. Trigger pulse 10 rolls V_1 into its flop state, and V_{1B} conducts. Its falling plate voltage flops V_{2B} into conduction. Plate voltage of V_{2B} decreases, flopping V_{3B} into conduction. Its negative-going plate voltage flops V_{4B} into conduction. All B tubes are now in conduction as they were at the beginning (0 pulse). This means that it required 10 input trigger pulses to reset all the stages, resulting in a divide-by-10 circuit.

Feedback signals are present at this tenth trigger pulse but have no effect. With V_{3B} starting to conduct, its negative-going plate voltage is a feedback signal. This signal is applied to the grid of V_{2A}. Since V_{2A} is already nonconducting, this signal has no effect. The other feedback signal is produced at V_{4A} plate. When V_{4A} became cut off, its plate voltage rose. This positive-going voltage is applied to the grid of V_{3B}. Since V_{3B} is already in heavy conduction, the feedback signal has no effect.

Figure 7·12 Decade counters. (*a*) Neon-tube decimal readout; (*b*) Nixie-tube decimal readout. (*Hewlett-Packard Co.*)

The modified binary-chain decade counter just described counts up to 10, with the count of ten resulting in a 0 count. This is the same as ordinary counting in the decimal system where, after the count of 9, the count becomes 0 and a 1 is used in the next column, which is the tens. With the decade counter, the output of the last stage, V_4, of the units decade is then fed into V_1 of the next decade, which becomes the tens decade. The output from this decade V_4 is then fed into V_1 of the next decade, which becomes the hundreds decade, etc. Figure 7·12 shows two models of a single decade counter, one using neon lights and the other using the Nixie readout tube. Figure 7·13 illustrates two complete counters, one employing eight decades and the other five decades.

(a)

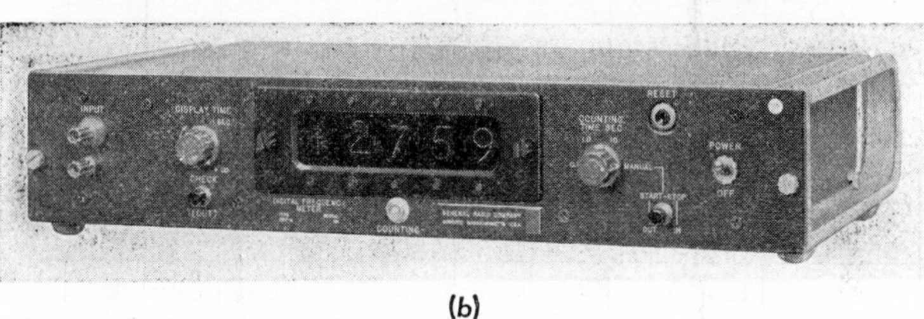

(b)

Figure 7·13 Time-interval counters. (*General Radio* Co.) (*a*) Eight-decade neon-tube decimal readout; (*b*) five-decade Nixie-tube decimal readout.

7·8 Decimal-readout System of Decade Counter

The decade counter, as explained in the previous section, is a binary chain modified with feedback. This advances the count of the chain from a divide by 16 (circuit resets on sixteenth pulse) to a divide by 10 (circuit resets on tenth pulse). Ten neon lamps are connected in the circuit to give a decimal readout. The lamps of one decade are placed in a column, as seen in the photographs Figs. 7·12*a* and 7·13*a*, one above the other. The numerals 0 through 9, one number in front of each lamp, are visible on the front panel. Duplicate columns, one for each decade, are placed to the left of the previous column with the first column numbers at the right being units, the next column being tens, etc.

The complete circuit diagram of the modified binary counter and the decimal-readout neon lamps is shown in Fig. 7·10. A partial diagram of the neon lamps is shown in Fig. 7·14, along with a table of operations. In the following discussion, reference should be made to Figs. 7·10 and 7·14 and Table 7·4. The ten neon lamps are numbered 0 through 9 and are connected, as shown, in pairs, 0 and 1, 2 and 3, 4 and 5, etc. The upper electrodes of all even-numbered lamps, 0, 2, 4, 6, and 8, are connected to a line called the "even bus." This line connects through R_7 to the plate

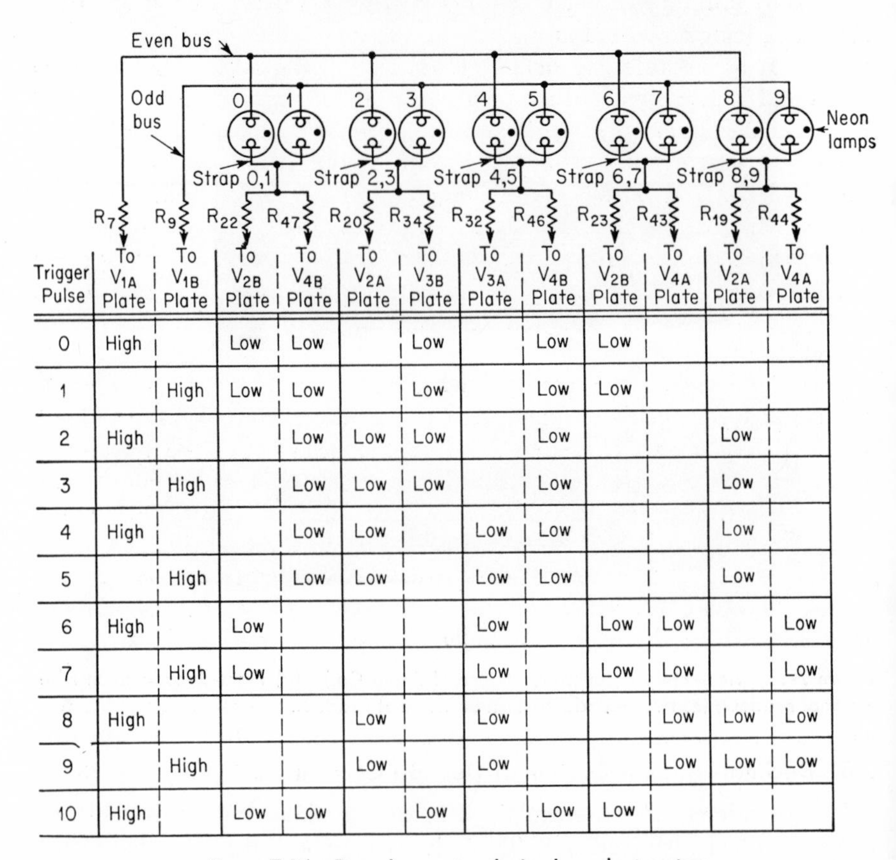

Trigger Pulse	To V_{1A} Plate	To V_{1B} Plate	To V_{2B} Plate	To V_{4B} Plate	To V_{2A} Plate	To V_{3B} Plate	To V_{3A} Plate	To V_{4B} Plate	To V_{2B} Plate	To V_{4A} Plate	To V_{2A} Plate	To V_{4A} Plate
0	High		Low	Low		Low		Low	Low			
1		High	Low	Low		Low		Low	Low			
2	High			Low	Low	Low		Low			Low	
3		High		Low	Low	Low		Low			Low	
4	High			Low	Low		Low	Low			Low	
5		High		Low	Low		Low	Low			Low	
6	High		Low				Low		Low	Low		Low
7		High	Low				Low		Low	Low		Low
8	High				Low		Low			Low	Low	Low
9		High			Low		Low			Low	Low	Low
10	High		Low	Low		Low		Low	Low			

Figure 7·14 Decade-counter decimal-readout system.

circuit of V_{1A}. When V_{1A} is cut off, as shown in Table 7·4, at pulses 0, 2, 4, 6, 8, and 10, its plate voltage is high, making the even bus a high voltage of +137 volts.

The upper electrodes of all odd-numbered neon lamps, 1, 3, 5, 7, and 9, are connected to the line called "odd bus." This line connects through R_9 to the plate circuit of V_{1B}. The odd line will get a high voltage of +137 volts on it when V_{1B} is cut off. This occurs at pulses 1, 3, 5, 7, and 9, as shown in Table 7·4 and Fig. 7·14.

The lower electrodes of the neon lamps are connected together in pairs by lines called "straps." Lamps 0 and 1 are tied together by strap 0, 1. Lamps 2 and 3, by strap 2, 3, as shown in the diagrams. The straps connect through resistors to the plate circuits of the various stages. For example, strap 0, 1 connects through R_{22} to the plate circuit of V_{2B} and through R_{47} to V_{4B}. Other straps connect as shown.

A strap will have any one of three voltages on it depending upon the condition (cutoff or conducting) of the vacuum tubes to which it is connected. If both tubes are cut off, their plate voltages are high, such as V_{2A} and V_{4A} at 0 pulse in the operations chart of Fig. 7·14, and strap 8, 9 is at a high voltage of +135 volts (Fig. 7·10). If only one tube is cut off, while the other conducts, one plate voltage is high and the other is low, such as V_{2A} and V_{3B} at 0 pulse in the operations chart of Fig. 7·14, and strap 2, 3 is at a medium voltage of +95 volts (Fig. 7·10). A strap is at a low voltage of +75 volts when both tubes are conducting, and *both* plate voltages are low. An example of this is at 0 pulse (Fig. 7·14) when V_{2B} plate and V_{4B} plate are both low because of these tubes conducting. Strap 0, 1 is then at a low voltage of +75 volts.

The neon lamps used here glow when about 62 volts is applied across their electrodes; smaller voltages cannot light them. The required 62 volts appear across a neon lamp only when the upper electrode has a high bus voltage of +137 volts on it, and the lower electrode has a low strap voltage of +75 volts, producing a voltage difference of 137 − 75 or 62 volts. In the operations chart of Fig. 7·14, this occurs to lamp 0 at pulse 0 since, at this instant, V_{1A} plate voltage is high (V_{1B} is conducting, Table 7·4), making the even bus +137 volts; at the same time, V_{2B} and V_{4B} are conducting (Table 7·4), with their plate voltages low, making strap 0, 1 a low voltage of +75 volts. Neon lamp 0, with 62 volts across it (137 − 75), lights. Neon lamp 1, on pulse 0, does not glow despite its low strap voltage of +75 volts, since its upper electrode is not at the required high voltage.

Pulse 1 sends V_{1A} into conduction (Table 7·4), and the plate voltage of V_{1B} is now high, making the odd bus line a high voltage. As shown in the chart of Fig. 7·14, low plate voltages now exist at V_{2B} and V_{4B}, producing a low voltage on strap 0, 1. Neon bulb 1 now lights.

Pulse 2 produces a high voltage at V_{1A} plate and on the even bus line. Strap 2, 3 now has the low voltage since V_{2A} and V_{3B} are both conducting (Table 7·4) and have low plate voltages (chart, Fig. 7·14). Neon lamp 2 now glows.

Pulse 3 produces a high voltage on V_{1B} plate and on the odd bus. Since V_{2A} and V_{3B} are still conducting (Table 7·4), their low plate voltages puts a low voltage on strap 2, 3. With a high voltage on the upper electrode and a low voltage on the other, neon lamp 3 now lights.

Pulse 4 brings the high voltage back to the plate of V_{1A}, making the even bus voltage high. As shown in Table 7·4, V_{3A} and V_{4B} are now conducting, and these plate voltages are low. Now strap 4, 5 has the low voltage, causing lamp 4 to glow.

Pulse 5 results in the high voltage appearing on the plate of V_{1B} and the odd bus. Strap 4, 5 is still at a low voltage since V_{3A} and V_{4B} are both still conducting (Table 7·4). Neon lamp 5 glows.

Referring to the operations chart of Fig. 7·14, it can be noticed that, starting with 0 pulse and progressing to pulse 9, the high voltage switches up and back from the even bus to the odd bus. This sets up the possibility for either an even- or odd-numbered lamp to light up. Which lamp in a pair *does* glow is then determined by the "low-low" voltage as shown in the chart of Fig. 7·14. Note that starting with 0 pulse, the "low-low" voltage is repeated in the same double column of the chart twice, then moving to the next double column to the right, and again repeating twice (for pulses 2 and 3). This pattern is followed through pulse 9, causing first the even-numbered lamp in a pair to glow and then the odd-numbered one in that pair to glow. Then the action repeats for each succeeding pair of lamps.

Pulse 10 produces the same condition as 0 pulse, where the 0 lamp is lit. However, the 1 lamp in the *next* decade will also light up since an output signal from V_{4B} of the first decade counter (the units column) is sent into V_1 of the next decade (the tens column). At pulse 10 therefore, two lamps are lit, lamp 1 in the tens column and lamp 0 in the units column.

In Fig. 7·10, an output is shown at the extreme lower-right corner being taken from the plate of V_{4B} in the schematic of the decade counter. This is a negative-going voltage (waveshapes, Fig. 7·11) which is produced for the first time at pulse 10. It is this output signal from V_{4B} which is used to trigger V_1 of the next decade.

Another available output voltage, called the "staircase-output voltage," is shown at the extreme left side of the diagram of Fig. 7·10. This is discussed in a later section under "Digital-to-analog conversion."

PROBLEMS

7·1 Convert the following decimal numbers into binary equivalents: (*a*) 47; (*b*) 61; (*c*) 189; (*d*) 77.

7·2 Convert the above decimal numbers into octal equivalents.

7·3 Convert the following binary numbers into decimal equivalents: (*a*) 10001; (*b*) 10001111; (*c*) 11101; (*d*) 1000101.

7·4 Convert the following octal numbers into decimal equivalents: (*a*) 103; (*b*) 256; (*c*) 400; (*d*) 625.

7·5 Convert the following binary numbers into octal equivalents: (*a*) 11010110; (*b*) 110010000; (*c*) 1111111; (*d*) 10101100.

MORE CHALLENGING PROBLEMS

7·6 Convert the following binary numbers into trinary (base of 3) equivalents: (*a*) 1001101; (*b*) 101111; (*c*) 10111101.

7·7 Referring to the unmodified binary chain of Fig. 7·8 and the operations Table 7·3, which neon bulbs would glow on the following input pulses if V_3 were defective and could not flip (remains in its *flop* or *zero* state: (*a*) third pulse; (*b*) fifth pulse?

7·8 Refer to schematic of decade counter (modified binary chain) of Fig. 7·10, to the operations of Table 7·4, and also to the decimal readout system of Fig. 7·14. If tube V_{3A} were dead so that V_3 remained in its *flop* or *zero* state, which decimal readout lights would glow on the third pulse?

7·9 If, in the previous problem, V_{4B} were dead (V_4 therefore remains in its *flip* or *one* state) instead of V_{3A}, which decimal readout lights would glow on: (*a*) pulse 2; (*b*) pulse 10?

7·10 If, in Prob. 7·8, all stages were functioning except that the feedback from V_{3B} plate to V_{2A} grid is open, then which *decimal* readout lamps glow on (*a*) pulse number ten; (*b*) pulse number twelve?

CHAPTER 8

GATES

8·1 Gates, General

A logic gate may be defined as a circuit having two or more input signals but only one output. One or more input signals may function as control or selector pulses. The output signal is produced only at a time chosen by one or more of the control pulses. The gate circuits described here are called *logical* or *switching* gates and are used in all applications of pulse circuitry such as radar, digital computers, and missiles. The circuits discussed here are OR gate, NOR gate, AND gate, NOT-AND gate, and INHIBITOR.

8·2 Diode OR Gates

An OR gate produces an output signal when any *one* or more inputs of the correct polarity are applied. In Fig. 8·1 negative-going input signals are applied to the cathodes of the diodes V_1, V_2, or V_3, or any combination of them. The diodes as shown could be either a thermionic type (vacuum tube) or a semiconductor type (crystal diode) as shown by the dashed-line connections. Without any input being applied, the diodes are nonconducting. No current flows through the load resistor R_L, and E_{out} is zero.

When a negative pulse is applied to a cathode, that diode conducts. As shown in Fig. 8·1*b*, a negative input pulse (at time t_1) is applied to the cathode of V_1 alone. V_1 conducts and current flows, as shown, down through R_L. E_{out} becomes a negative voltage. V_2 and V_3 are still nonconducting. If R_L is much larger than the resistance of the conducting diode (forward resistance) plus the generator impedance, then practically the full input pulse appears across R_L as output signal.

At time t_2 of Fig. 8·1*b*, a -10-volt input pulse is applied to the cathode of V_2. V_2 current now flows down through R_L, and E_{out} is again a -10-volt pulse. Now, V_1 and V_3 are nonconducting. A similar thing happens at time t_3, when a -10-volt input pulse at V_3 cathode makes V_3 conduct, producing a -10-volt output signal. V_1 and V_2 are now nonconducting. Note that an input to V_1 *or* V_2 *or* V_3 produces an output signal, giving the circuit its name of OR gate.

An output is similarly produced when inputs are applied to more than one diode at the same time. At time t_4, -10-volt input pulses are applied to both V_1 and V_2, making these diodes conduct, while V_3 is now cut off.

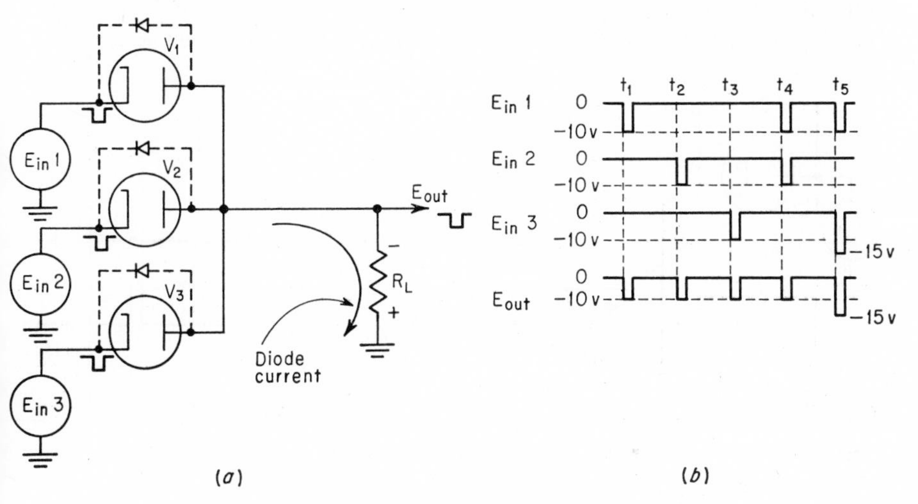

Figure 8·1 **Diode OR gate.**

V_1 and V_2 current flowing down through R_L produces a -10-volt output pulse.

The output signal is equal to the amplitude of the input signal when there is only that one input. When there are two or more simultaneous inputs of equal amplitude, the output is again equal to the amplitude of the input pulse. At time t_5 of Fig. 8·1*b*, a -10-volt pulse is applied to V_1 cathode while a -15-volt pulse is applied to the cathode of V_3. V_3 conducts and E_{out} becomes -15 volts. This -15 volts reverse biases V_1 so that, despite the -10-volt input to the cathode, V_1 is cut off along with V_2. V_3 conducts alone, producing a -15-volt output. An OR circuit with simultaneous input pulses of different amplitude produces an output which is equal in amplitude to the *largest* input signal.

Figure 8·1, as explained above, works with negative input pulses only. An OR gate that requires positive input pulses is shown in Fig. 8·2. Here the diodes are simply reversed from those shown previously. As shown in Fig. 8·2*b*, a $+10$-volt input pulse is applied either to the plate of V_1 (at time t_1), making it conduct, or to the plate of V_2 (at time t_2), making V_2 conduct, or to both plates (at time t_3), making both conduct. For each of these inputs, the output is equal to the amplitude of the input, or +10 volts. At time t_4 of Fig. 8·2*b*, unequal amplitude inputs are applied; +15 volts to V_1 and +10 volts to V_2. Conduction of V_1 up through R_L produces an E_{out} of +15 volts. This reverse biases V_2, preventing V_2 from conducting.

In Fig. 8·3*a*, an OR circuit is shown capacitively coupled from the input circuits. V_1 and V_2 (vacuum tubes or crystal diodes) and R_L comprise the OR gate. V_3 and V_4 (also either thermionic or semiconductor diodes) are clamper diodes required for d-c restoration (DCR) as explained in Sec. 4·8.

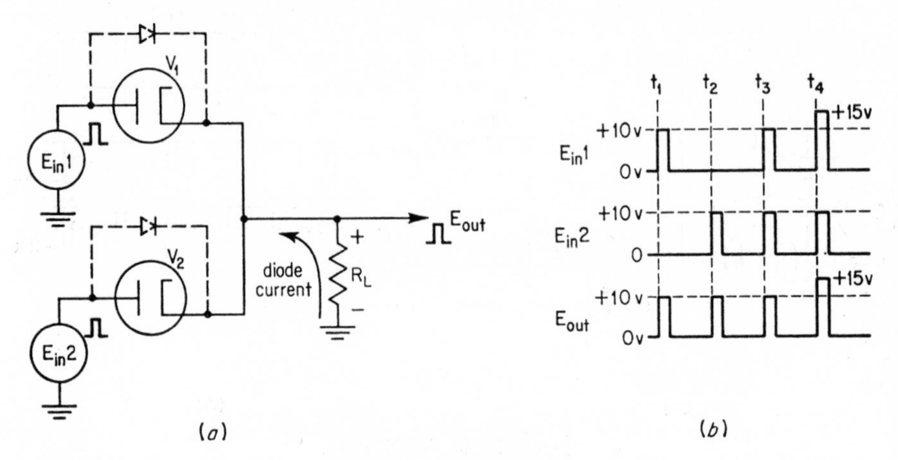

Figure **8·2 Diode OR gate.**

The input signals, as shown in Fig. 8·3*b*, are fluctuating positive d-c voltages. Coupling capacitors C_1 and C_2 block the d-c voltage component, passing only the 10-volt a-c peak-to-peak component. When the input (at 1 or 2) is +200 volts, the associated shunt diode DCR (V_3 or V_4) conducts, clamping the voltage at points X and Y to ground (zero volts). When either input (at 1 or 2) decreases 10 volts to +190 volts, the negative-going signal causes either V_3 or V_4 to cut off. Voltage at points X or Y now becomes −10 volts, as shown in Fig. 8·3*b*. This negative pulse at either V_1 or V_2 cathode, or both, causes one or both of these diodes to conduct down through R_L. An E_{out} of −10 volts is produced whenever either V_1 or V_2, or both, conduct. R_1 and R_2 should be much larger than R_L. The backward resistance of the crystal diodes V_3 and V_4 (shown in the dashed-line connections of Fig. 8·3*a*) may be used instead of R_1 and R_2, with R_1 and R_2 being actually omitted. When the input (1 or 2) is up at +200 volts, the shunt DCR diode (V_3 or V_4) charges up the coupling capacitor (C_1 or C_2). The negative-going input signal causes the OR

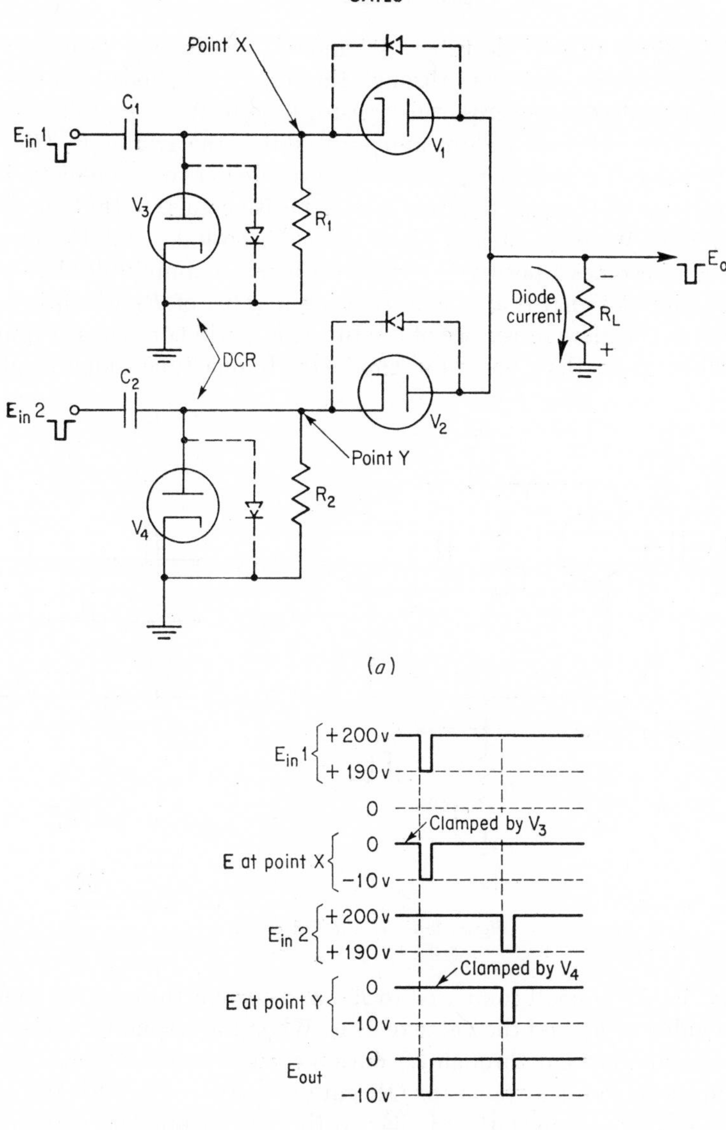

Figure 8·3 **Diode OR gate and DCR.**

diodes (V_1 or V_2) to conduct, permitting the capacitor (C_1 or C_2) to discharge partially.

8·3 Triode, Pentode, and Transistor OR Gates

Triodes, both vacuum tubes and transistors, and multigrid tubes may be employed as OR gates as shown in Figs. 8·4 to 8·7.

In Fig. 8·4*a*, two triode tubes, V_1 and V_2, are shown biased beyond cutoff. Since no current flows through the common cathode-load resistor, R_L, the output voltage across this resistor is zero. When a positive-going pulse of sufficient amplitude is applied to either the grid of V_1 (E_{in} 1 of Fig. 8·4*b*) *or* to the grid of V_2 (E_{in} 2), that particular tube conducts. Plate current flows up through R_L, producing a positive pulse output, as shown in Fig. 8·4*b*. When E_{in} 1 and E_{in} 2 coincide, both V_1 and V_2 conduct. E_{out} again becomes a positive pulse and is larger in amplitude than when only one input signal is applied. By using a value of R_L which is much larger than the plate resistance of the tubes, E_{out} will not increase appreciably when more than one tube conducts. If one tube alone conducts

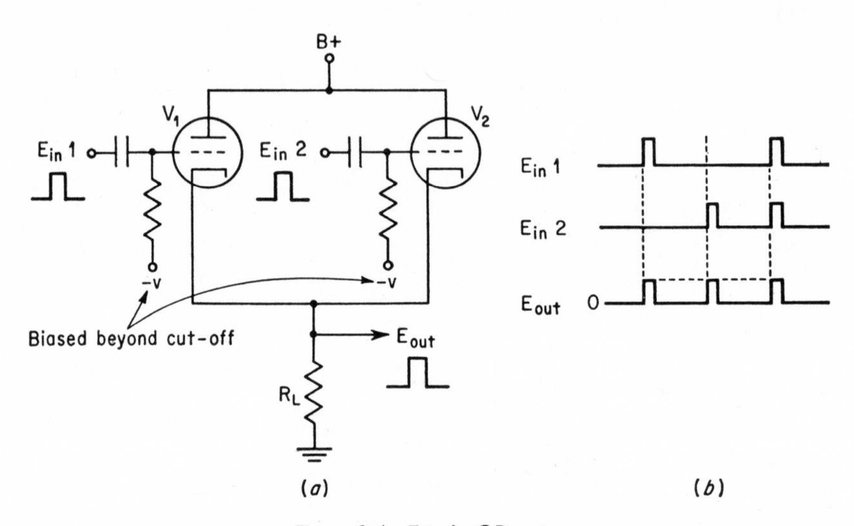

Figure 8·4 Triode OR gate.

heavily, its R_P is small compared to R_L, and current through R_L depends on the value of R_L and the value of B+. When the second tube also conducts heavily, current through R_L does not materially increase.

Figure 8·5*a* shows a transistor OR gate circuit using a PNP transistor using a common base (CB). Similar to the triode tube OR gate of Fig. 8·4*a*, the transistors, T_1 and T_2, are inoperative without an input signal. The bias voltage applied between emitter and base, as shown in Fig. 8·5*a*, is *reversed* with the + voltage applied to the N base. To conduct, *forward* bias must be applied between these elements, with the P emitter positive and the N base negative with respect to each other. As a result of the reverse bias, neither T_1 nor T_2 will conduct. With no current flow through R_L, the lower end of R_L is the same potential as the upper end which connects to the applied negative d-c voltage, $-V$. E_{out} is therefore $-V$.

To conduct, the P emitter must be driven sufficiently positive to overcome the applied reverse bias voltage. When the positive-going input signal, E_{in} 1, is applied to the P emitter of transistor T_1, current flows down through R_L, as shown. Note that transistor emitter-collector current flow, as shown in Fig. 8·5*a*, is backward to the direction of the emitter arrow. (Electron flow is always backward to the arrow direction of the emitter. This arrow indicates the movement of the holes in a transistor.) T_1 current, flowing down through R_L, makes the lower end of R_L the more positive end, and E_{out} goes positive from its normal negative voltage level. This is shown in Fig. 8·5*b*. Transistor T_2 conducts when E_{in} 2, a

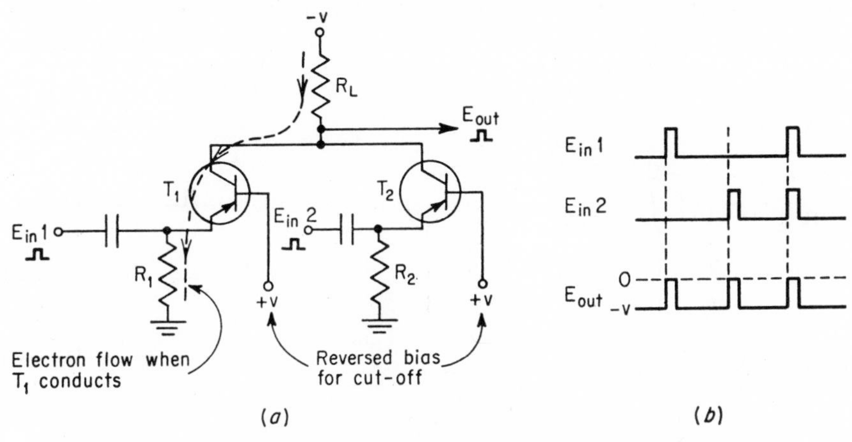

Figure 8·5 **PNP-transistor (CB) OR gate.**

positive-going pulse, is applied to the emitter of this transistor. T_2 current, like T_1 current, flows down through R_L, and E_{out} again is a positive-going pulse. When E_{in} 1 and E_{in} 2 coincide, as shown in Fig. 8·5*b*, both T_1 and T_2 conduct. E_{out} is, as before, a positive-going pulse. If R_L is much larger than R_1, R_2, and the emitter-to-collector resistance of the conducting transistors, then E_{out} will rise to zero from its normal negative voltage. As a result, E_{out} will be the same amplitude whether E_{in} 1 *or* E_{in} 2, *or* both are applied. Note that there is no inversion between the input and output signals here. Only a common emitter (CE) transistor can produce inversion.

Figure 8·6*a* shows another transistor OR gate. This circuit uses NPN transistors with common collectors (CC). As in the previous transistor OR gate of Fig. 8·5*a*, reverse bias between base and emitter is employed to keep the transistors cut off. The P base of Fig. 8·6*a* which should be forward biased (positive with respect to the N emitter) if transistor current is to flow, is connected to a negative voltage (reverse bias), keeping the transistor inoperative.

The positive-going input signal, E_{in} 1, as shown in Fig. 8·6, is applied to the base of T_1. Driving the P base sufficiently positive overcomes the reverse bias and produces T_1 current. Current flows, as shown, up through R_L and from T_1 emitter to collector to the applied positive d-c voltage. Note that this electron flow, as in all transistors, is in the reverse direction from that indicated by the emitter arrow. When T_1 conducts, due to the positive pulse E_{in} 1, a positive-going output pulse is produced across R_L, as shown in Fig. 8·6*a* and *b*. A positive pulse E_{out} is also produced when E_{in} 2 is applied to the base of T_2, as shown. When both inputs E_{in} 1 and E_{in} 2 coincide, both T_1 and T_2 conduct. E_{out} is again a positive pulse, as shown in Fig. 8·6*b*. Since the input pulses drive T_1 and T_2 to saturation, the upper end of R_L (E_{out}) rises from its normal zero value

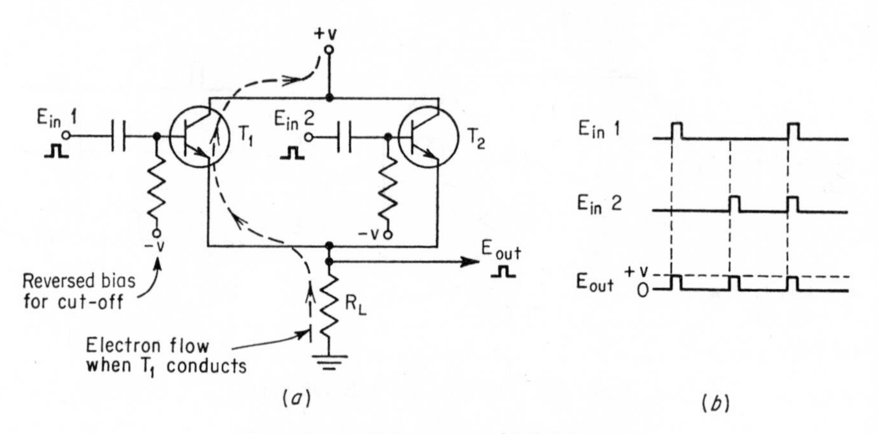

Figure 8·6 **NPN-transistor (CC) OR gate.**

(when T_1 and T_2 are inoperative) to the applied positive d-c voltage. As a result, E_{out} pulse has the same amplitude when E_{in} 1 is applied alone, or E_{in} 2 is applied alone, or when both are fed in simultaneously, as shown in Fig. 8·6*b*.

Figure 8·7*a* shows a pentode tube operating as an OR gate and also as a NOR gate. NOR gates are discussed in the next section (Sec. 8·4). The negative d-c bias voltages applied to the control grid and to the suppressor grid still permit plate current flow. That is, the tube is biased less than cut off. This plate current, flowing up through R_{L_1}, produces a positive voltage at the upper end of R_{L_1}. When a negative pulse (E_{in} 1) of sufficient amplitude is applied to the suppressor grid, plate current is cut off. E_{out} at the *cathode* falls to zero, as shown in Fig. 8·7*b*. This output also falls to zero if a negative pulse (E_{in} 2) is applied to the control grid, again cutting off plate current. If E_{in} 1 coincides with E_{in} 2, plate current cuts off, and the output at the *cathode* decreases from its normal positive value down to zero. The output at the cathode is a *negative*-going pulse whenever a *negative* pulse is applied to either the suppressor grid *or* to the control grid, *or* to both, thereby operating as an OR gate. The output from the pentode *plate* of Fig. 8·7*a* is called the NOR output and is discussed in the following section.

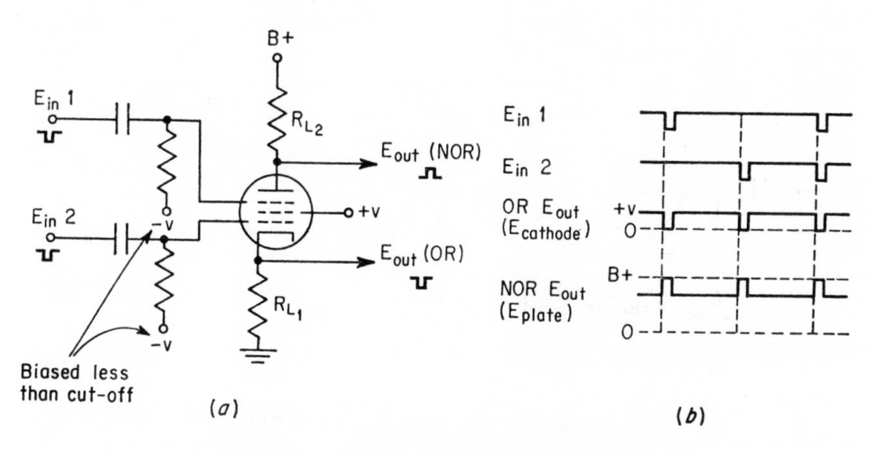

Figure 8·7 **Pentode OR gate, also NOR gate.**

OR gates are often used to prevent feedback or interaction between the output circuits of commonly coupled generators.

8·4 Triode, Pentode, and Transistor NOR Gates

When an OR-gate *output* signal is the reverse polarity of the *input*, the circuit is called a NOT-OR gate, or a NOR gate.

In the OR gates of Figs. 8·4 to 8·6, a *positive* input signal produced a *positive* output. Similarly, in the OR gate of Fig. 8·7, a *negative* input pulse produces the same type of output at the cathode, that is, a *negative*-going pulse. The output signal produced at the plate, however, is of the *opposite* polarity, being a *positive*-going pulse. This is called a NOT-OR or a NOR gate and is accordingly labelled in Fig. 8·7*a* and *b*.

Figure 8·8*a* is a triode NOR gate. This circuit is the same as the triode OR gate of Fig. 8·4*a* except that the output is taken from the plates rather than from the cathodes. As shown in Fig. 8·8*a* and *b*, the tubes V_1 and V_2 are biased beyond cutoff. E_{in} 1 is a positive pulse which causes V_1 to conduct. Plate current flows through R_L, producing a voltage drop across it, and the output at the plate falls from the B+ value. E_{in} 2 would also produce current flow through R_L and a resultant negative-going output pulse. If E_{in} 1 coincides with E_{in} 2, both tubes conduct, and, as shown in Fig. 8·8*b*, a negative pulse output is again produced. If R_L is much larger than the R_P of the conducting tubes, then E_{out} is about the same amplitude when one tube or both conduct.

Figure 8·9*a* is the diagram of a transistor NOR gate. This PNP transistor is using a common emitter (CE), which is the only transistor circuit

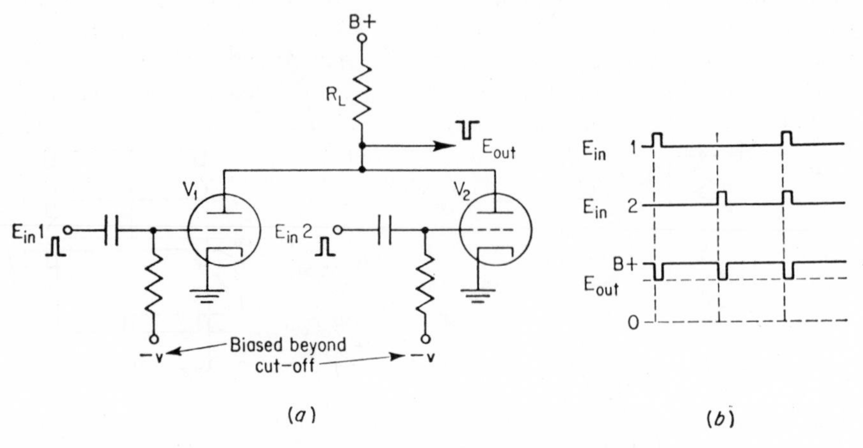

Figure 8·8 Triode NOR gate (NOT-OR).

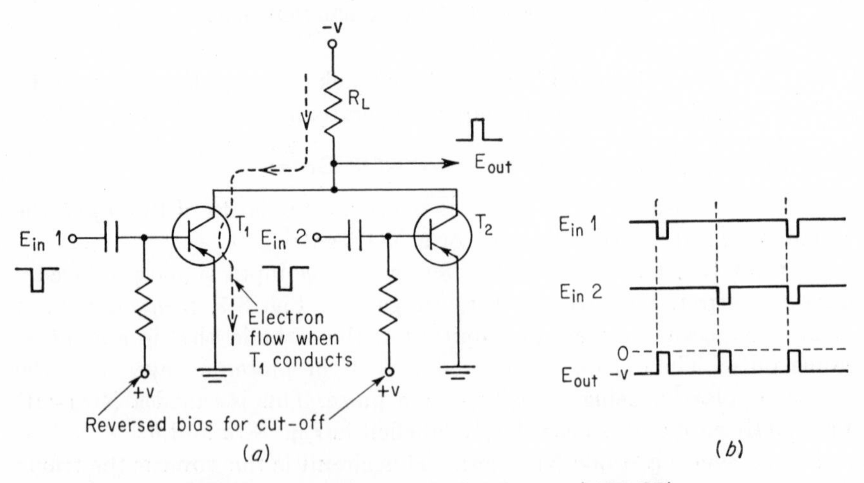

Figure 8.9 PNP-transistor (CE) NOR gate (NOT-OR).

producing phase inversion between input and output signals. The transistors, T_1 and T_2, are reverse biased so that both are cut off. *Forward* bias between emitter and base is required to have transistor current. A P emitter must be positive with respect to the N base, which must be negative. As shown in Fig. 8·9*a*, the reverse bias makes the N base positive with respect to the P emitter. As a result, no current flows through R_L, and E_{out} is the negative d-c voltage applied at the upper end of R_L. Actually zero bias may be employed between base and emitter, instead of the reverse bias shown, to cut off the transistor.

To produce collector current, the N base must be driven negative to overcome the reverse-bias voltage. E_{in} 1, as shown in Fig. 8·9*b*, is a negative-going pulse. Collector current, due to E_{in} 1, flows down through R_L, as shown in Fig. 8·9*a*. Note that, as usual, electron flow in a transistor is in the reverse direction from that indicated by the emitter arrow symbol. Since the negative-going E_{in} 1 pulse drives the transistor into saturation, R_L is much larger than the internal collector-to-emitter transistor resistance, and the entire applied d-c voltage is dropped across R_L. As a

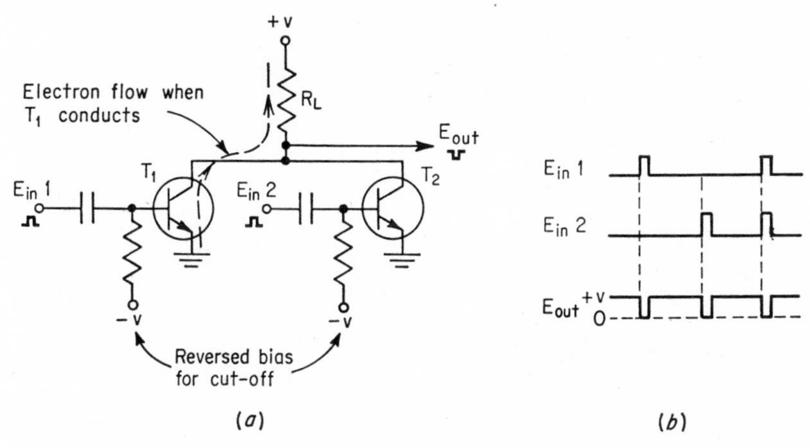

Figure 8.10 NPN-transistor (CE) NOR gate (NOT-OR).

result, E_{out} rises from its normal negative voltage to zero, as shown in Fig. 8·9*b*. Note that a *negative* input signal produces a *positive* output pulse. The same thing occurs if E_{in} 2 were applied instead of E_{in} 1, or if both were applied simultaneously as shown. The circuit is therefore a NOT-OR or a NOR gate.

Figure 8·10*a* is another transistor NOR gate, identical to that shown in Fig. 8·9*a* except that the transistors of Fig. 8·10*a* are NPN type, requiring therefore opposite polarities of the d-c voltages. As in the previous diagram, reverse bias is used to keep the transistors cut off in the absence of an input signal. The reverse bias, as shown in Fig. 8·10*a*, makes the P base negative with respect to the N emitter. In this circuit, a positive pulse input to the base is required to produce collector-current flow. When E_{in} 1, *or* E_{in} 2, *or* both are applied as shown, either T_1, or T_2, or both conduct. This collector current flows up through R_L, as shown in Fig. 8·10*a*, and E_{out} falls from its normal positive voltage down to zero. As shown in Fig. 8·10*b*, *positive* input signals, E_{in} 1 *or* E_{in} 2, produce *negative* output pulses. The circuit therefore functions as a NOT-OR or NOR gate.

The NOR gates of Figs. 8·9*a* and 8·10*a* use transistors in parallel; that is, all collectors are connected together, and all emitters are grounded. Several transistors may be parallel-connected in such a circuit, the number being limited by the leakage currents of the normally inoperative stages. The total current is, of course, the sum of each and may be sizeable. A circuit where the total current is about equal to that of only one transistor is the series-transistor circuit of Fig. 8·11*a*. Forward bias is used on

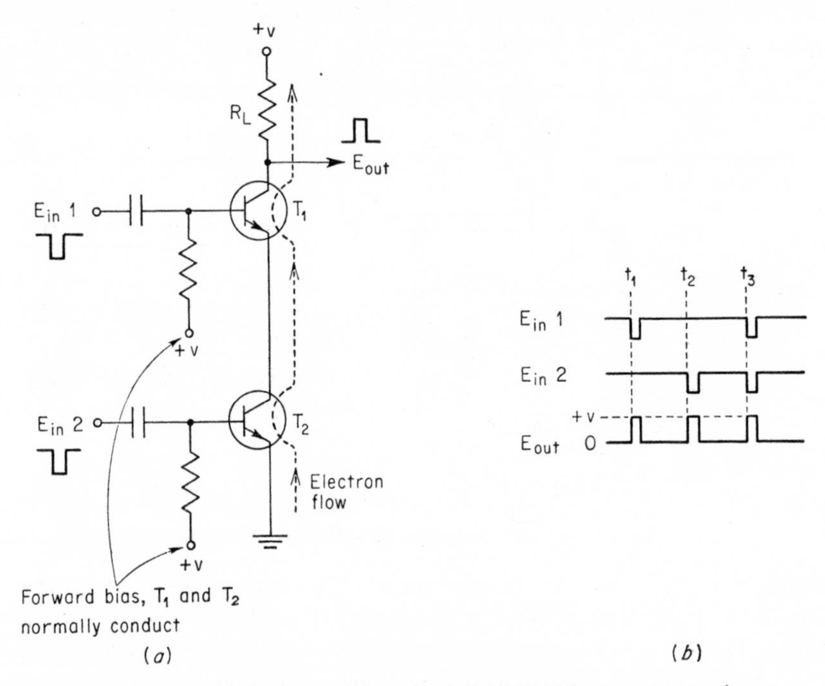

Figure **8·11** **NPN-transistor NOR gate (NOT-OR) (series transistors).**

each base of the NPN transistors so that T_1 and T_2 normally conduct. Current flow through the transistors, as indicated by the dotted-line arrows, produces a voltage drop across R_L. E_{out} will just about be at zero volts before any input signal is applied, as shown before instant t_1 of Fig. 8·11*b*.

At instant t_1, a negative pulse is applied to the base of transistor T_1. This produces a reverse bias on this base, cutting off transistor T_1 current and resulting in T_2 also becoming inoperative. As a result, no current flows through R_L, and E_{out} rises to the applied positive d-c voltage. The negative-going input pulse at instant t_1 of Fig. 8·11*b* has therefore produced a positive-going output pulse, and the circuit is a NOT-OR or NOR gate. Similar output pulses are produced at instant t_2, when transistor T_2 is cut off, and at instant t_3, when input pulses cut off both transistors directly. Output signals are therefore produced, with phase inversion, when an input is applied to T_1, *or* to T_2, *or* to both.

8·5 Diode AND Gates

An AND gate produces an output signal only when *all* the input signals of the correct polarity occur at the same time. If one of the input signals is not present, there will be no output signal produced even though the

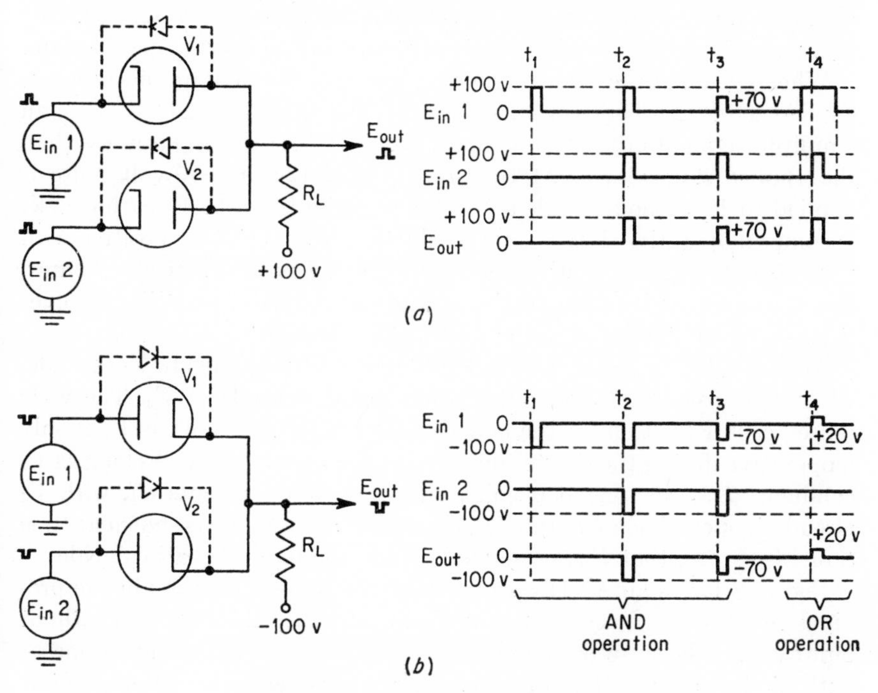

Figure 8·12 Simple diode AND gates.

other inputs are being applied. If the input signals have different pulse durations, the output signal appears only during the time that *all* inputs *coincide*. Another name for an AND gate is therefore a *coincidence* circuit.

Figure 8·12 shows two simple diode AND gates, with diagram (a) operating with positive-going input signals and (b) working with negative input pulses. In Fig. 8·12*a* and *b* it should be assumed that the diodes, V_1 and V_2 (either vacuum tubes or crystals), have practically zero ohms forward resistance. Also, the input generators are assumed to have zero ohms internal impedance. Later in this section a more detailed discussion will be given assigning ohmic values to these, and the need for a clamping diode with the AND gate will be established.

The diodes, V_1 and V_2, in Fig. 8·12*a*, without input signals, are both conducting due to the applied +100 volts d-c, making the plates positive with respect to the cathodes. Assuming practically zero ohms for the diode forward resistance and for the input generator, the applied 100 volts d-c all appears across R_L. As a result, E_{out} is zero volts. At time t_1 of Fig. 8·12*a*, a +100-volt pulse is applied to the cathode of V_1, cutting off this diode. V_2, however, with no input, continues to conduct, and E_{out} remains at zero volts.

At time t_2 of Fig. 8·12*a*, +100-volt pulses are applied to both V_1 *and* V_2 cathodes. Both diodes become inoperative, no voltage is dropped across R_L, and E_{out} becomes +100 volts, as shown in the waveshape diagram.

When inputs of the correct polarity, but with different amplitudes, are applied to *all* the diodes of an AND gate, the output signal will have an amplitude equal to the *smallest* input, provided that this input is less than the applied d-c voltage. At time t_3 of Fig. 8·12*a*, a +70-volt pulse is applied to V_1 cathode, while a +100-volt pulse is fed to V_2 cathode. V_2 becomes inoperative, but V_1 still conducts, since its plate is positive with respect to its cathode. With V_1 conducting and having practically 0 ohms, the +70 volts at its cathode also appears at its plate, and E_{out} becomes +70 volts, as illustrated in the waveshape diagram.

At time t_4 in Fig. 8·12*a*, a wide +100-volt pulse is applied to V_1 cathode, while a narrow +100-volt pulse is fed to the cathode of V_2. The wide pulse keeps V_1 cut off for the long duration of the pulse, while V_2 is only inoperative during the shorter duration of the narrow pulse. As long as V_2 conducts, E_{out} remains at 0 volts. However, when V_2 cuts off, with V_1 already inoperative, an output signal is produced. Note, as shown at time t_4 of Fig. 8·12*a*, that E_{out} occurs only when the input signals coincide.

Figure 8·12*b* is an AND gate for negative-going pulses. As in the previous circuit of diagram (a), V_1 and V_2 are normally in conduction without input signals because of the applied −100 volts d-c. As long as one or both diodes conduct, E_{out} remains at zero volts, as shown before, and at time t_1. At t_1, a −100-volt pulse is applied to the plate of V_1, cutting off this diode. V_2, however, continues to conduct, and E_{out} remains clamped to zero.

At time t_2, −100-volt pulses are applied to V_1 *and* V_2 anodes. Both diodes become inoperative, and no current flows through R_L. With no voltage across R_L, E_{out} becomes −100 volts, as shown.

The output becomes equal to the *smaller* input at time t_3 where a −70-volt input pulse is applied to V_1 anode, while a −100-volt pulse is fed to V_2 anode. V_2 cuts off, but V_1 continues to conduct. With −70 volts at the anode of the conducting V_1, its cathode, and E_{out}, becomes clamped to this figure, as shown in the waveshapes of Fig. 8·12*b*.

An AND gate would operate as an OR circuit when an input pulse of opposite polarity is applied; that is, opposite in polarity to that required by the AND circuit. This OR operation is illustrated at time t_4 of Fig. 8·12*b*. Here a +20-volt pulse is applied to the anode of V_1, with no input applied to V_2. With the forward resistance of V_1 considered to be zero, the +20 volts at the anode also appears at its cathode. V_2 becomes inoperative since its cathode, connected to V_1 cathode, is +20 volts, while its anode is zero. E_{out}, as shown, is +20 volts. The circuit therefore produces a positive-going output signal when a positive input signal is applied to V_1 *or* V_2, making the circuit function as an OR gate.

The discussion of the diode AND gates of Fig. 8·12*a* and *b* assumed that the diodes, V_1 and V_2, had zero ohms forward resistance and also that the input generators had zero ohms. With this theoretical setup, output signals are only produced when inputs of the correct polarity are applied to V_1 *and* to V_2. However, actually both the diodes and the generators have an ohmic value which is not zero. As a result, the AND gate produces a small output signal when only *one* input is applied and a large output when *all* inputs are applied. A clamper diode is therefore added to the AND gate to prevent the small output from being produced.

The following discussion of Fig. 8·13 will show why the clamper (V_3 of Fig. 8·13*a*) is necessary in a practical AND gate.

In Fig. 8·13*a*, small values of 100 ohms are assumed for the forward resistance of V_1 and V_2 and for each of the input generators internal resistances, R_1 and R_2. A value of 900 ohms will, at first, be used for R_L in the various calculations to determine E_{out} without the clamper V_3. Then a larger R_L of 9,900 ohms will be used to show the effect on E_{out}. Finally, E_{out} will be determined using the large R_L and the clamper V_3.

Without an input signal, and without V_3 of Fig. 8·13*a*, V_1 and V_2 both conduct because of the −100 volts d-c applied. As shown in (c), V_1 and R_1 in series act as a 200-ohm resistor, and V_2 and R_2 in series act like another 200-ohm resistor. The two 200-ohm resistors are in parallel, acting like a 100-ohm resistor. This 100-ohm resistor is, in turn, in series with the 900-ohm R_L, making a total resistance of 100 + 900 or 1 kilohm. As shown in (*c*), the voltage across the 100 ohms (the two 200 ohms in parallel) is $^{100}/_{1,000} \times 100$ volts = 10 volts. Since the current flow is down through the 200-ohm resistors, the top of these, which is E_{out}, is a negative voltage or −10 volts. This is shown in Fig. 8·13*b*, waveshape (3) before time t_1.

At time t_1, a −100-volt pulse is applied to the anode of V_1, cutting off this diode. In the previous diagram of Fig. 8·12*b*, the output was shown as zero when only one input was applied. However, because of the actual finite values (not zero) of the diodes and R_1 and R_2, the output will not be zero. As shown in Fig. 8·13*d*, V_2 will conduct alone (since V_1 has been cut off by the −100-volt input pulse). V_2 and R_2 make up a 200-ohm

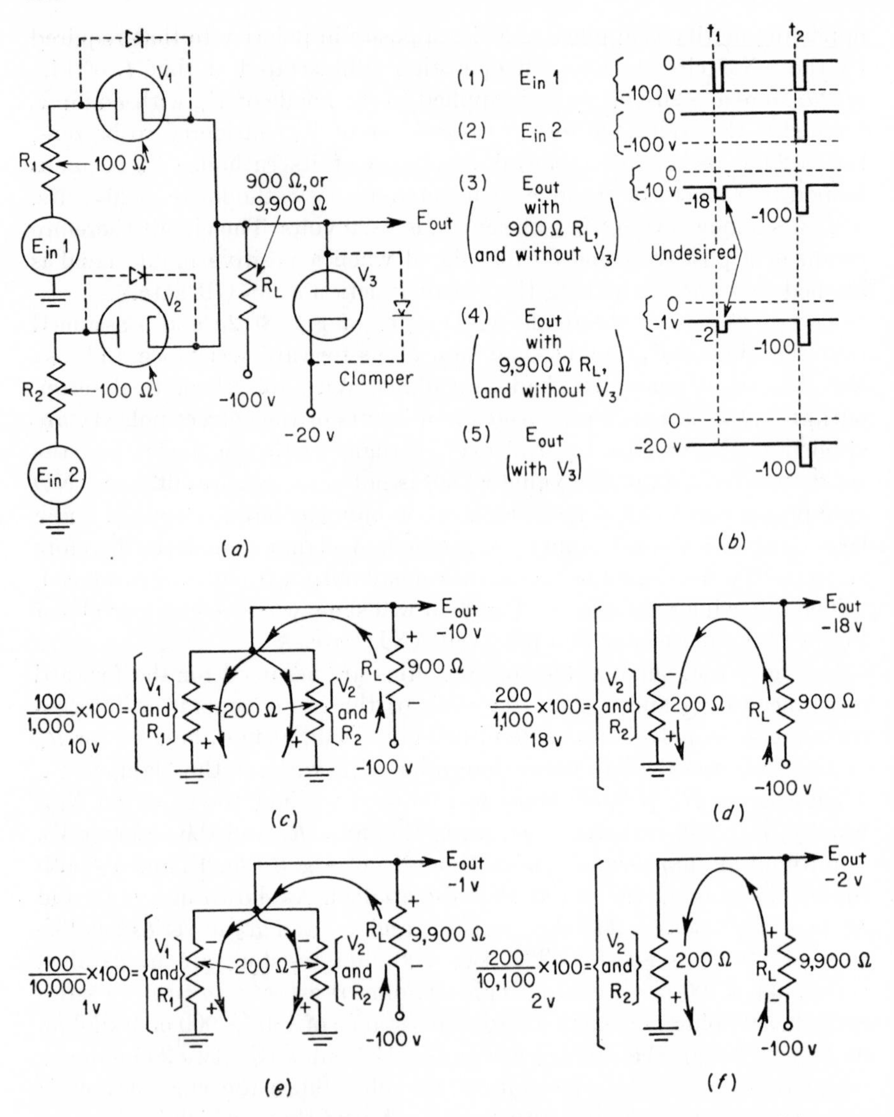

Figure 8·13 **Details of diode AND gate, with and without clamper. (a) Circuit; (b) waveshapes; (c) V_1 and V_2 conducting; (d) V_2 conducting alone; (e) V_1 and V_2 conducting; (f) V_2 conducting alone.**

resistor in series with the 900-ohm R_L, for a total resistance of 200 + 900 or 1,100 ohms. As shown in (*d*), voltage across the 200-ohm resistor is $^{200}/_{1,100} \times 100$ volts = 18 volts (approx.). Since current is flowing down through the 200 ohms, the upper end is negative, and E_{out} is -18 volts. Therefore, as illustrated in diagram (*b*) at time t_1, an input of

-100 volts to V_1 only produces an output pulse of -18 volts, as shown in waveshape (3). This is undesirable in an AND gate.

At time t_2 of Fig. 8·13*b*, -100-volt pulses are applied to V_1 and V_2, cutting off both diodes. With no current flowing through R_L (still assuming that V_3 is out of the circuit), E_{out} becomes -100 volts, as shown in waveshape (3) and is the usual operation of an AND gate.

The undesirable output pulse of -18 volts [waveshape (3) at time t_1] due to the input pulse applied to V_1 only can be decreased by using a much larger value of R_L. If R_L is made 9,900 ohms (still omitting V_3), then at a time before t_1 when no input pulse is applied, V_1 and V_2 conduct normally. The circuit is as shown in Fig. 8·13*e*. V_1 and R_1, a total of 200 ohms, are in parallel with V_2 and R_2, another 200 ohms. This parallel combination amounts to 100 ohms and is in series with the 9,900-ohm R_L, making a total resistance of $100 + 9{,}900 = 10{,}000$ ohms. As shown in diagram (*e*), voltage across the 100-ohm resistor (the two 200 ohms in parallel) is $^{100}/_{10{,}000} \times 100$ volts $= 1$ volt. Since current flow is down through the 200-ohm resistors, their upper ends are a negative voltage, and E_{out} is -1 volt. This output is shown in diagram (*b*) waveshape (4) just before time t_1.

At time t_1, a -100-volt input pulse is applied to V_1 anode, cutting off that diode. V_2, however, still conducts, and the circuit is as shown in (f). V_2 and R_2 make up a 200-ohm resistor which is in series with the 9,900-ohm R_L, for a total resistance of $200 + 9{,}900 = 10{,}100$ ohms. Voltage across the 200-ohm resistor is $^{200}/_{10{,}100} \times 100$ volts $= 2$ volts (approx.). As shown in (f), current flow is down through the 200-ohm resistor, making the upper end negative, and E_{out} is -2 volts. This very small output pulse is shown in diagram (*b*) waveshape (4) at time t_1. Note that this output pulse, using R_L of 9,900 ohms, is much smaller than the one shown at t_1 of waveshape (3), where an R_L of 900 ohms was used. At this time t_1, where only one input is being applied, no output pulse is desired.

At time t_2, when -100-volt pulses are applied to V_1 *and* V_2, both diodes cut off, and E_{out}, as usual, becomes -100 volts, as shown at t_2 waveshape (4).

Now consider the output with the clamper V_3 added to the circuit of Fig. 8·13*a*. V_3 will conduct whenever the output voltage at its anode is positive with respect to its cathode. As shown in diagram (*a*), if the cathode of V_3 is connected to -20 volts d-c, V_3 will conduct at all times during waveshape (4) except when the output becomes more negative than -20 volts. Before t_1, the output of waveshape (4) is at -1 volt. This makes V_3 anode positive with respect to its cathode. V_3 conducts, and E_{out} is clamped at -20 volts, as shown in waveshape (5), assuming zero ohms for V_3. The actual low forward resistance of V_3 simply results in clamping at slightly less than -20 volts.

Clamping by V_3 also occurs at time t_1 when E_{out} of waveshape (4) tries to become -2 volts. Again V_3 conducts, clamping E_{out} to -20 volts, as shown in waveshape (5).

At time t_2 of Fig. 8·13*b*, E_{out} becomes -100 volts. Since this makes V_3 anode negative with respect to its cathode, V_3 becomes inoperative, and E_{out} is permitted to become -100 volts, as shown in waveshape (5).

The clamper V_3 also prevents the AND gate from functioning as an OR circuit for opposite-polarity input pulses. If a positive input pulse were applied to the anode of V_1, the conduction of V_1 would then make E_{out} positive. V_3 would then conduct, clamping E_{out} to -20 volts and preventing the positive input pulse from appearing in the output.

8·6 Triode, Pentode, and Transistor AND Gates

Triode AND gates, like the diode AND circuits, require input signals to *all* stages to produce an output.

Figure 8·14*a* shows a triode AND gate. V_1 and V_2 are normally conducting. Their plate currents flow through the common cathode load resistor R_L. Before instant t_1 of Fig. 8·14*b*, no input signal is yet applied. E_{out} at point *A* is a small positive d-c voltage, say $+10$ volts as an example, as shown in Fig. 8·14*b*. At instant t_1, a negative-going pulse (E_{in} 1) is applied to the grid of V_1, cutting off this tube. V_2, however, still conducts. With less current flow through R_L, E_{out} at point *A* (the top of resistor R_L) becomes less positive, say $+6$ volts as an example, as shown in Fig. 8·14*b*. This negative-going output at t_1 is undesired in an AND gate since only one input is being applied at this time. This output at point *A* will, however, be quite small since this negative-going output pulse on the cathode of V_2 effectively acts like a positive-going pulse at its grid (less bias). V_2 conducts more heavily, making the top of R_L go more positive. This opposes, somewhat, the falling voltage at the upper end of R_L with the result that E_{out} at point *A* does not decrease as much.

To prevent even this small pulse from appearing in the output, a series-diode clamper V_3 is added. The output is now taken from point *B* in Fig. 8·14*a*. When the cathode of diode V_3 is either $+10$ volts (before instant t_1 of Fig. 8·14*b*) or $+6$ volts (at instant t_1), V_3 does not conduct since its anode is less positive, $+3$ volts in the diagram. Since no current flows, as yet, through R_3, E_{out} at point *B* is $+3$ volts, and the undesirable negative-going output pulse at point *A* does not appear at point *B*.

A large negative-going pulse is produced at point *A* (top of R_L) at instant t_2 of Fig. 8·14*b*, when negative input pulses (E_{in} 1 and E_{in} 2) are applied to the grids of V_1 *and* V_2. Both triodes are cut off, and no current flows through R_L. E_{out} at point *A* now becomes 0 volts. This negative-going output pulse at instant t_2 is desirable since it is due to inputs at V_1 *and* V_2. Series-diode clamper V_3 conducts when its cathode (point *A*) becomes zero volts. Current now flows through R_3, and E_{out} at point B becomes clamped to zero at instant t_2.

Another triode AND gate is shown in Fig. 8·15*a*. Here input signals are applied to the cathodes of V_1 and V_2, with the output taken off at the common plate connection. Both V_1 and V_2 are normally conducting through the common plate-load resistor R_L. With a large R_L, most of the +300 volts applied is dropped across R_L, producing a small voltage, say, +50 volts, at each plate. This is shown in Fig. 8·15*b*, before instant t_1, as E_{out} at point A.

At instant t_1, a positive-going pulse (E_{in} 1) is applied to the cathode of V_1. This has the effect of driving the grid negative, and V_1 cuts off, and

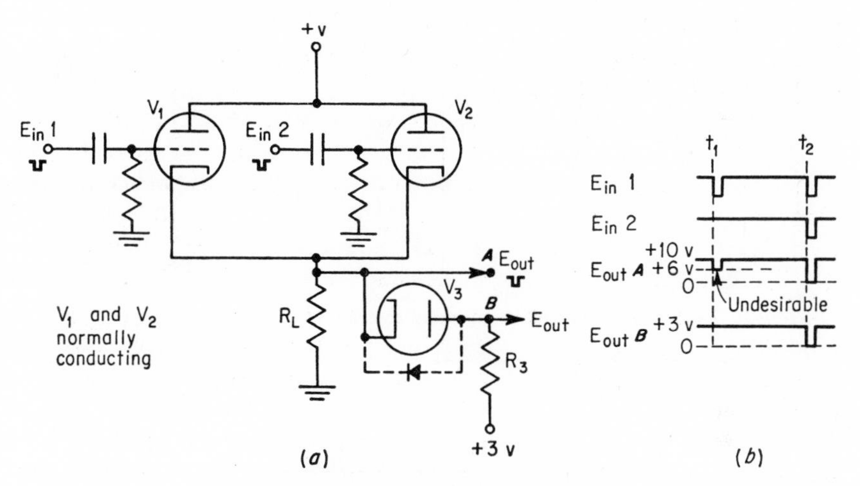

Figure 8·14 Triode AND gate.

V_2 continues to conduct. With less current through R_L, less voltage is dropped across it, and E_{out} at point A rises to, say, +70 volts, as shown in Fig. 8·15*b*. The amount of this rise is reduced by the action of V_2. When V_1 becomes cut off, because of E_{in} 1 at instant t_1, the rising plate voltage (E_{out} at point A) causes V_2 to conduct more heavily. This tends to lower its plate, thus partially reducing the rise of voltage.

This rise of output voltage at point A is undesirable in an AND gate since only one input is applied at this time. Series diode clamper V_3 in Fig. 8·15*a* prevents this small positive-going output at point A from appearing as output at point B. With +100 volts d-c applied through R_3 to the cathode of V_3, the diode cannot conduct when its anode is +50 volts (before instant t_1) or at +70 volts (at instant t_1). With no current flow through R_3, E_{out} at point B is at the +100 volts d-c applied, preventing the undesirable positive-going output pulse from appearing at point B.

At instant t_2 of Fig. 8·15*b*, positive-going input pulses (E_{in} 1 and E_{in} 2) are applied to the cathodes of V_1 *and* V_2, cutting off both triodes. With no current flow through R_L, there is zero voltage drop across it, and E_{out} at point A rises to the full +300 volts d-c applied. This +300 volts on V_3 anode causes the diode to conduct, producing a voltage across R_3, and

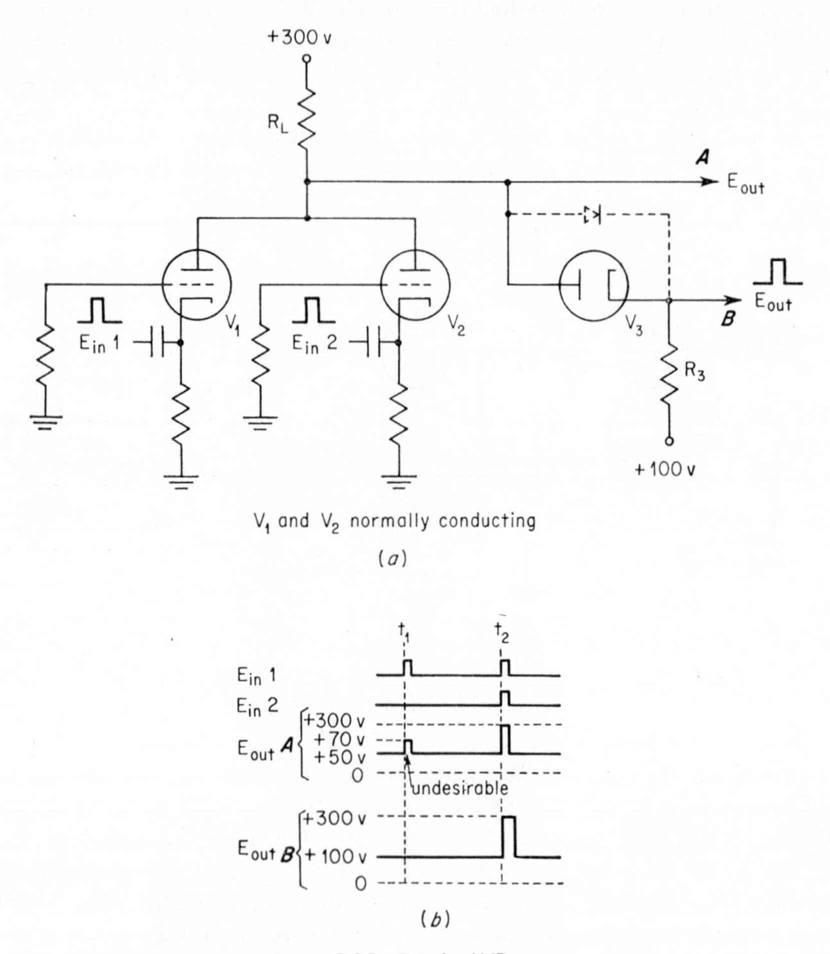

Figure 8·15 Triode AND gate.

E_{out} at point B becomes clamped to +300 volts. As shown in Fig. 8·15*b*, E_{out} at point B consists of a positive-going pulse only at instant t_2, when inputs are applied to V_1 *and* V_2. No output signal appears at point B at instant t_1, when an input is applied only to V_1.

A pentode AND gate is shown in Fig. 8·16*a*. This stage is normally inoperative due to the choice of d-c voltages applied to the various elements. Plate current can only flow when both the control grid *and* the suppressor grid are driven positive simultaneously.

At instant t_1 of Fig. 8·16*b*, a positive-going pulse (E_{in} 1) is applied to the control grid. The tube still remains cut off due to the negative d-c voltage still present on the suppressor grid. E_{out} at the *cathode*, with no current flow through R_K, remains at zero volts, as shown in Fig. 8·16*b*.

At instant t_2, positive-going pulses (E_{in} 1 and E_{in} 2) are applied to the control grid *and* to the suppressor. Plate current now flows through R_K, and the output at the *cathode* now goes positive, producing the output of an AND gate. The output from the plate, shown as a *negative*-going pulse, is of the reverse polarity compared to the input signals. This is the output of a NOT-AND gate and is discussed in Sec. 8·7.

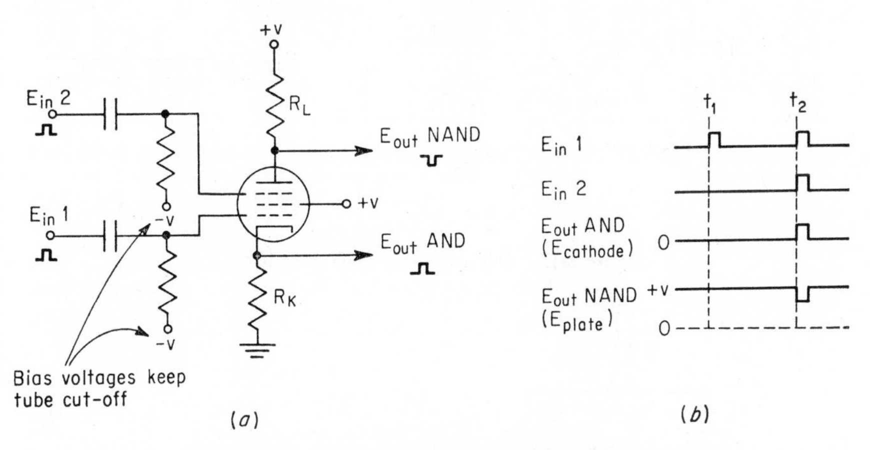

Figure 8·16 Pentode AND gate and NAND gate.

Transistor AND gates are shown in Figs. 8·17*a* and 8·18*a*. In Fig. 8·17*a*, a PNP-type transistor is shown using forward biasing between the P emitter and the N base, with the N base being connected to a negative d-c voltage. The transistors T_1 and T_2 are normally heavily conducting. Electron flow is indicated by the dotted-line arrows of Fig. 8·17*a*, flowing down through R_L. E_{out} is taken from the positive end of R_L. When both transistors T_1 and T_2 are conducting, most of the negative d-c applied voltage is dropped across R_L, and E_{out} is practically at zero volts, provided that R_L is very much larger than the collector-to-emitter resistance of each transistor and also much larger than R_1 and R_2.

At instant t_1 of Fig. 8·17*b*, a negative-going pulse (E_{in} 1) is applied to the emitter of T_1. This negative pulse reverse biases the P-type emitter, cutting off T_1. T_2 continues to conduct however, and again practically all the negative d-c applied voltage is dropped across R_L. E_{out}, as shown in Fig. 8·17*b*, remains at zero volts.

At instant t_2 of Fig. 8·17*b*, negative-going pulses (E_{in} 1 and E_{in} 2) are applied to the emitters of T_1 *and* T_2. This reverse biases both transistors, and they cut off. With no current flowing through R_L, there is no voltage drop across it, and E_{out} becomes the full applied negative d-c voltage. As

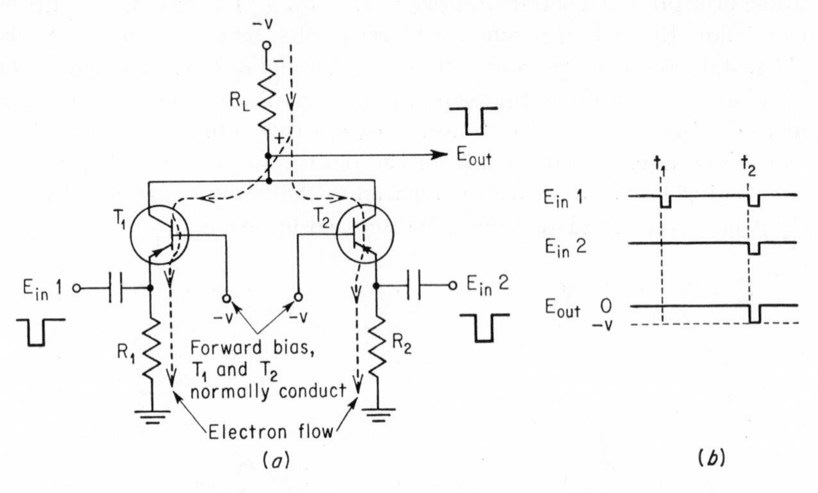

Figure 8·17 **PNP-transistor AND gate (CB).**

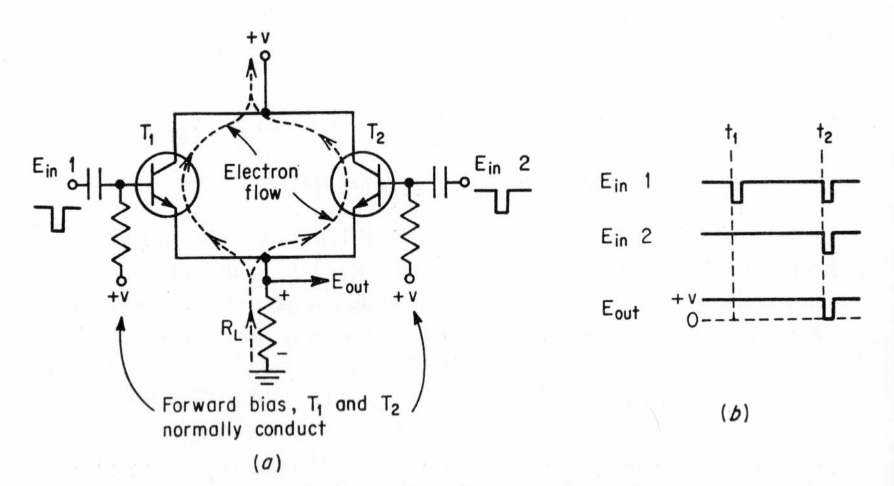

Figure 8·18 **NPN-transistor AND gate (CC).**

shown in Fig. 8·17*b*, a negative-output pulse is produced by this AND gate only when negative pulses are applied to *all* inputs. Note that there is no phase inversion in the circuit of Fig. 8·17*a* since this diagram uses a common base (CB). Only a common emitter (CE) circuit produces phase inversion.

The NPN transistor AND gate of Fig. 8·18*a* uses a common collector (CC) and similarly produces no phase inversion. Operation is very similar to the circuit of Fig. 8·17, just described.

In Fig. 8·18*a*, the NPN transistors are forward biased by virtue of the applied positive d-c voltage applied to the bases. Transistors T_1 and T_2 are normally in heavy conduction. Electron flow is shown by the dotted-line arrows and is up through R_L. E_{out} is therefore normally a positive d-c voltage, just about equal to that of the applied d-c, since R_L is much larger than the emitter-to-collector transistor resistance.

When transistor T_1 is cut off by the negative input pulse at instant t_1 of Fig. 8·18*b*, which reverse biases the P-type base, T_2 continues to conduct. E_{out} remains at its positive d-c voltage level. When both transistors become inoperative at instant t_2, due to the negative pulses (E_{in} 1 and E_{in} 2) reverse biasing both bases, no current flows through R_L. E_{out} now becomes zero, as shown in Fig. 8·18*b*, and a negative-going output pulse is produced only when negative pulses are applied to *all* inputs.

8·7 NOT-AND (NAND) Gates

When an AND gate produces an output which is phase inverted compared to the input pulses, the circuit is called a NOT-AND gate and is also referred to as a NAND gate. Triode, pentode, and transistor AND gates which produce phase inversion are therefore called NOT-AND or NAND gates.

In the pentode coincidence gate of Fig. 8·16, positive-going input signals (E_{in} 1 and E_{in} 2) at instant t_2 drive the tube into conduction. The pulse of plate current flows up through R_K and R_L. This produces a *positive*-going output pulse at the cathode (an AND gate) and a *negative*-going pulse at the plate (a NOT-AND gate or NAND), as shown in Fig. 8·16*b*.

A triode NOT-AND gate (NAND) circuit is shown in Fig. 8·19*a*. This circuit is identical to the triode AND gate of Fig. 8·15 except that the inputs are applied to the control grids instead of to the cathodes. To cut the tubes off therefore, the inputs must be negative-going, as shown at instants t_1 and t_2 of Fig. 8·19*b*. When no input is applied, E_{out} at point A (plate voltage) is low, say +50 volts, as shown in Fig. 8·19*b*, because of the large value of R_L.

A small, *undesirable*, positive pulse is produced at instant t_1 at point A when a negative pulse (E_{in} 1) is applied to the grid of V_1. This output pulse is not desired in a NAND (NOT-AND) gate since only one input pulse has been applied at instant t_1. At this instant, E_{in} 1 cuts off V_1, and the plate voltage (E_{out} at point A) rises to say +70 volts. The amplitude of this rise is reduced by V_2 which conducts more heavily when the voltage at its plate rises. The heavier conduction of V_2 tends to lower the voltage at point A, preventing a large rise.

The series-diode clamper V_3 prevents this small, positive output pulse at point A from appearing at point B. The diode is biased with +100 volts applied through R_3 to its cathode, as shown in Fig. 8·19*a*. When the

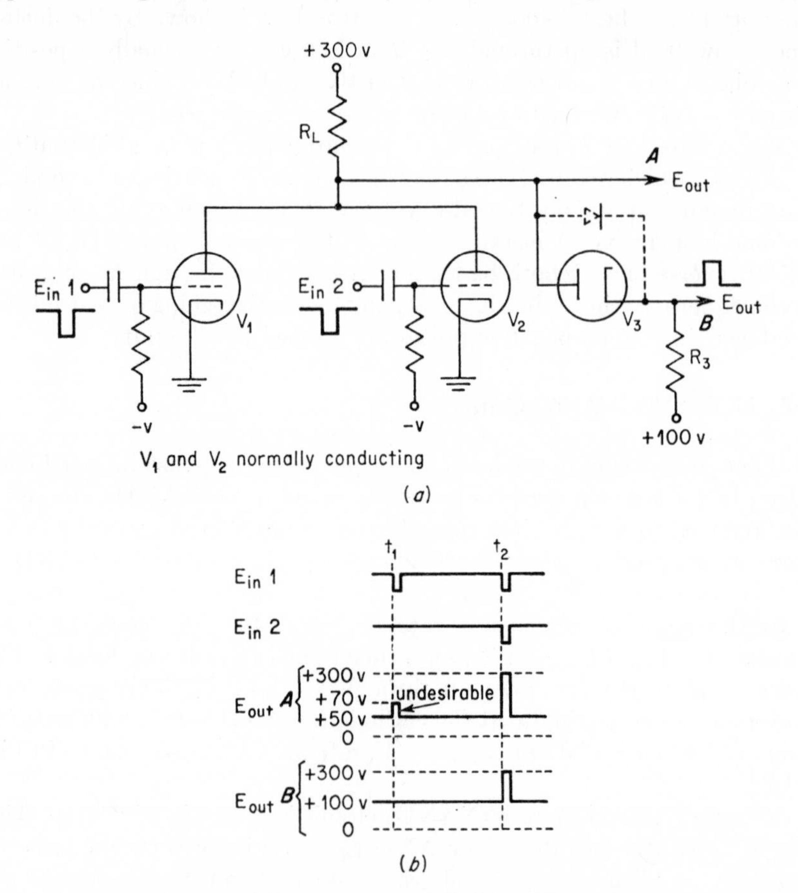

Figure 8·19 **Triode-NAND (NOT-AND) gate.**

voltage at point A is +50 volts (before t_1) and then rises to +70 volts (at t_1), V_3 remains inoperative. No current flows through R_3, and E_{out} at point B is +100 volts as shown in Fig. 8·19*b*.

At instant t_2, V_1 and V_2 are both cut off because of negative input pulses (E_{in} 1 *and* E_{in} 2) being applied to both control grids. With no triode current flow through R_L, E_{out} at point A rises to the full applied d-c voltage of +300. V_3, with +300 volts on its anode, conducts, clamping point B to +300 volts. As a result, E_{out} at point B consists of a *positive-*

going pulse only when *negative*-going pulses are applied to *all* inputs. The circuit therefore operates as a *phase inverting* AND gate, or a NOT-AND or NAND gate.

Transistor NAND gates are shown in Figs. 8·20*a* and 8·21*a*. Figure 8·20*a* uses two NPN transistors in a parallel common emitter (CE) circuit. A CE circuit is a phase inverter. Transistors T_1 and T_2 are forward biased (positive d-c voltage connected to the P-material base), and both conduct heavily. Since R_L is very much larger than the transistor emitter-to-collector resistance, practically the entire applied positive d-c voltage is dropped across R_L. E_{out} (at the collector) is therefore zero volts, before instant t_1, as shown in Fig. 8·20*b*.

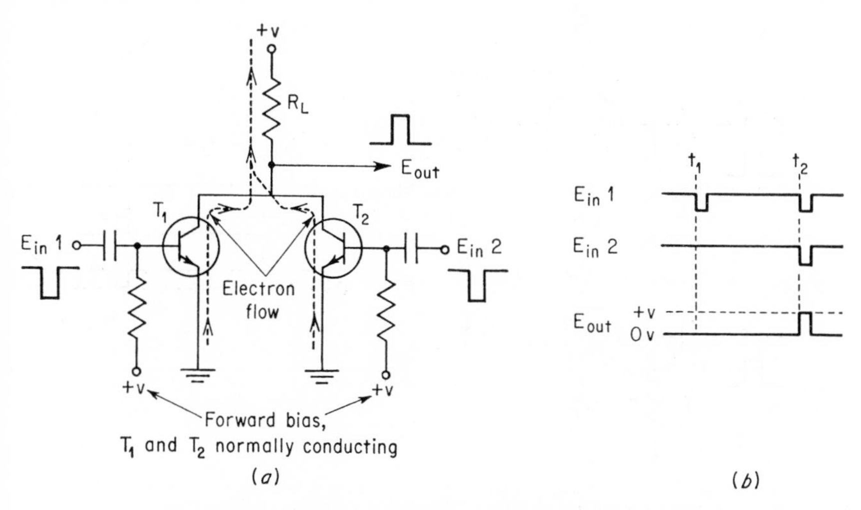

Figure **8·20 NPN-transistor NAND (NOT-AND) gate (CE).**

At instant t_1, a negative pulse (E_{in} 1) is applied to the P-type base, reverse biasing T_1, and this transistor cuts off. However, T_2 still conducts, and E_{out} remains at about zero volts, as shown in Fig. 8·20*b*.

At instant t_2, negative pulses (E_{in} 1 *and* E_{in} 2) are applied to both bases. This reverse biases both T_1 and T_2, cutting them off. With no current flow through R_L, E_{out} rises to the full applied positive d-c voltage. Since a *positive*-pulse output is produced only when *negative* input signals coincide at *all* inputs, the circuit is a *phase-inverting* AND gate, or a NOT-AND or NAND gate.

The transistor NAND gate of Fig. 8·21*a* uses two NPN transistors in a series circuit. The transistors T_1 and T_2 are reverse biased (negative d-c voltage connected to the P material base) and are therefore inoperative. Since no current flows through R_L, E_{out} is the full applied positive d-c voltage, as shown in Fig. 8·21*b*.

At instant t_1, a positive pulse (E_{in} 1) is applied to the P-type base of transistor T_1, forward biasing T_1. However, since T_2 is still reverse biased, T_2 cannot conduct and being in series with T_1, T_1 is likewise kept inoperative. E_{out} remains at the same positive voltage as it was before the input pulse. At instant t_2, a positive pulse (E_{in} 2) is applied to the P-type base of transistor T_2, forward biasing T_2. However, now T_2 is prevented from conducting by T_1 which is reverse biased. Again, E_{out} remains at the same positive potential.

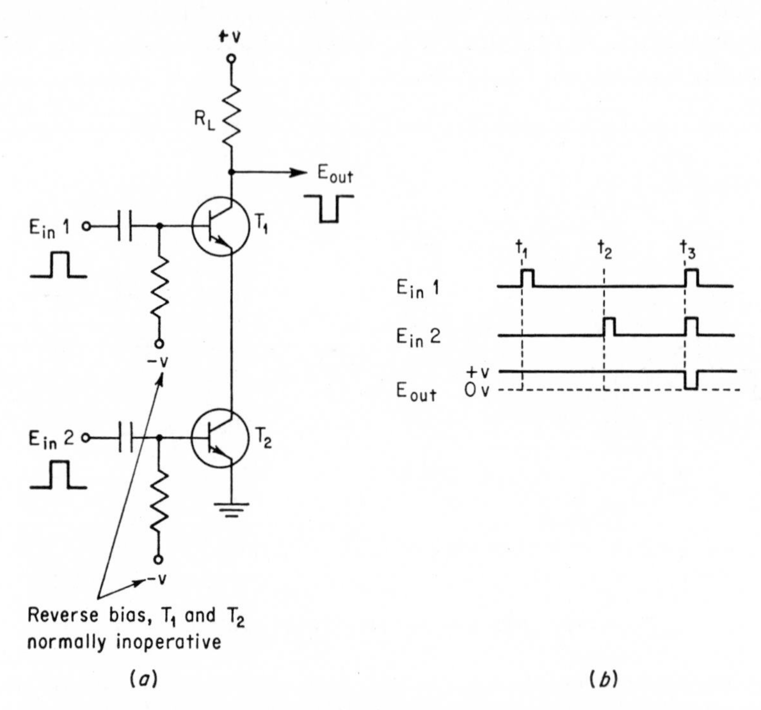

Figure 8·21 NPN-transistor NAND gate (series transistors).

At instant t_3 of Fig. 8·21*b*, positive pulses (E_{in} 1 *and* E_{in} 2) are applied to the P-type bases of transistors T_1 *and* T_2, forward biasing them. Now, T_1 and T_2 conduct in series. The current flow through the large R_L causes E_{out} to drop to practically zero, producing a negative-going output pulse. The circuit is therefore a *phase inverting* AND gate, or a NOT-AND or NAND gate since a *negative* output pulse is produced only when *positive* input signals are applied to *all* inputs.

8·8 INHIBITION Gates

An INHIBITION gate or INHIBITOR is essentially an *anticoincidence* circuit. When *all* input signals coincide, no output signal is produced. This circuit is therefore sometimes referred to as a NOT-AND gate but should not be confused with the other type of NOT-AND gate (NAND) of Sec. 8·7 which is simply an AND gate with a phase-inverted output.

A diode INHIBITION-gate circuit is shown in Fig. 8·22*a*. V_1 and V_2, vacuum tubes or crystal diodes (shown in the dotted-lines), normally conduct since their anodes are grounded, and their cathodes are connected to the -100 volts through R_L. V_3 is normally inoperative due to the -100 volts on its anode. Conduction of V_1 and V_2, without an input signal (before instant t_1 of Fig. 8·22*b*) makes the upper end of R_L about zero volts, causing the clamper tube V_4 to conduct. Conduction of V_4 clamps the output voltage to about -20 volts, as shown in Fig. 8·22*b*.

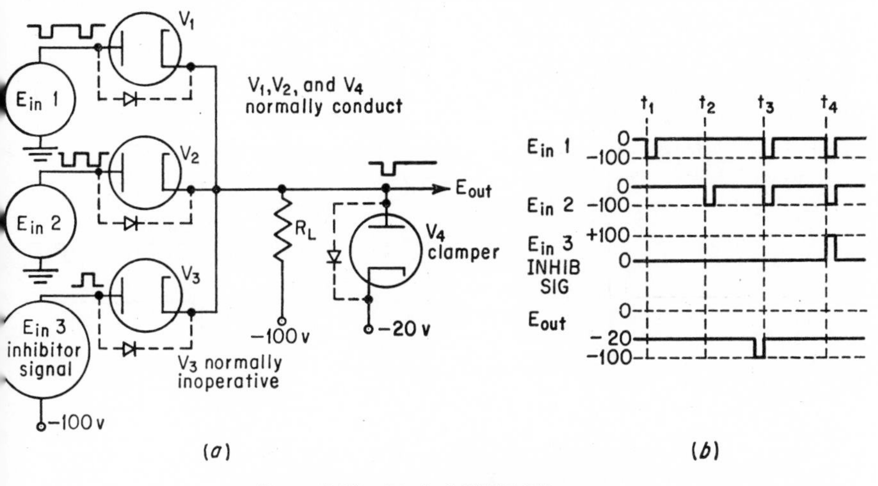

Figure 8·22 Diode INHIBITION gate.

At instant t_1, a negative pulse (E_{in} 1) is applied to the anode of V_1, cutting off this diode. V_2, however, continues to conduct, as does V_4, and E_{out} is still clamped to -20 volts. A similar situation exists at instant t_2, when E_{in} 2 cuts off V_2, and V_1 and V_4 continue to conduct. E_{out}, as shown, remains clamped at -20 volts.

At instant t_3 of Fig. 8·22*b*, negative pulses (E_{in} 1 *and* E_{in} 2) cut off V_1 and V_2. No current flows through R_L, and the upper end of R_L becomes -100 volts. V_4, with -100 volts on its anode, also becomes inoperative, and E_{out} becomes -100 volts, as shown. The circuit up to now acts like an AND gate, producing a negative output pulse only when negative pulses E_{in} 1 *and* E_{in} 2 are applied simultaneously (instant t_3 of Fig. 8·22*b*).

The inhibiting action occurs at instant t_4. At this instant, negative pulses (E_{in} 1 and E_{in} 2) are applied to the anodes of V_1 and V_2, cutting them off. The inhibitor signal (E_{in} 3), a positive-going pulse of sufficient amplitude, is applied to the anode of V_3, driving this diode into conduction. Plate current of V_3 flows up through R_L, making the anode of V_4 positive with respect to its cathode. V_4 conducts, clamping E_{out} to -20 volts, as shown, and preventing the formation of a negative output pulse. E_{in} 3, applied to the anode of V_3, is called the *inhibitor* input since its presence prevents an output signal from being produced. Without this inhibitor input signal, as at instant t_3, an output signal is produced.

A transistor INHIBITOR circuit is shown in Fig. 8·23*a*. The PNP transistor is reverse biased at the N-type base, preventing transistor current. Since the collector has zero volts applied to it, it also prevents

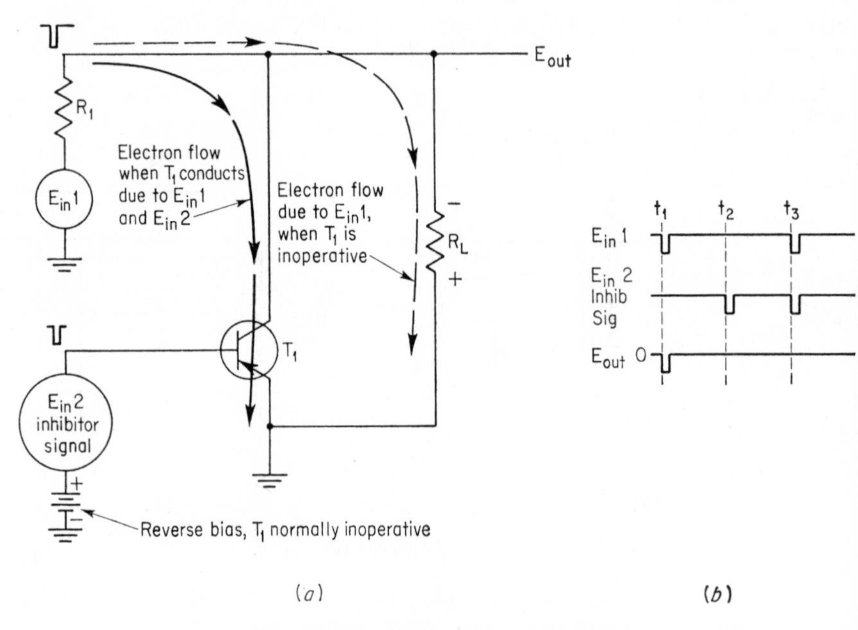

Figure 8·23 **PNP-transistor INHIBITOR.**

transistor current. An output signal will appear only when input signal E_{in} 1 is applied, but the inhibitor signal (E_{in} 2) is not present.

At instant t_1 of Fig. 8·23*b*, a negative pulse (E_{in} 1) is applied to the collector, but the transistor T_1 cannot conduct because of the reverse bias applied to the base (Fig. 8·23*a*). As shown, with the transistor inoperative, current flows down through R_L because of the applied negative signal E_{in} 1. If R_L is made much larger than R_1 (the resistance of the input circuit of E_{in} 1), practically the entire E_{in} 1 appears across R_L as a negative E_{out} pulse.

At instant t_2 of Fig. 8·23*b*, a negative pulse (E_{in} 2) is applied to the base of T_1. If this pulse has sufficient amplitude, it overcomes the reverse bias voltage, forward biasing T_1. However, despite this forward bias, T_1 cannot conduct since there is zero volts at the collector without E_{in} 1 being applied. No current flows through T_1 (collector-emitter), and no current flows through R_L. E_{out}, at instant t_2, is therefore zero, as shown in Fig. 8·23*b*.

INHIBITOR action occurs at instant t_3. At this instant, as shown in Fig. 8·23*b*, negative pulses (E_{in} 1 and E_{in} 2) are applied to the collector and to the base of transistor T_1. E_{in} 1 reverse biases the P-type collector, which is required for transistor conduction. E_{in} 2 forward biases the N-type base, which, of course, is also required for transistor operation. As a result of both input signals being present, T_1 conducts. This current flows, as indicated by the solid-line arrows of Fig. 8·23*a*, from the E_{in} 1 generator through R_1 and from collector to emitter. (Electron flow through a transistor is in the opposite direction from the emitter arrow symbol.)

With T_1 conducting, its emitter-to-collector internal resistance is much smaller than R_1. As a result, practically the full pulse E_{in} 1 is dropped across R_1, leaving zero volts across T_1, and E_{out} is zero, as shown at instant t_3 of Fig. 8·23*b*.

E_{in} 2 is the inhibitor signal, since its presence, along with E_{in} 1 signal, causes T_1 to conduct, preventing E_{in} 1 from appearing in the output.

PROBLEMS

8·1 For the diagram accompanying this problem, draw the output voltage waveshapes with values at (*a*), (*b*), and (*c*). (*d*) Is this an AND or an OR gate?

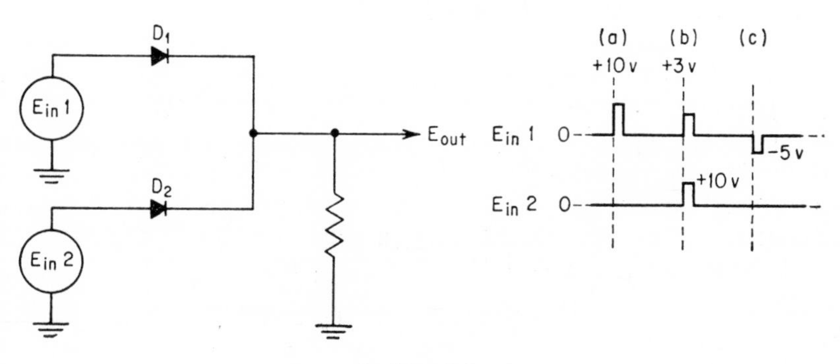

Problem 8·1

8·2 For the accompanying diagram draw the output voltage waveshapes with values at (*a*), (*b*), and (*c*).

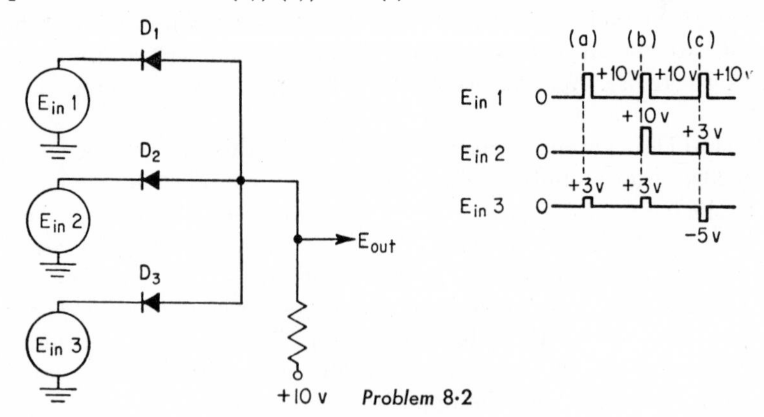

Problem 8·2

8·3 For the accompanying diagram, draw the output voltage waveshapes with values at (*a*), (*b*), (*c*), and (*d*).

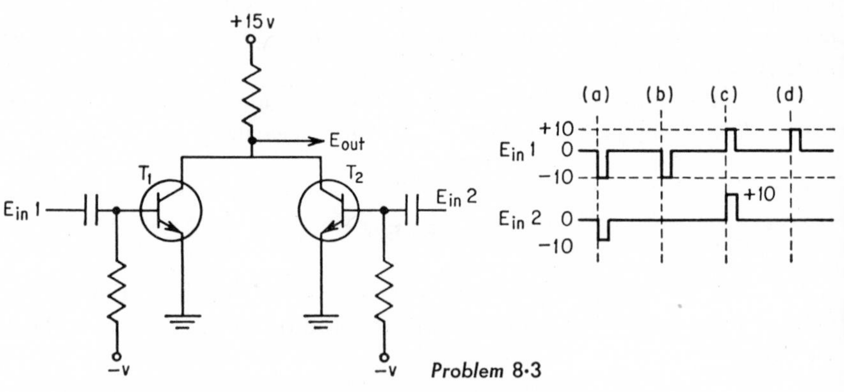

Problem 8·3

8·4 For the accompanying diagram, draw the output voltage waveshapes with values at (*a*), (*b*), and (*c*).

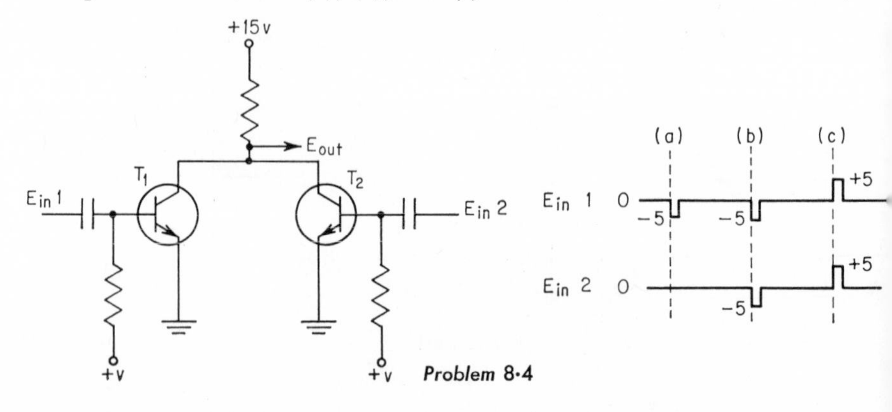

Problem 8·4

MORE CHALLENGING PROBLEMS

8·5 For the accompanying diagram, draw the output voltage waveshapes with values and the D_1 and D_2 plate current waveshapes with values at (*a*), (*b*), (*c*), and (*d*), neglecting diode resistance.

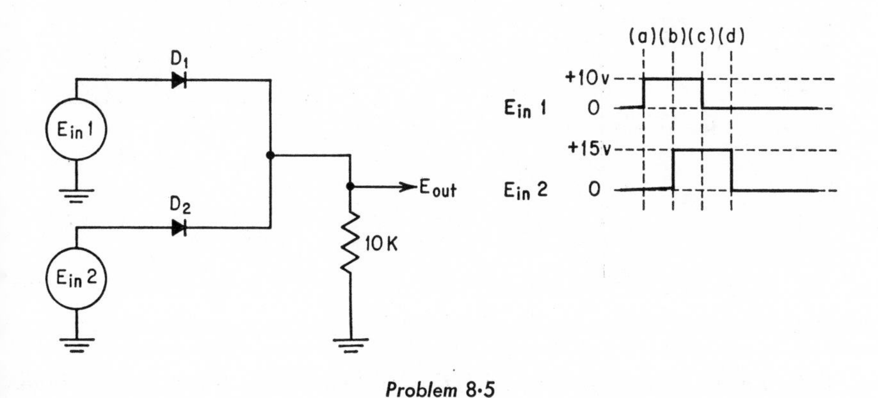

Problem 8·5

8·6 For the diagram shown with this problem, draw the output signals, E_{out} 1 and E_{out} 2, with values at (*a*), (*b*), (*c*), (*d*), (*e*), and (*f*). Assume that $R_L = R_K$, and neglect the resistance of T_1 and T_2.

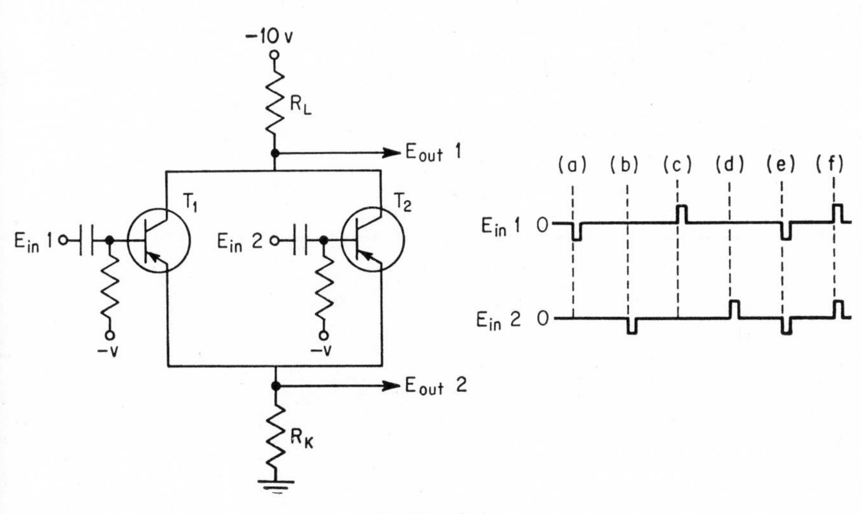

Problem 8·6

8·7 For the diagram accompanying this problem, draw the output voltage waveshapes with the switch open and then with the switch closed at (*a*), (*b*), (*c*), and (*d*).

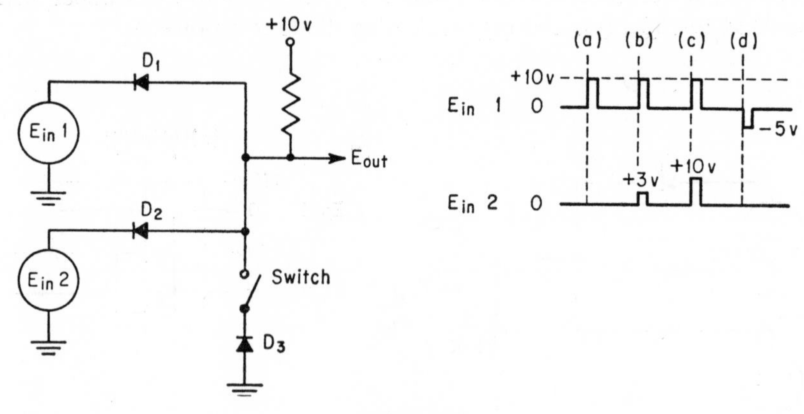

Problem 8·7

8·8 For the diagram given with this problem, draw the output voltage at (*a*) through (*f*).

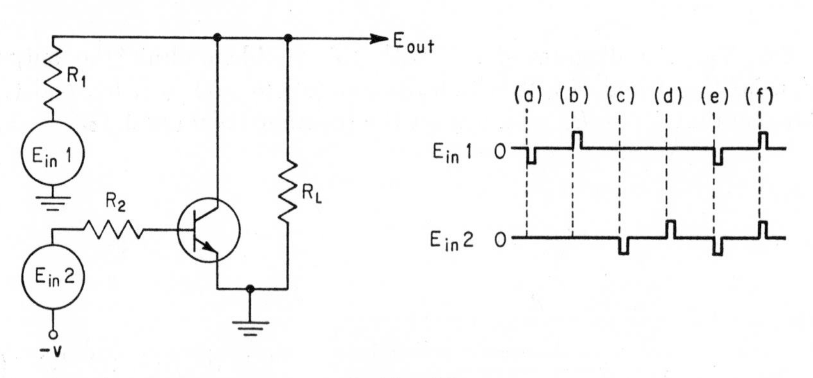

Problem 8·8

CHAPTER 9

SOME APPLICATIONS OF PULSE CIRCUITS

9·1 General

In some of the previous chapters, circuit theory was discussed covering many topics. In this chapter, various equipments employing these circuits are explained. It will be assumed here that the reader understands such things as *mesh circuits* (Chap. 1), *multivibrators* (Chap. 5), *binary numbers* (Chap. 7), and *gates* (Chap. 8). The discussions in this chapter are held to the block-diagram level since it is assumed that the reader is sufficiently sophisticated in these circuits so that he could, if necessary, substitute circuitry for each block. A thorough understanding, therefore, of those previous chapters mentioned above is a prerequisite for the discussions given here.

9·2 Digital Multiplier

One of the many circuits of an electronic computer is known as a *digital multiplier*. As its name implies, this circuit performs multiplication. A simplified block diagram is shown in Fig. 9·1, together with the operations chart. Note that this diagram employs AND gates, OR gates, and Eccles-Jordan or flip-flop stages.

The numbers to be multiplied must be in the binary system. This is discussed in detail in Chap. 5 and will be only briefly reviewed here. The *binary* system uses the number 2 as its *base*, although the system only uses two digits, 0 and 1. The first position or right-hand number has a value of 2^0 or 1 but is omitted or ignored if the digit 0 appears in this position. The value of each position or column is doubled as you progress to the left so that the second column from the right has a value of 2^1 or 2, the third is 2^2 or 4, the fourth position from the right is 2^3 or 8. Again the number *0* in any position makes that position or column value *zero*. The

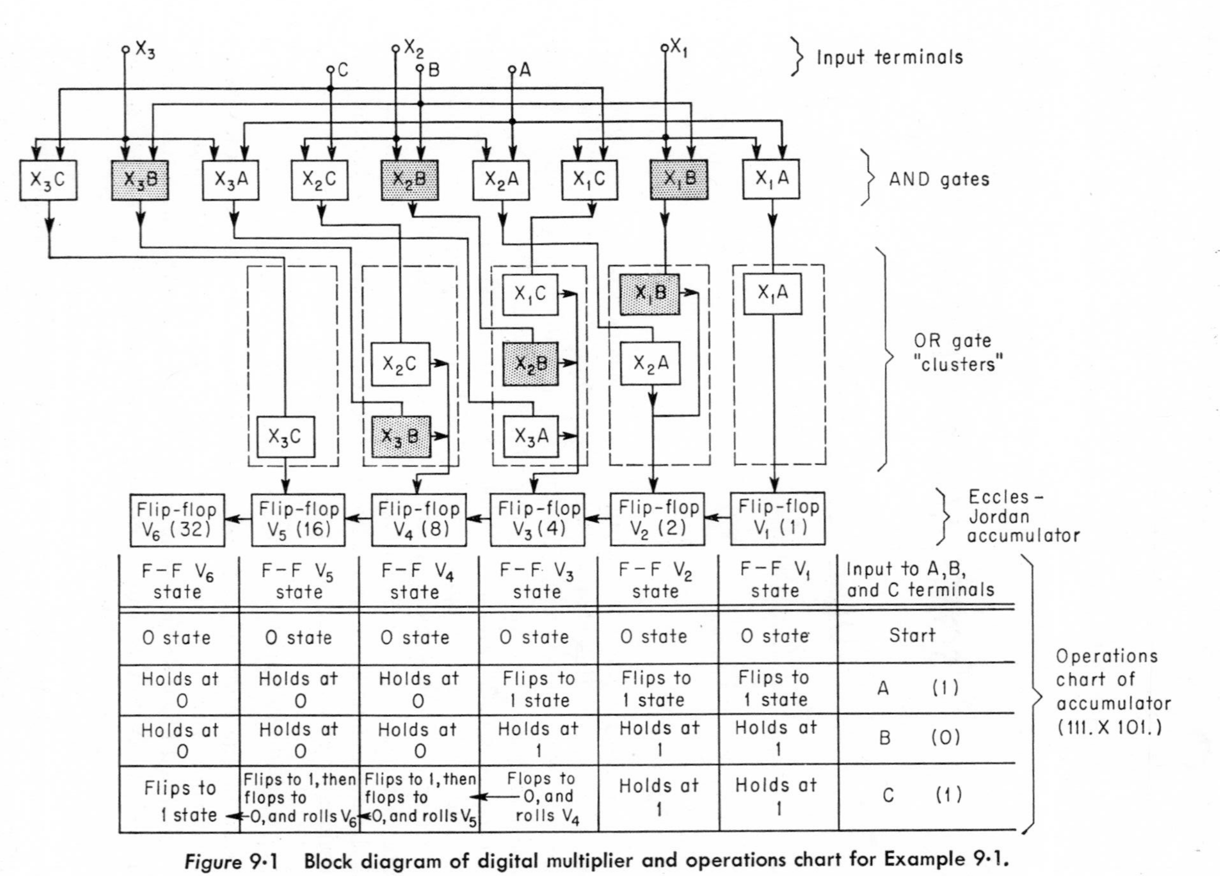

F–F V_6 state	F–F V_5 state	F–F V_4 state	F–F V_3 state	F–F V_2 state	F–F V_1 state	Input to A, B, and C terminals
0 state	0 state	0 state	0 state	0 state	0 state	Start
Holds at 0	Holds at 0	Holds at 0	Flips to 1 state	Flips to 1 state	Flips to 1 state	A (1)
Holds at 0	Holds at 0	Holds at 0	Holds at 1	Holds at 1	Holds at 1	B (0)
Flips to 1 state	Flips to 1, then flops to 0, and rolls V_6	Flips to 1, then flops to 0, and rolls V_5	Flops to 0, and rolls V_4	Holds at 1	Holds at 1	C (1)

Figure 9·1 **Block diagram of digital multiplier and operations chart for Example 9·1.**

numbers in parentheses in the blocks of the Eccles-Jordan accumulator stages, V_1, V_2, V_3, etc. in Fig. 9·1 indicate these binary values. As examples, the decimal number 5 is the binary 101. (which is $4 + 0 + 1$); decimal number 14 is the binary 1110. (which is $8 + 4 + 2 + 0$).

In the digital multiplier, the binary numbers to be multiplied are therefore a series of 1 or zero digits. The digit 1 will represent a voltage which will be applied to an input terminal in Fig. 9·1, while a zero digit represents no voltage to be applied.

Example 9·1 Let us multiply the decimal number 7 (binary 111.) by the decimal number 5 (binary 101.). The decimal answer, of course, is 35, with the binary answer being 100011. (which is $32 + 0 + 0 + 0 + 2 + 1$). Now, let us follow the general operation of this rather complicated piece of equipment as it produces the binary product.

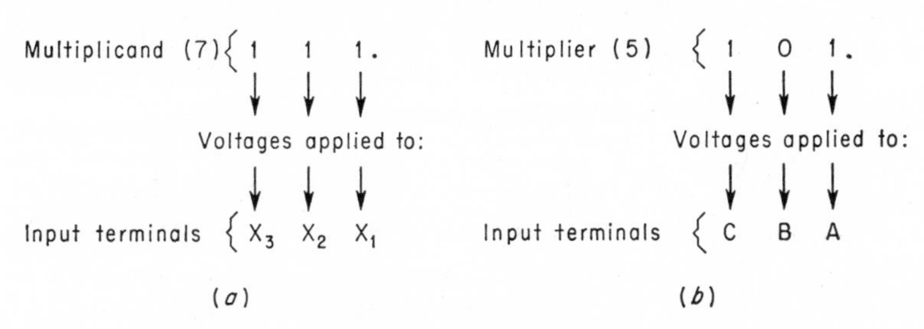

Figure 9·2 Example 9·1.

Either one of the numbers to be multiplied is called the *multiplicand*, while the other is called the *multiplier*. The multiplicand, say, the binary number 111., is applied to the X input terminals in the form of applied voltages, with the right-hand digit 1, applying a voltage to the X_1 input terminal; the second digit from the right, another 1, applying a voltage to the X_2 terminal; and the left-hand 1 applying a voltage to the X_3 terminal. This is shown in Fig. 9·2*a*. The other number 101. causes voltages to be applied to the appropriate A input terminal, B terminal, and C terminal as shown in Fig. 9·2*b*. In this example, 101., voltages are applied to the A and C terminals, but nothing is applied to terminal B.

In the block diagram of Fig. 9·1, note that the X_1 terminal connects to all the X_1 AND gates (X_1A, X_1B, and X_1C), while the X_2 terminal connects to all the X_2 AND gates (X_2A, X_2B, and X_2C), and the X_3 terminal connects to all the X_3 AND gates (X_3A, X_3B, and X_3C). Similarly, the A, B, and C terminals connect to their respective AND gates, with the A terminal connected to the X_1A, X_2A, and X_3A AND gates, while the B terminal is connected to the X_1B, X_2B, and X_3B AND gates, and the C terminal is connected to the X_1C, X_2C, and X_3C AND gates. An AND gate, as discussed in Chap. 8, produces an output signal only when *all* inputs to that gate are present. If there are two input connections to an AND gate, as there are here, then one input can be called the *enabling signal,* which, in effect, opens the gate to allow the other input to go through. With either input absent, there is no output signal.

In the multiplication of binary 111. (to X_1, X_2, and X_3 terminals) and binary 101. (to A, B, and C terminals), the B input terminal alone gets no applied voltage, as shown in Fig. 9·2*b*. This means that all AND gates which connect to the B input terminal are not getting *both* input voltages. As a result, AND gates X_1B, X_2B, and X_3B cannot produce an output signal and are shown shaded in the block diagram to indicate this lack of output signal.

The output from each AND gate is connected, as shown in Fig. 9·1, to the corresponding OR gate. For example, the X_1A AND-gate output is applied to the X_1A OR gate, and the X_1B AND-gate output is connected to the X_1B OR-gate input, etc. An OR gate, as explained in Chap. 8, produces an output signal when either one or the other input signals are applied, or whenever there is an input signal.

The OR gates in Fig. 9·1 are arranged in groups or *clusters.* The first OR cluster at the right in Fig. 9·1 is made up of only the X_1A OR gate. The second cluster from the right comprises the X_1B OR gate and the X_2A OR gate. The outputs of both of the OR gates in this cluster are combined to form the output from this cluster which connects to flip-flop V_2 in the E-J accumulator. The next cluster, the third from the right in Fig. 9·1, consists of the X_1C OR, the X_2B OR, and the X_3A OR. The outputs from these three OR gates in this cluster are connected together to form the input to flip-flop stage V_3 in the E-J accumulator. The next cluster contains the X_2C OR and the X_3B OR, and their common output is connected to flip-flop V_4. The last cluster at the left in Fig. 9·1 contains only the X_3C OR gate. Its output connects to flip-flop V_5 in the E-J accumulator. The OR gates shown shaded indicate that they are not getting an input signal from their associated AND gates (due to the lack of applied voltage which corresponds to the 0 digit of the multiplier 101). These *shaded* OR gates, as a result, produce no output signal to the flip-flop stages.

The E-J accumulator, as shown in Fig. 9·1, consists of the flip-flop or bistable multivibrator stages. A discussion of these circuits is given in Chap. 5. A *flip-flop* or *bistable multivibrator* consists of two triodes (vacuum tube or semiconductor). At the start, all flip-flops in the E-J accumulator are made to operate in the same manner; that is, the corresponding halves of each stage are conducting, while the other halves are cut off. This is achieved by means of a reset switch which applies the proper d-c voltages to each flip-flop. The stages are said to be in their *flop* or 0 state now.

A flip-flop stage will *roll*, that is, go from its *flop* or 0 state into its *flip* or 1 state, or vice versa, whenever an input signal is applied to that flip-flop stage from its associated OR-gate cluster. In the E-J accumulator, a flip-flop stage will roll also whenever the previous stage goes *into* or *through* its zero state. This means that if V_3 is in its 1 state or *flip* condition and it is rolled into its zero state or *flop* condition (either by an input from the OR cluster that connects to it or because of the previous flip-flop V_2 going into its zero state), then V_4 will be rolled by the signal from V_3. This is a case of a flip-flop going *into* its zero state and rolling the succeeding stage. An example of a flip-flop going *through* its zero state and rolling the next stage is the following: Assume that V_3 is in its 1 or *flip* state, and an input signal from its associated OR cluster is applied to V_3. The stage rolls into its zero or *flop* state. If another input, this time from the previous flip-flop V_2 going into its zero state, now enters V_3, it rolls V_3 from 0 back to 1. V_3 therefore went from 1 to zero, and back to 1. The moment that V_3 went into its zero state, it rolled the next following stage, V_4, even though V_3 zero state was short-lived, and V_3 went back to 1. V_3 went *through* its zero state and rolled V_4.

Now, let us go through the detailed operations chart at bottom of Fig. 9·1, step by step, and follow the digital multiplier manipulations as it multiplies the binary numbers 111. and 101.

1. The multiplicand (binary 111.) is applied to the X_1, X_2, and X_3 input terminals of Fig. 9·1, in the form of voltages, *enabling* all the X_1, X_2, and X_3 AND gates.

2. Before the multiplier (binary 101.) is applied, all flip-flop stages in the E-J accumulator are in their zero or *flop* state at the *start.* This is indicated in the operations chart at the lower section of Fig. 9·1.

3. The multiplier (101.) may be applied to the A, B, and C input terminals in any order at all, as long as the correct digit or voltage is applied at the correct terminal, as shown in Fig. 9·2*b*.

4. In the operations chart, the order used here starts at the right-hand digit and the A terminal first, progressing to the next digit to the left and the B terminal, etc.

5. When an input voltage is applied to the A terminal, an output is produced from *all* AND gates having the letter A in their name (X_1A, X_2A, and X_3A) since *all* AND gates have been previously enabled by the multiplicand (111.), and the A-input terminal signal now goes through these AND gates.

6. Each similarly named OR gate (X_1A, X_2A, and X_3A) now gets an input signal, each thereby producing an output signal.

7. X_1A OR feeds flip-flop V_1, and this stage flips to its 1 state, as shown in the operations chart.

8. X_2A OR similarly flips V_2, and X_3A OR flips V_3, as shown.

9. V_4, V_5, and V_6 are not rolled during this A input terminal signal, and these flip-flop stages *hold* at their zero state as shown.

10. Since there is no signal at the B input terminal (corresponding to the 0 digit in the multiplier 101.), there are *no* output signals from any AND-OR gates, and all flip-flop stages *hold* or remain in their previous state, as shown in the operations chart of Fig. 9·1.

11. When an input signal is applied to the C terminal, outputs are produced from X_1C, X_2C, and X_3C AND gates since these gates have been previously enabled by the multiplicand (111.).

12. The corresponding OR gates, X_1C, X_2C, and X_3C ORs, now each produces an output signal to their associated flip-flops, V_3, V_4, and V_5 respectively.

13. V_1 and V_2, during this C input terminal signal, are not affected, and they hold in their previous 1 states, as shown.

14. On this C input signal, V_3 is flopped into its zero state.

15. A signal from V_3 now is produced and applied to V_4, as indicated by the arrow in the last line of the operations chart.

16. V_4 will now receive a signal from X_2C OR gate and also a signal from V_3. V_4 therefore first flips to its 1 state (from its previous zero state) and then flops back to its zero state.

17. A signal is now developed from V_4 and applied to V_5, as indicated by the arrow in the last line of the operations chart.

18. V_5 now gets a signal from V_4 and also a signal from X_3C OR gate.

19. V_5 now first flips to its 1 state (from its previous zero state), and then flops back to its zero state, as shown.

20. V_5 going into its zero state produces a signal which is applied to V_6, as indicated by the arrow in the last line of the chart.

21. Flip-flop V_6 now gets an input from V_5 which flips V_6 into its 1 state.

22. Since there are no further input terminal signals in this problem of 111. multiplied by 101., the *states* of the flip-flop stages now express the binary answer, reading from left to right, starting with V_6 (count value of 32), then to V_5 (value of 16), etc. This binary answer, as shown in the last line of the chart of Fig. 9·1, is 100011., which agrees with the answer stated at the beginning of this discussion.

The block diagram of the digital multiplier shown in Fig. 9·1, because of lack of space, only used *three* X terminals (X_1, X_2, and X_3) and *three* others (A, B, and C), requiring *nine* AND gates and *nine* OR gates, and *six* flip-flops. This simple circuit could, of course, only multiply up to three-digit binary numbers. In the following discussion, additional input terminals and additional gates and flip-flops are employed to provide multiplication by larger numbers.

Example 9·2 Using the partial block diagram of the larger digital multiplier of Fig. 9·4, multiply the binary number 11101. (decimal number 29) by the binary number 10110. (decimal number 22).

Solution. As shown in Fig. 9·3*a*, the digits of the multiplicand of the binary number 11101., starting with the digit 1 at the right, are applied as voltages to the X_1, X_2, X_3, X_4, and X_5 input terminals. The zero digit does not cause any voltage to be applied to the X_2 input terminal. The 1 digits of the multiplicand do cause voltages to be applied to all the other X terminals. These X input terminals each connect to the corresponding numbered AND gates (not shown in Fig. 9·4 for simplicity). The X_1 input terminal connects to each of the X_1 AND gates, the X_2 terminal feeds each X_2 AND gate, etc., as shown in Fig. 9·1. Since, in this example, as shown in Fig. 9·3*a*, the X_2 terminal does not get an applied voltage, all

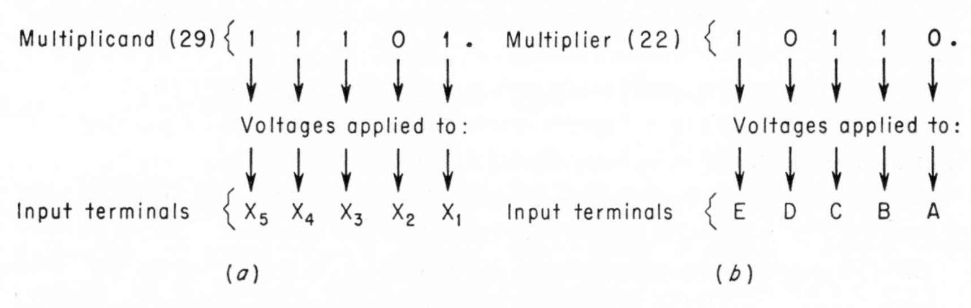

Figure 9·3 Example 9·2.

the X_2 AND gates are not enabled and cannot produce an output signal. As a result, each of the associated OR gates (X_2A, X_2B, X_2C, X_2D, and X_2E ORs) do not receive an input and consequently produce no output signal to the flip-flop stages. Note that in the block diagram of Fig. 9·4, all the X_2 ORs are shown *shaded* to indicate their inability to produce output signals in this example.

The multiplier, binary number 10110., as shown in Fig. 9·3*b*, causes voltages to be applied to input terminals A, B, C, D, and E. A zero digit, such as the one at the far right and also the second from the left, produces no voltages to the A and D input terminals. As a result, all A and D AND gates (X_1A, X_2A, X_3A, X_4A, and X_5A, and X_1D, X_2D, X_3D, X_4D, and X_5D ANDs) are incapable of producing output signals. The associated OR gates, with these same names, therefore receive no inputs and are inoperative. In the block diagram of Fig. 9·4, all these OR gates are *shaded* to indicate their inability, in this example, to produce output signals to the flip-flop stages.

In the following step-by-step discussion of the operations chart of the flip-flop stages of the E-J accumulator, it will be assumed that the multiplicand, 11101., has been previously applied, in the form of voltages, to the X terminals, enabling the AND gates. Now, the multiplier, 10110., will be applied as voltages to the A, B, C, etc., terminals. The application of these voltages may be in any order at all, just as long as the far-right digit is associated with the A terminal, the next digit to the left with the B terminal, etc. In the operations chart and discussion, the order is as shown in Fig. 9·4, with first the far right zero digit of the 10110. (to A terminal), then the next digit to the left, 1 (to B terminal), etc. The discussion of the operations will be simplified by the following. When an input voltage is applied to, say, the B terminal, every OR gate containing the letter B in its name (except those shown *shaded*) produces an output which rolls that associated flip-flop. Refer to the operations chart of Fig. 9·4.

1. At the start, before the multiplier binary number is applied, all flip-flop stages are placed in their zero or flop state, as shown in the first line of the chart.

2. The digit zero does not cause any voltage to be applied to the A terminal, and no output signals are forthcoming from the OR gates. All flip-flop stages hold or remain in their previous zero states.

3. The input voltage to B terminal (corresponding to the digit 1) produces output signals from the following OR gates: X_1B, X_3B, X_4B, and X_5B. (The X_2B is inoperative, as indicated by the shading in Fig. 9·4.) These OR gate signals roll their associated flip-flops. V_1 holds at zero state. V_2 flips to its 1 state, due to X_1B OR signal. V_3 holds at zero state. V_4 flips to its 1 state because of signal from X_3B OR gate. V_5 flips to its 1 state due to X_4B OR signal. V_6 flips to its 1 state due to X_5B OR gate signal. V_7, V_8, V_9, and V_{10} hold at zero.

4. Applied voltage to C input terminal (corresponding to the digit 1 of the multiplier) produces outputs from X_1C, X_3C, X_4C, and X_5C OR gates. This rolls the associated flip-flop stages. As shown in the fourth line of the operations chart, V_1 holds at zero state. V_2 holds at its previous 1 state. V_3 flips to 1 state because of X_1C OR gate signal. V_4 holds at its previous 1 state. V_5 flops to zero state because of signal from X_3C OR gate. When V_5 rolls to zero state, it produces a signal which rolls V_6, as indicated by the arrow in the fourth line of the operations chart. V_6 now receives signals from V_5 and also from X_4C OR gate. V_6 first flops to its zero state and then flips back to its 1 state. As V_6 goes through zero, it sends a signal on to V_7, as indicated by the arrow. V_7 now receives signals from X_5C and also from V_6. V_7 first flips to its 1 state and then flops back to its zero state. This sends a signal on to V_8, as again shown by the arrow in the chart. V_8 is flipped to its 1 state by signal from V_7. V_9 and V_{10} hold at their zero states.

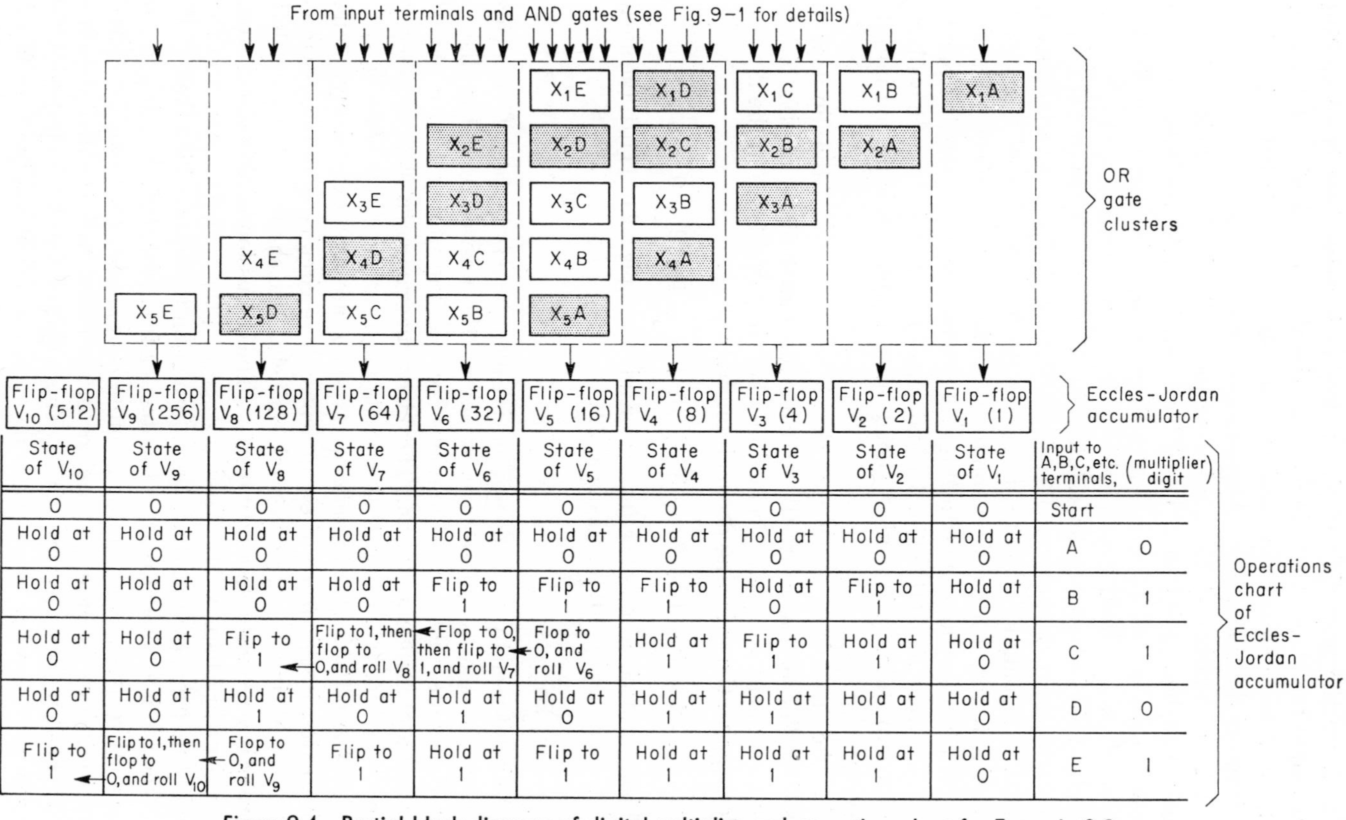

State of V_{10}	State of V_9	State of V_8	State of V_7	State of V_6	State of V_5	State of V_4	State of V_3	State of V_2	State of V_1	Input to A, B, C, etc. terminals,	(multiplier digit)
0	0	0	0	0	0	0	0	0	0	Start	
Hold at 0	Hold at 0	Hold at 0	Hold at 0	Hold at 0	Hold at 0	Hold at 0	Hold at 0	Hold at 0	Hold at 0	A	0
Hold at 0	Hold at 0	Hold at 0	Hold at 0	Flip to 1	Flip to 1	Flip to 1	Hold at 0	Flip to 1	Hold at 0	B	1
Hold at 0	Hold at 0	Flip to 1	Flip to 1, then flop to 0, and roll V_8	Flop to 0, then flip to 1, and roll V_7	Flop to 0, and roll V_6	Hold at 1	Flip to 1	Hold at 1	Hold at 0	C	1
Hold at 0	Hold at 0	Hold at 1	Hold at 0	Hold at 1	Hold at 0	Hold at 1	Hold at 1	Hold at 1	Hold at 0	D	0
Flip to 1	Flip to 1, then flop to 0, and roll V_{10}	Flop to 0, and roll V_9	Flip to 1	Hold at 1	Flip to 1	Hold at 1	Hold at 1	Hold at 1	Hold at 0	E	1

Figure 9·4 **Partial block diagram of digital multiplier and operations chart for Example 9·2.**

5. Since the next digit of the multiplier is zero, there is no applied voltage to the D input terminal and no signals from any of the OR gates. All flip-flops hold in their previous states, as shown by the next to the last line of the chart.

6. The last digit at the far left of the multiplier is 1. This causes a voltage to be applied to E input terminal. Outputs are now produced from X_1E, X_3E, X_4E, and X_5E OR gates to their associated flip-flop stages V_5, V_7, V_8, and V_9. V_1 holds at its previous zero state. V_2, V_3, and V_4 also hold in their previous states of 1. V_5 flips to its 1 state because of X_1E OR signal. V_6 holds at its 1 state. V_7 flips to its 1 state because of X_3E OR signal. V_8 flops to its zero state because of X_4E OR signal. V_8 going to zero sends a signal on to V_9, as shown by the arrow in the last line of the operations chart of Fig. 9·4. V_9 now receives signals from V_8 and also from X_5E OR gate. V_9 first flips to its 1 state, then flops back to its zero state. This sends a signal on to V_{10}, which is flipped into its 1 state.

7. The final binary answer, as shown by the states of the flip-flops in the last line of the chart, is 1001111110. The 1 digits of this binary number, as shown by the decimal numbers in parentheses in the flip-flop blocks, give the following decimal-number values:

$$512 + 0 + 0 + 64 + 32 + 16 + 8 + 4 + 2 + 0 = 638$$

This is the decimal-number answer for this Example 9·2 of 11101. (decimal 29) multiplied by 10110. (decimal 22).

9·3 Shift Register

A circuit used by many computers is called the *shift register*. This device is used for short-term storage, removing the answer from the digital multiplier (described in the previous Sec. 9·2) and storing it until needed for some other computation. The block diagram of a shift register is shown in Fig. 9·5. Note that circuits such as the *Eccles-Jordan*, also called the *bistable multivibrator* or *flip-flop*, are used extensively, along with gates, in this piece of equipment. These circuits were discussed at some length in Chaps. 5 and 8 respectively. Also shown at the lower-left corner is a monostable multivibrator. The discussion of this circuit too, was given in Chap. 5. The flip-flop stages, V_1 through V_5, are part of the digital multiplier from which the information to be stored is taken.

When it is desired to move the answer from the digital multiplier to the shift register, a commutation pulse is applied through an *L-C* delay line to each of the gates, A through E, associated with the digital multiplier flip-flop stages V_1 through V_5. This single commutation pulse is produced when the key or switch is depressed which starts the transferral from digital multiplier to shift register. This pulse could be produced by

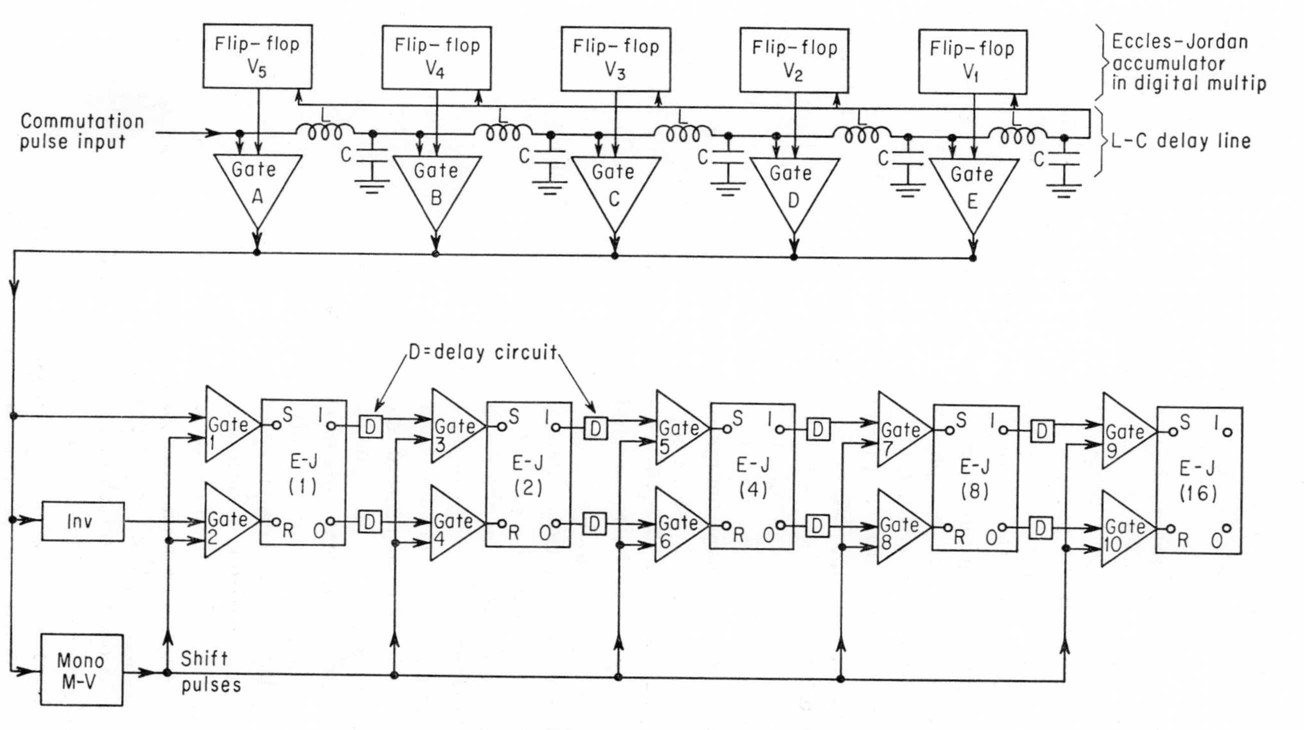

Figure 9·5 **Block diagram of shift register.**

a monostable (one-shot) multivibrator, or blocking oscillator. As shown in the block diagram of Fig. 9·5, the commutation pulse reaches gate A first, allowing that gate to produce an output signal. This output signal depends on the state of V_5. If V_5 is in its 1 state, one of the plate or collector voltages of this double-triode flip-flop is high, while the other is low. This high voltage goes through gate A, during the commutation pulse period, in the form of a positive pulse. If V_5 is in its zero state, then that particular plate or collector voltage is low. In this condition, the output from gate A is a negative pulse. The same is true for all of the flip-flops V_5 down through V_1 and gates A through E. That is, a *positive* pulse comes out of a gate if that particular flip-flop is in its 1 state, while a *negative* pulse is produced if that flip-flop is in its zero state.

The applied commutation pulse opens the gates in sequence, starting with gate A, then moving on to gate B, etc. The delay between the opening of gate A and the opening of gate B and then the opening of the next is determined by the inductance and capacitance, L and C. The time delay in seconds is approximately equal to the square root of the product of L and C, where L is in henrys, and C is in farads.

The bistable multivibrators, or E-J circuits, shown as E-J (1), E-J (2), E-J (4), E-J (8), and E-J (16) are two-line grid triggered. They can be rolled only by a positive pulse applied to the triode grid. The four terminals marked S, 1, R, and zero denote the following: S represents the grid or transistor base of that triode which conducts when the stage (both triodes together) is in its 1 state, also referred to as the *set* state; the 1 terminal represents the plate or transistor collector of that triode which conducts when the stage is in its 1 state; the R terminal represents the grid or transistor base of that triode which conducts when the stage (both triodes together) is in its zero state, also referred to as *reset;* the zero terminal represents the plate or transistor collector of that triode which conducts during the zero state of the stage.

Gates 1 and 2 are enabled only by a positive pulse coming from the digital-multiplier flip-flops via gates A through E. For example, a positive pulse will enable gate 1, while a negative one will not. However, a negative pulse is reversed by the inverter stage, and the resulting positive pulse enables gate 2. The remaining gates, 3 to 10, are enabled by the *low* d-c voltage at the plate (or collector), terminal 1 or zero, of the previous E-J stage. A delay circuit, shown as the blocks marked D in Fig. 9-5, prevents this *low* d-c voltage from immediately reaching a gate. For example if E-J (1) is flipped to its 1 state, gate 3 will not *immediately* become enabled. Gate 3 enables after the delay circuit, D, permits the *low* voltage at E-J (1) terminal 1 to reach gate 3. Similarly, if E-J (1) is flopped into its zero state, gate 4 will enable shortly after. Gates 3 to 10 are each said to be enabled by the *previous* state of the E-J stage associated with that gate.

The last bit of necessary theory, before the complete operation of the shift register can be gone into, is the monostable multivibrator, shown in Fig. 9·5 as *mono M-V*. This stage generates a signal called the *shift pulse* for *every* input to this stage from the digital-multiplier flip-flops via gates A through E. These shift pulses, as shown, are fed to gates 1 to 10, going through those gates which are in an enabled condition. A shift pulse going through a gate could roll an E-J stage if that stage is in an opposite state. For example, if E-J (1) is in its zero state and a shift pulse gets through gate 1 to the terminal marked S, it flips E-J (1) to its 1 state, or if E-J (1) is in its 1 state and a shift pulse gets through gate 2 to terminal R, it flops E-J (1) to its zero state. However, if E-J (1) were already in its zero state, and a shift pulse went through gate 2 to terminal R, E-J (1) stage would be unaffected and would hold in its zero state. Now let us follow the detailed step-by-step operations of the shift register by referring to the block diagram of Fig. 9·5 and to the operations chart of Fig. 9·6 in the following example.

Example 9·3 The binary number 10011. is to be transferred from the digital multiplier to the shift register. Go through the detailed operations that take place for this transfer.

Solution. Refer to both the shift-register block diagram and to the operations chart, Figures 9·5 and 9·6, while going through each of the following steps:

1. At the start, line 1 of the chart before the commutation pulse is applied to gates A through E, all E-J stages of the shift register are in their zero states. The low d-c voltage now present at each E-J stage zero terminal enables the associated gates, gates 4, 6, 8, and 10.

2. When the commutation pulse is applied, line 2 of the chart, gate A allows the high d-c voltage at V_5, which is in its 1 state, to cause a positive pulse to come from gate A. This positive pulse enables gate 1. This pulse also causes the monostable multivibrator to produce a shift pulse. The shift pulse goes through gate 1 and flips E-J (1) to its 1 state. This does not immediately enable gate 3, since the low d-c voltage at terminal 1 of E-J (1) cannot get to gate 3 at this instant because of the delay circuit, shown as D. Gate 4 is still enabled, as are gates 6, 8, and 10. The shift pulse gets through all these enabled gates (4, 6, 8, and 10) but does not effect E-J (2), E-J (4), E-J (8), or E-J (16), since they are already in their zero states. Note that the first digit 1 (the far *left* digit) of the binary number 10011. has now been applied to the shift register, causing E-J (1)

	Input from digital multiplier				Gate 1 or 2 enabled	E − J (1) state	Gate 3 or 4 enabled	E − J (2) state	Gate 5 or 6 enabled	E − J (4) state	Gate 7 or 8 enabled	E − J (8) state	Gate 9 or 10 enabled	E − J (16) state
	From flip-flop	Via gate	Binary digit	Pulse polarity										
(Line 1)		Start		None	None	0	4	0	6	0	8	0	10	0
(Line 2)	V_5	A	1	+	1	Flips to 1	4	Holds at 0	6	Holds at 0	8	Holds at 0	10	Holds at 0
(Line 3)	V_4	B	0	−	2	Flops to 0	3	Flips to 1	6	Holds at 0	8	Holds at 0	10	Holds at 0
(Line 4)	V_3	C	0	−	2	Holds at 0	4	Flops to 0	5	Flips to 1	8	Holds at 0	10	Holds at 0
(Line 5)	V_2	D	1	+	1	Flips to 1	4	Holds at 0	6	Flops to 0	7	Flips to 1	10	Holds at 0
(Line 6)	V_1	E	1	+	1	Holds at 1	3	Flips to 1	6	Holds at 0	8	Flops to 0	9	Flips to 1

Figure 9·6 **Operations chart of shift register for Example 9·3.**

to be in its 1 state. As subsequent digits are applied, this first digit 1 (or 1 state) will move across to E-J (2), then to E-J (4), etc.

3. On line 3 of the operations chart, when the commutation pulse reaches gate B, it allows the low d-c voltage at V_4 (of the digital multiplier), which is in its zero state, to cause a negative pulse to come out of gate B. This corresponds to the zero digit second from the left in the binary number 10011. The negative pulse does not enable gate 1 but comes out of the inverter stage (Fig. 9·5) as a positive pulse. This pulse enables gate 2, allowing the shift pulse from the monostable M-V to go through gate 2 into the R terminal of E-J (1). This flops E-J (1) back to its zero state. The low d-c voltage now at the zero terminal of E-J (1) is delayed from reaching gate 4. However, the *previous* condition of E-J (1), line 2 of the chart, was its 1 state. Now, the low d-c voltage which had been at the 1 terminal of E-J (1) gets to gate 3, enabling this gate. This is shown in the chart, where an arrow goes from the column *state of the E-J (1) stage*, 1 (line 2) to the column *gate 3 or 4 enabled* (line 3). Gates 3 to 10, as has been mentioned before, are enabled by the *low* d-c voltage because of the state of the associated E-J stage from the *previous* line of the operations chart. Therefore, on line 3 of the chart, gate 3 is enabled because of the *previous* 1 state of E-J (1). The monostable M-V shift pulse now goes through gate 3 to the S terminal of E-J (2), flipping E-J (2) to its 1 state. Since this stage had previously (line 2) been in its zero state, it now (line 3) enables gate 6. The shift pulse goes through gate 6 to the R terminal of E-J (4) but does not affect this stage since it is already in its zero state. Gate 8 is still enabled, and the shift pulse goes through it but again does not affect E-J (8) which is already in its zero state. Gate 10 is still enabled, and the shift pulse goes through it but does not affect E-J (16) which is already in its zero state.

4. On line 4 of the chart, the commutation pulse has reached gate C, allowing the low d-c voltage at V_3, which is in its zero state, to produce a negative pulse from gate C. The zero state of V_3 is the zero digit, third from the left of the binary number 10011. The negative pulse is reversed in polarity by the inverter, and the resulting positive pulse enables gate 2. The monostable M-V shift pulse goes through gate 2 but does not affect E-J (1) which is already in its zero state. The previous condition of this stage (line 3) was zero state, and now (line 4) gate 4 is enabled. The shift pulse goes through this gate to the R terminal of E-J (2), flopping this stage to its zero state. Since the previous condition (line 3) of E-J (2) was 1 state, gate 5 now (line 4) enables. The shift pulse goes through gate 5 to the S terminal of E-J (4), flipping this stage into its 1 state. Since the previous condition of this stage (line 3) was zero state, gate 8 is still enabled. The shift pulse goes through gate 8 but does not affect E-J (8) since this stage is already in its zero state. The previous zero state of

E-J (8) keeps gate 10 enabled, and the shift pulse goes through it to E-J (16) but does not affect this stage since it is already in its zero state. Note that up to now, the digits 1, zero, and zero have been applied to the shift register, and the digit 1 in the form of the 1 state has moved along from the E-J (1) stage (line 2), to the E-J (2) stage (line 3), to the E-J (4) stage (line 4), followed by the zero digits, in the form of zero states in the E-J (2) and in the E-J (1) stages.

5. When the commutation pulse reaches gate D (Fig. 9·5), it permits the high d-c voltage at V_2, which is in its 1 state (corresponding to the 1 digit next to the far right in the binary number 10011.), to cause a positive pulse to come from gate D. This positive pulse enables gate 1, as shown in line 5 of the operations chart. The monostable M-V shift pulse goes through gate 1 to the S terminal of E-J (1), flipping this stage to its 1 state. Since the previous condition (line 4) of this stage was zero state, it now (line 5) keeps gate 4 enabled. The shift pulse passes through gate 4 to the R terminal of E-J (2) but does not affect this stage since it is already in its zero state. Since the previous state (line 4) of E-J (2) was zero, it now (line 5) enables gate 6. The shift pulse goes through gate 6 to the R terminal of E-J (4), flopping this stage to its zero state. The previous condition (line 4) of E-J (4) was its 1 state, which now (line 5) enables gate 7. The shift pulse goes through gate 7 to the S terminal of E-J (8), flipping this stage to its 1 state. Since the previous state of E-J (8), line 4, was zero, it now (line 5) enables gate 10. The shift pulse passes through gate 10 but does not affect E-J (16) since this stage is already in its zero state.

6. When the commutation pulse reaches gate E, it permits the high d-c voltage at V_1, which is in its 1 state (corresponding to the last digit at the right in the binary number 10011.), to produce a positive pulse from gate E, as tabulated in line 6 of the operations chart. This positive pulse enables gate 1, permitting the monostable M-V shift pulse to reach the S terminal of E-J (1), not affecting this stage since it is already in its 1 state. With the previous condition of E-J (1), line 5, in its 1 state, gate 3 is now (line 6) enabled. The shift pulse goes through gate 3 to the S terminal of E-J (2), flipping this stage to its 1 state. The previous state (line 5) of this stage was zero, and now (line 6) gate 6 stays enabled. The shift pulse goes through gate 6 to the R terminal of E-J (4) but does not affect this stage since it is already in its zero state. The previous state (line 5) of E-J (4) was zero, and now (line 6) it enables gate 8. The shift pulse passes through gate 8 to the R terminal of E-J (8), flopping this stage to its zero state. Since the previous condition (line 5) of E-J (8) was its 1 state, then it now (line 6) enables gate 9. The shift pulse passes through gate 9 to the S terminal of E-J (16). This stage flips, for the first time, to its 1 state.

7. The transfer of the binary number 10011. from the digital multiplier to the shift register is now complete. Note that the first digit, at the left, of 10011. went through each of the E-J stages in the shift register, ending up in the highest-value E-J stage, the E-J (16). The binary number, as shown in line 6 of the operations chart for each E-J stage reading from *right* to *left* (highest to lowest values), is 10011., the original number to be transferred.

8. After the commutation pulse reaches gate E, making the transfer complete, it is delayed by the last section of the delay line, finally entering all the flip-flop stages in the E-J accumulator of the digital multiplier, resetting or flopping them all back to their zero states. This "erases" the answer in the digital multiplier.

9·4 Forward-backward Counter

An electronic pulse counting device which produces a *difference* readout is the *forward-backward counter* shown in the block diagram of Fig. 9·7. This counter, as shown, employs flip-flop or bistable multivibrators (discussed in detail in Chap. 5), used in conjunction with AND and OR gates (discussed in Chap. 8). The flip-flop stages used in counters have numerical count values as shown by the figures in parentheses in each block, with V_1 having a count value of 1, V_2 a value of 2, V_3 a value of 4, etc. This is explained in greater detail in Chap. 7.

The forward-backward counter, as its name implies, can actually count in the usual forward manner, going progressively higher, or can count backwards or in the reverse manner, going progressively lower. Input pulses applied to one point cause a count in one direction, while pulses fed to another input result in an opposite-direction count. The pulses fed to each input, in practice, may be produced by a photo-cell amplifier device set up on each of two factory production lines, for the purpose of comparison. By applying a series of pulses first to one input and then another series to the other input, the counter will count first in one direction and then in reverse. The pulses may be applied either as a *series* fed to one input first, followed by a *series* fed to the other input, or pulses may be intermittently mixed, with one or several fed to one input, then with one or several fed to the other, followed with one or several again applied to the first input terminal, etc. As shown in Fig. 9·7, the input terminals connect to the *backward line* and to the *forward line*. If the counter is started with a setting of zero (all flip-flops in their zero states) and the number of pulses applied to the forward line is equal to that applied to the backward line, then the final result is that the counter again reads zero. If the number of pulses fed to the forward line exceeds the number fed to the reverse line, say by 5, then the counter will indicate this difference as simply a count of 5. This will be discussed in detail in Example 9·4.

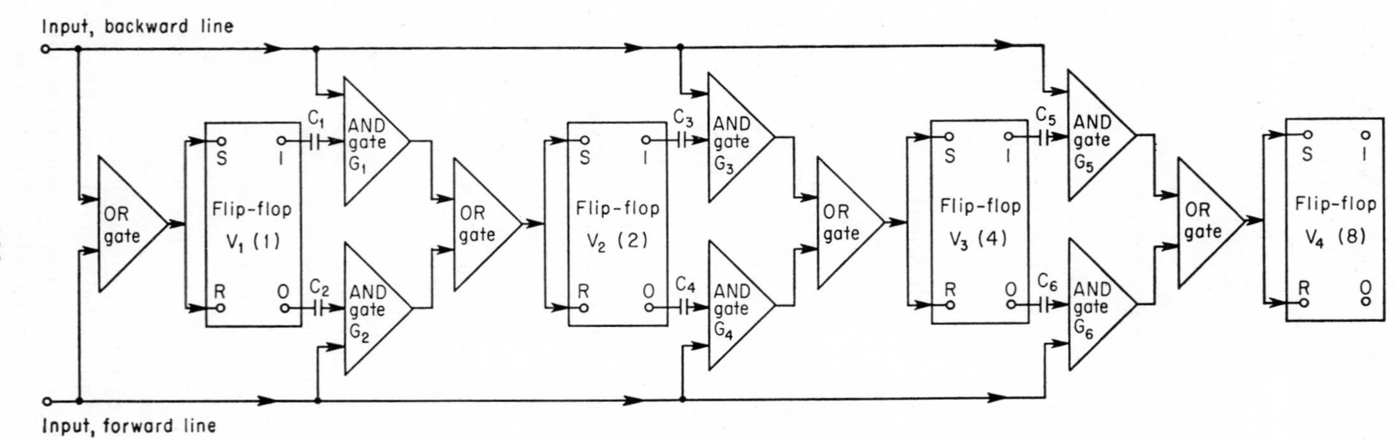

Figure 9·7 **Block diagram of forward-backward counter.**

The device is said to produce a *difference* readout. If the possibility exists that the number of pulses to be fed to the backward line is larger than that to be applied to the forward line, the counter should be started off at some advanced count, with several or all of the flip-flops in their 1 states. This will be detailed later in Example 9·5. The following Example 9·4 illustrates the details, using the operations chart of Fig. 9·8 and the block diagram of Fig. 9·7 where the forward-line pulses are greater in number than those applied to the backward line.

Example 9·4 If eight pulses are consecutively applied to the forward-line input terminal followed by three applied to the backward line, work out the detailed operation of the forward-backward counter shown in Fig. 9·7.

Solution. In the following detailed discussion steps, reference should be made to both the block diagram of Fig. 9·7 and to the operations chart of Fig. 9·8. The AND gates in this circuit are enabled only *momentarily* by a negative-going voltage from the 1 or zero terminal of the associated flip-flop stage. The two upper terminals of each flip-flop stage block marked S and 1 represent the grid and plate of a triode tube (or base and collector of a transistor), while the two lower terminals marked R and zero represent the same elements of the other triode in that flip-flop stage. The S stands for set, while R is reset. These stages, as shown, are single-line grid (or base) triggered. Whenever an input pulse (of the correct polarity) enters a single-line triggered flip-flop, that stage rolls. If the stage were in its *flop* or zero state (low voltage at zero terminal), an input pulse applied to the connected terminals S and R would roll the stage to its *flip* or 1 stage (low voltage now appears at the 1 terminal). If the stage were in its 1 state, the input to terminals S and R would flop it to its zero state. The coupling capacitors, C_1 through C_6, prevent the d-c voltage at the output terminals (1 and zero) of a flip-flop from getting to the AND gates. When the voltage at a terminal falls, the coupling capacitor passes only the *negative-going* pulse which *momentarily* enables that associated gate. A steady d-c voltage, or a rising voltage at a flip-flop terminal, will not enable an AND gate. The step-by-step discussion follows.

1. At the start, before any input is applied, all flip-flops (V_1, V_2, V_3, and V_4) are reset to their flop or zero state. Except for the instant of reset, the AND gates are all disabled. Line 1 of the operations chart, Fig. 9·8, illustrates this.

2. As shown in line 2, the *first* pulse applied to the forward line flips V_1 to its 1 state. The decrease of voltage at terminal 1 of V_1 enables gate G_1, but no signal goes through this gate since there isn't any on the backward line. The forward-line pulse cannot get through the disabled gate G_2, and

	Input Forward	Input Backward	F-F V_1 (1) state	G_1 or G_2 enabled	F-F V_2 (2) state	G_3 or G_4 enabled	F-F V_3 (4) state	G_5 or G_6 enabled	F-F V_4 (8) state	Count V_1	Count V_2	Count V_3	Count V_4	Count Total
Line 1	Start		0		0		0		0					0
Line 2	1		Flip to 1	G_1	Hold at 0		Hold at 0		Hold at 0	(1)				1
Line 3	2		Flop to 0	G_2	Flip to 1	G_3	Hold at 0		Hold at 0		(2)			2
Line 4	3		Flip to 1	G_1	Hold at 1		Hold at 0		Hold at 0	(1)	(2)			3
Line 5	4		Flop to 0	G_2	Flop to 0	G_4	Flip to 1	G_5	Hold at 0			(4)		4
Line 6	5		Flip to 1	G_1	Hold at 0		Hold at 1		Hold at 0	(1)		(4)		5
Line 7	6		Flop to 0	G_2	Flip to 1	G_3	Hold at 1		Hold at 0		(2)	(4)		6
Line 8	7		Flip to 1	G_1	Hold at 1		Hold at 1		Hold at 0	(1)	(2)	(4)		7
Line 9	8		Flop to 0	G_2	Flop to 0	G_4	Flop to 0	G_6	Flip to 1				(8)	8
Line 10		1	Flip to 1	G_1	Flip to 1	G_3	Flip to 1	G_5	Flop to 0	(1)	(2)	(4)		7
Line 11		2	Flop to 0	G_2	Hold at 1		Hold at 1		Hold at 0		(2)	(4)		6
Line 12		3	Flip to 1	G_1	Flop to 0	G_4	Hold at 1		Hold at 0	(1)		(4)		5

Figure 9·8 **Operations chart of forward-backward counter for Example 9·4.**

V_2 holds in its zero state. With no decrease of voltage from V_2, neither gate G_3 nor G_4 is enabled, and the forward line pulse cannot get into V_3, which holds in its zero state. Neither gate G_5 nor G_6 is enabled, and V_4 holds at zero. As shown in line 2, only V_1 is in its 1 state, giving a total count of 1.

3. On line 3, the *second* pulse applied to the forward line flops V_1 back to its zero state. The decreasing voltage at this terminal of V_1 enables gate G_2, permitting the forward-line input pulse to pass through this gate and flip V_2 to its 1 state. This enables gate G_3, but without a backward line pulse, no signal can reach V_3, and V_3 holds at zero state. Neither gate G_5 nor G_6 is enabled, preventing a signal from reaching V_4, which holds at zero. Only V_2 is now in the 1 state, giving a total count of 2.

4. As indicated on line 4, the *third* forward-line pulse flips V_1 to its 1 state, enabling gate G_1, but no signal can reach V_2 which holds in its previous 1 state. Neither G_3 nor G_4 is enabled, and no signal reaches V_3, which holds at zero. V_4 similarly holds at zero since no signal can get through disabled gates G_5 and G_6. Now, as shown on line 4, V_1 and V_2 are both in their 1 states, giving a count value of 3, or $1 + 2$.

5. The *fourth* input pulse on the forward line, as shown on line 5 of the operations chart, flops V_1 to its zero state, enabling gate G_2. This permits the forward-line pulse to reach V_2, flopping this stage to its zero state. This enables gate G_4, allowing the pulse to enter V_3, flipping it to its 1 state. Gate G_5 is now enabled, but the forward-line pulse cannot go through G_6, and V_4 holds in its zero state. The count is now 4 since V_3 is the only stage in the 1 state.

6. Forward line pulse 5, as shown on line 6 of the chart, flips V_1 to its 1 state, enabling gate G_1. However, the forward-line pulse cannot go through G_2, and V_2 holds at zero state. Neither gate G_3 nor G_4 is enabled, and no pulse reaches V_3, which holds at its previous 1 state. Similarly, neither gate G_5 nor G_6 is enabled, and no pulse reaches V_4, which holds in its zero state. V_1 and V_3 are both in their 1 states, giving a count of 5 or $1 + 4$.

7. On line 7 of the chart, the *sixth* forward-line pulse flops V_1 to its zero state, enabling gate G_2. This allows the forward-line pulse to enter V_2, flipping it to its 1 state. This enables gate G_3, but the forward-line pulse cannot get through G_4, and V_3 holds in its previous 1 state. Since neither gate G_5 nor G_6 is enabled, no pulse reaches V_4, which holds at its zero state. V_2 and V_3 are now in their 1 states, giving a count of 6 or $2 + 4$.

8. The *seventh* forward-line pulse, as shown on line 8 of the chart, flips V_1 to its 1 state, enabling gate G_1. However, the forward-line pulse cannot pass through G_2 to V_2, which remains in its previous 1 state.

Neither G_3 nor G_4 is enabled, and no pulse reaches V_3, which holds in its previous 1 state. V_4 also holds in its previous zero state, since no pulse reaches it due to gates G_5 and G_6 being disabled. Now, V_1, V_2, and V_3 are all in their 1 states, giving a count of 7, or $1 + 2 + 4$.

9. The *eighth* forward-line pulse, as shown on line 9 of the chart, flops V_1 to its zero state, enabling gate G_2. This allows the forward-line pulse to reach and roll V_2 into its flop or zero state. Gate G_4 is now enabled, allowing the pulse to reach, and roll V_3 into its flop or zero state. Gate G_6 is now enabled, and the forward-line pulse reaches and rolls V_4 into its flip or 1 state. Now, only V_4 is in the 1 state, giving a count of 8.

10. As shown on line 10 of the operations chart, the *first* backward-line input pulse flips V_1 to its 1 state. This enables gate G_1, and the backward-line pulse now passes through G_1 to V_2, flipping it to its 1 state. Gate G_3 is now enabled, and the backward-line pulse reaches and rolls V_3 to its flip or 1 state. Gate G_5 is now enabled, and the pulse reaches and rolls V_4 to its flop or zero state. V_1, V_2, and V_3 are now in their 1 states, giving a count of 7 or $1 + 2 + 4$. The previous count (line 9) was *8*, after 8 forward-line pulses. Now, with 1 backward-line pulse, the count was rolled back to *7*.

11. On the *second* backward-line pulse, as indicated on line 11 of the chart, V_1 is flopped to its zero state, enabling gate G_2. However, the backward-line pulse cannot get through G_1, and V_2 holds at its 1 state. Neither G_3 nor G_4 is enabled, and no pulse reaches V_3, which remains in its previous 1 state. Similarly, neither gate G_5 nor G_6 is enabled, and no pulse reaches V_4, which holds at its zero state. The count value is now 6 or $2 + 4$, since V_2 and V_3 are in their 1 states. This count of 6 is correct, of course, since the previous count (line 10) was 7, and the backward line pulse of line 11 rolled the count back to 6.

12. The *third* backward-line pulse, as shown on line 12 of the chart, flips V_1 to its 1 state. Gate G_1 is enabled, passing the backward-line pulse to V_2. This flops V_2 into its zero state, enabling G_4. However, the backward-line pulse cannot get through G_3, and V_3 therefore holds in its previous 1 state. Neither gate G_5 nor G_6 is enabled, and no pulse reaches V_4, which holds at its zero state. With V_1 and V_3 in their 1 states, the final count is now 5 or $1 + 4$. This final result is correct since a total of 8 forward-line pulses and 3 backward-line pulses were applied. The counter reads this difference of 5.

In Example 9·4, the forward-backward counter was started off with a count of zero. In the next example, the backward-line pulses are more numerous, and the counter is started off at an advanced count.

Example 9·5 With the forward-backward counter of Fig. 9·7 started off at a count of 15, follow the step-by-step operations if 3 forward-line pulses and 5 backward-line pulses are applied intermittently as shown in the operations chart of Fig. 9·9.

Solution. Refer to the block diagram of Fig. 9·7, and to the operations chart of Fig. 9·9 during the following operational steps. As shown, two backward-line pulses are first applied, followed by two forward-line signals. Then two more backward-line pulses are followed by a forward-line pulse and one last backward-line signal. This comprises a total of 5 backward-line and 3 forward-line pulses, or a net effect of 2 backward-line pulses. Since the counter, in this example, is being started at an advanced setting of 15, then the final result should give a reading of 13, or 2 less than 15. The operation steps are as follows:

1. At the start, as shown in line 1, all flip-flops, V_1, V_2, V_3, and V_4, are in their 1 states. With steady d-c voltages on all flip-flop output terminals, none of the AND gates are enabled. The count is now 15 or $1 + 2 + 4 + 8$.

2. The *first* backward-line input pulse, as shown on line 2 of the chart, Fig. 9·9, flops V_1 into its zero state. Gate G_2 enables, but the backward-line pulse cannot pass G_1, and V_2 holds at its 1 state. Gates G_3 and G_4 are disabled, and no pulse reaches V_3, which holds at its 1 state. V_4 also holds at its 1 state since no pulse can come through the disabled G_5 and G_6 gates. V_2, V_3, and V_4 are now in their 1 states, making the count 14 or $2 + 4 + 8$. This is correct since the first backward-line pulse rolled the count back from its starting point of 15 to 14.

3. The *second* backward-line pulse, as shown in line 3 of the chart, Fig. 9·9, flips V_1 to its 1 state, enabling gate G_1. The backward-line pulse goes through G_1 to V_2, flopping it to its zero state. This enables G_4, but the backward-line pulse cannot pass the disabled G_3, and V_3 holds at its 1 state. Since neither gate G_5 nor G_6 is enabled, no pulse gets to V_4, and it holds at its 1 state. V_1, V_3, and V_4 are now in their 1 states, making the count 13 or $1 + 4 + 8$.

4. The *first* forward-line input pulse, as shown in line 4 of the chart, Fig. 9·9, flops V_1 to its zero state, enabling gate G_2. The forward-line pulse passes through G_2, reaching and rolling V_2 to its flip or 1 state. This enables gate G_3, but the forward-line pulse cannot pass the disabled gate G_4, and V_3 receives no input pulse. V_3 holds in its 1 state. Neither gate G_5 nor G_6 is enabled, and no pulse reaches V_4, which holds in its 1 state. Now V_2, V_3, and V_4 are in their 1 states, making the count 14 or $2 + 4 + 8$. This is correct since the previous count (line 3) was 13, and the forward-line pulse brought the count now (line 4) to 14.

5. The *second* forward-line pulse, as shown in line 5 of the chart of Fig. 9·9, flips V_1 to its 1 state, enabling gate G_1. The forward-line pulse, however, cannot pass through G_2, and V_2 holds in its previous 1 state.

	Input		F–F V_1 (1) state	G_1 or G_2 enabled	F–F V_2 (2) state	G_3 or G_4 enabled	F–F V_3 (4) state	G_5 or G_6 enabled	F–F V_4 (8) state	Count				
	Forward	Backward								V_1	V_2	V_3	V_4	Total
Line 1	Start		1		1		1		1	(1)	(2)	(4)	(8)	15
Line 2		1	Flop to 0	G_2	Hold at 1		Hold at 1		Hold at 1		(2)	(4)	(8)	14
Line 3		2	Flip to 1	G_1	Flop to 0	G_4	Hold at 1		Hold at 1	(1)		(4)	(8)	13
Line 4	1		Flop to 0	G_2	Flip to 1	G_3	Hold at 1		Hold at 1		(2)	(4)	(8)	14
Line 5	2		Flip to 1	G_1	Hold at 1		Hold at 1		Hold at 1	(1)	(2)	(4)	(8)	15
Line 6		3	Flop to 0	G_2	Hold at 1		Hold at 1		Hold at 1		(2)	(4)	(8)	14
Line 7		4	Flip to 1	G_1	Flop to 0	G_4	Hold at 1		Hold at 1	(1)		(4)	(8)	13
Line 8	3		Flop to 0	G_2	Flip to 1	G_3	Hold at 1		Hold at 1		(2)	(4)	(8)	14
Line 9		5	Flip to 1	G_1	Flop to 0	G_4	Hold at 1		Hold at 1	(1)		(4)	(8)	13

Figure 9·9 **Operations chart of forward-backward counter for Example 9·5.**

Neither G_3 nor G_4 is enabled, and no pulse reaches V_3, which holds in its 1 state. Gates G_5 and G_6 are disabled, preventing the pulse from getting to V_4, which holds in its 1 state. V_1, V_2, V_3, and V_4 are all in their 1 states. The count is now 15 or $1 + 2 + 4 + 8$.

6. Backward-line pulse 3, as shown in line 6 of the chart, flops V_1 to its zero state, enabling gate G_2. The backward-line pulse cannot go through gate G_1, and V_2 holds in its 1 state. Neither gate G_3 nor G_4 is enabled, and no pulse reaches V_3, which holds in its 1 state. Since neither gate G_5 nor G_6 is enabled, no pulse reaches V_4, which holds in its 1 state. V_2, V_3, and V_4 are in their 1 state, giving a count of 14 or $2 + 4 + 8$. This is correct since the previous count (line 5) was 15, and the backward-line pulse (line 6) rolls the count back to 14.

7. The *fourth* backward-line input pulse, line 7, flips V_1 to its 1 state, enabling gate G_1. The backward-line pulse passes through G_1, reaching V_2, flopping V_2 to its zero state. This enables gate G_4, but the backward-line pulse cannot get through G_3. V_3 therefore holds in its 1 state. Neither gate G_5 nor G_6 is enabled, and without an input pulse, V_4 holds in its 1 state. V_1, V_3, and V_4 are in their 1 states, and the count is now 13 or $1 + 4 + 8$.

8. Forward-line pulse 3, line 8 of Fig. 9·9, flops V_1 to its zero state, enabling gate G_2. The forward-line pulse passes through G_2 and flips V_2 to its 1 state, enabling gate G_3. However, the forward-line pulse cannot get through G_4, and no pulse reaches V_3. V_3 holds in its 1 state. Neither gate G_5 nor G_6 is enabled, and no pulse gets to V_4, which holds in its 1 state. V_2, V_3, and V_4 are in their 1 states, giving a count of 14 or $2 + 4 + 8$. This is correct since the previous count (line 7) was 13, and the forward-line pulse (line 8) advances the count to 14.

9. The *fifth* and last backward-line pulse, as shown on line 9 of Fig. 9·9, flips V_1 to its 1 state, enabling gate G_1. The backward pulse passes through G_1 to V_2, rolling it to its flop or zero state. This enables gate G_4, but the backward pulse cannot get through G_3, and without an input pulse, V_3 holds in its previous 1 state. Neither gate G_5 nor G_6 is enabled, and no pulse reaches V_4, which holds in its previous 1 state. V_1, V_3, and V_4 are in their 1 states, giving a count of 13 or $1 + 4 + 8$. This final result checks with the fact that the original setting of the counter was 15, and with 5 backward- and 3 forward-line pulses, the count is rolled back 2 steps to 13.

9·5 Television Synchronizing Pulse Generator

A complicated piece of equipment that produces the pulses required by the television system is called the *television synchronizing pulse generator*. To keep the scanning electron beam in the receiver picture tube moving exactly in step with the camera tube scanning beam, synchronizing pulses are necessary. The pulses, shown in Fig. 9·10, are the *horizontal synchronizing pulses*, the *equalizing pulses*, and the *vertical synchronizing pulses.* These are generated by the circuits of the block diagram of Fig. 9·11, which is one method of producing them. These circuits, as shown, are flip-flop or Eccles-Jordan (E-J) bistable multivibrators (discussed in detail in Chap. 5), gates (discussed in Chap. 8), and pulse-frequency dividers (discussed in Chaps. 5 to 7).

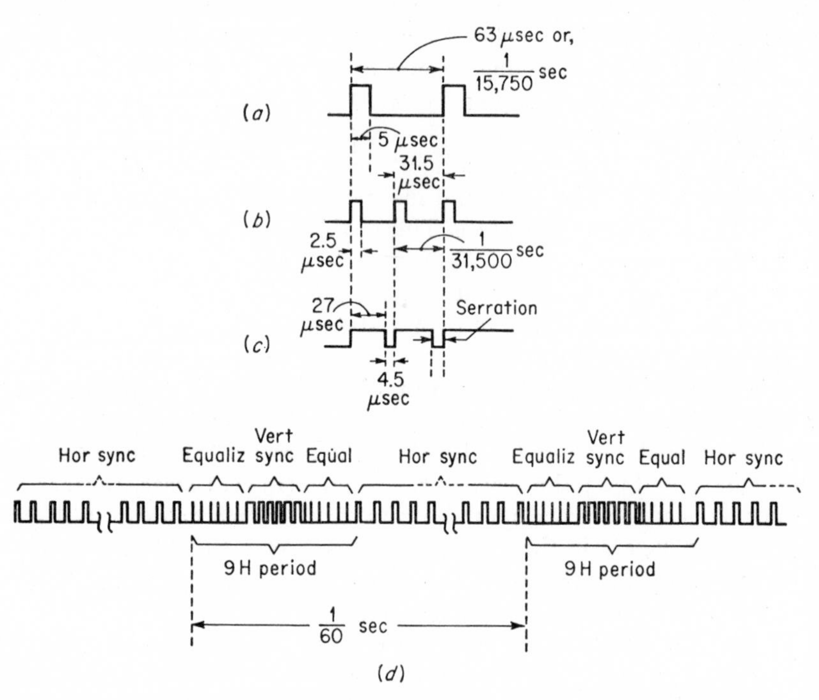

Figure 9·10 TV sync pulses. (*a*) Horizontal sync pulses 15,750 cps; (*b*) equalizing pulses 31,500 cps; (*c*) vertical sync-pulse serrations 31,500 cps; (*d*) composite sync pulses.

The synchronizing pulses, also referred to as *sync* pulses, are shown in their individual details in Fig. 9·10*a*, *b*, and *c* and are shown in sequence in part (*d*). The horizontal sync pulses (Fig. 9·10*a*) are approximately 5 μsec in duration, at a frequency of 15,750 cps. The equalizing pulses (Fig. 9·10*b*) are about 2.5 μsec wide, having a frequency of 31,500 cps. The vertical sync pulses (Fig. 9·10*c*) are approximately 27 μsec wide, with 4.5-μsec serrations between the wider pulses, and have a frequency of 31,500 cps. As shown in Fig. 9·10*d*, numerous (actually 262) horizontal sync pulses are generated at first. These are followed by a group of six equalizing pulses, then six vertical sync pulses, and then six more equalizing pulses. This group of 18 pulses occurs during a period called *9H.*

After the last equalizing pulse, at the end of the 9H period, the horizontal sync pulses are again produced. After many of these (again 262), the next 9H period, comprising the equalizing and vertical sync pulses, occurs. As shown, the recurrent rate of the 9H period is 60 cps or 1/60 of a second from the *beginning* of one 9H period to the *beginning* of the next.

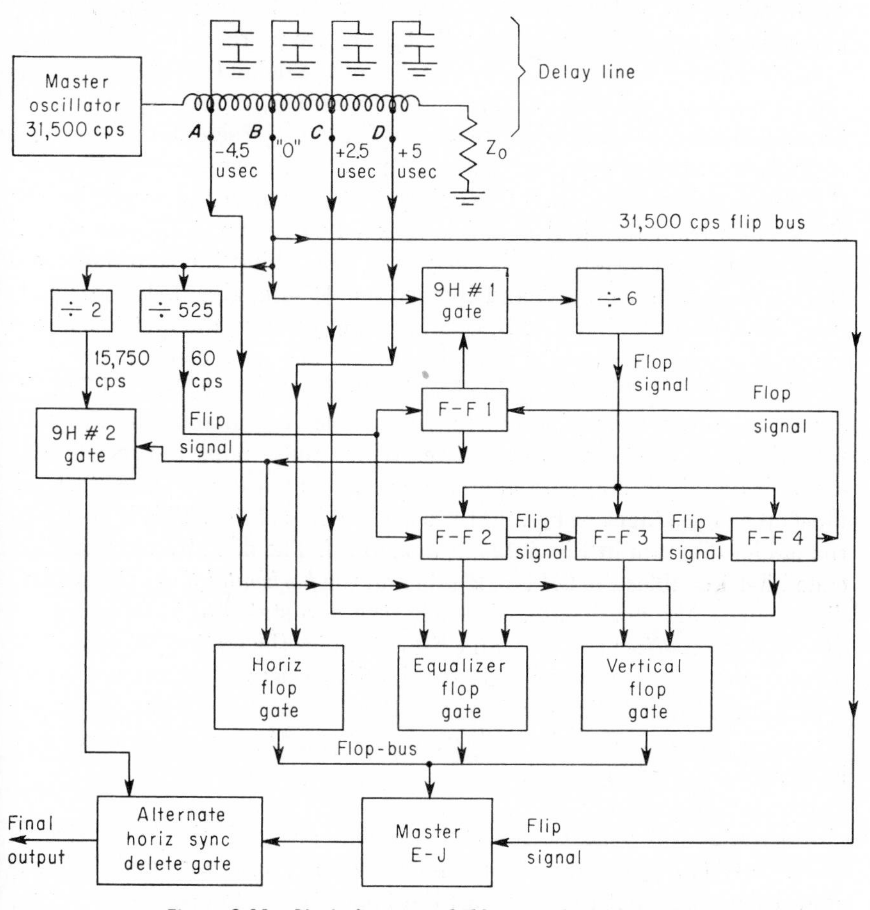

Figure 9·11 Block diagram of TV sync-pulse generator.

To understand how this complex sync signal, Fig. 9·10*d*, is generated, the block diagram of Fig. 9·11 will be discussed in detail, stage by stage. The ladder diagram of the waveshapes of many of these stages, shown in Fig. 9·12, will aid in the discussion of the block diagram.

The master oscillator produces narrow pulses at a frequency of 31,500 cps. These signals are applied to a tapped delay line, as shown at the top of the block diagram of Fig. 9·11. The delay line usually consists of several coils in series, with shunt capacitors connected as shown in the block diagram of the shift register of Fig. 9·5. The time delay is approximately equal to the square root of the product of L and C. In Fig. 9·11, an output pulse is developed first at terminal A of the delay line. Then, 4.5 μsec later, a signal appears at terminal B. 2.5 μsec after that, a pulse appears at terminal C and another 2.5 μsec later, a pulse appears at terminal D of the line. The line is terminated in an impedance, Z_0, which is equal to the line's characteristic impedance so that there are no reflected pulses back down the line. The pulse from terminal B of the delay line is used here as the *reference* or *clock* pulse and is called the zero time pulse and occurs at the 31,500-cps frequency of the master oscillator. The signal from terminal A precedes this by 4.5 μsec and is referred to as the −4.5 μsec pulse. Output signal from terminal C is delayed 2.5 μsec after the zero reference clock pulse from terminal B and is referred to as the +2.5-μsec signal. The output from terminal D is delayed 5 μsec behind the zero reference-clock pulse from terminal B and is called the +5-μsec signal. These signals from delay-line terminals A, B, C, and D are shown in the first line of the ladder diagram of waveshapes, Fig. 9·12*a*. The zero clock pulse from terminal B *always* flips the *master Eccles-Jordan* (master E-J), while only *one* of the other delay line pulses, from either terminal A, or C, or D, rolls the stage back to its flop state. Which pulse flops the master E-J depends on which flop gate is in the enabled condition. At the start, *all* flip-flop or E-J stages of Fig. 9·11 are in their flop states. This is shown in the ladder diagram in Fig. 9·12*b*, *c*, *d*, *e*, and *j*. The flop state of the stage called F-F1 enables the 9H No. 2 gate and the *horizontal flop gate* and also disables the 9H No. 1 gate. Similarly, the flop states of F-F2 and F-F4 disable the *equalizer flop gate*, and F-F3 disables the *vertical flop gate*. The following discussions explain the production processes of the horizontal sync pulses, the equalizing pulses, and the vertical sync pulses. Reference should be made to the block diagram of Fig. 9·11 and to the waveshape ladder diagram of Fig. 9·12.

The horizontal sync pulses. All flip-flops are in their flop states to start. Only two gates are enabled now, the 9H No. 2 gate and the horizontal flop gate, both enabled by the flop-state voltage of F-F1. A zero clock pulse from terminal B of the delay line flips the master E-J. Five μsec later, a pulse from terminal D (the +5-μsec pulse) goes through the enabled horizontal flop gate, and flops the master E-J, producing a 5-μsec-wide pulse from the master E-J. This process can be seen from the first few pulses of Fig. 9·12*a* and *j*.

The delay-line output pulse from terminal A, the −4.5-μsec signal, cannot reach the master E-J since the vertical flop gate is disabled, because of the flop state of F-F3. Likewise, the delay line output pulse from terminal C, the +2.5-μsec signal, cannot reach the master E-J since

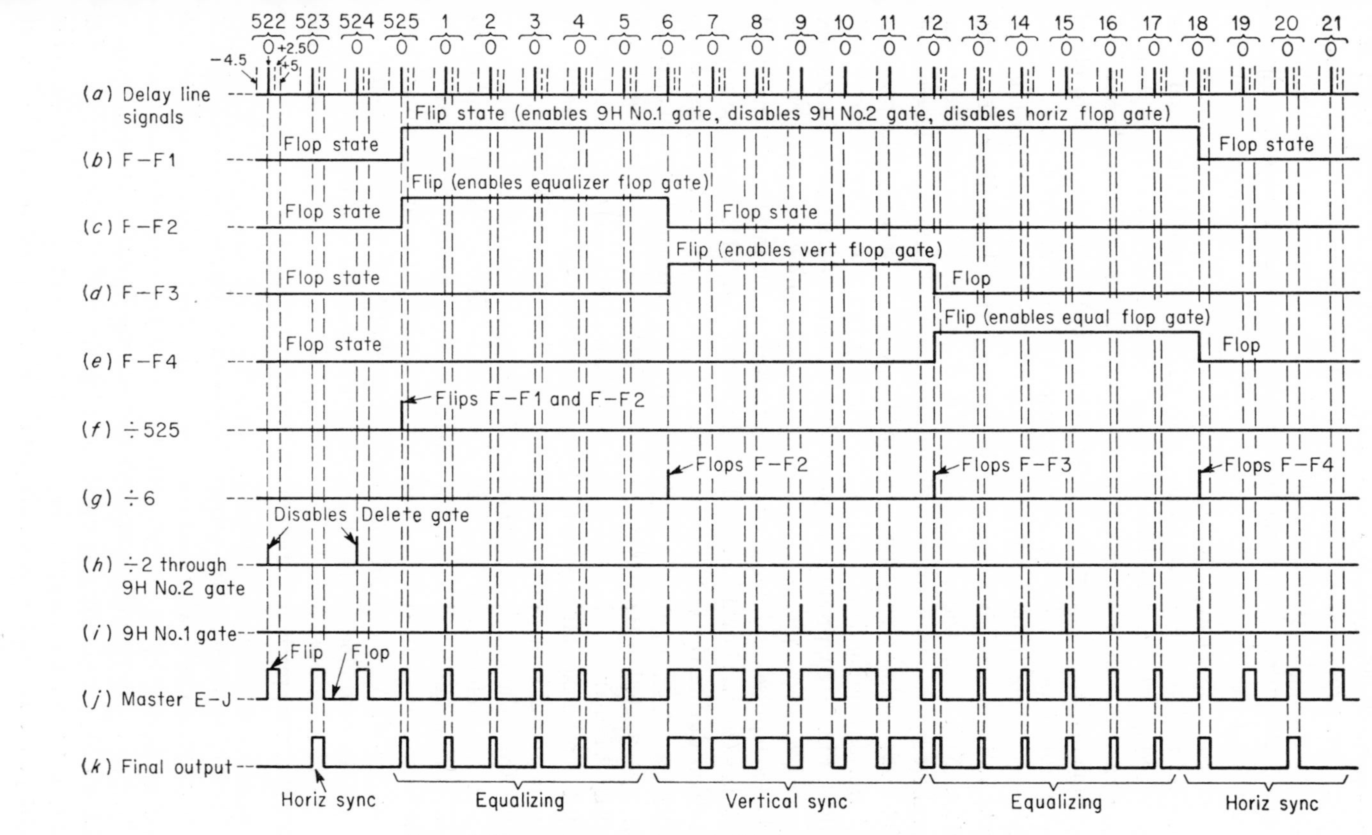

Figure 9·12 Ladder diagram of TV sync generator waveshapes.

the equalizer flop gate is disabled, due to the flop states of F-F2 and F-F4. This process of the zero reference pulse from terminal B flipping the master E-J, and the $+5$-μsec pulse from terminal D, through the horizontal flop gate, flopping the master E-J, is repeated over and over again 525 times.

The zero reference pulse from delay-line terminal B also is applied to the $\div 2$ and to the $\div 525$ stages, as well as to the 9H No. 1 gate. As described in Chaps. 5 and 7, a series of bistable multivibrators (binary chain) acts as a division circuit, where a series of input signals are required in sequence to produce one output signal. For every *second* zero clock pulse from terminal B, one pulse appears at the output of the $\div 2$ stage. This signal goes through the 9H No. 2 gate, which is enabled by the flop-state voltage of F-F1. This signal is shown in the ladder diagram of Fig. 9·12*h*. Note that the $\div 2$ and 9H No. 2 pulses appear for the even-numbered zero clock pulses (Fig. 9·12*a*) only. The ladder diagram of Fig. 9·12*a* is shown starting with the 522nd pulse. Actually, the previous 521 pulses are identical to that shown. The 522nd and 524th zero clock pulses, Fig. 9·12*a*, produce output pulses from the $\div 2$ and 9H No. 2 gate, Fig. 9·12*h*. These pulses are applied to the *alternate horizontal sync delete* gate, each pulse disabling the gate. Output from the master E-J is applied to this gate, as shown in the block diagram, Fig. 9·11. Since the gate is disabled during the even numbered pulses (2, 4, 6, . . . , 522, and 524), every *second* 5-μsec pulse from the master E-J is prevented from appearing or is deleted in the final output of the generator. Note that, as shown in Fig. 9·12*a* and *k*, only the 523rd zero pulse and $+5$-μsec pulse actually result in a horizontal (H) sync pulse in the final output. The frequency of the horizontal sync pulses is therefore half that of the original 31,500-cps signal from the *master oscillator*, or 15,750 cps.

The zero reference pulse from delay-line terminal B is also applied to the $\div 525$ stage, as shown in the block diagram, Fig. 9·11. This stage, of course, produces an output signal only after 525 pulses have been applied in sequence. As shown in the ladder waveshapes of Fig. 9·12*a* and *f*, at the 525th zero clock pulse, an output pulse is produced from the $\div 525$ stage. This, as will be explained next, is done to halt any further generation of the 5-μsec horizontal sync pulses.

Equalizing pulses, first group. As shown in Fig. 9·10*b*, the equalizing pulses are 2.5 μsec wide and occur at a 31,500-cps frequency. They are only generated during the 9H period, as shown in Fig. 9·10*d*, and then, as groups of six during the first third and last third of the 9H period.

The 525th zero reference pulse from terminal B of the delay line (Fig. 9·11) produces an output pulse from the $\div 525$ stage (Fig. 9·12*f*). This signal, as shown in the block diagram of Fig. 9·11, is applied to F-F1 and

F-F2, flipping these stages. The flip-state voltage of F-F2 enables the equalizer flop gate, while the flip of F-F1 disables both the 9H No. 2 gate and the horizontal flop gate and also enables the 9H No. 1 gate. The 9H No. 2 gate being disabled prevents the ÷2-stage signal from reaching the *alternate horizontal sync delete* gate. This means that there will be no deletion of signals produced by the master E-J. The disabling of the horizontal flop gate now prevents any +5-μsec pulse, from terminal *D* of the delay line, from reaching and flopping the master E-J.

Now, as shown in the ladder waveshapes of Fig. 9·12*a* and *j*, the 525th zero clock pulse, as usual, flips the master E-J. The +2.5-μsec pulse from delay-line terminal *C* now goes through the *equalizer flop* gate (enabled by the flip condition of F-F2, Fig. 9·12*c*) and flops the master E-J. This produces a 2.5-μsec-wide pulse from the master E-J, and this signal goes through the *delete* gate (which has not been disabled due to any ÷2 stage pulse) to the final output. This is repeated for the next five zero- and +5-μsec pulses, numbers 1 through 5 of Fig. 9·12*a*. The result is the 2.5-μsec, 31,500-cps equalizing pulses shown in Fig. 9·12*j* and *k*.

As shown in the block diagram of Fig. 9·11, the zero clock pulses from delay line terminal *B* are also applied to the 9H No. 1 gate. Starting with the zero clock pulse number 1 of Fig. 9·12*a*, a series of outputs from the 9H No. 1 gate are produced as shown in Fig. 9·12*i*. These pulses are applied to the ÷6 stage of the block diagram of Fig. 9·11. At the zero-clock pulse number 6 (Fig. 9·12*a*), an output pulse is produced from the ÷6 stage (Fig. 9·12*g*). This output pulse from the ÷6 stage flops F-F2, which disables the *equalizer flop* gate. This, as will be explained next, is done to halt the generation of any further 2.5-μsec equalizing pulses and allow the vertical sync pulses to begin.

Vertical sync pulses. As shown in Fig. 9·10*d* during the 9H period, the first group of six equalizing pulses are followed by the six vertical sync-pulse serrations. The output from the ÷6 stage (see Fig. 9·12*g* and the block diagram of Fig. 9·11) flops F-F2 but has no effect on F-F3 or F-F4 since these latter two are already in their flop state. When F-F2 goes into its flop state, it flips F-F3, as shown in Fig. 9·11 and in the ladder waveshape diagram at the zero clock pulse number 6 (Fig. 9·12*a*) and for the F-F2 and F-F3 waves (Fig. 9·12*c* and *d*).

The flop of F-F2 disables the equalizer flop gate, while the flip of F-F3 enables the vertical flop gate. The master E-J is, as usual, flipped by the zero-clock pulse (in group number 6 now, in Fig. 9·12*a*). The +2.5-μsec delay line pulse from terminal *C*, in this group number 6, cannot get through the disabled equalizer flop gate (Fig. 9·11) to reach the master E-J. Neither can the +5-μsec delay-line pulse from terminal *D*, in this group number 6, get through the disabled horizontal flop gate to reach the master E-J. The -4.5-μsec delay-line pulse from terminal A, in the next group of pulses (number 7 of Fig. 9·12a), now goes through the enabled vertical flop gate and flops the master E-J. This produces the first vertical sync pulse of the group of six of Fig. 9·12j and k. Each of the remaining five vertical sync pulses is produced in the same manner with the zero clock pulse flipping the master E-J and then the -4.5-μsec pulse flopping this stage.

The vertical sync pulses occur at the same frequency as the zero clock pulses, which is 31,500 cps. The duration of one entire cycle is therefore $1/31{,}500$ sec, or approximately 31.5 μsec. As shown in Fig. 9·10c and in Fig. 9·12j and k, the period between the vertical sync pulses, called serrations, is 4.5 μsec, leaving approximately 27 μsec for the wide pulse. Each vertical sync pulse produced by the master E-J goes through the delete gate, since this gate is in its enabled state, and these pulses become the final output as shown in Fig. 9·12k.

After six vertical sync pulses, the generator discontinues these and again resumes the production of equalizing pulses.

Equalizing pulses, second group. As shown in Fig. 9·10d and Fig. 9·12j and k, a second group of six equalizing pulses is generated during the last third of the 9H period, following the six vertical sync pulses. The 9H No. 1 gate, which was enabled by the flip state of F-F1 (Fig. 9·12b), passes the zero reference-clock pulses (Fig. 9·12i). Every sixth 9H No. 1 gate pulse produces a pulse from the $\div 6$ stage, as shown in Fig. 9·12g. The $\div 6$-stage output, which corresponds to the zero clock pulse in group number 12 (Fig. 9·12a), flops F-F3 (Fig. 9·12g and block diagram Fig. 9·11). This pulse does not affect F-F2 nor F-F4, which are already in their flop states.

The flop of F-F3 disables the vertical flop gate and flips F-F4, which, in turn, enables the equalizer flop gate. This is indicated in the block diagram of Fig. 9·11 and the ladder waveshapes of Fig. 9·12d and e. Now, the master E-J is again flopped by the $+2.5$-μsec pulse from terminal C of the delay line, after the zero-clock pulse has flipped the master E-J. Six 2.5-μsec-wide equalizing pulses are produced in this manner, corresponding to the zero clock pulses and the $+2.5$-μsec pulses in groups 12 to 17 of Fig. 9·12a and j.

In group number 18 of Fig. 9·12a, the zero clock pulse produces an output pulse from the $\div 6$ state (Fig. 9·12g), which flops F-F4. This, in turn, disables the equalizer flop gate, and also flops F-F1, as shown in Fig. 9·11. The flop of F-F1 returns the gates to their original operation during the production of horizontal sync pulses; the 9H No. 1 gate is disabled; the horizontal flop gate is enabled; and the 9H No. 2 gate is enabled, allowing the ÷2 stage signal to alternately disable and enable the alternate horizontal sync delete gate. This blocks every *other* 5-μsec-wide pulse produced by the master E-J (flipped by zero clock pulse, and flopped by +5-μsec delay-line pulse) from reaching the final output. The signals which do are the horizontal sync pulses, occurring at the 15,750-cps frequency.

9·6 Digital Voltmeter, Analog-to-Digital Converter

A fluctuating voltage, whether d-c or a-c, changes from its initial value to some other amplitude by going through all the in-between values. This is an example of an *analog* quantity. Conversely, a *digital* quantity is one which has no in-between values. When counting items moving along a

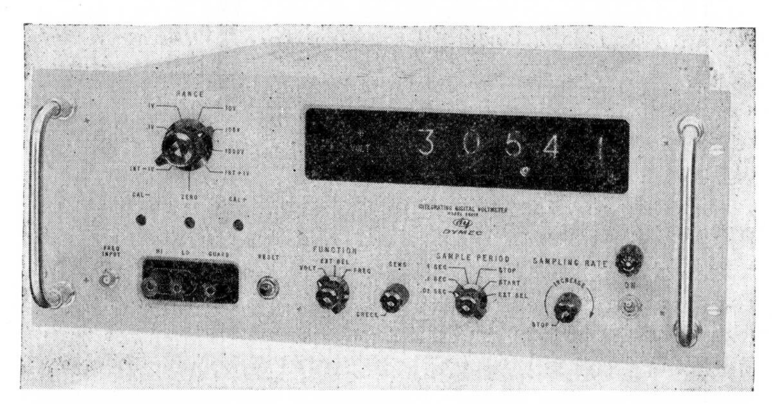

Figure 9·13 Digital voltmeter. *(DYMEC, a subsidiary of Hewlett-Packard Co.)*

conveyor belt, the count jumps sharply from one value to the next higher one in steps, without going through the in-between values. Counting a group of equalizing pulses in a television signal, the count advances from zero to one, to two, etc., in discrete steps, without any fractional in-between values. To apply the digital procedures of pulse circuits to analog quantities, analog-to-digital conversion must occur. A *digital voltmeter* of the type described here is an *analog-to-digital converter device.* An example of such a piece of equipment is shown in Fig. 9·13. Numeral indication may be displayed by columns of neon lights, or by a row of single tubes, each one of which replaces a column of lights. Two types of these tubes are known by their trade names of Nixie and Numerik.

A block diagram of a simple digital voltmeter is shown in Fig. 9·14. As shown, this unit employs flip-flop or bistable multivibrators and a monostable multi (discussed in Chap. 5) as well as AND gates (Chap. 8). Before the operation of the entire digital voltmeter is discussed, some of the individual blocks will first be presented as a simple theory review.

Constant-current generator. The block called *constant-current source* in Fig. 9·14 is also known as a *Norton generator*. This is simply a voltage source having a high internal resistance, much larger than that of the load. A pentode vacuum tube may be considered to be a *constant-current generator* since its internal resistance is much larger than the load resistor.

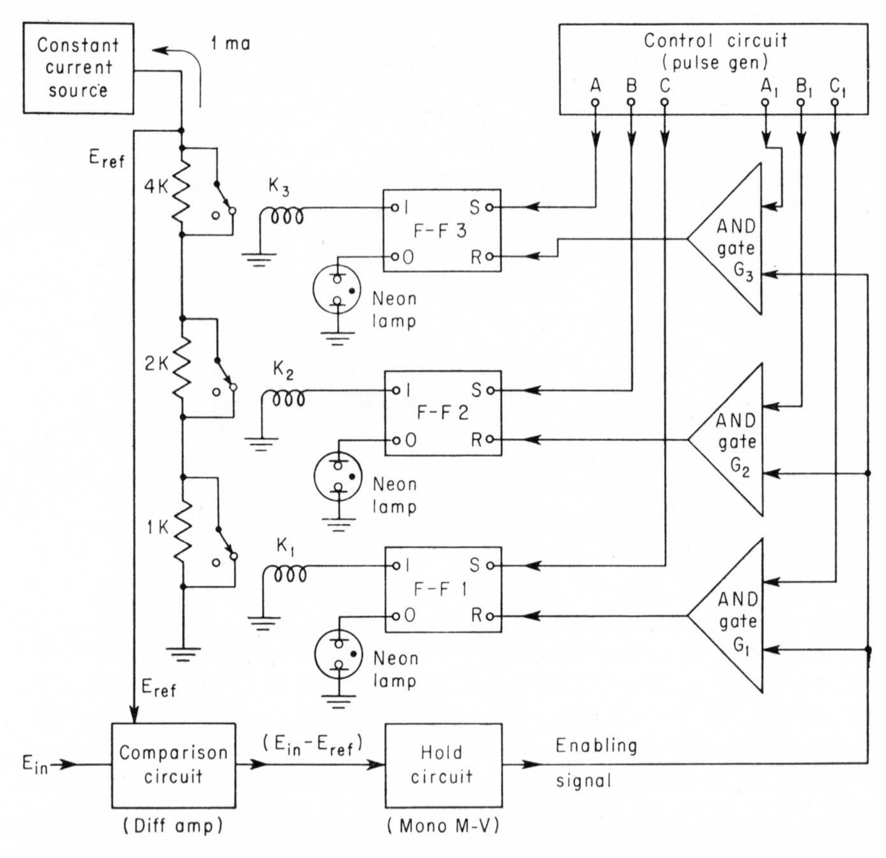

Figure 9·14 **Block diagram of digital voltmeter.**

In Fig. 9·15*a*, a 1,000-volt d-c generator or battery, having a 1-megohm internal resistance (R_{int}), is shown connected in series with a variable load resistor (R_{load}). As shown in the chart of Fig. 9·15*b*, the current remains practically 1 ma even though the load is changed from zero ohms to 7 kilohms. This is comparable to the resistance variation that the relays (K_1, K_2, and K_3) produce in the block diagram of Fig. 9·14.

Comparison circuit. Another part of the digital-voltmeter block diagram of Fig. 9·14 is called the *comparison circuit*. Some simple schematic diagrams are shown in Fig. 9·16. These are also known as *difference amplifiers* or *summing devices.*

In Fig. 9·16*a*, a triode class *A* amplifier is shown with two input signals, E_2 applied to the grid and E_1 applied to the cathode, with both signals going positive. E_1 driving the cathode *positive,* has the same effect as if the grid were being driven negative. When the amplitude of E_1 is equal to E_2, the effect is complete cancellation, resulting in the same amount of electron flow through the tube as if no input signals were being applied. If 100 volts were dropped across R_L without any input signal, then E_{out} is +200 volts, as shown in Fig. 9·16*a*.

When E_2 is greater than E_1, the net effect is that the grid is being driven in a *positive* direction (actually, a decrease in the bias). With more plate

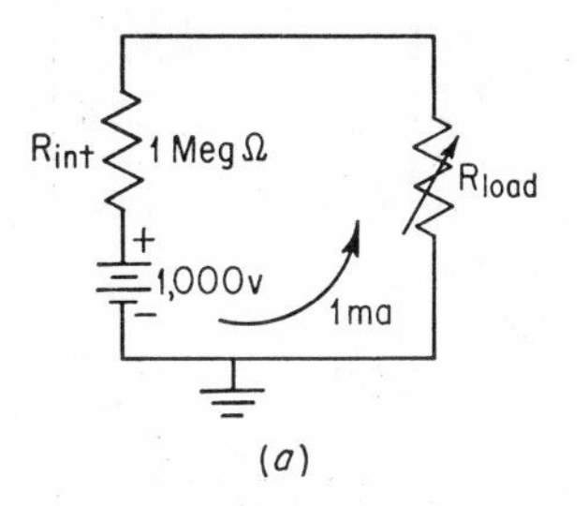

(*a*)

E_{total}	R_{int}	R_{load}	I
1,000v	1 Meg	0 Ω	1 ma
"	"	1 K	0.999 ma
"	"	7 K	0.993 ma

(*b*)

Figure 9·15 **Example of constant-current generator (Norton generator).**

current flowing, a larger voltage drop takes place across R_L, say, 120 volts. E_{out}, as shown, is now +180 volts (300 − 120).

When E_2 is less than E_1, the net effect is that the grid is being driven in a *negative* direction (an increase in bias). With less plate current flowing, a smaller voltage drop occurs across R_L, say, 80 volts. E_{out}, as shown, is now +220 volts (300 − 80).

A PNP transistor circuit, common emitter diagram (C-E), is shown in Fig. 9·16*b*. Electron flow, as shown, is backward to the emitter-symbol arrow and is *down* through R_L. In this circuit, E_1 and E_2 are negative inputs. When E_2 goes negative, it increases the forward bias at the N-material base, and more current flows through the transistor. When E_1 goes negative, an opposite effect is created. Driving the P-type emitter negative decreases the forward bias between base and emitter, and less current flows. With no input signal, current flows, and a voltage drop (say 5 volts) exists across R_L, with the polarity shown in Fig. 9·16*b*. E_{out} is −5 volts (−10 + 5), as shown. When $E_1 = E_2$, their effect is as if no input were applied, and E_{out} remains the same.

When the negative-going E_2 is larger than E_1, the effect is that the N-type base is driven more negative, increasing the forward bias. The transistor conducts more heavily, and voltage across R_L increases to, say,

7 volts. E_{out} now becomes -3 volts $(-10 + 7)$. An opposite result occurs when E_2 is less than E_1. Now, the negative-going E_1 decreases the forward bias between emitter and base, and less current flows. Voltage across R_L decreases to say 3 volts, and E_{out} becomes -7 volts $(-10 + 3)$, as shown.

Another triode tube difference amplifier circuit is shown in Fig. 9·16*c*, with an NPN-transistor diagram shown in Fig. 9·16*d*. In both diagrams,

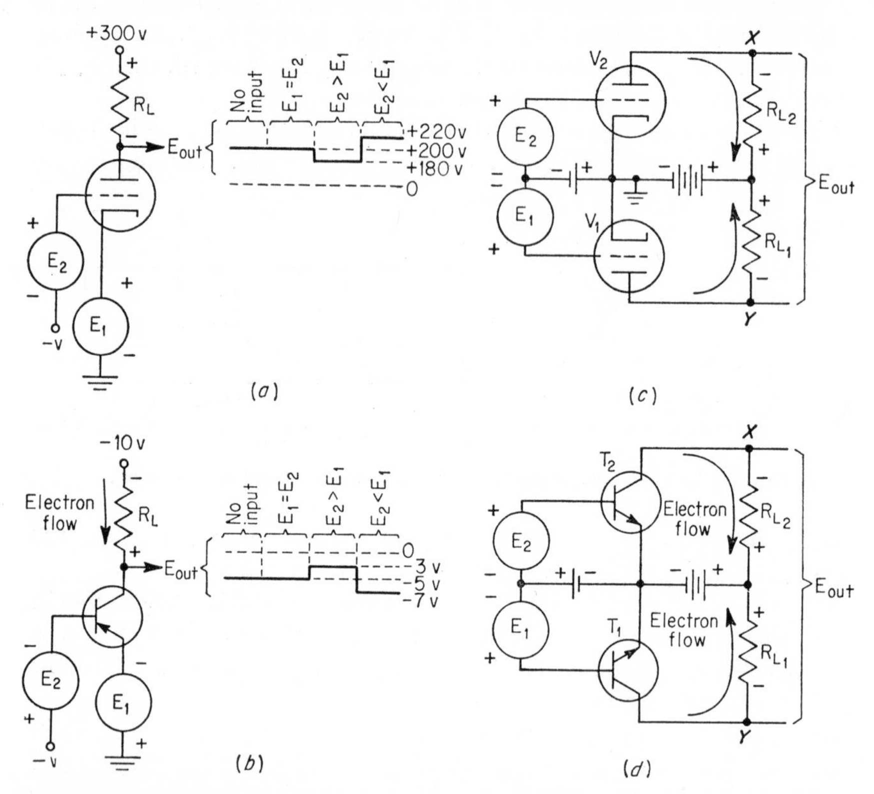

Figure 9·16 **Difference amplifiers. (*a*) Triode tube; (*b*) PNP transistor; (*c*) triode tubes; (*d*) NPN transistors.**

the input signals E_1 and E_2 are positive-going, causing the tubes V_1 and V_2 to conduct more heavily and the transistors T_1 and T_2 to conduct more. Without any input signals, the tubes conduct equally, producing equal voltage drops across R_{L_1} and R_{L_2}. Point X is therefore at the same potential, with respect to ground (cathode), as point Y. E_{out} (between X and Y) is therefore zero. The same occurs when $E_1 = E_2$.

When E_2 is greater than E_1, V_2 (and T_2) conducts more heavily than V_1 (and T_1). $E_{R_{L_2}}$ is larger than $E_{R_{L_1}}$, and point X is now less positive than point Y. E_{out} (point X with respect to Y) is now *negative*. E_{out} would become *positive* if E_2 were less than E_1. V_1 (and T_1 also) would conduct more than V_2 (and T_2). $E_{R_{L_1}}$ would be larger than $E_{R_{L_2}}$, and point X would now become more positive than point Y.

Digital-voltmeter operation. The block diagram of the digital voltmeter, Fig. 9·14, should be referred to during the following discussions of the theory operation of the device. With the simple diagram shown, only three flip-flop stages are used. As a result, this circuit can only measure up to a maximum voltage equal to the binary number 111., or the decimal number 7. Additional flip-flops would enable the voltmeter to read higher voltages. For example, eight stages could produce a maximum reading equal to the binary 11111111. (eight digits), or the decimal number 255. (Binary numbers are discussed in Chap. 7.)

The comparison circuit in the block diagram of Fig. 9-14 is a difference amplifier of the type shown in Fig. 9·16. The input signals are the voltage to be measured, called E_{in}, and a voltage called the reference or E_{ref}, produced by the current from the *constant-current source* flowing through the 4-, 2-, and 1-kilohm resistors. The output from the *comparison circuit* will be zero volts if $E_{in} = E_{ref}$; or a plus voltage if E_{in} is larger than E_{ref}; or a minus voltage if E_{ref} is larger than E_{in}. This output is fed to the *hold circuit.*

The hold circuit of Fig. 9·14 is designed to produce an enabling signal to all the AND gates only when the signal from the comparison circuit is negative. This occurs only when E_{ref} is larger than E_{in}. The hold circuit could be simply a monostable multivibrator which is triggered by a negative input.

The *control circuit* of the block diagram of Fig. 9·14 consists primarily of a pulse generator which produces pulses delayed from each other out of terminals A, B, C, etc. The sequence of pulses from these terminals is first A, then A_1, then B, B_1, C, and finally C_1.

In the following examples, reference should be made to the block diagram of the digital voltmeter, Fig. 9·14, and to the operations chart for that particular example.

Example 9·6 If the voltage to be measured (E_{in}) is 6 volts, follow the operations of the stages of the digital voltmeter to produce this 6-volt reading.

Solution. Refer to the operations chart of Fig. 9·17 and to the block diagram of Fig. 9·14 in the following step-by-step discussion.

1. At the start, all flip-flop stages are in the flop or zero state. This means that high plate voltages exist at the terminals marked 1 of each flip-flop. These high voltages cause relays K_3, K_2, and K_1 to energize, closing the switch contacts which short out the 4-, 2-, and 1-kilohm resistors. The 1-ma current from the constant-current source is shunted around the resistors through the relay switches. This produces zero volts E_{ref}, and with an E_{in} of 6 volts, the output of the comparison circuit is a positive voltage which does not affect the hold circuit. The AND *gates* are not enabled.

2. When a pulse is produced at terminal A of the control circuit, flip-flop stage F-F3 is flipped to its set or 1 state because of the A pulse being applied to the S terminal. The voltage at terminal 1 of F-F3 is now low, and relay K_3 becomes de-energized, and its switch contacts open. F-F2 and F-F1 remain in their zero states, and their associated relays K_2 and K_1 are still energized. Relay K_3 being de-energized removes the short across the 4-kilohm resistor. The 1 ma from the constant-current source now flows through the 4-kilohm resistor, producing an E_{ref} of 4 volts (1 ma $\times$ 4 kilohms). Since E_{in} (voltage to be measured) is 6 volts, it is larger than the 4 volts E_{ref}, and the output from the comparison circuit is positive. This does not affect the hold circuit (which requires a negative voltage), and the AND gates are still disabled, as shown in the chart, Fig. 9·17, line 2.

3. Pulse A_1 from the control circuit cannot go through AND gate G_3 since it is disabled. F-F3 holds in its 1 state, as shown in the operations chart, Fig. 9·17, line 3. With a low voltage at F-F3 terminal 1, relay K_3 is still de-energized, and the 4-kilohm resistor is in the circuit, not shorted. F-F2 and F-F1 remain in their zero states, and their relays, K_2 and K_1, remain energized, shorting out the 2- and 1-kilohm resistors. E_{ref} is still 4 volts, and the output from the comparison circuit is still positive. This does not affect the hold circuit, and the AND gates remain disabled.

4. Pulse B from the control circuit rolls F-F2 to its set or 1 state since the pulse is applied to terminal S. The low voltage at terminal 1 de-energizes relay K_2, and the short is removed from across the 2-kilohm resistor. As shown in line 4 of the operations chart, Fig. 9·17, F-F3 holds in its 1 state, relay K_3 is still de-energized, F-F1 holds in its zero state, and relay K_1 is still energized. Only the 1-kilohm resistor is shorted. The 1-ma current from the constant-current source flows through the 4- and 2-kilohm resistors, and E_{ref} is 6 volts (1 ma $\times$ 6000 ohms). Since $E_{ref} = E_{in}$ now, the comparison circuit produces zero output, and the hold circuit remains unaffected. The AND gates are still disabled.

5. Pulse B_1 from the control circuit cannot get through the disabled AND gate G_2, and F-F2 holds in its 1 state. F-F3 and F-F1 likewise hold in their previous states, 1 and zero respectively. The relays remain as they were previously, with only K_1 energized. E_{ref} is still 6 volts, and the comparison circuit output of zero still does not affect the hold circuit. All AND gates remain disabled, as shown in line 5, Fig. 9·17.

	Pulse	F-F 3 state	K_3 relay	F-F 2 state	K_2 relay	F-F 1 state	K_1 relay	E_{ref}	Comparison circuit out ($E_{in} - E_{ref}$)	Gates
Line 1	Start	0	Energized	0	Energized	0	Energized	0 V	6 − 0 = +6	Disabled
Line 2	A	Set, 1	De-energ	Holds at 0	Energized	Holds at 0	Energized	4 V	6 − 4 = +2	Disabled
Line 3	A_1	Holds at 1	De-energ	Holds at 0	Energized	Holds at 0	Energized	4 V	6 − 4 = +2	Disabled
Line 4	B	Holds at 1	De-energ	Set, 1	De-energ	Holds at 0	Energized	6 V	6 − 6 = 0	Disabled
Line 5	B_1	Holds at 1	De-energ	Holds at 1	De-energ	Holds at 0	Energized	6 V	6 − 6 = 0	Disabled
Line 6	C	Holds at 1	De-energ	Holds at 1	De-energ	Set, 1	De-energ	7 V	6 − 7 = −1	Enabled
Line 7	C_1	Holds at 1	De-energ	Holds at 1	De-energ	Reset, 0	Energized	6 V	6 − 6 = 0	Disabled

Neon lights read binary number 1 1 0., or decimal number 6

Figure 9·17 **Operations chart of digital voltmeter, Example 9·6.**

6. Pulse C from the control circuit flips F-F1 to its set or 1 state, lowering the voltage at terminal 1, and de-energizing relay K_1. This removes the short from across the 1-kilohm resistor. As shown in line 6 of the chart, Fig. 9·17, F-F2 and F-F3 hold in their 1 states, and all relays are de-energized. All three resistors, 4-, 2-, and 1-kilohm, are now in the circuit. The 1 ma from the constant-current source flows through these three resistors, with a total resistance of 7 kilohms, producing E_{ref} of 7 volts (1 ma $\times$ 7 kilohms), which is larger than the E_{in} of 6 volts. As a result, the output from the comparison circuit is now a negative voltage. This triggers the hold circuit, which, in turn, enables the AND gates.

7. Pulse C_1 from the control circuit goes through the enabled AND gate G_1 to the R terminal of F-F1. This *resets* F-F1 to its zero state. The high voltage at terminal 1 now energizes relay K_1, closing the switch contacts and shorting out the 1-kilohm resistor. F-F2 and F-F3 hold in their 1 states, and relays K_2 and K_3 remain de-energized. The 4- and 2-kilohm resistors are still in the circuit, and the 1 ma from the constant-current source, flowing through these resistors, produces an E_{ref} of 6 volts. Since E_{ref} is now equal to E_{in}, the comparison circuit produces zero volts output. This does not affect the hold circuit, which now disables the AND gates.

As shown in line 7 of the chart, Fig. 9·17, at the end of the last pulse C_1, F-F1 is in its zero state, F-F2 in its 1 state, and F-F3 in its 1 state. The neon bulbs connected to the zero terminals of the flip-flop stages glow when that terminal has a high voltage. When a flip-flop or bistable multivibrator is in its zero state, the zero terminal of the block diagram has a low voltage, while the 1 terminal has a high voltage. Conversely, when in its 1 state, the 1 terminal is low, while the zero terminal is high. As a result, with F-F1 in its zero state, its associated neon bulb (with a binary count value of 1) is not glowing. F-F2 in its 1 state causes its neon bulb (with a binary count of 2) to glow. Similarly, with F-F3 in its 1 state, its neon bulb (with a binary count value of 4) also glows. The reading of the digital voltmeter is therefore the binary number 110. or the decimal number 6. Thus, with 6 volts applied as the voltage to be measured (E_{in}), the device reads 6. As discussed in Chap. 7, under electronic counters, the binary numbers indicated by the neon lights are usually converted to a decimal-readout system indicated by either the columns of neon lights, or by the more convenient NIXIE-type tubes, one of which replaces a column of ten neon lamps.

Example 9·7 In the digital voltmeter of Fig. 9·14, 1 volt (E_{in}) is to be measured. Go through the operations of each stage to produce a final reading of 1 volt.

Pulse	F-F 3 state	K_3 relay	F-F 2 state	K_2 relay	F-F 1 state	K_1 relay	E_{ref}	Comparison circuit out ($E_{in} - E_{ref}$)	Gates
Start	0	Energized	0	Energized	0	Energized	0 v	1 − 0 = +1	Disabled
A	Set, 1	De-energ	Holds at 0	Energized	Holds at 0	Energized	4 v	1 − 4 = −3	Enabled
A_1	Reset, 0	Energized	Holds at 0	Energized	Holds at 0	Energized	0 v	1 − 0 = +1	Disabled
B	Holds at 0	Energized	Set, 1	De-energ	Holds at 0	Energized	2 v	1 − 2 = −1	Enabled
B_1	Holds at 0	Energized	Reset, 0	Energized	Holds at 0	Energized	0 v	1 − 0 = +1	Disabled
C	Holds at 0	Energized	Holds at 0	Energized	Set, 1	De-energ	1 v	1 − 1 = 0	Disabled
C_1	Holds at 0	Energized	Holds at 0	Energized	Holds at 1	De-energ	1 v	1 − 1 = 0	Disabled

Neon lights read binary number 0 0 1., or decimal number 1

Figure 9·18 **Operations chart of digital voltmeter, Example 9·7.**

Solution. In this example, the operations chart is shown in Fig. 9·18. The step-by-step procedure is omitted here to have the reader make up his own, using the solutions in the operations chart.

PROBLEMS

9·1 Refer to the digital-multiplier block diagram and operations chart of Fig. 9·1. What two conditions could roll a flip-flop stage, such as V_4, in the Eccles-Jordan (E-J) accumulator?

9·2 The binary numbers 1011 (decimal number 11) and 1101 (decimal number 13) are to be multiplied in the digital multiplier. Use which numbers you desire as the multiplicand and the multiplier, and make up the block diagram and operations chart as shown in Figs. 9·1 and 9·4. Perform the required steps, and then check whether the final product in the binary form is equal to the decimal number 143 (11×13).

9·3 Refer to the block diagram and operations chart of the shift register in Figs. 9·5 and 9·6. Assume that the binary number 10110 is a product answer resulting in the E-J accumulator stages. This product is to be transferred to the shift register. Make up the operations chart, and go through the required steps to set these digits up in the shift register. [Note that the first digit 1 (of 10110) is the highest column value and should *end* up in the E-J 16 stage of the operations chart.]

9·4 Perform the required steps, as shown in the block diagram (Fig. 9·7), and the operations chart (Fig. 9·8) of the forward-backward counter, if five forward-line pulses and then three backward-line pulses were applied. Check to see that the final result should, of course, equal a count of *two*.

9·5 Referring to the digital-voltmeter block diagram (Fig. 9·14) and to the operations charts (Figs. 9·17 and 9·18), perform the required steps if 5 *volts* were applied as the voltage to be measured, E_{in} to the device.

MORE CHALLENGING PROBLEMS

9·6 The decimal numbers 27 and 30 are to be multiplied in the digital multiplier. What is the binary answer in the device (give the highest column value first) if the *coupling* from the *OR* cluster to flip-flop V_5 is *open*?

9·7 The decimal numbers 14 and 29 are to be multiplied in the digital multiplier. What is the binary answer in the device (give highest column value first) if flip-flop V_4 *is defective* and *cannot flip* (remains in zero state) and X_4B AND gate cannot enable? (Use 29 as multiplicand.)

9·8 The decimal number 762 is to be transferred from the E-J accumulator of the digital multiplier into the shift register. What is the *binary* answer in the shift register (give highest column value first) if gate G_5 is defective and cannot enable?

9·9 The decimal number 14 is to be transferred from the digital multiplier E-J accumulator into the shift register. What is the *binary* answer in the shift register (give the highest column value first) if gate G_3 is defective and cannot enable?

9·10 A forward-backward counter is fed four backward-line pulses and then two forward-line pulses. What is the binary answer (starting with the highest column value) if the coupling capacitor to gate G_2 is open?

9·11 A forward-backward counter is started off at a count of 1111. What is the result after two forward-line, and then four backward-line pulses are applied, if the coupling capacitor to gate G_3 is open?

9·12 In the TV sync-pulse generator (Figs. 9·11 and 9·12), what is the output signal if F-F1 does not flip?

9·13 In the TV sync-pulse generator of Figs. 9·11 and 9·12, what is the output signal if F-F2 does not flop?

9·14 In the difference amplifier circuit shown for this problem, $E_{\text{in}}\ 1 = E_{\text{in}}\ 2$, T_1 and T_2 are conducting equally, and points A and B are at equal potentials. If $E_{\text{in}}\ 2$ becomes greater than $E_{\text{in}}\ 1$, which point (A or B) becomes more *negative*?

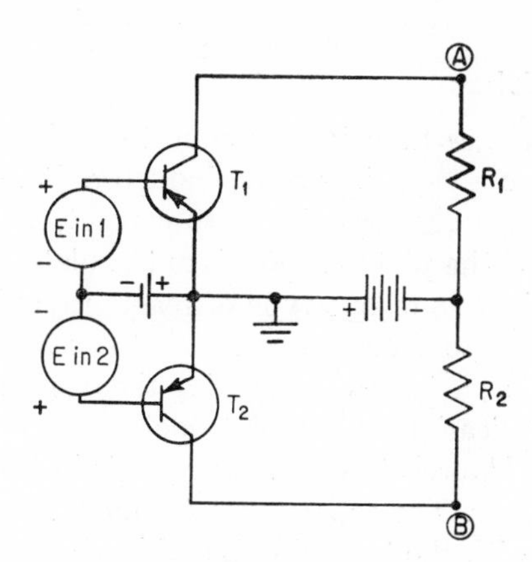

Problem 9·14

CHAPTER 10

MISCELLANEOUS CIRCUITS

10·1 Miscellaneous Circuits

The topics included in this chapter, although used extensively in pulse equipment, have, in the main, no direct affiliation with the circuits of other chapters of this book and therefore appear here as unrelated miscellany.

10·2 Reactance Tubes

General. A *reactance* tube is a vacuum-tube circuit which acts electrically as if it were a capacitor or an inductor. It exhibits a capacitive effect if the plate voltage lags the plate current, whereas an inductive effect is produced if the plate current lags the plate voltage. In any one circuit, the effect is that of one type of reactance only. The circuit diagram must be altered to change from an X_C to an X_L effect. A reactance-tube circuit is used as a means of changing the *frequency* of a tank-circuit oscillator. One application of this is a method of producing frequency modulation (f-m). The reactance-tube circuit is in parallel with the tank circuit of an oscillator. The capacitance (or inductance) of the reactance tube is in parallel with the tank circuit and helps determine the resonant frequency. The audio modulating signal, applied to the reactance tube, causes the capacitance (or inductance) of this circuit to vary. This changes the oscillator tank-circuit resonant frequency, producing f-m.

Another application of the reactance tube is in an automatic frequency control (afc) circuit, discussed in a later part of this chapter. Here too, the reactance tube changes the oscillator frequency. In this afc circuit, the oscillator signal and a reference signal are applied to a phase-comparator stage. As the oscillator frequency drifts, the phase-comparator circuit applies a d-c correction voltage to the reactance tube. This changes the capacitance (or inductance) of the circuit, changing the oscillator frequency and bringing it back to its correct value.

Basic reactance tube circuit. As shown in Fig. 10·1, the reactance-tube circuit usually consists of a pentode tube, an impedance, Z_{PG}, connected between plate and grid and another impedance, Z_{GK}, between grid and cathode. As shown, the circuit is in parallel with the resonant tank circuit of the oscillator.

A pentode is usually employed since its plate current is practically independent of its plate voltage. The a-c voltage, E, Fig. 10·1, is across the tank circuit and is also the tube plate voltage, E_P. Current, I_1, is the a-c current through Z_{PG} and Z_{GK} because of oscillator-tank voltage E. Z_{PG} in all reactance tube circuits is always much larger than Z_{GK} and is

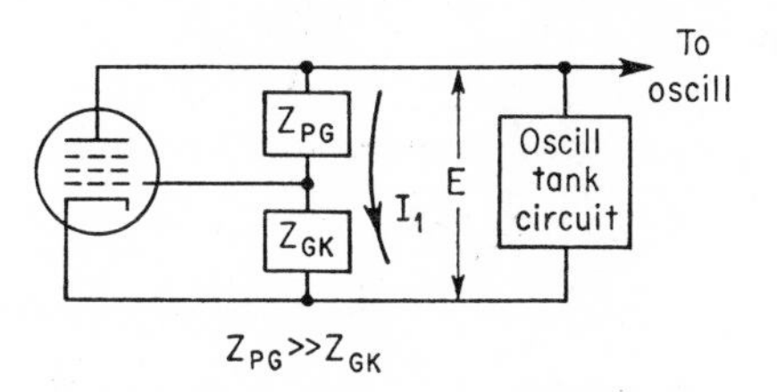

Figure 10·1 Basic reactance-tube diagram.

usually large in order to prevent loading down the oscillator tank circuit. Z_{PG} and Z_{GK} must be a combination of a resistor and a reactance. That is, either one may be a resistor, but the other must then be either a capacitor or an inductor. As will be shown in the discussions to follow, which of the two, Z_{PG} or Z_{GK}, is the resistor and which is the reactance determines whether the circuit is an inductive or a capacitive reactance device.

Capacitive reactance tube circuits. A capacitive reactance-tube circuit is shown in Fig. 10·2*a*. X_{C_1} is the Z_{PG} component of Fig. 10·1, and R_1 is the Z_{GK} of Fig. 10·1. As shown, X_{C_1} is much larger than R_1. To explain how the vacuum-tube circuit of Fig. 10·2*a* is a capacitive reactance, two methods will be employed. The first, and simpler, discussion will analyze the circuit using vectors. Then a second more rigorous discussion will be given using several equations.

A simple approach using vectors to explain a reactance tube is the subject of the following discussion, which refers to the circuit of Fig. 10·2*a* and to the vectors of Fig. 10·2*b*. The discussion is presented as a series of steps, each number of which also coincides with the numbers of the vectors.

1. The voltage E of Fig. 10·2*a* causes a current I_1 to flow through C_1 and R_1. The vector current I_1 is shown as vector 1 in Fig. 10·2*b*.

2. Voltage across resistor R_1, E_{R_1}, is in phase with current I_1 and is small since R_1 is small. Vector voltage E_{R_1} is shown as vector 2. This voltage is also the grid-input signal, e_G.

3. Voltage across capacitor C_1, E_{C_1}, lags the current through it, I_1, by 90°. Since vectors rotate counterclockwise, E_{C_1}, vector 3, is shown *lagging* I_1, vector 1, in Fig. 10·2*b*.

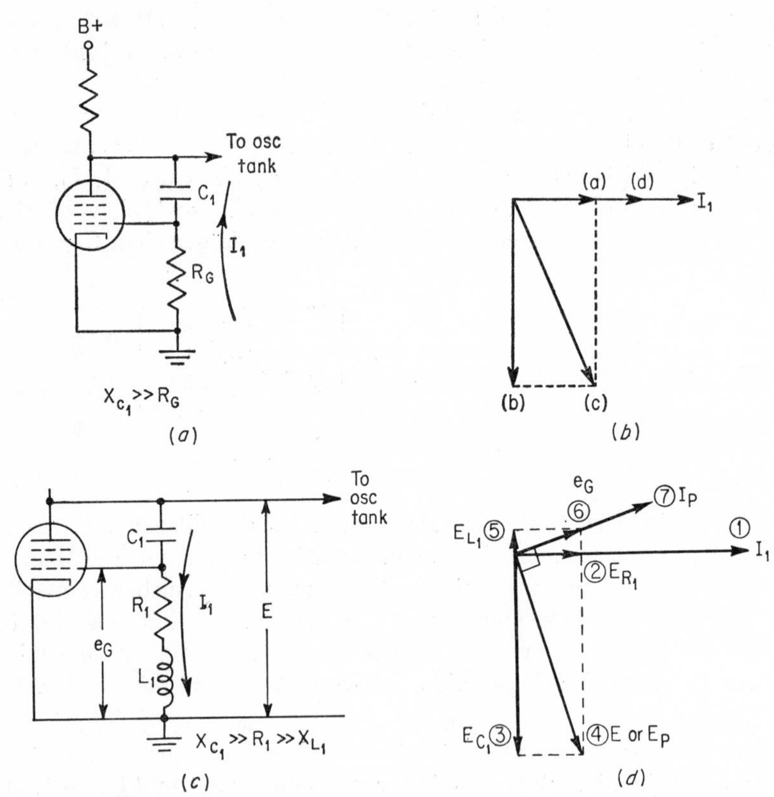

Figure 10·2 Capacitive reactance-tube circuits and vectors. (*a*) Circuit; (*b*) E_P lags I_P; (*c*) circuit; (*d*) E_P lags I_P by 90°.

4. The vector sum of E_{C_1} and E_{R_1} is the voltage E, which is the tank-circuit voltage of Fig. 10·2*a*, and also the tube plate voltage, E_P, vector 4, Fig. 10·2*b*.

5. Plate current, I_P, vector 5, is in phase with grid-input signal, e_G, vector 2.

6. From the vector diagrams of Fig. 10·2*b*, it can be seen that the *plate voltage*, E_P, vector 4, is *lagging* the *plate current* I_P, vector 5. The tube therefore acts like a capacitive reactance.

Note that from the vectors of Fig. 10·2*b*, plate voltage, E_P, lags plate current, I_P, by slightly *less* than 90°. A *pure* capacitor has voltage lagging current by exactly 90°. To bring E_P closer to a 90° angle of lag, making the tube act more like a pure capacitor, a small *inductor* L_1 is added in series with R_1, as shown in Fig. 10·2*c*. As before, X_{C_1} is much larger than R_1, and R_1 is much larger than X_{L_1}. The following brief discussion steps refer to the diagram of Fig. 10·2*c* and to the vectors of Fig. 10·2*d*.

1. Current I_1, in Fig. 10·2*c*, flows through C_1, R_1, and L_1 and is shown as vector 1, in Fig. 10·2*d*.

2. E_{R_1}, vector 2, is in phase with I_1.

3. E_{C_1}, vector 3, is large (since X_{C_1} is large) and lags I_1 by 90°.

4. The vector sum of E_{R_1} and E_{C_1} is E, vector 4. E, as shown in the diagram of Fig. 10·2*c*, is also the tube plate voltage E_P.

5. Voltage across the small inductor, L_1, E_{L_1}, vector 5, leads I_1 (or I_1 lags E_{L_1}) by 90°.

6. The vector sum of E_{R_1}, vector 2, and E_{L_1}, vector 5, is the grid-input signal, e_G, vector 6. Note that e_G is now, in Fig. 10·2*d*, slightly leading I_1. Previously, in Fig. 10·2*b*, e_G was in phase with I_1.

7. As usual, plate current, I_P, vector 7, is in phase with grid signal e_G.

8. From the vectors of Fig. 10·2*d*, it can be seen that plate voltage, E_P, vector 4, lags plate current, I_P, vector 7 by exactly 90°.

A second method, and a more rigorous one, of explaining how the circuit of Fig. 10·2*a* acts as a capacitive reactance follows.

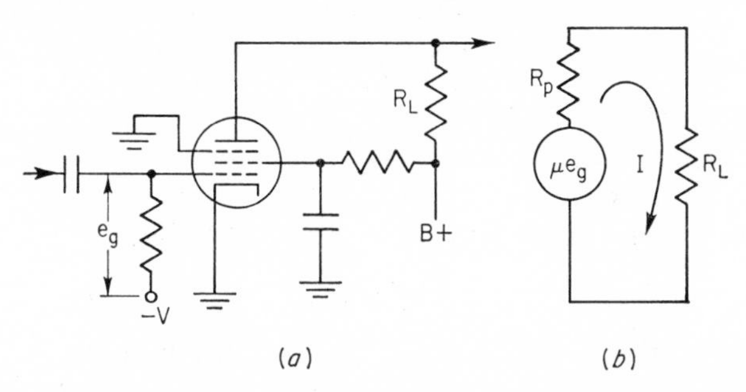

Figure 10·3 Pentode circuit. (*a*) Pentode amplifier; (*b*) Thévenin equivalent circuit.

Before continuing with the discussion of the reactance-tube circuit, some pertinent facts on pentodes must be reviewed. A typical pentode vacuum-tube circuit is shown in Fig. 10·3*a*, with its Thévenin equivalent circuit shown in Fig. 10·3*b*. The voltage of the equivalent generator is μe_G, as shown. Current, I, from Ohm's law, is E/R, where E is μe_G, and R is $R_P + R_L$, or

$$I = \frac{E}{R} = \frac{\mu e_G}{R_P + R_L} \tag{10·1}$$

In a pentode, R_L is much smaller than R_P and can therefore be dropped from the denominator of Eq. (10·1) giving

$$I = \frac{\mu e_G}{R_P} \tag{10·2}$$

The transconductance, G_M, is $\frac{\mu}{R_P}$. Therefore, substituting G_M for μ/R_P in Eq. (10·2) gives

$$I = G_M e_G \tag{10·3}$$

This is the current equation for a pentode tube, which is often referred to as a constant-current source.

Now, returning to the reactance-tube circuit of Fig. 10·2*a*, the *feedback* current, I_1, due to oscillator tank voltage, E is

$$I_1 = \frac{E}{Z_1} \tag{10·4}$$

where Z_1 is the vector sum of X_{C_1} and R_1.

However, since X_{C_1} is much larger than R_1, then Z_1 is just about equal to X_{C_1}. Therefore,

$$I_1 = \frac{E}{X_{C_1}} \tag{10·5}$$

and

$$I_1 = \frac{E}{1/2\pi f C_1} \tag{10·6}$$

and

$$I_1 = \frac{E}{1/j\omega C_1} \tag{10·7}$$

$$I_1 = j\omega C_1 E \tag{10·8}$$

Grid signal, e_G, of Fig. 10·2*a* is

$$e_G = I_1 R_1 \tag{10·9}$$

Substituting for I_1 from Eq. (10·8) in Eq. (10·9) gives

$$e_G = j\omega C_1 E R_1 \tag{10·10}$$

Plate current of the pentode tube, I_P, from Eq. (10·3) is

$$I_P = G_M e_G \tag{10·3}$$

Substituting for e_G from Eq. (10·10) in Eq. (10·3) gives

$$I_P = G_M j\omega C_1 E R_1 \tag{10·11}$$

The oscillator-tank circuit "sees" the pentode tube as an impedance, $Z = E/I_P$. Substituting for I_P gives

$$Z = \frac{E}{G_M j\omega C_1 E R_1} \tag{10·12}$$

Cancelling the Es in numerator and denominator of Eq. (10·12) and moving the j term from the denominator to the numerator, where it now becomes a *negative* j, gives

$$Z = \frac{-j}{G_M \omega C_1 R_1} \tag{10·13}$$

The *negative* j term means that Z of the pentode tube is a *capacitive* reactance. This is the same conclusion that is shown by the vectors I_P and E_P of Fig. 10·2*b*.

The actual *capacitance* that the circuit of Fig. 10·2*a* places in parallel with the oscillator tank is

$$C = G_M C_1 R_1 \tag{10·14}$$

To vary this capacitance, an input signal voltage is applied to one of the pentode tube elements, causing a change in the transconductance, G_M.

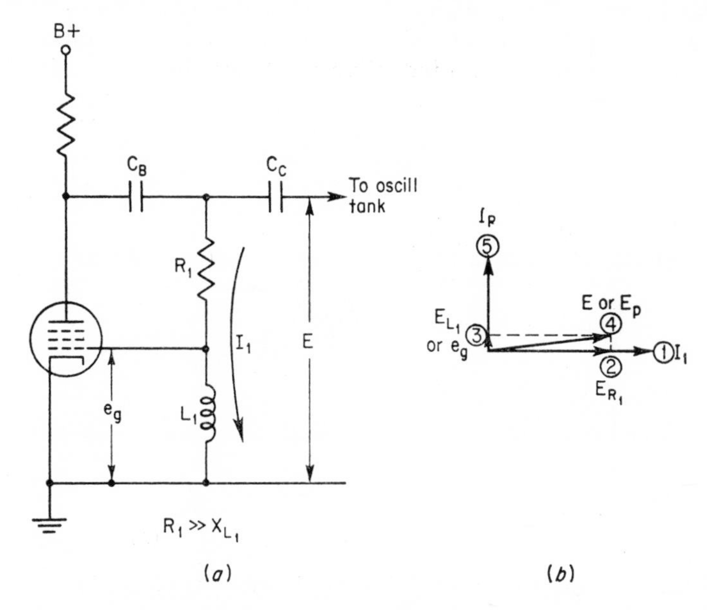

Figure 10·4 Another capacitive reactance-tube circuit and vectors. (*a*) Circuit; (*b*) E_P lags I_P.

To accomplish this, of course, the tube must be operated nonlinearly (cannot be class A) so that G_M is variable.

Another *capacitive-reactance* tube circuit is shown in Fig. 10·4*a*. Note that here R_1 is the Z_{PG} of Fig. 10·1, and X_{L_1} (Fig. 10·4*a*) is the Z_{GK} of Fig. 10·1. Compare these with the diagram of Fig. 10·2*a*. Capacitors C_B and C_C of Fig. 10·4*a* are simply for d-c blocking purposes and have negligible impedance at the oscillator frequency. The discussion of how this circuit acts as if it were a capacitive reactance is given as a series of steps, using the vectors shown in Fig. 10·4*b*.

1. Current, I_1, flows through R_1 and L_1 of Fig. 10·4*a*. I_1 is shown as vector 1 in Fig. 10·4*b*.
2. E_{R_1}, vector 2, is in phase with I_1.

3. E_{L_1}, vector 3, *leads* I_1 (or I_1 *lags* E_{L_1}) by 90°. Since X_{L_1} is small, then E_{L_1} is small. E_{L_1} is also the grid-input signal, e_G.

4. The vector sum of E_{R_1} (vector 2) and E_{L_1} (vector 3) is voltage E or plate voltage E_P, vector 4.

5. Plate current, I_P (vector 5), is in phase with grid signal e_G (vector 3).

6. From the vector diagram, it can be seen that plate voltage, E_P (vector 4), is lagging plate current, I_P (vector 5), making the tube circuit act like a capacitive reactance.

Inductive reactance tube circuits. By changing the circuits described just previously, plate current can be made to lag the plate voltage, causing the circuit to act like an inductive reactance. In the circuit of Fig.

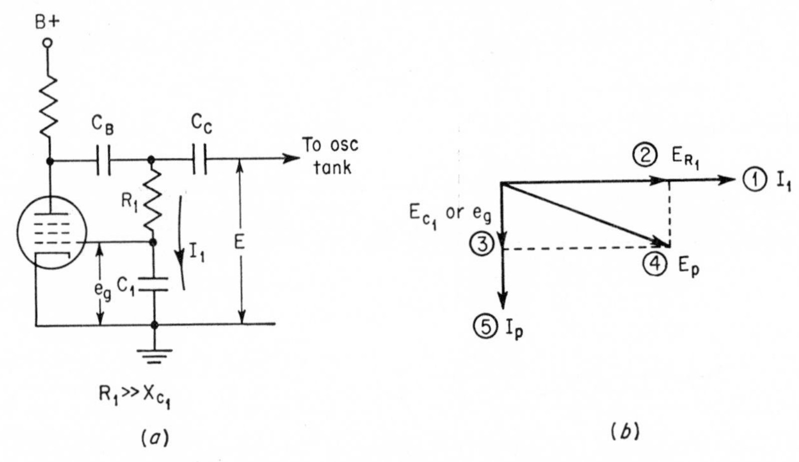

Figure 10·5 **Inductive reactance-tube circuit and vectors. (a) Circuit; (b) I_P lags E_P.**

10·5*a*, R_1 acts as the large Z_{PG} (of Fig. 10·1), and X_{C_1} acts as the much smaller Z_{GK} (of Fig. 10·1). The circuit of Fig. 10·5*a* will be discussed first using the vectors of Fig. 10·5*b* to show that plate *current* lags plate *voltage*. Then a second more rigorous explanation is given involving some simple equations.

The following steps refer to Fig. 10·5*a* and *b*, with each numbered step coinciding with the numbers assigned to the vectors.

1. Current I_1, because of oscillator-tank voltage E, flows through R_1 and C_1. Current vector I_1 is shown as vector 1 in Fig. 10·5*b*.

2. E_{R_1}, vector 2, is in phase with I_1.

3. E_{C_1}, vector 3, lags I_1 by 90° and is small since X_{C_1} is small. E_{C_1} is also the grid-input signal, e_G.

4. The vector sum of E_{R_1} (vector 2) and E_{C_1} (vector 3) is tank voltage E and is also plate voltage E_P, vector 4.

5. Plate current I_P, vector 5, is in phase with grid signal, e_G (vector 3).

6. From the vector diagram of Fig. 10·5*b*, it can be noted that plate *current*, I_P (vector 5), is lagging plate *voltage*, E_P (vector 4), making the pentode-tube circuit of Fig. 10·5*a* act like an inductive reactance.

The following is a more rigorous explanation of the inductive reactance-tube circuit (Fig. 1 0·5*a*). Since R_1 is much larger than X_{C_1} then current, I_1, is approximately equal to E/R_1. Grid signal, e_G, is then

$$e_G = I_1 X_{C_1} \tag{10·15}$$

$$e_G = \frac{E}{R_1}\left(\frac{1}{j2\pi f C_1}\right) \tag{10·16}$$

$$e_G = \frac{E}{R_1}\left(\frac{1}{j\omega C_1}\right) \tag{10·17}$$

where $\omega = 2\pi F$

$$e_G = \frac{E}{j\omega C_1 R_1} \tag{10·18}$$

Plate current I_P in a pentode is

$$I_P = G_M e_G \tag{10·3}$$

Substituting for e_G from Eq. (10·18) gives

$$I_P = G_M\left(\frac{E}{j\omega C_1 R_1}\right) \tag{10·19}$$

The impedance, Z, that the oscillator tank circuit "sees" is

$$Z = \frac{E}{I_P} \tag{10·20}$$

Substituting for I_P from Eq. (10·19) gives

$$Z = \frac{E}{G_M\left(\dfrac{E}{j\omega C_1 R_1}\right)} \tag{10·21}$$

$$Z = E\left(\frac{j\omega C_1 R_1}{G_M E}\right) \tag{10·22}$$

Cancelling the Es in numerator and denominator gives

$$Z = \frac{j\omega C_1 R_1}{G_M} \tag{10·23}$$

The positive j term in the numerator of Eq. (10·23) signifies that the impedance, Z, of the pentode circuit of Fig. 10·5*a* is *inductive*. The *actual inductance* that this circuit places in parallel with the oscillator tank is taken from Eq. (10·23) and is

$$L = \frac{C_1R_1}{G_M} \tag{10·24}$$

This inductance, L, can be varied by operating the pentode over the nonlinear region of its characteristic curve, thus having a variable transconductance, G_M.

Another variation of an *inductive reactance* tube circuit is shown in Fig. 10·6*a*, with the explanatory vectors shown in Fig. 10·6*b*. The following

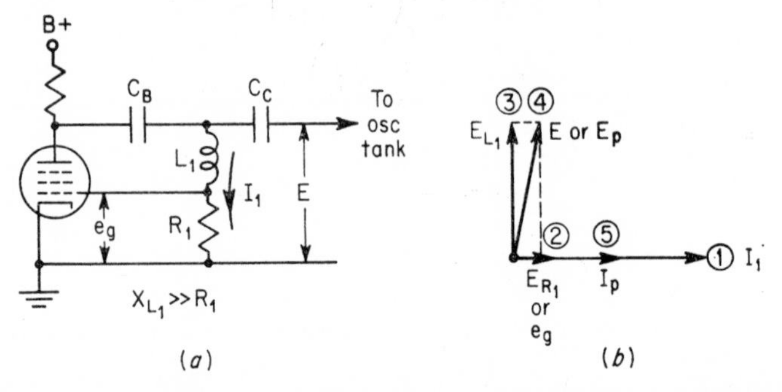

Figure 10·6 Another inductive reactance-tube circuit and vectors. (a) Circuit; (b) I_P lags E_P.

step-by-step discussion refers to this diagram, and each step is numbered the same as the related vector.

1. Current, I_1, flows through L_1 and R_1 because of oscillator-tank voltage, E. I_1 is shown as vector 1.
2. E_{R_1}, vector 2, is in phase with I_1. E_{R_1} is also the grid signal e_G and is small, since R_1 is small.
3. E_{L_1}, vector 3, leads I_1 (or I_1 lags E_{L_1}) by 90°.
4. The vector sum of E_{R_1} (vector 2) and E_{L_1} (vector 3) is tank voltage E, which is also plate voltage, E_P, vector 4.
5. Plate current, I_P, vector 5, is in phase with grid signal, e_G (vector 2).
6. From the vector diagram of Fig. 10·6*b*, it can be seen that plate *current*, I_P (vector 5), is *lagging* the plate *voltage*, E_P (vector 4), making the pentode-tube circuit act like an inductive reactance.

A more complex reactance tube circuit. The circuit of Fig. 10·7*a* is that of a reactance-tube circuit that has been used extensively in the automatic frequency control (afc)-horizontal-synchronization circuit of numerous television receivers. This afc circuit, described in the next section of this chapter, is known as sine wave-pulse afc, also as *synchrolock*.

The circuit is very similar to that of Fig. 10·2*a*, a capacitive-reactance tube circuit, the chief difference being the feedback signal E_{R_1}. In Fig.

10·2*a* and in all the standard previous circuits shown (Figs. 10·2*a* and *c* and 10·6*a*), the voltage E_{R_1} is applied to the control grid. In Fig. 10·7*a*, E_{R_1} is fed to the *cathode*. As related in the following discussion, the effect is that the pentode *tube* acts like an *inductive* reactance, while the entire *circuit*, including the tube, acts as if it were a *capacitive* reactance that is in parallel with the oscillator tank.

As was done previously, two discussions are presented; the first one embraces the simpler method of the vectors of Fig. 10·7*b*, while the second is more rigorous, employing a series of derived equations. The following step-by-step discussion refers to the circuit of Fig. 10·7*a* and to the vectors

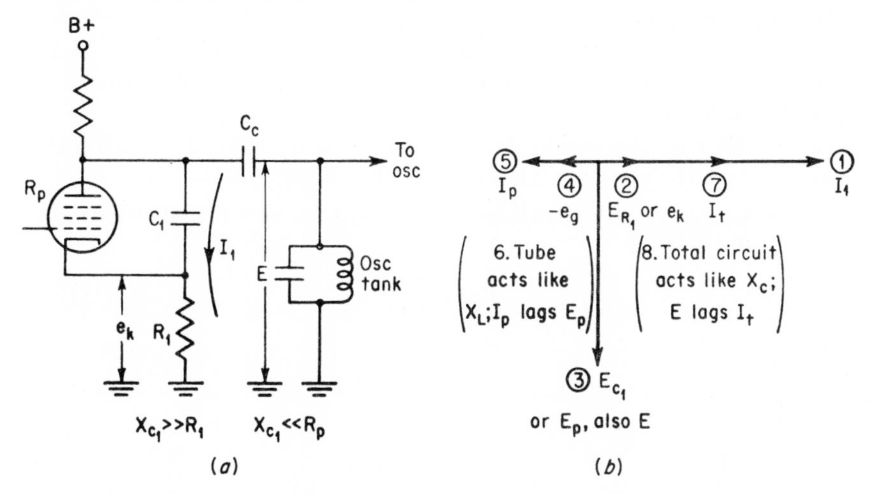

Figure 10·7 A more complex reactance-tube circuit and vectors.

of Fig. 10·7*b*. Each numbered step relates to the same number shown on the vectors.

1. Current I_1, vector 1, because of oscillator-tank voltage E, flows through C_1 and R_1.

2. E_{R_1}, vector 2, is in phase with I_1 and is small since R_1 is very small. E_{R_1} is also the input signal which is applied to the cathode and is called e_K.

3. E_{C_1}, vector 3, lags the current I_1 by 90°. E_{C_1} is also the plate-to-cathode voltage, E_P. Since X_{C_1} is very much larger than R_1, then E_{C_1}, or E_P, is just about equal to tank-voltage E.

4. There is actually no oscillator-tank voltage applied to the grid, since it is fed instead to the cathode. By driving a *cathode positive*, the effect is that the *grid* acts as if it were being driven more *negative*. This means that cathode-input signal, e_K (vector 2), effectively produces a signal at the grid that is 180° out of phase. This *effective* signal at the grid is shown as $-e_G$, vector 4.

5. Plate current I_P, vector 5, is in phase with the effective grid signal, $-e_G$ (vector 4), or I_P is 180° out of phase with cathode signal, e_K (vector 2).

6. As can be seen from the vector diagram, Fig. 10·7*b*, tube plate current, I_P (vector 5) is *lagging* plate voltage E_P (vector 3) by 90°. The pentode *tube* is therefore acting like an inductive reactance.

7. Since R_P of the pentode is much larger than X_{C_1}, then plate current I_P (vector 5) is smaller than current through C_1, I_1 (vector 1). From the vector diagram, it can be seen that I_P is 180° out of phase with I_1. The *total current*, I_T (vector 7), is the difference (vector sum) of I_P and I_1 and is in phase with I_1.

8. From the vectors, it can be seen that the *total* circuit acts like a *capacitive* reactance since voltage, E (vector 3), lags total current, I_T (vector 7), by 90°.

The following discussion of the circuit of Fig. 10·7*a* is more rigorous than that of the vectors which has just preceded.

Current, I_1, flows through the large X_{C_1} and the very small R_1. I_1 is then approximately

$$I_1 = \frac{E}{X_{C_1}} \tag{10·25}$$

Cathode *input* signal e_K is

$$e_K = I_1 R_1 \tag{10·26}$$

Substituting for I_1 gives

$$e_K = \frac{E}{X_{C_1}} R_1 \tag{10·27}$$

In the previous vector discussion of this circuit, it was pointed out that a cathode-input signal has the same effect as if a reverse-polarity signal were applied to the grid. That is, driving a cathode in the *positive* direction with an *input* applied to the cathode makes the grid appear to become more *negative*. Therefore,

$$e_K = -e_G \tag{10·28}$$

Plate current for a pentode is

$$I_P = G_M e_G \tag{10·3}$$

Substituting $-e_K$ from Eq. (10·27) for e_G in Eq. (10·3) gives

$$I_P = G_M \left(- \frac{E R_1}{X_{C_1}} \right) \tag{10·29}$$

The impedance of the pentode tube is

$$Z_{\text{tube}} = \frac{E}{I_P} \tag{10·30}$$

Substituting for I_P [from Eq. (10·29)] gives

$$Z_{\text{tube}} = \frac{E}{G_M\left(-\frac{ER_1}{X_{C_1}}\right)} \tag{10·31}$$

$$= E\left(-\frac{X_{C_1}}{G_M E R_1}\right) \tag{10·32}$$

Cancelling the E terms in numerator and denominator gives

$$Z_{\text{tube}} = -\frac{X_{C_1}}{G_M R_1} \tag{10·33}$$

Substituting $\frac{-j}{\omega C_1}$ for X_{C_1} gives

$$Z_{\text{tube}} = -\left(\frac{\frac{-j}{\omega C_1}}{G_M R_1}\right) \tag{10·34}$$

$$Z_{\text{tube}} = -\left(\frac{-j}{G_M R_1 \omega C_1}\right) \tag{10·35}$$

$$Z_{\text{tube}} = \frac{j}{G_M R_1 \omega C_1} \tag{10·36}$$

The positive j term in Eq. (10·36) denotes that the tube impedance is inductive. The total impedance of the circuit, including that of the tube, which is "seen" by the oscillator-tank circuit, consists of the tube impedance (Z_{tube}) in parallel with X_{C_1}. As indicated in Fig. 10·7a, X_{C_1} is much smaller than the pentode impedance R_P.

$$Z_{\text{total}} = \frac{\text{product of } Z_{\text{tube}} \text{ and } X_{C_1}}{\text{sum of } Z_{\text{tube}} \text{ and } X_{C_1}} \tag{10·36a}$$

substituting for Z_{tube}, from Eq. (10·33), gives

$$Z_{\text{total}} = \frac{-\frac{X_{C_1}}{G_M R_1} X_{C_1}}{-\frac{X_{C_1}}{G_M R_1} + X_{C_1}} \tag{10·37}$$

Combining the two parts of the denominator of Eq. (10·37) into a single fraction with a common denominator, $G_M R_1$, gives, as shown in Eq. (10·38),

$$Z_{\text{total}} = \frac{-\frac{X_{C_1}{}^2}{G_M R_1}}{\frac{-X_{C_1} + X_{C_1}(G_M R_1)}{G_M R_1}} \tag{10·38}$$

Inverting the large fraction which is the denominator of Eq. (10·38) and multiplying the numerator by the inverted fraction gives

$$Z_{\text{total}} = \left(-\frac{X_{C_1}{}^2}{G_M R_1}\right)\left(\frac{G_M R_1}{-X_{C_1} + X_{C_1} G_M R_1}\right) \tag{10·39}$$

Cancelling the terms $G_M R_1$ from numerator and denominator produces

$$Z_{\text{total}} = \frac{-X_{C1}{}^2}{-X_{C_1} + X_{C_1} G_M R_1} \tag{10·40}$$

Multiplying all terms in the numerator and denominator of above by -1 produces

$$Z_{\text{total}} = \frac{X_{C1}{}^2}{X_{C_1} - X_{C_1} G_M R_1} \tag{10·41}$$

Factoring out the X_{C_1} term in the denominator of above gives

$$Z_{\text{total}} = \frac{X_{C_1}{}^2}{X_{C_1}(1 - G_M R_1)} \tag{10·42}$$

Dividing numerator and denominator of above by X_{C_1} produces

$$Z_{\text{total}} = \frac{X_{C_1}}{1 - G_M R_1} \tag{10·43}$$

Substituting $\frac{-j}{\omega C_1}$ for X_{C_1} gives

$$Z_{\text{total}} = \frac{\frac{-j}{\omega C_1}}{1 - G_m R_1} \tag{10·44}$$

$$Z_{\text{total}} = \frac{-j}{\omega C_1}\left(\frac{1}{1 - G_M R_1}\right) \tag{10·45}$$

$$Z_{\text{total}} = \frac{-j}{\omega C_1(1 - G_M R_1)} \tag{10·46}$$

The *negative j* term in Eq. (10·46) denotes that the total impedance of the circuit, Z_{total}, is *capacitive*. The actual capacitance ($C_{\text{equivalent}}$) that this circuit injects in parallel with the oscillator tank is then derived as follows:

$$Z_{\text{total}} = X_{C_{\text{equiv}}} \tag{10·47}$$

Substituting $X_{C_{\text{equiv}}}$ for Z_{total} in Eq. (10·46) gives:

$$X_{C_{\text{equiv}}} = \frac{-j}{\omega C_1(1 - G_M R_1)} \tag{10·48}$$

Substituting $\frac{-j}{\omega C_{\text{equiv}}}$ for $X_{C_{\text{equiv}}}$ in above yields

$$\frac{-j}{\omega C_{\text{equiv}}} = \frac{-j}{\omega C_1(1 - G_M R_1)} \tag{10·49}$$

Cross multiplying produces

$$-j\omega C_{\text{equiv}} = -j\omega C_1(1 - G_M R_1) \tag{10·50}$$

Dividing both sides of equation by $-j\omega$ yields

$$C_{\text{equiv}} = C_1(1 - G_M R_1) \tag{10·51}$$

The term in parentheses is always less than unity, or 1, since the digit 1 (inside the parentheses) is larger than the product of G_M and R_1. The larger G_M is made (by driving the control grid in a positive direction, for example), the closer the $G_M R_1$ product approaches 1. The *difference* $(1 - G_M R_1)$ then *decreases*, making C_{equiv} decrease. This *decreases* the total capacitance of the oscillator tank, *increasing* the resonant frequency of the oscillator.

10·3 Phase Comparators in Automatic-frequency-control Circuits

General. An automatic frequency control, or afc, circuit is basically a servo system. A signal voltage from the oscillator, the frequency of which is to be controlled, is applied to the phase-comparator stage along with some reference signal. A correction voltage is then produced by the comparator circuit, if the oscillator and reference signals are not in correct phase relationship. Depending upon the circuit, this correction voltage is either fed back to the oscillator directly or applied to a reactance tube which is electrically part of the oscillator tank circuit. In either case, the correction voltage changes the oscillator frequency, pulling it back to the correct value. In this chapter, two types of phase comparators are discussed. The first type employs diodes, where the oscillator signal is a sine wave, and the reference signal used for comparison is a pulse. A second circuit uses a triode, where the oscillator signal is a sawtooth, and the reference is a pulse. Although there are numerous circuits of these types, an understanding of the two representative types presented here should enable the reader to intelligently follow any of the others that he might encounter.

Sine-wave and pulse phase comparator. A *phase-comparator* circuit used for sine waves and pulses is shown in Fig. 10·8, along with the oscillator and associated reactance tube. This circuit has been used in television receivers, usually referred to as the *synchrolock* circuit, and in other equipment requiring afc such as the electronic guidance circuits of missiles.

The oscillator circuit of Fig. 10·8 is an electron-coupled Hartley. The first three elements (cathode, control grid, and screen grid) of the pentode tube V_3 act as a triode. Together with the resonant tank, the circuit is a conventional *series-fed Hartley oscillator*. The pentode plate and suppressor grid are not part of the oscillator, simply acting as an amplifier stage. The frequency of the generated sine waves depends mostly on the inductance and capacitance in the oscillator-tank circuit. Paralleling the tank is a capacitive reactance tube circuit V_4 described in the previous section. The capacitance of the V_4 circuit adds to that of the resonant tank, determining the oscillator frequency. The sine waves from the

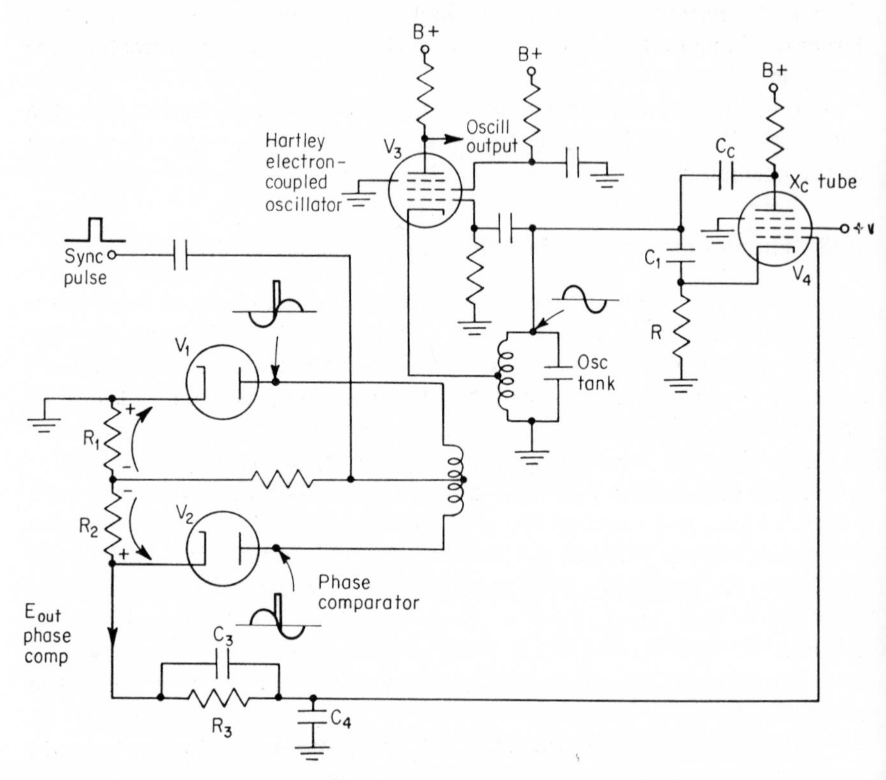

Figure 10·8 Phase-comparator circuit in afc, using sine wave and pulse.

oscillator tank are transformer coupled to the diode plates, V_1 and V_2, the *phase comparator*.

This phase comparator circuit has sine waves applied to its plates 180° out of phase from each other, which is usual when transformer coupling is employed. A positive-going pulse (in television it is the *horizontal synch pulse*), is used as the reference signal. It is applied to the center tap of the transformer secondary, and appears on each diode plate as if they were in parallel, as a positive-going pulse. The pulse combines with the sine waves at each plate. If the oscillator sine waves and the pulses are in the correct phase relationship, the composite signals appear, as shown in Fig. 10·9*a*, at the diode plates of V_1 and V_2. The pulse occurs at the instant that the sine waves are passing through their zero axes. Note in Fig. 10·9*a* that the sine waves on the diode plates are 180° out of phase, while the pulse appears positive-going on both plates.

In Fig. 10·8, it is shown that when V_1 conducts, electron flow is *up* through resistor R_1, producing the voltage polarity across R_1 as indicated, *negative* at the lower end of R_1.

When V_2 conducts, in Fig. 10·8, electron movement is *down* through R_2, producing the voltage polarity as indicated in the circuit, *positive* at the lower end.

Because of the sine waves alone, the diodes conduct alternately each time that their plates are driven positive. When V_1 conducts *alone*, there is a *negative* voltage developed across R_1 and none across R_2 since V_2 is nonconducting. The output from the phase comparator, taken across R_1 and R_2 (between the lower end of R_2 and ground), is then a negative voltage. When V_2 conducts alone, because of the sine wave driving its plate positive, there is no voltage across R_1 and a *positive* voltage across R_2. The output from the comparator is now a positive voltage. Resistor R_3 and capacitor C_4 act as a large *R-C* integrator circuit which filters out the negative and positive voltages, producing zero voltage across C_4 at the grid of the X_C tube, V_4.

Now consider the effect of the positive-going pulse (the reference signal). If the oscillator frequency is correct, and the phase relationship between sine wave and pulse is as shown in Fig. 10·9*a*, then the pulse drives *both* diode plates equally positive at the same time. If the pulse were 10 volts amplitude and it occurred at the zero axis of the sine wave, then each diode plate is driven to +10 volts. V_1 and V_2 now conduct equally through R_1 and R_2 respectively. Since R_1 and R_2 are equal in size, E_{R_1} is equal to E_{R_2} but is of opposite polarity. The output (the sum of E_{R_1} and E_{R_2}) from the comparator is therefore zero volts to the grid of the X_C tube, V_4. As a result, no change is produced in V_4, and no change is produced in oscillator frequency.

If the oscillator frequency now starts *decreasing*, the phase relationship between the sine waves and the pulses changes to that shown in Fig. 10·9*b*. Since the sine waves are now going lower in frequency, the pulses appear to be going higher in frequency, and the pulses start moving to the left with respect to the sine waves. At the plate of V_1, the 10-volt amplitude, positive-going pulse now occurs during the negative portion of a sine wave, say at −3 volts, as shown in Fig. 10·9*b*. The pulse and sine wave now drive the plate of V_1 to only +7 volts (−3 and +10). At the plate of V_2, the pulse now occurs during the positive portion of the sine wave, say, at +3 volts, as shown in Fig. 10·9*b*. The pulse and sine wave now drive V_2 plate to +13 volts (+3 and +10). V_2 now conducts more heavily than V_1, and E_{R_2} is larger than E_{R_1}. The output from the phase comparator ($E_{R_1} + E_{R_2}$) is now a *positive* voltage. This is the d-c correction voltage which will change the oscillator frequency, pulling it back to its correct value. This positive voltage is applied to the grid of the X_C tube, V_4, increasing the transconductance, G_M. The equivalent capacitance that V_4 places in parallel with the oscillator tank is

$$C_{\text{equiv}} = C_1(1 - G_M R) \qquad (10\cdot51)$$

The product $G_M R$ is less than 1, and the figures in parenthesis therefore come to less than 1. As G_M increases, the product $G_M R$ approaches 1, and the figures inside the parenthesis *decrease*. This *decreases* the capacitance

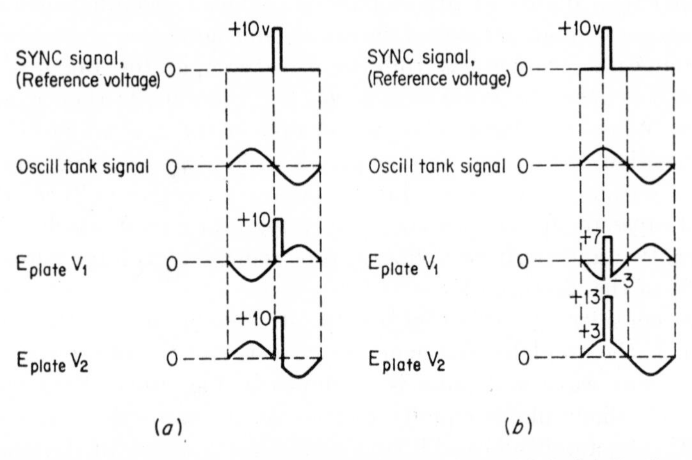

Figure 10·9 **Phase-comparator voltage waveshapes of Fig. 10·8. (*a*) Oscillator signal and sync pulse at correct phase relationship. V_1 and V_2 conduct equally. $E_{R_1} = E_{R_2}$. E_{out} from phase comparator is 0 volts. (*b*) Oscillator frequency too low. V_2 conducts more heavily than V_1. $E_{R_2} > E_{R_1}$. E_{out} from phase comparator is plus.**

that V_4 puts in parallel with the oscillator tank. With less capacitance, the resonant frequency of the oscillator increases, since $f = \dfrac{1}{2\pi\sqrt{LC}}$. Thus, the attempt by the oscillator to drift lower in frequency causes the action just described, and the oscillator frequency increases, offsetting the attempt to decrease.

Sawtooth and pulse phase comparator. A very commonly used *phase-comparator* circuit is shown in the simplified diagram of Fig. 10·10. Here, the phase relationship between a *sawtooth* and a *pulse* are compared. This circuit is called *pulse-width afc* and is also referred to as *synchroguide*.

The frequency of the *blocking oscillator-discharge tube*, V_2, is determined mainly by the R-C time constant in the grid circuit and by the amount of external bias which is applied. This circuit is discussed more fully in Chap. 6. The voltage across resistor R_K acts as external bias for the blocking oscillator, V_2. E_{R_K} is due to the conduction of the phase comparator tube, V_1. This current flows *up* through R_K, making the top of R_K *positive.* At the same time, capacitor C_1, in the grid of the blocking oscillator, sends a discharge current *down* through R_1 and R_K, making the top of R_K *negative.* Since the discharge current *down* through R_K is larger than the V_1 plate current flowing *up* through R_K, the upper end of R_K is negative. If V_1 conducts *less,* the upper end of R_K becomes more negative, *increasing* the bias on the blocking oscillator, V_2, grid. If V_1 conducts *more heavily,* the upper end of R_K goes in a positive direction and actually becomes less negative, *decreasing* the bias on V_2. Frequency of a blocking oscillator is

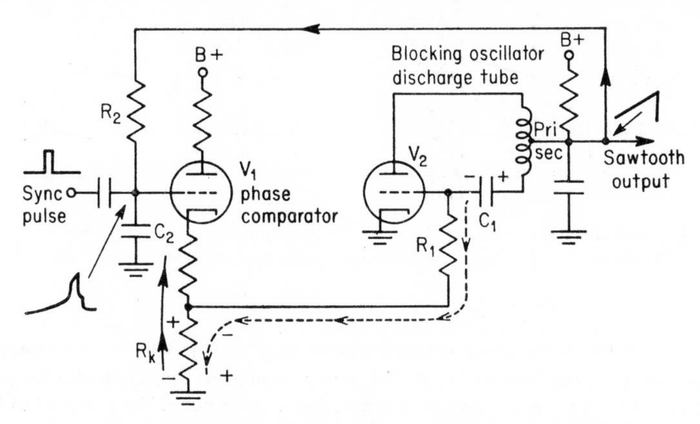

Figure 10·10 Phase comparator (partial circuit) using sawtooth and pulse in afc.

inversely proportional to the bias. A larger bias (more negative) produces a lower frequency, whereas a smaller bias (less negative) results in a higher frequency. The voltage, E_{R_K}, due to V_1 conduction, is the d-c correction voltage for the oscillator.

The blocking oscillator-discharge tube, V_2, produces a sawtooth-voltage waveshape output. This sawtooth is fed back to the grid of the phase comparator, V_1, through a large *R-C* integrator circuit, R_2-C_2. The integrated sawtooth is called a *parabolic* wave and is shown in Fig. 10·11. The sync pulse, which is the reference signal, is also applied to the grid of the phase-comparator tube, V_1, adding to the parabolic wave. The combined waveshape is shown in Fig. 10·11.

If the oscillator frequency is correct and the phase relationship is also correct, the sync pulse occurs so that it "sits" astride the sawtooth, as shown in Fig. 10·11*a*. That is, the left half of the pulse occurs during the *rise* time of the sawtooth, with the right half of the pulse occurring during the decay time of the sawtooth. The sum of the two results in the left half of the pulse, shown as the dark-shaded section, "riding" up high on the rise part of the integrated sawtooth. The right half of the pulse, shown as the lighter section, "slips" down low on the decay portion of the sawtooth. These are shown as the $E_{G_{V_1}}$ wave of Fig. 10·11*a*. Grid voltage cutoff for V_1 is such that the tube conducts only during the left half of the pulse of Fig. 10·11*a*. The pulse of plate current of V_1 flows up through R_K, producing an *average* d-c voltage across R_K which determines the bias, and the frequency, of the blocking oscillator, V_2.

When the oscillator frequency *increases*, the phase relationship between the sync pulse and the sawtooth changes as shown in Fig. 10·11*b*. The higher-frequency sawtooth now moves to the left with respect to the sync

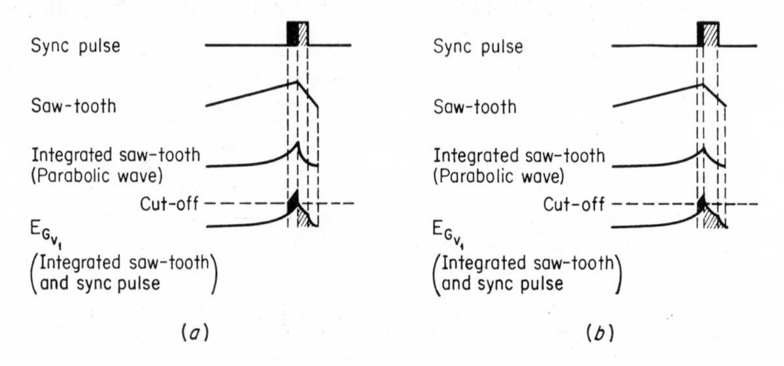

Figure 10·11 Phase-comparator voltage waveshapes of Fig. 10·10. (a) Oscillator frequency is correct. V_1 conducts during the left half of sync pulse. (b) Oscillator frequency is too high. V_1 conducts for less than half the duration of sync pulse. The plus part of E_{R_K} decreases, increasing the bias on oscillator V_2.

pulse so that only a small section, less than the left half of the pulse, now occurs during the rise time of the sawtooth. As shown in Fig. 10·11*b*, the grid of V_1 is now above (less negative than) the cutoff value only during the very short period which is less than the left half of the pulse. V_1 now conducts for a shorter time than normal, and the *average* d-c voltage across R_K is less positive, or actually more negative. This *increases* the bias on the grid of the blocking oscillator, V_2, *decreasing* its frequency. Thus, the attempt by the oscillator to increase its frequency results in its being brought back down to the correct value.

10·4 Digital-to-analog Conversion Mesh

A digital device such as the electronic decade counter described in Chap. 7 usually has some provision for an output which can operate analog-type equipment. The decade counter, which is a modified binary chain, is discussed in Chap. 7, with the complete circuit shown in Fig.

7·10. The detailed discussion of the bistable multivibrators or binaries of that circuit is given in Chap. 5. The analog-type output signal in the counter of Fig. 7·10 is called the *staircase output* on that diagram. A partial circuit from Fig. 7·10 which produces the staircase-shaped signal is shown in Fig. 10·12 and is called the digital-to-analog conversion mesh. The chart of Fig. 10·13 lists the conducting tubes of the counter of Fig. 7·10 for each input pulse and the amplitudes of the staircase output voltage, which are each derived in the following discussions.

In the digital-to-analog conversion mesh of Fig. 10·12, each voltage shown represents the plate voltage of a vacuum tube which is one half of a bistable multivibrator. The voltages shown in parentheses are the plate voltages of conducting tubes, while the other voltages are those of

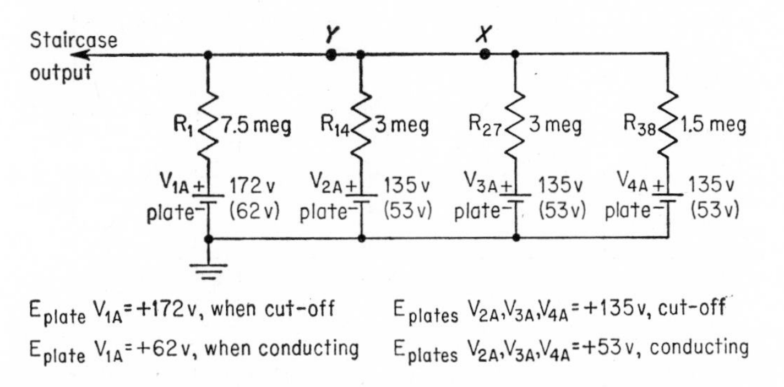

Figure 10·12 Digital-to-analog conversion mesh of decade-counter circuit of Fig. 7·10

cutoff tubes. The calculations to find the *amplitudes* of the staircase output voltage use the Thévenin equivalent circuit solutions covered in Chap. 1. The Thévenin solution for the circuit of Fig. 10·12 first opens the circuit just to the left of point X, solving for the *equivalent resistance* and *voltage* between X and ground, which includes the two branches at the right (R_{27} and R_{38}). Then the circuit is opened just to the left of point Y, and the Thévenin *equivalent resistance* and *voltage* between point Y and ground is now found. This includes the R_{14} branch and the previously found X-to-ground. Finally, the *equivalent voltage* at the *output* point (staircase output) is found, which includes the left branch, R_1, and the previously found Y-to-ground.

The solutions for the staircase output voltage are as follows:

For zero input pulse, that is, before a pulse is applied, all the B tubes of the multivibrators of Fig. 7·10 are conducting, and the A tubes are cut off, as shown in the chart of Fig. 10·13. The plate voltages of the A tubes are shown in Fig. 10·12 and are those which are not in parentheses.

Opening the circuit just to the left of point X as shown in Fig. 10·14a and solving for Thévenin equivalent resistance and voltage gives the following: Since the two voltage sources, V_{3A} plate and V_{4A} plate, are equal and of opposing polarities, no current will flow *inside* the loop of R_{27} and R_{36}. This means that there is zero volts across R_{27} and across R_{36}. The voltage therefore, from point X to ground is the sum of $E_{R_{27}}$ (zero volts) and V_{3A} plate (135 volts), or 135 volts. This is the *equivalent* voltage from X to ground and is shown in Fig. 10·14b.

Pulse input	Tubes conducting (from table of Fig. 7-4)				Staircase output voltage
0	1B,	2B,	3B,	4B	+ 138
1	1A,	2B,	3B,	4B	+ 128
2	1B,	2A,	3B,	4B	+ 119
3	1A,	2A,	3B,	4B	+ 109
4	1B,	2A,	3A,	4B	+ 101
5	1A,	2A,	3A,	4B	+ 91
6	1B,	2B,	3A,	4A	+ 83
7	1A,	2B,	3A,	4A	+ 73
8	1B,	2A,	3A,	4A	+ 64
9	1A,	2A,	3A,	4A	+ 54
10	1B,	2B,	3B,	4B	+ 138

Figure **10·13 Operation chart of conversion mesh of circuit of Fig. 10·12.**

The Thévenin equivalent resistance from X to ground in Fig. 10·12 assumes that the voltage sources are shorted and consists of R_{27} (3 megohms) and R_{36} (1.5 megohms) in parallel. This is then

$$R_{X \text{ to ground}} = \frac{(3 \text{ megohms})(1.5 \text{ megohms})}{3 \text{ megohms} + 1.5 \text{ megohms}} \tag{10·52}$$

$$= \frac{4.5 \text{ (megohms)}^2}{4.5 \text{ megohms}}$$

$$= 1 \text{ megohm}$$

The 1-megohm equivalent resistor between point X and ground is shown in Fig. 10·14b.

Now, the circuit of Fig. 10·12 is opened just to the left of point Y, as shown in Fig. 10·14c. The equivalent voltage and resistance between Y and ground includes the R_{14} branch and the X-to-ground equivalent circuit. Again, the two voltage sources, V_{2A} plate and the X-to-ground equivalent, are equal but opposite in polarity. No current flows inside the loop of R_{14} and the 1 megohm. There is no voltage across either of these resistors, and the Y-to-ground voltage is the sum of $E_{R_{14}}$ (zero volts) and V_{2A} plate (135 volts) or 135 volts. This equivalent voltage between Y and ground of 135 volts is shown in Fig. 10·14*d*.

The Thévenin equivalent *resistance* between point Y and ground of Fig. 10·14*c* assumes that the voltage sources are shorted. This places R_{14}

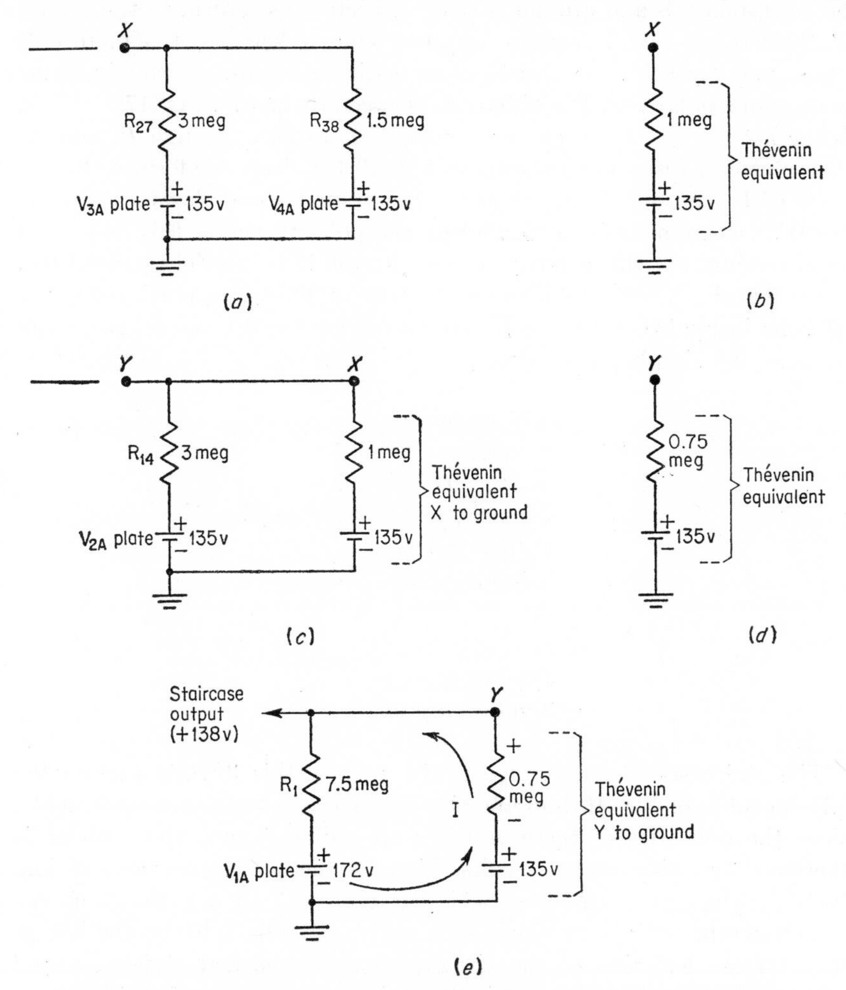

Figure 10·14 Thévenin equivalent circuits of Fig. 10·12 for input pulse zero.

(3 megohms) in parallel with the 1 megohm (X-to-ground equivalent resistance).

$$R_{Y \text{ to ground}} = \frac{(3 \text{ megohms})(1 \text{ megohm})}{3 \text{ megohms} + 1 \text{ megohm}} \qquad (10\cdot53)$$

$$= \frac{3 \text{ megohms}}{4 \text{ megohms}}$$

$$= 0.75 \text{ megohm}$$

This 0.75-megohm equivalent resistor between point Y and ground is shown in Fig. 10·14*d*.

Finally, the amplitude of the *staircase-output* voltage may be found from the circuit of Fig. 10·14*e*. The equivalent resistance and voltage between point Y and ground is now connected, as shown, to the branch at the left, R_1 and V_{1A} plate. The two voltage sources, V_{1A} plate (172 volts) and the 135 volts (Y-to-ground equivalent), are not equal and are of opposing polarities. The difference between them, 37 volts (172 − 135), is the *net* voltage which causes a current, I, to flow through R_1 and the 0.75-megohm resistor, as shown in Fig. 10·14*e*. Note that since the 172 volts of V_{1A} plate is larger than the 135 volts, current flows *up* through the 0.75 megohm resistor, producing the polarity across this resistor as shown. Note that this polarity is *adding* to the 135 volts. The current flow is through both resistors in series because of the *net* applied voltage of 37 volts inside this loop circuit. Voltage across the 0.75-megohm resistor is, as in any simple series circuit,

$$\begin{aligned} E_{0.75\text{ megohm}} &= \frac{0.75\text{ megohm}}{R_{\text{total}}} E_{\text{applied}} \qquad (10\cdot54) \\ &= \frac{0.75\text{ megohm}}{7.5\text{ megohms} + 0.75\text{ megohm}} \times 37 \\ &= \frac{0.75\text{ megohm}}{8.25\text{ megohms}} \times 37 \\ &= \frac{1}{11} \times 37 \\ &= 3\tfrac{4}{11}\text{ volts} \\ &= 3\text{ volts (approx.)} \end{aligned}$$

The staircase-output voltage is the *sum* of this 3 volts (across the 0.75 megohm of Fig. 10·14*e*) and the 135 volts (Y-to-ground equivalent), since the polarities of these voltages are aiding. The output voltage is therefore 3 + 135, or +138 volts. This is indicated in the chart of Fig. 10·13 for the zero pulse input. This figure of +138 is also shown on the complete schematic of the electronic counter of Fig. 7·10, on the line at the extreme left side of the circuit, which is marked *staircase-output voltage*. The complete diagram is shown in its start or reset state.

For input pulse 1 to the counter, multivibrator V_1 flips, and V_{1A} now conducts. V_{2B}, V_{3B}, and V_{4B} still continue to conduct. This is indicated in the chart of Fig. 10·13. The plate voltages of all tubes are therefore unchanged with the exception of V_{1A}. Plate voltage of V_{1A} is now 62 volts, the value shown in parentheses in Fig. 10·12. Since V_{2A}, V_{3A}, and V_{4A} plates are all unchanged, the Thévenin equivalent resistance and voltage between X and ground is unchanged and is the same as shown in Fig. 10·14*b*. Likewise, the equivalent circuit from Y to ground is still the same and is as shown in Fig. 10·14*d*.

The staircase output is now found from the circuit of Fig. 10·15, where the voltage V_{1A} plate is now 62 volts. The difference between the two voltage sources, V_{1A} plate (62 volts) and Y-to-ground equivalent voltage (135 volts), is 73 volts and is the *net* applied voltage *inside* the loop. As a result of this 73 volts, current will flow as indicated in Fig. 10·15, *down*

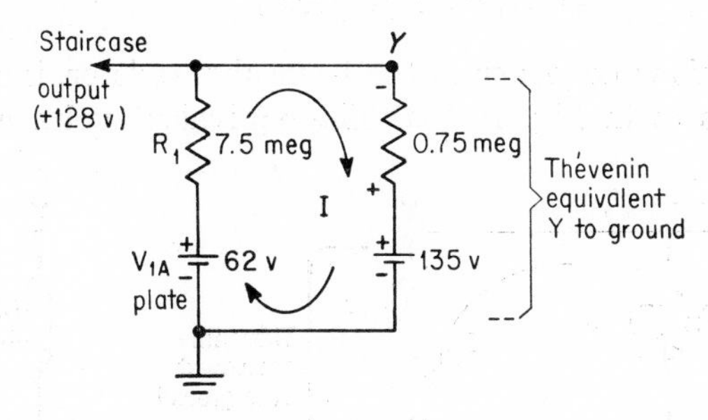

***Figure* 10·15 Thévenin equivalent circuit of Fig. 10·12 for input pulse 1.**

through the 0.75-megohm resistor, producing the voltage polarity across the resistor as shown. This voltage is opposing that of the 135 volts and is

$$E_{0.75\text{ megohm}} = \frac{0.75\text{ megohm}}{7.5\text{ megohms} + 0.75\text{ megohm}} \times 73 \qquad (10\cdot55)$$

$$= \frac{0.75\text{ megohm}}{8.25\text{ megohms}} \times 73$$

$$= \frac{1}{11} \times 73$$

$$= 6\tfrac{7}{11}$$

$$= 7\text{ volts (approx.)}$$

This 7 volts opposes the 135 volts, and the voltage output, the algebraic sum of the two, is +128 volts, as indicated in the chart of Fig. 10·13.

On *input pulse* 2, as shown in the chart of Fig. 10·13, V_{1B}, V_{2A}, V_{3B}, and V_{4B} are conducting. This means that the plate voltages of V_{3A} and V_{4A} are still at their high previous values of +135 volts, as shown in Fig. 10·12. The Thévenin equivalent voltage and resistance between point X and ground is still the same and is as shown in Fig. 10·14*b*. With V_{2A} now conducting, its plate voltage is down at +53 volts, the value shown in parentheses in Fig. 10·12. The Y-to-ground equivalent voltage and resistance is now determined by the circuit of Fig. 10·16*a*. Current flows inside the loop as shown because of the difference between the two bucking voltage sources, 53 and 135 volts. This difference of 82 volts is the *net* applied voltage *inside* the loop. Voltage across the 1-megohm resistor is

$$E_{1\text{ megohm}} = \frac{1\text{ megohm}}{3\text{ megohms} + 1\text{ megohm}} \times 82 \qquad (10\cdot56)$$
$$= \frac{1\text{ megohm}}{4\text{ megohms}} \times 82$$
$$= 20\tfrac{1}{2}\text{ volts}$$
$$= 21\text{ volts (approx.)}$$

The *polarity* of this voltage, across the 1-megohm resistor, is opposing the 135 volts, as shown in Fig. 10·16*a*. The equivalent voltage from Y to

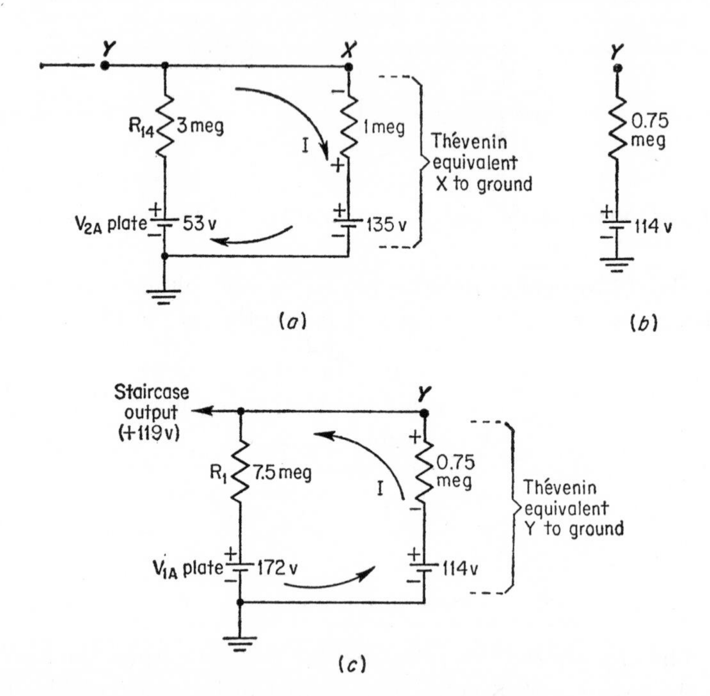

Figure 10·16 **Thévenin equivalent circuits of Fig. 10·12 for input pulse 2.**

ground is then the algebraic sum of the two voltages, or

$$135 - 21 = 114\text{ volts}$$

The equivalent resistance is still the 1 megohm and the 3 megohms in parallel, giving the same 0.75 megohm as before. The Y-to-ground Thévenin equivalent voltage and resistance is shown in Fig. 10·16*b*.

The staircase output voltage is determined by the circuit of Fig. 10·16*c*. The difference between the two voltages shown, the 172 volts (V_{1A} plate) and the 114 volts (Y-to-ground equivalent voltage), is 58 volts. This is the *net* applied voltage *inside* the loop which causes a current flow *up* through the 0.75-megohm resistor, as shown. Note that this voltage polarity across the resistor is aiding the 114 volts.

$$E_{0.75\text{ megohm}} = \frac{0.75\text{ megohm}}{7.5\text{ megohms} + 0.75\text{ megohm}} \times 58 \qquad (10\cdot57)$$
$$= \frac{0.75\text{ megohm}}{8.25\text{ megohms}} \times 58$$
$$= \frac{1}{11} \times 58$$
$$= 5\tfrac{3}{11}\text{ volts}$$
$$= 5\text{ volts (approx.)}$$

The staircase-output voltage, from Fig. 10·16*c*, is the sum of the 5 volts (across the 0.75-megohm resistor) and the 114 volts, or +119 volts, as indicated in the chart of Fig. 10·13, for pulse 2.

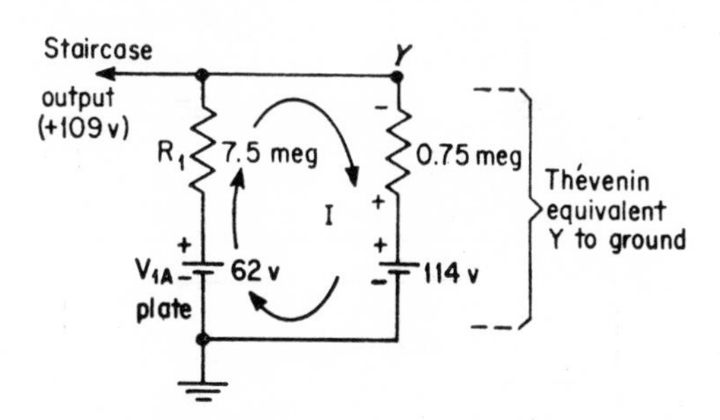

Figure 10·17 Thévenin equivalent circuit of Fig. 10·12 for input pulse 3.

Input pulse 3, as shown in the chart of Fig. 10·13, results in tubes V_{1A}, V_{2A}, V_{3B}, and V_{4B} conducting. Since the last three stages (V_{2A}, V_{3B}, V_{4B}) were already in conduction for the previous pulse 2, their plate voltages have not changed. As a result, the Thévenin equivalent circuit from X to ground is as it had been previously, shown in Fig. 10·14*b*, and the Y-to-ground equivalent, also unchanged from the previous pulse, is shown in Fig. 10·16*b*.

The staircase-output voltage can now be determined from Fig. 10·17. The *net* applied voltage inside the loop is the difference between the two voltages, 62 volts (V_{1A} plate voltage) and the 114 volts or 52 volts. This 52 volts causes a current to flow *down* through the 0.75-megohm resistor as shown, producing a voltage polarity across the resistor which opposes the 114-volt source. The resistor voltage is

$$E_{0.75\text{ megohm}} = \frac{0.75\text{ megohm}}{7.5\text{ megohms} + 0.75\text{ megohm}} \times 52 \qquad (10\cdot58)$$
$$= \frac{0.75\text{ megohm}}{8.25\text{ megohms}} \times 52$$
$$= \frac{1}{11} \times 52$$
$$= 4\tfrac{8}{11}\text{ volts}$$
$$= 5\text{ volts (approx.)}$$

The staircase-output voltage is then the algebraic sum of the 5 volts opposing the 114 volts, or an output of +109 volts, shown in the chart of Fig. 10·13 for pulse 3.

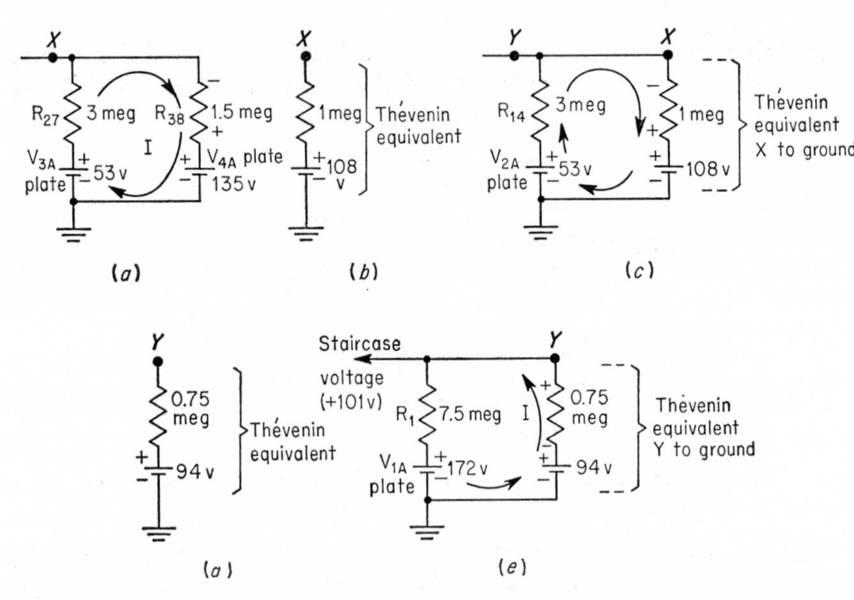

Figure 10·18 Thévenin equivalent circuits of Fig. 10·12 for input pulse 4.

Input pulse 4 results in V_{1B}, V_{2A}, V_{3A}, and V_{4B} conducting, as shown in the chart of Fig. 10·13. Since the V_{3A} tube of Fig. 10·12 is now conducting for the first time, its plate voltage is down at 53 volts for the first time. The equivalent voltage and resistance from point X to ground is now found from the circuit of Fig. 10·18*a*. The *net* applied voltage *inside* this loop is the difference between the 53 volts (V_{3A} plate voltage) and the 135 volts (V_{4A} plate voltage) or 82 volts. This 82 volts causes a current flow *down* through R_{38} as shown, producing a polarity of voltage across R_{38} which opposes the 135 volts (V_{4A} plate voltage).

$$E_{R_{38}} = \frac{R_{38}}{R_{27} + R_{38}} E_{\text{net applied}} \tag{10·59}$$
$$= \frac{1.5 \text{ megohms}}{3 \text{ megohms} + 1.5 \text{ megohms}} \times 82$$
$$= \frac{1.5 \text{ megohms}}{4.5 \text{ megohms}} \times 82$$
$$= \frac{1}{3} \times 82$$
$$= 27\tfrac{1}{3} \text{ volts}$$
$$= 27 \text{ volts (approx.)}$$

The equivalent voltage from X to ground is then the algebraic sum of the 27 volts ($E_{R_{38}}$) and the opposing 135 volts ($E_{V_{4A}}$ plate) or +108 volts. The equivalent resistance between point X and ground is unchanged, being the 3 megohms (R_{27}) and the 1.5 megohms (R_{38}) in parallel, giving 1 megohm. The X-to-ground *equivalent* circuit is shown in Fig. 10·18*b*.

The equivalent voltage and resistance from point Y to ground can now be calculated from the circuit of Fig. 10·18*c*. The *net* applied voltage inside this loop is the difference between the two voltages, 53 volts and 108 volts, or 55 volts. As a result of this 55 volts, current flows *down* through the 1-megohm resistor (the Thévenin equivalent X-to-ground), producing a voltage polarity which opposes the 108 volts (the Thévenin X-to-ground voltage). Voltage across the 1 megohm is

$$E_{1 \text{ megohm}} = \frac{1 \text{ megohm}}{3 \text{ megohms} + 1 \text{ megohm}} \times 55 \tag{10·60}$$
$$= \frac{1 \text{ megohm}}{4 \text{ megohms}} \times 55$$
$$= \frac{1}{4} \times 55$$
$$= 13\tfrac{3}{4} \text{ volts}$$
$$= 14 \text{ volts (approx.)}$$

The Y-to-ground equivalent voltage is then the 14 volts opposing the 108 volts, giving 94 volts. The Y-to-ground resistance, as before, consists of the 3 megohms and the 1 megohm in parallel, or 0.75 megohm, as shown in Fig. 10·18*d*.

The staircase output is calculated from the circuit of Fig. 10·18*e*. The *net* applied voltage *inside* the loop is the difference of the 172 volts ($E_{V_{1A}}$) and the 94 volts (Y-to-ground equivalent) or 78 volts. As a result of this 78 volts, current flows *up* through the 0.75-megohm resistor, producing a voltage polarity which aids that of the 94 volts, as shown.

$$E_{0.75\text{ megohm}} = \frac{0.75\text{ megohm}}{7.5\text{ megohms} + 0.75\text{ megohm}} \times 78 \qquad (10\cdot61)$$
$$= \frac{0.75\text{ megohm}}{8.25\text{ megohms}} \times 78$$
$$= \frac{1}{11} \times 78$$
$$= 7\tfrac{1}{11}\text{ volts}$$
$$= 7\text{ volts (approx.)}$$

The output is therefore the sum of this 7 volts and the 94 volts, or a total of +101 volts, as indicated in the chart of Fig. 10·13 for pulse 4.

Input pulse 5, as shown in the chart of Fig. 10·13, results in V_{1A}, V_{2A}, V_{3A}, and V_{4B} conducting. Since the latter three (V_{2A}, V_{3A}, and V_{4B}) also

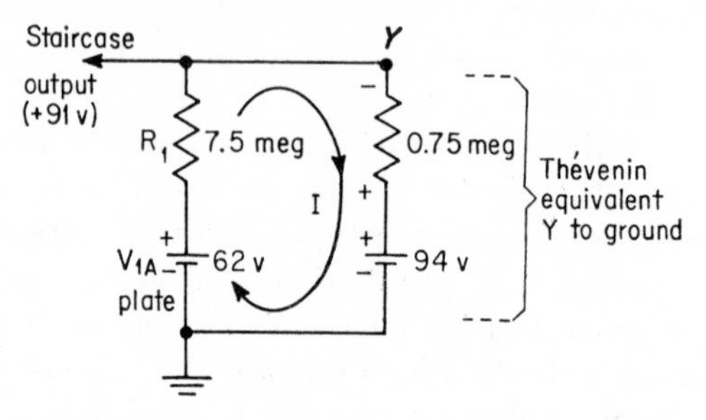

***Figure* 10·19 Thévenin equivalent circuit of Fig. 10·12 for input pulse 5.**

were conducting during the previous discussion of pulse 4, then the X-to-ground and the Y-to-ground equivalent circuits are the same as for pulse 4. The X-to-ground equivalent is still as shown in Fig. 10·18*b*, and the Y-to-ground remains as shown in Fig. 10·18*d*.

The *staircase output* however is now found from the circuit of Fig. 10·19. The difference between the 94 volts and the 62 volts is the *net* applied voltage *inside* the loop. This 32 volts causes a current flow *down* through the 0.75-megohm resistor, producing a voltage polarity that opposes the 94 volts. The voltage across the 0.75 megohm is

$$E_{0.75\text{ megohm}} = \frac{0.75\text{ megohm}}{7.5\text{ megohms} + 0.75\text{ megohm}} \times 32 \qquad (10\cdot62)$$
$$= \frac{0.75\text{ megohm}}{8.25\text{ megohms}} \times 32$$
$$= \frac{1}{11} \times 32$$
$$= 2\tfrac{10}{11}\text{ volts}$$
$$= 3\text{ volts (approx.)}$$

The output is therefore the algebraic sum of the 3 volts opposing the 94 volts, or +91 volts, as indicated in the chart of Fig. 10·13 for pulse 5.

Input pulse 6, as shown in the chart of Fig. 10·13, results in V_{1B}, V_{2B}, V_{3A}, and V_{4A} conducting. The equivalent voltage and resistance between point X and ground can be calculated from the circuit of Fig. 10·20*a*. With both voltages 53 volts, and bucking, there is no voltage *inside* the loop, and no current flows in the loop. $E_{R_{27}}$ and $E_{R_{38}}$ are each zero. Voltage from point X to ground is $E_{R_{36}}$ (0 volts), added to $E_{V_{4A}}$ plate (53 volts), or

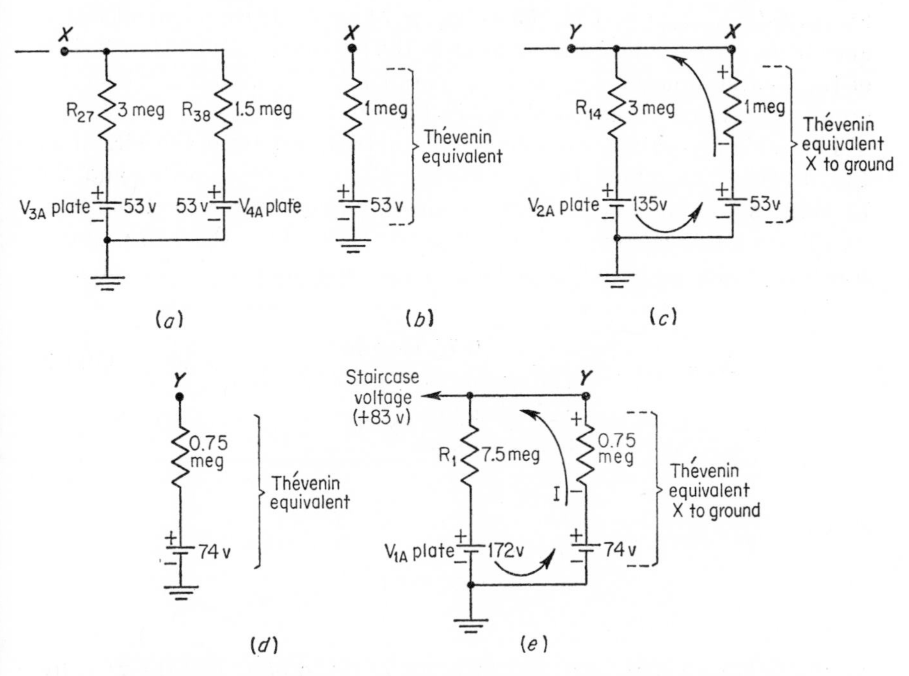

Figure 10·20 Thévenin equivalent circuits of Fig. 10·12 for input pulse 6.

53 volts. Resistance from point X to ground is the usual 3 megohms in parallel with 1.5 megohms, as in previous instances, and amounts to 1 megohm. The Thévenin equivalent voltage and resistance between X and ground is shown in Fig. 10·20*b*.

Equivalent voltage and resistance between Y and ground can be calculated from the circuit of Fig. 10·20*c*. The difference between the two voltages of 135 volts ($E_{V_{2A}}$ plate), and 53 volts (X-to-ground equivalent voltage), is 82 volts and is the *net* applied voltage *inside* the loop. As a result of this 82 volts, current flows, as shown, *up* through the 1-megohm resistor. The polarity of the voltage across the 1 megohm is aiding that of the 53 volts.

$$E_{1\text{ megohm}} = \frac{1\text{ megohm}}{3\text{ megohms} + 1\text{ megohm}} \times 82 \qquad (10\cdot63)$$
$$= \frac{1\text{ megohm}}{4\text{ megohms}} \times 82$$
$$= 20\tfrac{1}{2}\text{ volts}$$
$$= 21\text{ volts (approx.)}$$

Equivalent voltage from point Y to ground is therefore the sum of the 21 volts ($E_{1\text{ megohm}}$) and the 53 volts, or 74 volts. The equivalent resistance from Y to ground is the same as in the previous examples, consisting of the 3- and 1-megohm resistors in parallel or 0.75 megohm. The Y-to-ground equivalent circuit is shown in Fig. 10·20*d*.

The staircase-output voltage is now determined from the circuit of Fig. 10·20*e*. The difference between the 172 volts ($E_{V_{1A}}$ plate) and the 74 volts is 98 volts and is the net voltage applied *inside* the loop. This 98 volts causes a current to flow, as shown, *up* through the 0.75-megohm resistor, producing a voltage polarity which aids that of the 74 volts.

$$E_{0.75\text{ megohm}} = \frac{0.75\text{ megohm}}{7.5\text{ megohms} + 0.75\text{ megohm}} \times 98 \qquad (10\cdot64)$$
$$= \frac{0.75\text{ megohm}}{8.25\text{ megohms}} \times 98$$
$$= \frac{1}{11} \times 98$$
$$= 8\tfrac{10}{11}\text{ volts}$$
$$= 9\text{ volts (approx.)}$$

Staircase-output voltage is then the sum of this 9 volts and the 74 volts, or 83 volts, as shown in the chart of Fig. 10·13 for pulse 6.

Input pulse 7 to the counter results in V_{1A}, V_{2B}, V_{3A}, and V_{4A} conducting, as shown in the chart of Fig. 10·13. As also noted, the latter three tubes also conducted for the previous operation (pulse 6). This means that the plate voltages of V_{2A}, V_{3A}, and V_{4A} did not change. Therefore the X-to-ground equivalent circuit is the same as it had been previously and is shown in Fig. 10·20*b*. Similarly, the Y-to-ground equivalent circuit is unchanged and still appears as it had been in Fig. 10·20*d*.

The staircase-output voltage is now determined by the circuit of Fig. 10·21 which is identical to Fig. 10·20*e*, except for the V_{1A} plate which is now at 62 volts since the tube is conducting. The difference between the 62 volts and the 74 volts, or 12 volts, is the *net* applied voltage *inside* the loop. This 12 volts causes a current flow, as shown, *down* through the

0.75-megohm resistor. The polarity of the voltage across this resistor opposes that of the 74 volts.

$$E_{0.75\text{ megohm}} = \frac{0.75\text{ megohm}}{7.5\text{ megohms} + 0.75\text{ megohm}} \times 12 \qquad (10{\cdot}65)$$
$$= \frac{0.75\text{ megohm}}{8.25\text{ megohms}} \times 12$$
$$= \frac{1}{11} \times 12$$
$$= 1\tfrac{1}{11}\text{ volt}$$
$$= 1\text{ volt (approx.)}$$

Staircase-output voltage is therefore the algebraic sum of the 1 volt opposing the 74 volts, or 73 volts, as listed in the chart of Fig. 10·13 for pulse 7.

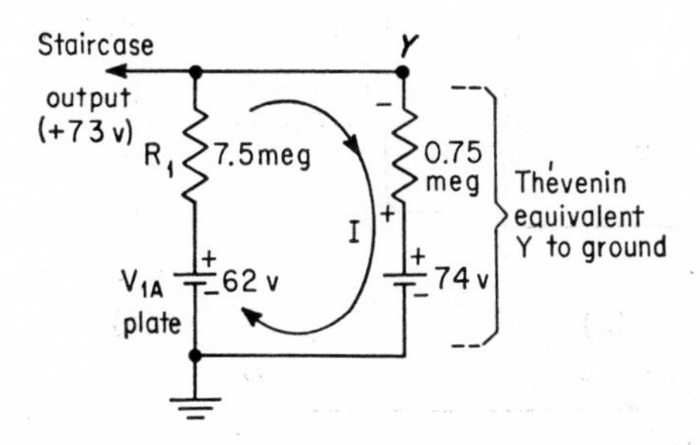

Figure **10·21 Thévenin equivalent circuit of Fig. 10·12 for input pulse 7.**

Input pulse 8, as shown in the chart of Fig. 10·13, results in tubes V_{1B}, V_{2A}, V_{3A}, V_{4A} conducting. The latter two, V_{3A} and V_{4A}, conducted also on the previous pulse, and their plate voltages therefore have remained unchanged. As a result, the X-to-ground equivalent voltage and resistance is the same as it had been and is shown in Fig. 10·20*b*.

The Y-to-ground equivalents are determined from the circuit of Fig. 10·22*a*. Since both voltages are 53 volts and opposing, there is zero volts *inside* the loop and no current flow. $E_{R_{14}}$ is zero volts, and the Y-to-ground voltage is therefore 53 volts. The Y-to-ground resistance is, as usual, the parallel combination of the 3 megohms and the 1 megohm or 0.75 megohm. The Y-equivalent circuit is shown in Fig. 10·22*b*.

The staircase-voltage output is now calculated from the circuit of Fig. 10·22*c*. The difference between the 172 volts ($E_{V_{1A}}$ plate) and the 53 volts, or 119 volts, is the *net* voltage applied *inside* the loop. This 119 volts causes a current to flow *up* through the 0.75-megohm resistor, producing the polarity of voltage that aids that of the 53 volts, as shown. The voltage across the resistor is

$$E_{0.75 \text{ megohm}} = \frac{0.75 \text{ megohm}}{7.5 \text{ megohms} + 0.75 \text{ megohm}} \times 119 \qquad (10\cdot66)$$
$$= \frac{0.75 \text{ megohm}}{8.25 \text{ megohms}} \times 119$$
$$= \frac{1}{11} \times 119$$
$$= 10^{9}/_{11} \text{ volts}$$
$$= 11 \text{ volts (approx.)}$$

The sum of this 11 volts and the 53 volts, which are aiding, is the staircase output of 64 volts, as listed in Fig. 10·13 for pulse 8.

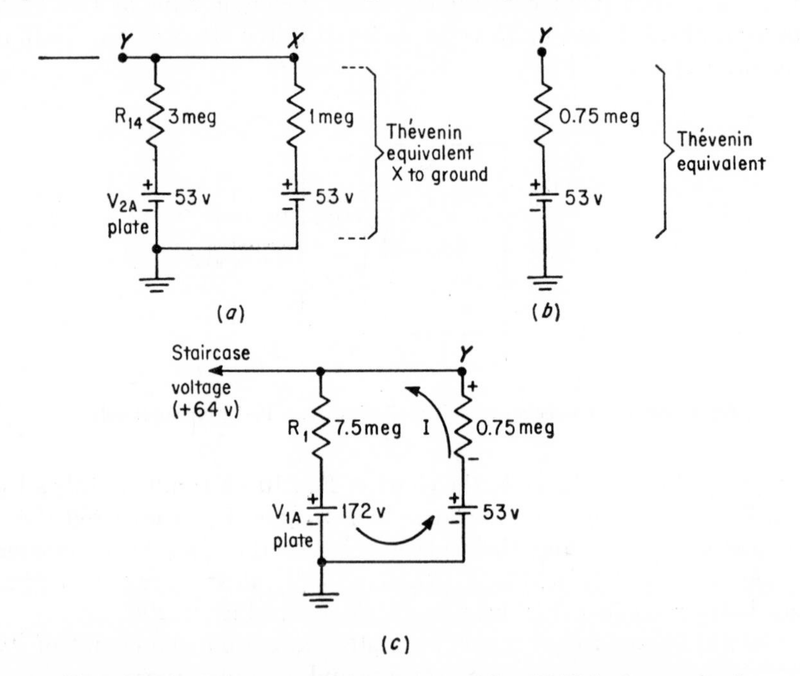

Figure 10·22 Thévenin equivalent circuits of Fig. 10·12 for input pulse 8.

Input pulse 9 to the counter results in tubes V_{1A}, V_{2A}, V_{3A}, and V_{4A} conducting. The latter three, V_{2A}, V_{3A}, and V_{4A}, conducted during the previous operation of pulse 8. As a result, the X-to-ground equivalent voltage and resistance and the Y-to-ground equivalents are unchanged from their last previous values. These are shown respectively in Fig. 10·20*b*, and Fig. 10·22*b*.

The staircase-output voltage is determined from the circuit of Fig. 10·23. The difference between the 62 volts and the 53 volts, or 9 volts, is the *net* applied voltage *inside* the loop. This 9 volts causes a current to flow, as shown, *up* through the 0.75-megohm resistor, producing a voltage polarity that is aiding that of the 53 volts.

$$
\begin{aligned}
E_{0.75 \text{ megohm}} &= \frac{0.75 \text{ megohm}}{7.5 \text{ megohms} + 0.75 \text{ megohm}} \times 9 \qquad (10\cdot67)\\
&= \frac{0.75 \text{ megohm}}{8.25 \text{ megohms}} \times 9\\
&= \frac{1}{11} \times 9\\
&= 9/11 \text{ volt}\\
&= 1 \text{ volt (approx.)}
\end{aligned}
$$

The sum of the 1 volt and the 53 volts is the 54 volts staircase output shown listed in Fig. 10·13 for pulse 9. *Input pulse* 10 resets the counter

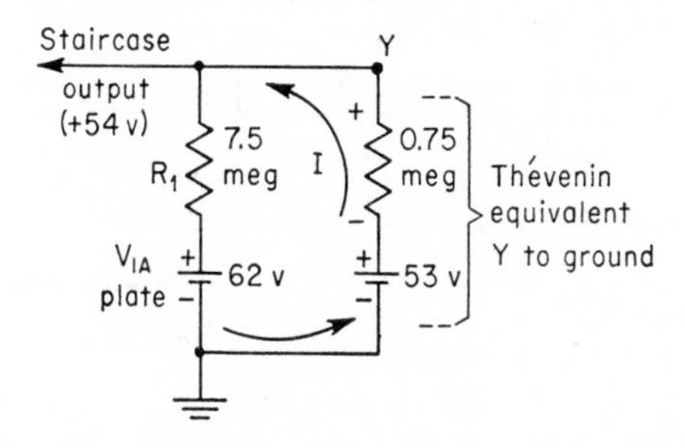

Figure 10·23 Thévenin equivalent circuit of Fig. 10·12 for input pulse 9.

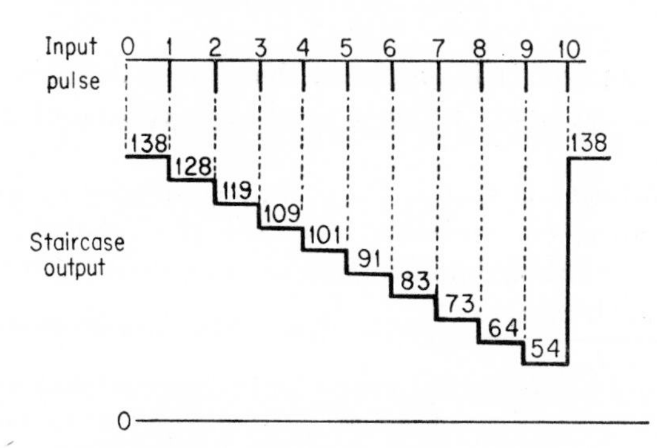

Figure 10·24 Digital and analog waveshapes.

circuits to its starting operation, the same as for the zero pulse, as shown in the chart of Fig. 10·13.

The diagram of Fig. 10·24 shows the input pulses (the digital waveshape) and the resultant staircase output voltage (the analog waveshape).

10·5 Step Counter

General. A circuit consisting basically of two diodes and two capacitors, as shown in Fig. 10·25, produces a staircase-output voltage when a series of input pulses are applied. The circuit, like the previous one described, is a *digital-to-analog converter*. This circuit is called a *step counter*, a *storage counter*, and also a *staircase-frequency divider*, when used in conjunction with a monostable blocking oscillator. The blocking oscillator is discussed in Chap. 6, and the combination circuit discussion is given later in this section.

In order to fully follow the operation of the step-counter circuit of Fig. 10·25, the reader should understand the action involving several capacitors in series when one or more has been previously charged, and a d-c

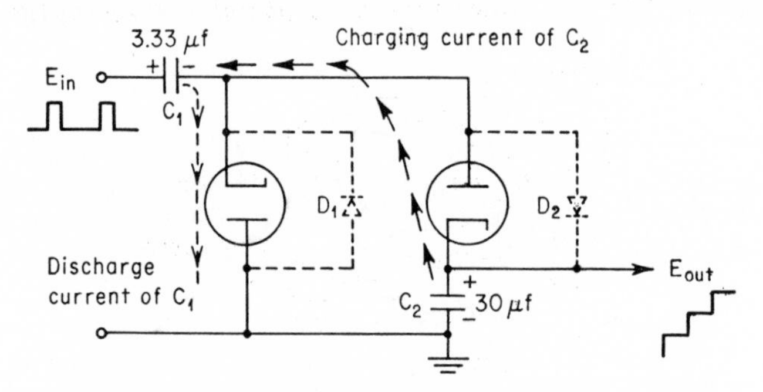

Figure **10·25 Step counter with positive-going input pulses.**

voltage is then applied. It is suggested that the reader review the following section before going on to the succeeding discussion of step counters.

Capacitor voltages in series. Capacitors connected in *series* result in a total capacitance, C_T, which is less than the smallest of the series capacitors. The equation, from elementary electricity, is the same as for resistors in parallel, being

$$C_T = \frac{1}{\dfrac{1}{C_1} + \dfrac{1}{C_2} + \dfrac{1}{C_3} + \text{etc.}} \tag{10·68}$$

Also, similar to the simplified equation for two resistors in series, the total capacitance, C_T, for two capacitors in series is

$$C_T = \frac{\text{product}}{\text{sum}} = \frac{C_1C_2}{C_1 + C_2} \tag{10·69}$$

As shown in Fig. 10·26*a*, the resistor-voltage divider, voltage across R_2 is a *fraction* of the applied voltage E.

$$E_{R_2} = \frac{R_2}{R_T} E \tag{10·70}$$

Voltage across a capacitor which is one of several in series is similar to that for a resistor [Eq. (10·70)], except that the *fraction* is reversed, as shown in the following:

$$E_{C_2} = \frac{C_T}{C_2} E \tag{10·71}$$

Example 10·1 (*a*) In the circuit of Fig. 10·26*a*, find the voltage across R_2 and across R_1. (*b*) In the circuit of Fig. 10·26*b*, find the voltage across C_2 and also across C_1. (*c*) In the circuit of Fig. 10·26*c*, find the voltage across C_2, across C_1, and also across C_3.

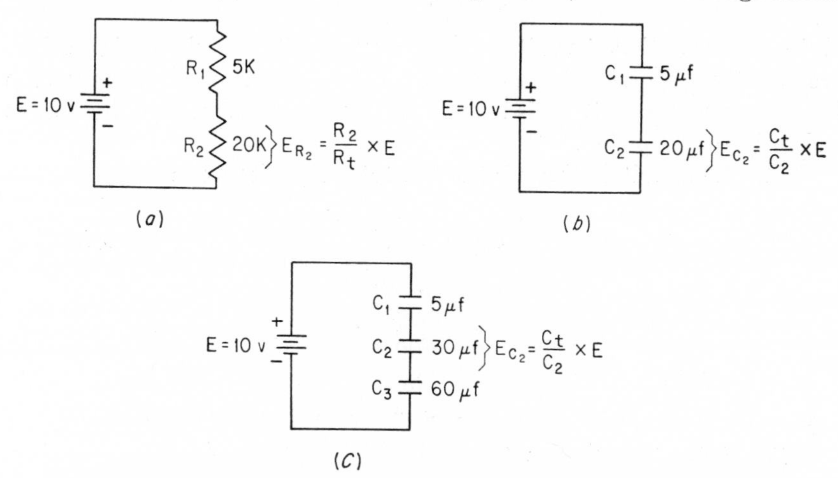

Figure 10·26 Resistor- and capacitor-voltage dividers for Example 10·1.

Solution. (*a*) Voltage across R_2 in Fig. 10·26*a* is

$$E_{R_2} = \frac{R_2}{R_T} E \tag{10·70}$$

where R_T = sum of R_1 and R_2.

$$\begin{aligned} E_{R_2} &= \frac{20{,}000}{5{,}000 + 20{,}000} \times 10 \\ &= \frac{20{,}000}{25{,}000} \times 10 \\ &= \frac{4}{5} \times 10 \\ &= 8 \text{ volts} \end{aligned}$$

With 10 volts applied, E, and 8 volts across R_2, E_{R_2}, then the remaining 2 volts must be across R_1. Solving for E_{R_1} in the same manner as the above, using Eq. (10·70), gives

$$\begin{aligned} E_{R_1} &= \frac{R_1}{R_T} E \\ &= \frac{5{,}000}{5{,}000 + 20{,}000} \times 10 \\ &= \frac{5{,}000}{25{,}000} \times 10 \\ &= \frac{1}{5} \times 10 \\ &= 2 \text{ volts} \end{aligned}$$

Note, of course, that the voltages in a resistor-voltage divider, Fig. 10·26*a*, are directly proportional to the resistor values. The larger resistor has the larger voltage across it. This is not true of a capacitor voltage divider, as shown in the following.

(*b*) In the circuit of Fig. 10·26*b*, the total capacitance is

$$\begin{aligned} C_T &= \frac{C_1 C_2}{C_1 + C_2} \qquad (10\cdot69) \\ &= \frac{5(20)}{5 + 20} \\ &= \frac{100}{25} \\ &= 4\ \mu\text{f} \end{aligned}$$

Voltage across capacitor C_2 is

$$\begin{aligned} E_{C_2} &= \frac{C_T}{C_2} E \qquad (10\cdot71) \\ &= \frac{4}{20} \times 10 \\ &= 2 \text{ volts} \end{aligned}$$

With 10 volts applied, E, and 2 volts across C_2, E_{C_2}, the remaining 8 volts must be across C_1. Solving for E_{C_1} in the same manner as for E_{C_2}, using Eq. (10·71), gives

$$\begin{aligned} E_{C_1} &= \frac{C_T}{C_1} E \\ &= \frac{4}{5} \times 10 \\ &= 8 \text{ volts} \end{aligned}$$

Note that in a capacitor voltage divider, unlike a resistor circuit, the *larger* capacitor gets the *smaller* voltage.

(c) In the circuit of Fig. 10·26c, the total capacitance C_T is

$$C_T = \frac{1}{\frac{1}{C_1} + \frac{1}{C_2} + \frac{1}{C_3}} \qquad (10\cdot68)$$

$$= \frac{1}{\frac{1}{5} + \frac{1}{30} + \frac{1}{60}}$$

$$= \frac{1}{\frac{12 + 2 + 1}{60}}$$

$$= \frac{1}{\frac{15}{60}}$$

$$= 1 \times \frac{60}{15}$$

$$= 4\ \mu\text{f}$$

Voltage across capacitor C_2 is then

$$E_{C_2} = \frac{C_T}{C_2} E \qquad (10\cdot71)$$

$$= \frac{4}{30} \times 10$$

$$= 1.33 \text{ volts}$$

Voltage across capacitor C_1 is then

$$E_{C_1} = \frac{C_T}{C_1} E \qquad \text{[from Eq. (10·71)]}$$

$$= \frac{4}{5} \times 10$$

$$= 8 \text{ volts}$$

With 10 volts applied, E, and 1.33 volts across C_2, E_{C_2}, and 8 volts across C_1, E_{C_1}, the remainder of 0.67 volts $[10 - (1.33 + 8)]$ must be across C_3. Solving for E_{C_3}, using Eq. (10·71), gives

$$E_{C_3} = \frac{C_T}{C_3} E$$

$$= \frac{4}{60} \times 10$$

$$= 0.67 \text{ volts}$$

Note again that in a capacitor voltage divider, Fig. 10·26c, the *largest* capacitor ($C_3 = 60\ \mu$f) develops the *smallest* voltage ($E_{C_3} = 0.67$ volts).

Similarly, the *smallest* capacitor ($C_1 = 5\ \mu f$), has the *largest* voltage ($E_{C_1} = 8$ volts).

When a capacitor, part of a series capacitor voltage divider, such as in Fig. 10·27*a*, has a voltage on it, e_C, and an additional new voltage, E, is applied, the capacitor charges up or discharges to a new value, E_C. The new capacitor voltage, E_C, depends upon the values of the capacitors and the *difference* between the applied voltage, E, and the original capacitor voltage, e_C. This *difference* is the *change* in voltage, or ΔE.

$$\Delta E = E - (e_{C_1} + e_{C_2}) \tag{10·72}$$

where ΔE is the *effective* applied voltage, or difference between the applied voltage, E, and the *original* capacitor voltages, e_{C_1} and e_{C_2}; E is the new *actual* applied voltage; and e_{C_1} and e_{C_2} are the voltages across each capacitor before the new voltage is applied.

The capacitor voltages, E_{C_1} and E_{C_2}, with the new voltage E applied are

$$E_{C_1} = e_{C_1} + \Delta e_{C_1} \tag{10·73}$$

and

$$E_{C_2} = e_{C_2} + \Delta e_{C_2} \tag{10·74}$$

where e_{C_1} and e_{C_2} = capacitor voltages *before* the new voltage E is applied

Δe_{C_1} and Δe_{C_2} = the *changes* in voltages across the capacitors.

$$\Delta e_{C_1} = \frac{C_T}{C_1} \Delta E \tag{10·75}$$

and

$$\Delta e_{C_2} = \frac{C_T}{C_2} \Delta E \tag{10·76}$$

where ΔE is the *change* in the circuit voltage when E is first applied; ΔE is the *effective* applied voltage or *difference* between the *actual* applied voltage, E, and the original capacitor voltages, e_{C_1} and e_{C_2}; ΔE is given in Eq. (10·72).

Example 10·2 In the circuit of Fig. 10·27*a*, if capacitor C_2 has been previously charged to 1 volt with the polarity shown ($e_{C_2} = 1$ volt), find the following when an applied voltage, E, of 10 volts, as shown, is connected by closing the switch: (a) ΔE, the *effective* applied voltage; (b) C_T,

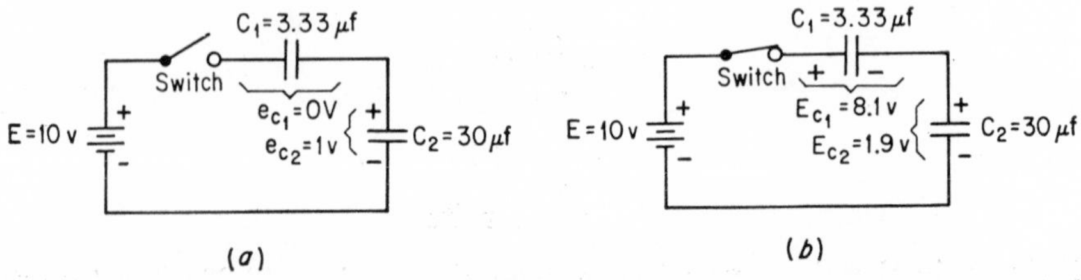

Figure 10·27 Circuit of Example 10·2. (a) Switch open; (b) switch closed.

the total capacitance; (c) Δe_{C_2}, the *change* of C_2 voltage; (d) E_{C_2}, the new voltage on C_2; (e) Δe_{C_1}, the change of C_1 voltage; (f) E_{C_1}, the new voltage on C_1.

Solution. (*a*) With 10 volts, E, applied through the switch and with C_2 previously charged to 1 volt, the two voltages, E and e_{C_2}, are opposing. The effect is that the difference voltage, ΔE, is being applied.

$$\begin{aligned} \Delta E &= E - (e_{C_1} + e_{C_2}) \qquad (10{\cdot}72) \\ &= 10 - (1 + 0) \\ &= 10 - 1 \\ &= 9 \text{ volts} \end{aligned}$$

(*b*)

$$\begin{aligned} C_T &= \frac{C_1 C_2}{C_1 + C_2} \qquad (10{\cdot}69) \\ &= \frac{3.33(30)}{3.33 + 30} \\ &= \frac{100}{33.33} \\ &= 3\ \mu\text{f} \end{aligned}$$

(*c*)

$$\begin{aligned} \Delta e_{C_2} &= \frac{C_T}{C_2}\Delta E \qquad (10{\cdot}76) \\ &= \frac{3}{30} \times 9 \\ &= 0.9 \text{ volt} \end{aligned}$$

(*d*)

$$\begin{aligned} E_{C_2} &= e_{C_2} + \Delta e_{C_2} \qquad (10{\cdot}74) \\ &= 1 + 0.9 \\ &= 1.9 \text{ volts} \end{aligned}$$

With 10 volts, E, applied and C_2 *now* charged to 1.9 volts ($E_{C_2} = 1.9$), C_1 voltage (E_{C_1}) must be equal to the remainder of 8.1 volts (10 − 1.9), as will be shown in the following:

(*e*)

$$\begin{aligned} \Delta e_{C_1} &= \frac{C_T}{C_1}\Delta E \qquad (10{\cdot}75) \\ &= \frac{3}{3.33} \times 9 \\ &= 8.1 \text{ volts} \end{aligned}$$

(*f*)

$$\begin{aligned} E_{C_1} &= e_{C_1} + \Delta e_{C_1} \qquad (10{\cdot}73) \\ &= 0 + 8.1 \\ &= 8.1 \text{ volts} \end{aligned}$$

The circuit with the switch closed and the final voltages across the capacitors, E_{C_1} and E_{C_2}, are shown in Fig. 10·27*b*.

Example 10·3 In the circuit of Fig. 10·28*a*, if C_1 has been previously charged up to 3 volts ($e_{C_1} = 3$) with the polarity shown and C_2 has been charged to 2 volts with the *opposite* polarity, as shown ($e_{C_2} = 2$) find the following when the switch is closed, applying 10 volts, E, to the circuit: (*a*) ΔE, the *effective* applied voltage; (*b*) C_T, the total capacitance; (*c*) Δe_{C_2}, the *change* of C_2 voltage; (*d*) E_{C_2}, the new voltage on C_2; (*e*) Δe_{C_1}, the *change* of C_1 voltage; (*f*) E_{C_1}, the new voltage on C_1.

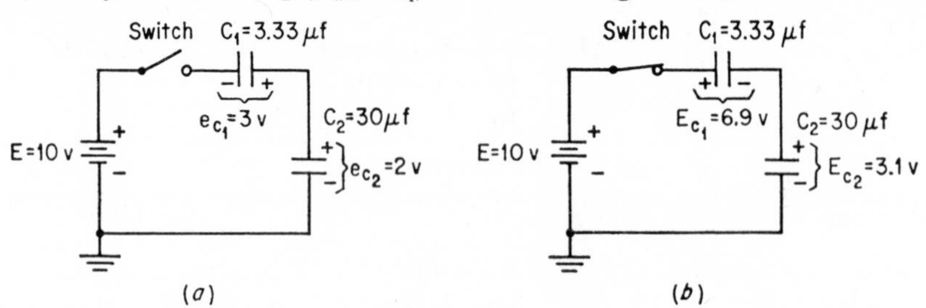

Figure **10·28 Circuit of Example 10·3. (a) Switch open; (b) switch closed.**

Solution. (*a*) The original charge on C_1 and C_2 are of opposite polarities and are bucking. The voltage across C_2, from the upper plate to ground, is positive, and e_{C_2} is +2 volts. Since voltage across C_2 is bucking that of C_1, then e_{C_1} is −3 volts. The sum of the two is therefore −1 volt. When the switch is closed, applying 10 volts, E, with the polarity shown, the 10 volts and the −1 volt are in series aiding, producing an *effective* applied voltage (ΔE) of 11 volts as shown in the following:

$$\begin{aligned} \Delta E &= E - (e_{C_1} + e_{C_2}) \qquad (10{\cdot}72) \\ &= 10 - (-3 + 2) \\ &= 10 - (-1) \\ &= 10 + 1 \\ &= 11 \text{ volts} \end{aligned}$$

(*b*)

$$\begin{aligned} C_T &= \frac{C_1 C_2}{C_1 + C_2} \qquad (10{\cdot}69) \\ &= \frac{3.33(30)}{3.33 + 30} \\ &= \frac{100.}{33.33} \\ &= 3\ \mu\text{f} \end{aligned}$$

(*c*)

$$\begin{aligned} \Delta e_{C_2} &= \frac{C_T}{C_2}\Delta E \qquad (10{\cdot}76) \\ &= \frac{3}{30} \times 11 \\ &= \frac{1}{10} \times 11 \\ &= 1.1 \text{ volts} \end{aligned}$$

(*d*)
$$E_{C_2} = e_{C_2} + \Delta e_{C_2} \tag{10·74}$$
$$= 2 + 1.1$$
$$= 3.1 \text{ volts}$$

(*e*)
$$\Delta e_{C_1} = \frac{C_T}{C_1} \Delta E \tag{10·75}$$
$$= \frac{3}{3.33} \times 11$$
$$= \frac{1}{1.11} \times 11$$
$$= 9.9 \text{ volts}$$

(*f*)
$$E_{C_1} = e_{C_1} + \Delta e_{C_1} \tag{10·73}$$
$$= -3 + 9.9$$
$$= +6.9 \text{ volts}$$

Note that the *original* charge on C_1 was 3 volts, *negative* at the *left* plate. When the switch was closed, C_1 and C_2 charged up to the applied voltage

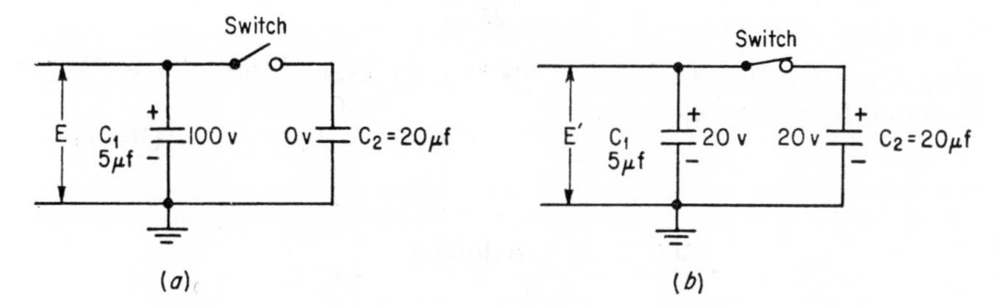

Figure **10·29 Circuit of Example 10·4. (a) Switch open; (b) switch closed.**

E, and C_1 actually reversed its polarity, as shown in comparing Fig. 10·28*a* with diagram *b*. Voltage across C_1 went from -3 volts, to $+6.9$ volts, or a change (Δe_{C_1}) of 9.9 volts.

As shown in Fig. 10·28*b*, with the switch closed, E_{C_1} is $+6.9$ volts, and E_{C_2} is $+3.1$ volts, making a total of $+10$ volts, which, of course, is equal to the applied 10 volts, E.

Example 10·4 In the circuit of Fig. 10·29*a*, C_1 is 5 μf and has been previously charged to 100 volts with the polarity shown. When the switch is closed, C_2, a 20-μf uncharged capacitor, is placed in parallel with C_1. Find the voltage, E', which will now appear across the parallel combination.

Solution. One method of solution, different from the previous examples, involves Q. The *quantity of charge*, Q_1, in coulombs on C_1 in Fig. 10·29*a* is

$$Q_1 = C_1E_1 \tag{10·77}$$

where Q is in *coulombs*, C is in *farads*, and E is in volts.

$$\begin{aligned} Q_1 &= C_1 E_1 \\ &= 5\ \mu\text{f (100 volts)} \\ &= (5 \times 10^{-6})\ 100 \\ &= 500 \times 10^{-6} \text{ coulombs} \end{aligned}$$

When the switch is closed, C_1 starts discharging, and E_{C_1} decreases. Discharge current from C_1 flows into C_2, and E_{C_2} increases from its starting value of zero. Electrons leave the lower plate of C_1 and move onto the lower plate of C_2. The total number of electrons (Q total) on both C_1 and C_2 in Fig. 10·29*b* is the same as that (Q_1) on C_1 in Fig. 10·29*a*, since the number of electrons (Q) cannot change without an *external* voltage source supplying them. Therefore, $Q_1 = Q_T$. The two capacitors in *parallel* are effectively a total capacitor, C_T.

$$\begin{aligned} C_T &= C_1 + C_2 \\ &= 5 + 20 \\ &= 25\ \mu\text{f} \end{aligned} \qquad (10{\cdot}78)$$

Since $Q_1 = Q_T$, then Q_T is 500×10^{-6} coulombs, and the voltage, E_T, can be found from the following:

$$\begin{aligned} Q_T &= C_T E_T \qquad (10{\cdot}77) \\ \frac{Q_T}{C_T} &= E_T \\ \frac{500 \times 10^{-6} \text{ coulombs}}{25\ \mu\text{f}} &= E_T \\ \frac{500 \times 10^{-6}}{25 \times 10^{-6}} &= E_T \\ 20 \text{ volts} &= E_T \end{aligned}$$

Since C_1 (5 μf) and C_2 (20 μf) in parallel act as C_T (25 μf), then the total voltage (E_T) across this combination is also the voltage across *each* capacitor in parallel. This is shown in Fig. 10·29*b*.

The result can easily be checked since the total quantity (Q_T) in Fig. 10·29*b* must be the same as the original Q_1 in Fig. 10·29*a*. In Fig. 10·29*b*, the quantity of electrons on C_1 will be called Q_1' to differentiate it from the quantity on C_1 in Fig. 10·29*a*.

$$Q_1' = C_1 E_1' \qquad (10{\cdot}77)$$

where E_1' = voltage across C_1 in Fig. 10·29*b* and is the same as E_T.

$$\begin{aligned} Q_1' &= C_1 E_1' \\ &= 5\ \mu\text{f (20 volts)} \\ &= (5 \times 10^{-6})\ 20 \\ &= 100 \times 10^{-6} \text{ coulombs} \\ Q_2 &= C_2 E_2' \qquad \text{[from Eq. (10·77)]} \end{aligned}$$

where $E_2' =$ voltage across C_2 in Fig. 10·29b and is the same as E_T.

$$\begin{aligned} Q_2 &= C_2 E_2' \\ &= 20 \ \mu\text{f} \ (20 \text{ volts}) \\ &= (20 \times 10^{-6}) \ 20 \\ &= 400 \times 10^{-6} \text{ coulombs} \end{aligned}$$

The total quantity (Q_T) of electrons in Fig. 10·29b is 500×10^{-6} coulombs. This checks with the following:

$$\begin{aligned} Q_T &= Q_1' + Q_2 \qquad (10\cdot79) \\ &= (100 \times 10^{-6}) + (400 \times 10^{-6}) \\ &= 500 \times 10^{-6} \text{ coulombs} \end{aligned}$$

Example 10·5 In the circuit of Fig. 10·30a, two capacitors C_1 and C_2 have each been previously charged up; C_1, a 12-μf capacitor, has 10 volts

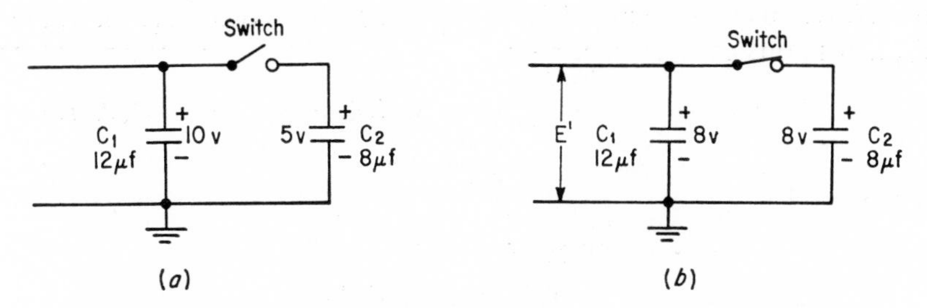

Figure 10·30 Circuit of Example 10·5. (a) Switch open; (b) switch closed.

across it with the polarity shown, while C_2, an 8-μf capacitor, has 5 volts as shown. When the switch is closed, Fig. 10·30b, find the voltage E' across the combination. Two methods of solution are presented in the following:

Solution. When the switch is closed, C_1, the capacitor with the larger voltage ($E_{C_1} = 10$ volts), discharges into the other capacitor, increasing its voltage, E_{C_2}. The quantity of electrons on C_1 in coulombs with the switch open (Fig. 10·30a) is

$$\begin{aligned} Q_1 &= C_1 E_1 \qquad (10\cdot77) \\ &= 12 \ \mu\text{f} \ (10 \text{ volts}) \\ &= (12 \times 10^{-6}) \ 10 \\ &= 120 \times 10^{-6} \text{ coulombs} \end{aligned}$$

The quantity of electrons, in coulombs, on C_2 with the switch open

(Fig. 10·30a) is

$$\begin{aligned} Q_2 &= C_2 E_2 \qquad \text{[from Eq. (10·77)]} \\ &= 8\ \mu\text{f (5 volts)} \\ &= (8 \times 10^{-6})\ 5 \\ &= 40 \times 10^{-6} \text{ coulombs} \end{aligned}$$

The total quantity of electrons (Q_T) on both capacitors is

$$\begin{aligned} Q_T &= Q_1 + Q_2 \qquad (10·79) \\ &= (120 \times 10^{-6}) + (40 \times 10^{-6}) \\ &= 160 \times 10^{-6} \text{ coulombs} \end{aligned}$$

The two capacitors in *parallel,* with the switch closed (Fig. 10·30b) are the equivalent of a single capacitor, C_T, and is

$$\begin{aligned} C_T &= C_1 + C_2 \qquad (10·78) \\ &= 12 + 8 \\ &= 20\ \mu\text{f} \end{aligned}$$

With C_T of 20 μf, and a quantity, Q_T, of 160×10^{-6} coulombs, the voltage, E_T, across this combination is

$$\begin{aligned} Q_T &= C_T E_T \qquad (10·77) \\ \frac{Q_T}{C_T} &= E_T \\ \frac{160 \times 10^{-6}}{20 \times 10^{-6}} &= E_T \\ 8 \text{ volts} &= E_T, \text{ or } E' \end{aligned}$$

This 8 volts is across *each* capacitor, C_1 and C_2, in the parallel circuit as shown in Fig. 10·30b and is the answer, E'. Note that with the switch closed, the new quantity of electrons on $C_1(Q_1')$, added to that on $C_2(Q_2')$, is still the *original* total quantity (Q_T) that was present in the open-switch circuit of Fig. 10·30a, as shown in the following:

$$\begin{aligned} Q_1' &= C_1 E_1' \qquad (10·77) \\ &= 12 \times 10^{-6}(8) \\ &= 96 \times 10^{-6} \text{ coulombs} \end{aligned}$$

and

$$\begin{aligned} Q_2' &= C_2 E_2' \qquad (10·77) \\ &= 8 \times 10^{-6}(8) \\ &= 64 \times 10^{-6} \text{ coulombs} \end{aligned}$$

The sum is still 160×10^{-6} coulombs

$$\begin{aligned} Q_T &= Q_1' + Q_2' \qquad (10·79) \\ &= (96 \times 10^{-6}) + (64 \times 10^{-6}) \\ &= 160 \times 10^{-6} \text{ coulombs} \end{aligned}$$

Alternate solution. Consider the circuit of Fig. 10·30*a*, where E_{C_2} is 5 volts across the 8-μf capacitor, C_2. When the switch is closed, Fig. 10·30*b*, a voltage E_{C_1} of 10 volts is placed across C_2. As far as C_2 is concerned, it "sees" 5 *additional* volts applied to it. This additional voltage is ΔE and is

$$\begin{aligned}\Delta E &= E_{C_1} - E_{C_2} \qquad (10\cdot 80)\\ &= 10 - 5\\ &= 5 \text{ volts}\end{aligned}$$

With the switch closed, C_1 and C_2, because of the current flow around the loop, could be considered to be in *series*. The total capacitance, C_T, of this *series* circuit is

$$\begin{aligned}C_T &= \frac{C_1C_2}{C_1 + C_2} \qquad (10\cdot 69)\\ &= \frac{12(8)}{12 + 8}\\ &= \frac{96}{20}\\ &= 4.8\ \mu\text{f}\end{aligned}$$

The *additional* voltage or *change* of voltage picked up by C_2, Δe_{C_2}, is then

$$\begin{aligned}\Delta e_{C_2} &= \frac{C_T}{C_2}\Delta E \qquad (10\cdot 76)\\ &= \frac{4.8}{8} \times 5\\ &= 3 \text{ volts}\end{aligned}$$

Voltage across C_2, with the switch closed, is now E'_{C_2} and is the sum of the original C_2 voltage, E_{C_2} or e_{C_2}, and the change, Δe_{C_2}.

$$\begin{aligned}E'_{C_2} &= e_{C_2} + \Delta e_{C_2} \qquad (10\cdot 74)\\ &= 5 + 3\\ &= 8 \text{ volts}\end{aligned}$$

Voltage across C_1 must be the same and may be calculated as follows: As far as C_1 is concerned it starts out with 10 volts on it (Fig. 10·30*a*). When the switch is closed, Fig. 10·30*b*, E_{C_2} placed across C_1 *effectively* tries to *lower* or *change* the voltage by 5 volts. This ΔE could then be called a *negative* 5 volts, to indicate a decrease of voltage. The *change* of voltage across $C_1(\Delta e_{C_1})$ is then

$$\begin{aligned}\Delta e_{C_1} &= \frac{C_T}{C_1}\Delta E \qquad (10\cdot 75)\\ &= \frac{4.8}{12}(-5)\\ &= -2 \text{ volts}\end{aligned}$$

The voltage across C_1 with the switch closed, E'_{C_1}, is then the sum of the original voltage, E_{C_1} or e_{C_1}, and the change across C_1, Δe_{C_1}.

$$E'_{C_1} = e_{C_1} + \Delta e_{C_1} \qquad (10\cdot73)$$
$$= 10 + (-2)$$
$$= 8 \text{ volts}$$

Storage counter. The circuit of Fig. 10·25 is that of a *storage counter* or *step counter*. As shown in the waveshape diagrams of Fig. 10·31, the circuit produces a staircase-shaped output voltage for either positive-going or negative-going input pulses. The operation for the former type is discussed first.

On the first positive pulse input, diode D_2 in Fig. 10·25 conducts, charging up the series capacitors C_1 and C_2. The values of C_1 and C_2, shown in the diagram, are 3.33 and 30 μf respectively. The total capacitance, C_T, is

$$C_T = \frac{C_1C_2}{C_1 + C_2} \qquad (10\cdot69)$$
$$= \frac{3.33(30)}{3.33 + 30}$$
$$= \frac{100}{33.33}$$
$$= 3\ \mu\text{f}$$

With a +10-volt pulse applied as pulse number 1, the output voltage across C_2 is

$$E_{C_2} = \frac{C_T}{C_2} E \qquad (10\cdot71)$$
$$= \frac{3}{30} \times 10$$
$$= 1 \text{ volt}$$

This 1-volt output is shown as the first step voltage of E_{out} in Fig. 10·31*a*. With 1 volt across C_2, and with 10 volts applied, E_{C_1} is therefore 9 volts. At the end of the first pulse, when E_{in} goes back down to zero, C_1 discharges through diode D_1. Voltage on C_2 remains steady.

At the second +10-volt pulse of Fig. 10·31*a*, diode D_2 again conducts. With 1 volt still across C_2, the *effective* applied voltage is the difference between the two, or 9 volts. This is also referred to as the change in voltage or ΔE. The *change* in voltage across C_2, Δe_{C_2}, is

$$\Delta e_{C_2} = \frac{C_T}{C_2} \Delta E \qquad (10\cdot76)$$
$$= \frac{3}{30} \times 9$$
$$= 0.9 \text{ volt}$$

This 0.9 volt, Δe_{C_2}, adds to the previous 1-volt charge on C_2, e_{C_2}, giving

$$\begin{aligned} E_{C_2} &= e_{C_2} + \Delta e_{C_2} \\ &= 1 + 0.9 \\ &= 1.9 \text{ volts} \end{aligned} \tag{10·74}$$

This 1.9 volts, E_{out}, is shown as the second-step voltage of Fig. 10·31*a*. With +10 volts applied, and E_{C_2} = 1.9 volts, the remainder of 8.1 volts is across C_1. At the end of pulse number 2, when E_{in} goes back to zero, C_1 discharges through diode D_1. Voltage on C_2 remains constant at 1.9 volts, since C_2 cannot discharge.

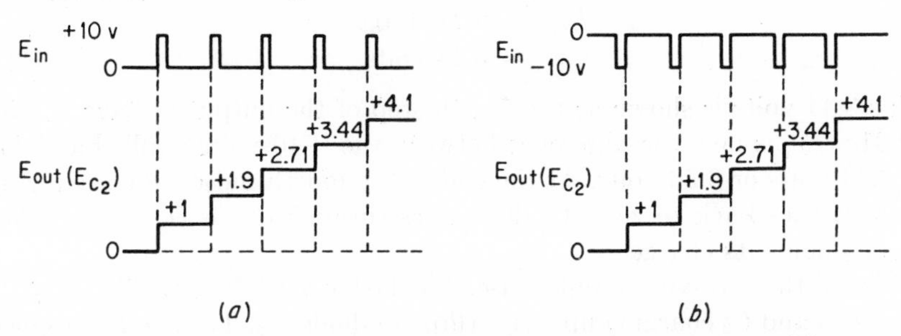

Figure 10·31 Step-counter waveshapes. (*a*) Positive-going input pulses; (*b*) negative-going input pulses.

At the third +10-volt pulse of Fig. 10·31*a*, D_2 (Fig. 10·25) conducts. With +10 volts applied and with 1.9 volts across C_2, the difference is the *effective* applied voltage, ΔE of 8.1 volts. The change of voltage across C_2, Δe_{C_2}, is

$$\begin{aligned} \Delta e_{C_2} &= \frac{C_T}{C_2}\,\Delta E \\ &= \frac{3}{30} \times 8.1 \\ &= 0.81 \text{ volt} \end{aligned} \tag{10·76}$$

This 0.81 volt is the additional charge placed on C_2, and E_{C_2} now becomes:

$$\begin{aligned} E_{C_2} &= e_{C_2} + \Delta e_{C_2} \\ &= 1.9 + 0.81 \\ &= 2.71 \text{ volts} \end{aligned} \tag{10·74}$$

This 2.71 volts is shown in Fig. 10·31*a* as the third step of the output voltage. With +10 volts applied and E_{C_2} = 2.71 volts, the difference of 7.29 volts is across C_1. At the end of pulse 3, when E_{in} returns to zero, C_1 discharges through diode D_1 (Fig. 10·25), but C_2 holds its charge.

At the fourth +10-volt pulse of Fig. 10·31*a*, diode D_2 again conducts. With +10 volts input and with C_2 charged (e_{C_2}) from the previous signals to 2.71 volts, the difference between the two, 7.29 volts, is the *effective* applied voltage or the *change* of voltage, ΔE. The change of voltage across C_2, Δe_{C_2}, is

$$\Delta e_{C_2} = \frac{C_T}{C_2}\,\Delta E \qquad (10\cdot76)$$
$$= \frac{3}{30} \times 7.29$$
$$= 0.729 \text{ volt or } 0.73 \text{ volt (approx.)}$$

This voltage (Δe_{C_2}) adds to the previous voltage on $C_2(e_{C_2})$ giving

$$E_{C_2} = e_{C_2} + \Delta e_{C_2} \qquad (10\cdot74)$$
$$= 2.71 + 0.73$$
$$= 3.44 \text{ volts}$$

This 3.44 volts is shown as the fourth step of the output voltage of Fig. 10·31*a*. E_{C_1} is now the difference between the +10 volts applied and E_{C_2} of 3.44 volts or 6.56 volts. At the end of the fourth pulse, when E_{in} (Fig. 10·31*a*) goes back to zero, C_1 discharges through diode D_1 (Fig. 10·25), but C_2 holds its charge.

The fifth +10-volt input pulse, the last one shown in Fig. 10·31*a*, starts C_1 and C_2 charging up again through diode D_2. The *effective* applied voltage, ΔE, is the difference between the actual applied voltage of +10 volts and the charge on $C_2(e_{C_2})$ of 3.44 volts because of the previous input signals. This difference, ΔE, is then 6.56 volts (10 − 3.44). C_2 will gain an additional voltage, Δe_{C_2}, which is

$$\Delta e_{C_2} = \frac{C_T}{C_2}\,\Delta E \qquad (10\cdot76)$$
$$= \frac{3}{30} \times 6.56$$
$$= 0.656 \text{ volt, or } 0.66 \text{ volt (approx.)}$$

The new voltage across C_2, E_{C_2} is now

$$E_{C_2} = e_{C_2} + \Delta e_{C_2} \qquad (10\cdot74)$$
$$= 3.44 + 0.66$$
$$= 4.1 \text{ volts}$$

The waveshapes of Fig. 10·31*a* only show five input pulses, with the output voltage rising higher and higher in a staircase fashion. If more input pulses were shown, E_{out} would keep rising with each pulse. Voltage across C_2 rises with each input pulse since C_2 has no path through which to discharge, in Fig. 10·25. Later, a discussion is given where an additional stage, a monostable blocking oscillator, provides such a discharge path for C_2.

The previous discussion of the step counter of Fig. 10·25 was for positive-going input pulses, as shown in Fig. 10·31*a*. Note that each step of the staircase output rises at the *leading* edge of the positive-going input pulse. The circuit of Fig. 10·25 also operates for negative-going input pulses, as shown in Fig. 10·31*b*. Note that in these waveshapes, each step voltage rises at the *lagging* edge of the negative-going pulse. The operation of the circuit for this type of pulse is as follows.

When the first input pulse becomes −10 volts, C_1 charges up to this voltage through diode D_1, as indicated in Fig. 10·32. Note that the polarity across C_1 is reversed from that of Fig. 10·25. At the completion of the first pulse, when E_{in} (Fig. 10·31*b*) has returned to zero, C_1 discharges through diode D_2 and capacitor C_2. The discharging of C_1 charges up C_2. The voltage that now appears across C_2 may be calculated similarly

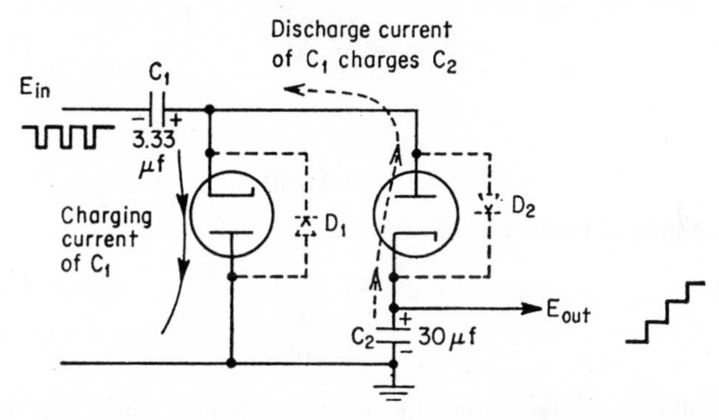

Figure 10·32 Step counter with negative-going input pulse.

to that of Example 10·4, Fig. 10·29, or as in Example 10·5, Fig. 10·30. The following computations follow the alternate solution method of Example 10·5.

C_1, a 3.33-μf capacitor charged to 10 volts, discharges through diode D_2 and C_2, a 30-μf capacitor. With zero volts across C_2, the 10 volts across C_1 act as an applied voltage. C_2 charges up to

$$E_{C_2} = \frac{C_T}{C_2} E \qquad (10\cdot71)$$

where C_T = total capacitance in the *series* discharge path of C_1, consisting of C_1 and C_2 in *series*.

$$C_T = \frac{C_1 C_2}{C_1 + C_2} \qquad (10\cdot69)$$

$$= \frac{3.33(30)}{3.33 + 30}$$

$$= 3 \ \mu\text{f}$$

and

$$E_{C_2} = \frac{3}{30} \times 10 \qquad (10\cdot71)$$

$$= 1 \text{ volt}$$

C_2 has now charged up to 1 volt because of the discharge of C_1, as shown in Fig. 10·32. This 1-volt value of E_{C_2} is shown as the first-step voltage of Fig. 10·31*b*.

When the second -10-volt input pulse is applied, C_1 recharges fully through diode D_1, and C_2 holds its 1-volt charge.

When pulse number 2, Fig. 10·31*b*, is ended, and E_{in} is back at zero, C_1 starts discharging its 10 volts through diode D_2 and capacitor C_2. The 10 volts E_{C_1} acts as an applied voltage which opposes the 1 volt e_{C_2}, producing an *effective* applied voltage or change of voltage, ΔE, of 9 volts. C_2 now charges up to a higher value (E_{C_2}) from its previous value (e_{C_2}). The additional voltage gained by C_2 is Δe_{C_2} and is

$$\Delta e_{C_2} = \frac{C_T}{C_2}\,\Delta E \qquad (10\text{·}76)$$

$$= \frac{3}{30} \times 9$$

$$= 0.9 \text{ volt}$$

C_2 is now charged to an E_{C_2} of

$$E_{C_2} = e_{C_2} + \Delta e_{C_2} \qquad (10\text{·}74)$$

$$= 1 + 0.9$$

$$= 1.9 \text{ volts}$$

This 1.9 volts E_{C_2} is shown as the second-step voltage in Fig. 10·31*b*.

The -10-volt input pulse number 3 causes C_1 to again charge fully through diode D_1, and C_2 remains charged at its 1.9 volts. At the end of pulse 3, the 10 volts across C_1 starts discharging, as shown in Fig. 10·32, through C_2 and diode D_2, causing C_2 to charge higher. The difference between the 10 volts across C_1, and the 1.9 volts across $C_1(e_{C_1})$, is the *effective* applied voltage or *change* of voltage, ΔE, applied of 8.1 volts $(10 - 1.9)$. C_2 now charges up an additional voltage, Δe_{C_2}, of

$$\Delta e_{C_2} = \frac{C_T}{C_2}\,\Delta E \qquad (10\text{·}76)$$

$$= \frac{3}{30} \times 8.1$$

$$= 0.81 \text{ volt}$$

Voltage on C_2 is now the sum of its previous charge (e_{C_2}) and its additional charge (Δe_{C_2}).

$$E_{C_2} = e_{C_2} + \Delta e_{C_2} \qquad (10\text{·}74)$$

$$= 1.9 + 0.81$$

$$= 2.71 \text{ volts}$$

This voltage across C_2 of 2.71 volts is indicated as the third-step output voltage in Fig. 10·31*b*.

The fourth -10-volt input pulse causes C_1 to again charge fully through diode D_1 with the polarity shown in Fig. 10·32. C_2 remains charged at 2.71 volts. At the end of this fourth pulse, C_1 discharges through C_2 and diode D_2. The voltage across C_2 now becomes $+3.44$ volts, as shown in the fourth step of the output voltage in Fig. 10·31*b*. The calculation of this $+3.44$ volts, and then the $+4.1$ volts (on the fifth step voltage), is left as an exercise for the reader.

From the values of the staircase-voltage output of Fig. 10·31, it should be noted that the *amplitude* of each step voltage, starting with the first, *decreases*. The first *increment* or increase is 1 volt (since the output went from zero volts to $+1$ volt). The second *increment* is 0.9 volt (the output went from $+1$ volt to $+1.9$ volts). The third increase is 0.81; the fourth is 0.73; and the fifth is 0.66. As these changes indicate, the staircase is not linear but is exponential such as that of a capacitor charging curve. To produce steps of more equal amplitudes, the circuit may be altered somewhat to that shown in Fig. 10·33*a*.

The anode of diode D_1 is connected to a variable voltage which is made equal to E_{C_2}. This is achieved through the use of feedback from an amplifier having unity gain. This feedback is a form of *bootstrap* circuit. The following discussion illustrates how this circuit theoretically produces a staircase with equal amplitude steps, using positive-going input pulses.

Before the first $+10$-volt pulse is applied, assume that point X, the feedback voltage, is zero. Neither diode conducts yet, and neither C_1 nor C_2 is charged. When the first $+10$-volt pulse is applied, C_1 and C_2 charge up through diode D_2. The total capacitance, C_T, of C_1 and C_2 in series is

$$\begin{aligned} C_T &= \frac{C_1 C_2}{C_1 + C_2} \qquad (10\cdot69) \\ &= \frac{3.33(30)}{3.33 + 30} \\ &= 3\ \mu\text{f} \end{aligned}$$

Voltages across these capacitors become

$$\begin{aligned} E_{C_2} &= \frac{C_T}{C_2}E \qquad (10\cdot71) \\ &= \frac{3}{30} \times 10 \\ &= 1\ \text{volt} \end{aligned}$$

and

$$\begin{aligned} E_{C_1} &= \frac{C_T}{C_1}E \\ &= \frac{3}{3.33} \times 10 \\ &= 9\ \text{volts} \end{aligned}$$

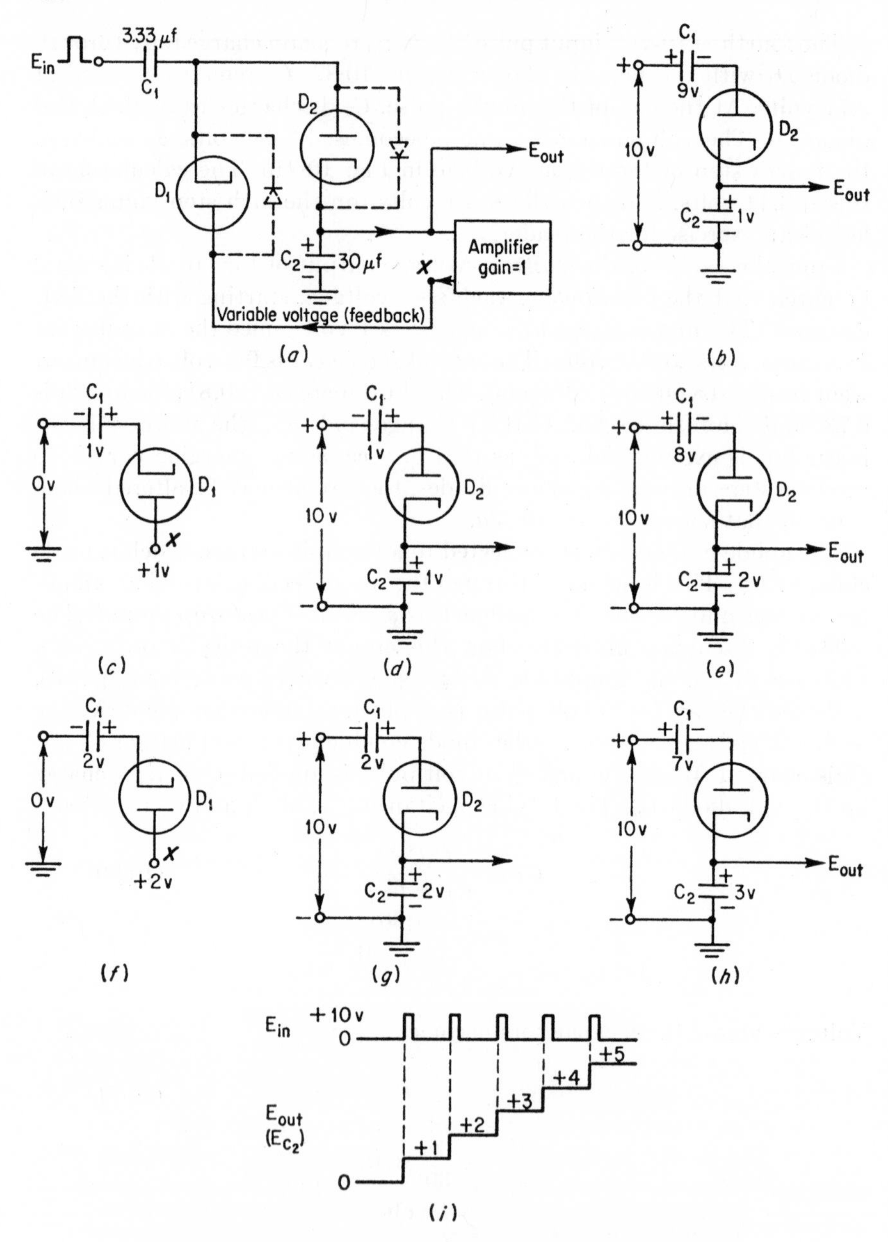

Figure 10·33 Step counter producing equal amplitude steps.

These voltages are shown in Fig. 10·33*b*, when E_{C_2} (+1 volt) is the output as shown in the first step of E_{out} in Fig. 10·33*i*.

If point $X = E_{C_2}$, then the plate of diode D_1 is now +1 volt. When the first +10-volt pulse has passed, E_{in} is at zero volts, and diode D_1 will first discharge C_1 and then charge up C_1 to 1 volt with the polarity shown in Fig. 10·33*c*. C_1 has gone from 9 volts with the polarity shown in Fig. 10·33*b* to 1 volt of the opposite polarity, a change in C_1 voltage of 10 volts. C_2 remains charged at 1 volt.

When the second +10-volt pulse first appears, it "sees" the circuit as shown in Fig. 10·33*d*. E_{C_1} and E_{C_2} are each 1 volt of opposite polarities, or zero volts across them in series. The +10 volts applied is therefore the *effective* applied voltage, or the *change*, ΔE. C_2 charges up through D_2 an additional voltage, Δe_{C_2}, which is

$$\Delta e_{C_2} = \frac{C_T}{C_2}\,\Delta E \qquad (10\cdot76)$$

$$= \frac{3}{30} \times 10$$

$$= 1 \text{ volt}$$

C_2 is now charged to the sum of the previous charge of 1 volt (e_{C_2}) and the additional change of 1 volt (Δe_{C_2}), or a total of 2 volts, as shown in Fig. 10·33*e*. This +2 volts is the output as shown in the second step of E_{out}, Fig. 10·33*i*. C_1 charges up through D_2 an additional voltage, Δe_{C_1}, which is

$$\Delta e_{C_1} = \frac{C_T}{C_1}\,\Delta E \qquad (10\cdot75)$$

$$= \frac{3}{3.33} \times 10$$

$$= 9 \text{ volts}$$

C_1, because of the +10 volts applied, now reverses its polarity from the 1-volt charge as shown in Fig. 10·33*d* to the 8 volts as shown in Fig. 10·33*e*, and C_1 has undergone a change (Δe_{C_1}) of 9 volts. Point X, the feedback applied to D_1 plate, is now +2 volts since it is equal to E_{C_2}.

When the second input pulse has been completed, E_{in} becomes zero, and diode D_1, with +2 volts on its plate, discharges C_1 and charges it up to 2 volts, with the polarity as shown in Fig. 10·33*f*.

The third +10-volt input pulse, when first applied, "sees" the circuit as shown in Fig. 10·33*g*. C_1 and C_2 are each charged to 2 volts, but of opposing polarities or zero volts across them in series. The +10 volts applied is therefore the *effective* applied voltage or the *change*, ΔE. C_2, as before, now charges up an additional voltage, Δe_{C_2}, of 1 volt as shown in the following:

$$\Delta e_{C_2} = \frac{C_T}{C_2} \Delta E \tag{10·76}$$

$$= \frac{3}{30} \times 10$$

$$= 1 \text{ volt}$$

E_{C_2} is now the sum of its previous charge (e_{C_2}) of 2 volts from Fig. 10·33*e*, and the new change (Δe_{C_2}) of 1 volt, or a total of +3 volts as shown in Fig. 10·33*h* and also as the third step of E_{out} in Fig. 10·33*i*. C_1 changes its voltage by

$$\Delta e_{C_1} = \frac{C_T}{C_1} \Delta E \tag{10·75}$$

$$= \frac{3}{3.33} \times 10$$

$$= 9 \text{ volts}$$

C_1, which had been charged to 2 volts with the polarity shown in Fig. 10·33*g*, now reverses its polarity because of the 10 volts applied and

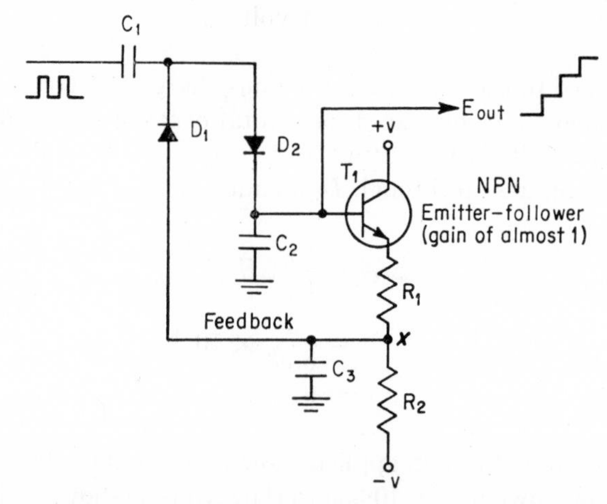

Figure 10·34 Step counter producing nearly equal amplitude steps.

becomes charged to 7 volts with the polarity shown in Fig. 10·33*h*, a *change* of 9 volts.

Note, as shown in the staircase-output voltage of Fig. 10·33*i*, that each step is of equal amplitude. The 1-volt equal increments occur, as explained, because of diode D_1 plate being connected to a variable voltage which is equal to E_{C_2}.

A circuit which could deliver a feedback voltage approximately equal to E_{C_2} is shown in Fig. 10·34. An *emitter follower*, NPN transistor T_1, with a gain of almost 1, provides the means for feeding back a voltage to the plate of diode D_1 which is directly proportional to E_{C_2}.

Before an input pulse is applied, T_1 conducts, and the junction of R_1 and R_2 (point X) is zero volts. As E_{C_2} charges, it drives the P-type base of the NPN transistor T_1 positive. Transistor current increases, and point X goes positive, although not quite as much as E_{C_2} because of the gain of less than 1. C_3 is a bypass capacitor which allows C_1 a low impedance path through which to discharge, bypassing R_2. If a PNP transistor emitter-follower were used, its base would require an increasing *negative*-going voltage. By inverting diodes D_1 and D_2, the staircase output across C_2 would be the required negative-going steps.

Frequency divider. The step-counter diode circuits of Fig. 10·25 or Fig. 10·32, combined with a monostable blocking oscillator, as shown in

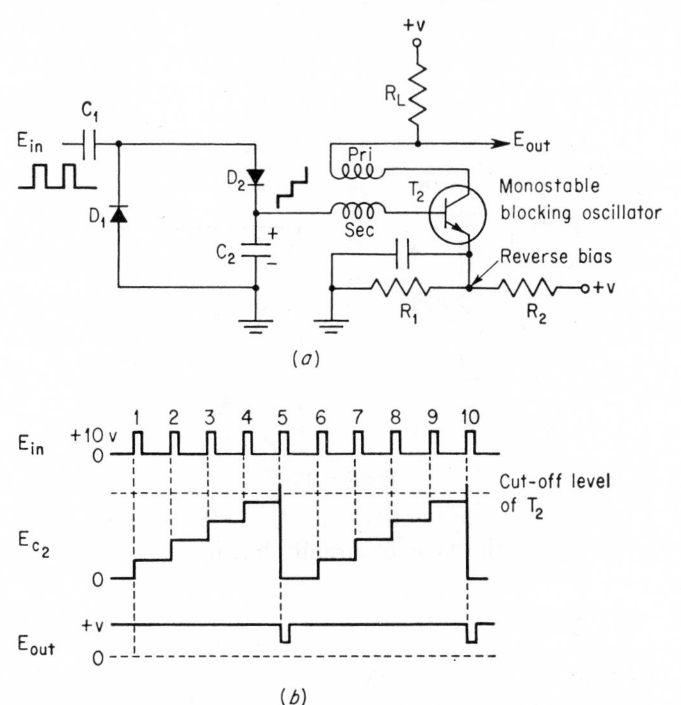

Figure **10·35 Step counter and monostable blocking oscillator; frequency division by 5.**

Fig. 10·35*a*, comprise a *frequency-divider* circuit. The monostable blocking oscillator, discussed in greater detail in Chap. 6, is kept inoperative by the positive voltage on the N-type emitter, or reverse biased. NPN transistor T_2 conducts whenever E_{C_2} rises sufficiently, making the P-type base positive. T_2 base current discharges C_2. Collector current flows as a short pulse up through resistor R_L, producing an E_{out} which is a negative-going pulse, shown in Fig. 10·35*b*. In the waveshapes shown here, it is assumed that the fifth step of E_{C_2} brings T_2 into conduction. Comparing the E_{in}

pulses with those of E_{out}, the frequency division can be observed since one output pulse is produced for five inputs, or a frequency division by five. By changing the values of R_1 and R_2, the frequency-division ratio can be varied. For example if R_1 were decreased, E_{R_1} would be smaller and the reverse bias would be less. As a result, T_2 would be triggered into conduction by a smaller positive voltage on its base, and instead of the fifth step of E_{C_2} doing this, an earlier step would, making it a division by four or by three.

PROBLEMS

10·1 In the accompanying diagram, identify the vectors (*a*), (*b*), (*c*), (*d*), and the type of circuit shown.

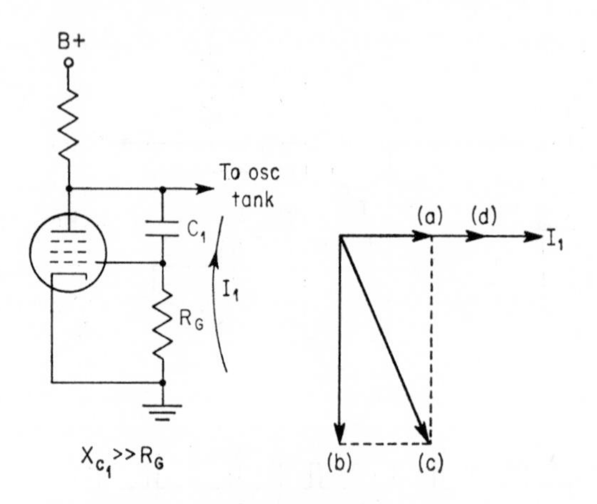

Problem 10·1

10·2 In the diagram accompanying this problem, identify the vectors (*a*), (*b*), (*c*), (*d*), and the type of circuit shown.

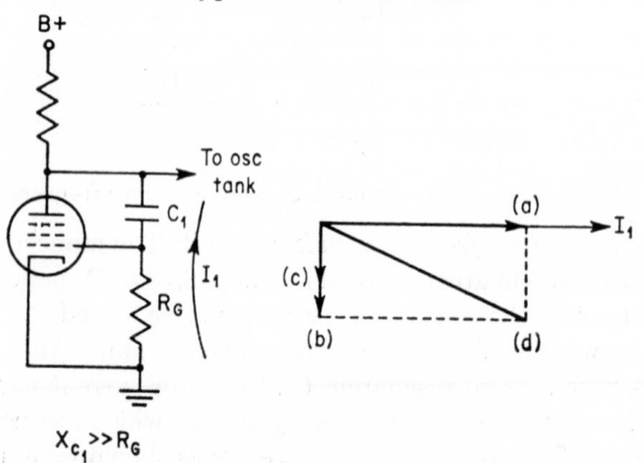

Problem 10·2

10·3 In the sine-wave and pulse phase comparator of Fig. 10·8, draw the voltage waveshapes at V_1 and V_2 plates, showing phase relationship, if the oscillator frequency started to increase. (See Fig. 10·9*a* for normal phase relationship.)

10·4 In the sawtooth and pulse phase comparator of Fig. 10·10, draw the voltage waveshape at V_1 grid if the oscillator frequency started to decrease. (See Fig. 10·11*a* for normal phase relationship.)

MORE CHALLENGING PROBLEMS

10·5 (*a*) From Fig. 10·4*a* derive an expression for the impedance Z, "seen" by the oscillator tank circuit. Follow the form of Eqs. (10·4) to (10·13) shown in the text for the circuit of Fig. 10·2*a*.

(*b*) Is the j term (in the numerator of Z of the final equation) positive or negative, and what does this denote?

10·6 (*a*) From Fig. 10·6*a* derive an expression for the impedance Z, "seen" by the oscillator-tank circuit. Follow the form of Eqs. (10·15) to (10·23) shown in the text for the circuit of Fig. 10·5*a*. (*b*) Is the j term (in the numerator of Z of the final equation) positive or negative, and what does this denote?

10·7 In the sine-wave and pulse phase-comparator circuit of Fig. 10·8, with the waveforms of Fig. 10·9, determine the following if the oscillator frequency starts increasing: (*a*) Does diode V_1 or V_2 conduct more heavily; (*b*) is E_{R_1} or E_{R_2} the larger; (*c*) what is the polarity of the phase comparator output; (*d*) what does this voltage do to the X_C tube, V_4; (*e*) what does this do to the oscillator tank frequency?

10·8 In the sawtooth and pulse phase-comparator circuit of Fig. 10·10, with the waveforms of Fig. 10·11, determine the following if the blocking oscillator frequency starts decreasing: (*a*) does the phase comparator tube V_1 conduct more heavily or less; (*b*) How does this affect the bias on the oscillator V_2; (*c*) How does this affect oscillator frequency?

10·9 Referring to the digital-to-analog conversion-mesh circuit of Fig. 10·12 and to the decade-counter schematic of Fig. 7·10, what is the *amplitude of the staircase-output voltage* at pulse number 3, if V_{4B} *is inoperative?* (V_{4A} is therefore permanently conducting.)

10·10 With the conditions the same as the previous problem, except that V_{3A} is inoperative instead of V_{4B}, find the amplitude of the staircase at the *seventh* input pulse.

10·11 In the circuit accompanying this problem, find the voltage across each capacitor after the switch is closed.

10·12 In the circuit given with this problem, find the voltage across each capacitor (and the polarity) after the switch is closed.

10·13 In the accompanying diagram, find the voltage across each capacitor (and the polarity) after the switch is closed.

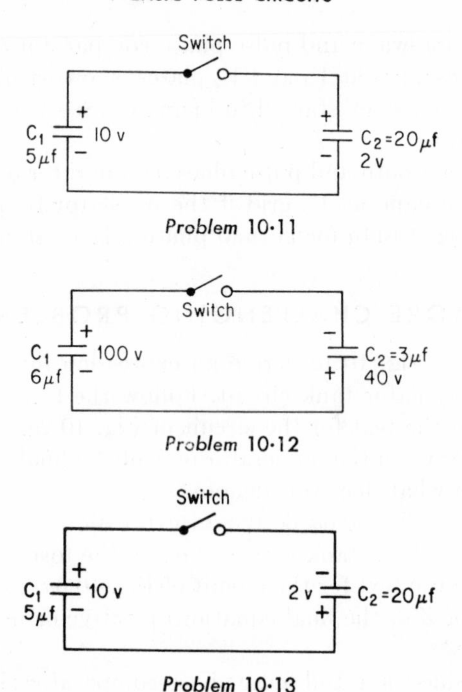

Problem 10·11

Problem 10·12

Problem 10·13

10·14 From the accompanying circuit, draw the output-voltage waveshapes for E_{in} (a) and also for E_{in} (b), showing the input and output signal in proper phase relationship.

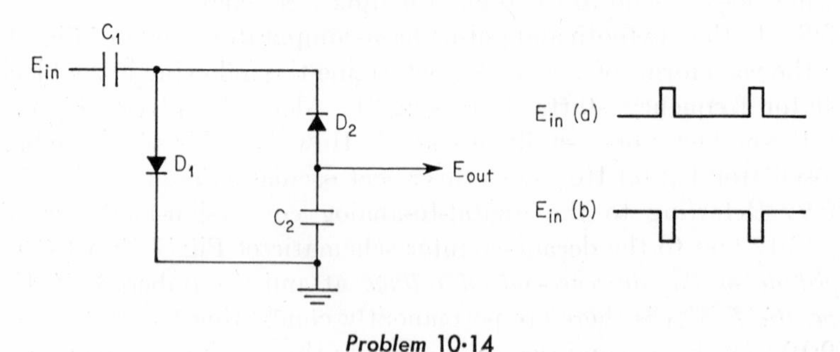

Problem 10·14

CHAPTER 11

TRANSIENT ANALYSIS

11·1 General

In Chap. 3, a discussion is given on the effects that a resistor and capacitor (an *R-C* circuit) and a resistor and inductor (*R-L* circuit) have upon an input-signal voltage. The solutions in that chapter are based on the universal time-constant chart, shown in that chapter, Fig. 3·1. The reader should recall that in an *R-C* circuit where a steady voltage is applied, voltage across the capacitor rises to 63 per cent of the applied voltage in the first *R-C* time constant (the product of *R* in ohms and *C* in farads). In each succeeding *R-C* time constant, the process repeats. That is, the capacitor in each *R-C* charges up 63 per cent of the difference between the applied voltage and the voltage across the capacitor. In the *first* *R-C*, the capacitor charges up 63 per cent of the applied voltage of 100 per cent. The difference is then 37 per cent. In the *second* *R-C*, the capacitor charges up 63 per cent of this 37 per cent difference, or about another 23 per cent. The capacitor is now at 86 per cent (63 + 23) of the full applied voltage and becomes practically fully charged in about five *R-C* time constants. This is all shown in the universal time-constant chart of Fig. 3·1. In Chap. 3, by referring to the chart, satisfactory approximate results could be obtained by using only some simple arithmetic.

In the present chapter, more precise answers will be obtained by using a basic formula and some algebra and logarithms. The mathematics involved will be reviewed in detail as it is employed.

11·2 *R-C* Circuit

If a steady d-c voltage, such as a battery, is applied to a resistor in series with a capacitor, as shown in Fig. 11·1*a*, then the voltages across the resistor (E_R) and capacitor (E_C) are as shown in Fig. 11·1*b* during the period of time T_1. If the battery is then removed and a short circuit substituted for the battery, E_R and E_C are as shown during the time T_2.

In order to find E_R and E_C, a basic equation is generally used. This basic equation is

$$\frac{E_2}{E_1} = \epsilon^{-\frac{T}{RC}} \tag{11·1}$$

where E_1 is the voltage across the resistor at *any* instant during a period of time T, and E_2 is the resistor voltage at some *later* instant of this same time period. However, both E_1 and E_2 *must* occur during the *same* time period, T_1 or T_2 of Fig. 11·1*b*. T in Eq. (11·1) is the time between the

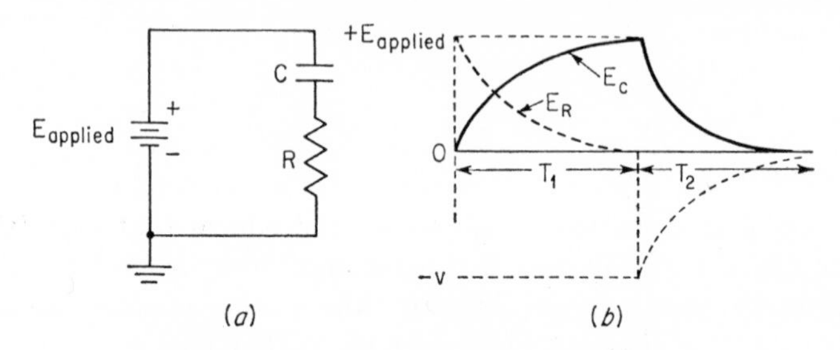

Figure **11·1 D-C applied to *R*-C circuit.**

instant of E_1 and the instant of E_2. Note from Fig. 11·1*b* that the resistor voltage, E_R, is maximum at first and decreases toward zero during T_1 and also during T_2. As a result, E_1 is *always* larger than E_2. In Eq. (11·1), the term ϵ is the Greek letter, pronounced *epsilon* and, like π, has a given mathematical value; $\epsilon = 2.718$.

Logarithms. Using Eq. (11·1) involves the use of a system called *natural logarithms,* or logarithms to the base ϵ. This involves the use of *natural logarithm* tables, or the log-log (*LL*) and *D* scales of the slide rule. Another system of logarithms (or logs) is more popular and is called the *common logs.* These use logs to the base number 10, where the log of any number is simply the *power* or *exponent* of 10 which makes it equal to that number. As an example, the log of 100 is 2.00, since $10^2 = 100$. Also, since $1{,}000 = 10^3$, log 1,000 = 3.00, and since $100{,}000 = 10^5$, log 100,000 = 5.00. Note in the above examples that the log of each number is shown containing a decimal part too, even though in these previous examples the decimal part is only zero. The decimal part of the log is called the *mantissa,* while the number to the left of the decimal point is called the *characteristic number.* The characteristic number part of the log is always one less than the number of digits which are to the left of the decimal point in the original number, as shown in the following:

The number 1,000. has *four* digits to the left of the decimal point; the *characteristic number* part of its log is therefore 3. The number 14,000. has *five* digits to the left of the decimal point; the *characteristic number* part of its log is therefore 4.

To find the common log of a number, look up that number in the log tables or on the D scale of a slide rule. The *mantissa* or decimal part of the log is found in the log tables or on the L scale of the slide rule. For example, the *mantissa* part of the log for the number 14,000 is 0.146 (from the slide rule), while the *characteristic number* is 4. Therefore, the log of 14,000. is 4.146. In the work of this chapter, the numbers whose logs are required will always be greater than 1, thereby avoiding negative characteristic numbers with their resulting additional confusion. A few additional examples are as follows.

The log of 24.2 is 1.384. The number 24.2 has *two* digits to the left of the decimal point; its characteristic number is therefore 1. The number 24.2 on the D scale of the slide rule is opposite the *mantissa* 0.384 on the L scale.

The log of 555. is 2.744. The number 555. has *three* digits to the left of the decimal point; therefore its characteristic number is 2. The number 555. on the D scale is opposite 0.744 (the mantissa) on the L scale.

Antilogarithms. When the log of a number is known, while the number itself is not, the number may be found by a method which is just the reverse of finding the log. For example, if the *log* of a number is 1.5, the mantissa 0.5 must be looked up first, either in the tables, or on the L scale of the slide rule. 0.5 on the L scale is opposite the digits 316 on the D scale. The characteristic number 1, of the original log 1.5, is one less than the number of digits to the left of the decimal point of the answer. The number of these digits is therefore 2. The digits 316 on the D scale therefore should be the number 31.6, with two digits to the left of the decimal point. When a number is to be found and its log is known, the method is usually referred to as the *antilog* and is often written with the exponent -1 following the word log, for example, $\log^{-1} 1.5$. This is read as the antilog of 1.5.

Resistor-voltage equations. With the brief review of logs and antilogs just presented, let us now return to the basic equation given previously.

$$\frac{E_2}{E_1} = \epsilon^{-\frac{T}{RC}} \qquad (11\cdot1)$$

The negative exponent of ϵ can be changed to a positive one by simply moving the ϵ term from the numerator of a fraction to the denominator, as shown in the following:

$$\frac{E_2}{E_1} = \frac{1}{\epsilon^{\frac{T}{RC}}} \tag{11·2}$$

As stated previously, E_1 is larger than E_2, and it would be more convenient to have the larger term in the numerator. Inverting both fractions in Eq. (11·2) (same as dividing both sides of equation into 1), gives

$$\frac{E_1}{E_2} = \epsilon^{\frac{T}{RC}} \tag{11·3}$$

As stated previously, ϵ is equal to 2.718. The number 1,000 is, of course, equal to 10^3. From the natural log tables (to the base ϵ), it can be proven that 1,000 is also equal to $\epsilon^{6.9}$. Restating these gives

$$1{,}000 = 10^3 \tag{11·4}$$

and

$$1{,}000 = \epsilon^{6.9} \tag{11·5}$$

therefore

$$\epsilon^{6.9} = 10^3 \tag{11·6}$$

From Eq. (11·6), it can be seen that to change from one term using a power of ϵ to an equal-value term using a power of 10, simply divide the exponent of ϵ by 2.3. Therefore, dividing the ϵ exponent of 6.9 in Eq. (11·6) by 2.3 results in 10^3, $(6.9 \div 2.3 = 3)$.

Returning to Eq. (11·3),

$$\frac{E_1}{E_2} = \epsilon^{\frac{T}{RC}} \tag{11·3}$$

By dividing the ϵ exponent T/RC by 2.3, the ϵ term is replaced by the number 10, as shown.

$$\frac{E_1}{E_2} = 10^{\frac{T}{2.3RC}} \tag{11·7}$$

Equation (11·7) is similar in form to the following:

$$1{,}000 = 10^3 \tag{11·8}$$

therefore

$$\log 1{,}000 = 3 \tag{11·9}$$

Equation (11·7), like (11·8) and (11·9), can therefore be changed to the following, where the number 10 is omitted, but its exponent $(T/2.3RC)$ is equal to the log of E_1/E_2, as shown.

$$\log \frac{E_1}{E_2} = \frac{T}{2.3RC} \tag{11·10}$$

Equation (11·10) is the basic equation that will be used in the transient analysis discussion that will follow. Other useful forms of this equation are, solving for the time, T, from Eq. (11·10):

$$T = 2.3RC \log \frac{E_1}{E_2} \tag{11·11}$$

Also, solving Eq. (11·10) for E_2 gives

$$\frac{E_1}{E_2} = \log^{-1} \frac{T}{2.3RC} \tag{11·12}$$

and

$$E_2 = \frac{E_1}{\log^{-1} \dfrac{T}{2.3RC}} \tag{11·13}$$

Simple R-C Circuit Examples

Example 11·1 In the circuit of Fig. 11·1*a*, the applied voltage is 100 volts, C is 0.01 μf, and R is 10 kilohms. Find the voltage across the resistor and across the capacitor if the voltage is applied for 200 μsec.

Solution. Equation (11·13) will be used in this example. The voltage across the resistor at the start of the 200-μsec period is the full applied voltage, as shown by the E_R curve in Fig. 11·1*b*. This voltage is E_1 in Eq. (11·13). The voltage across the resistor at the end of the 200-μsec period is the unknown, E_2.

$$\begin{aligned} E_2 &= \frac{E_1}{\log^{-1} \dfrac{T}{2.3RC}} \qquad (11·13) \\ &= \frac{100}{\log^{-1} \left(\dfrac{200\ \mu\text{sec}}{2.3(10\ \text{kilohms})0.01\ \mu\text{f}} \right)} \\ &= \frac{100}{\log^{-1} \left(\dfrac{200 \times 10^{-6}}{2.3(10 \times 10^3)0.01 \times 10^{-6}} \right)} \\ &= \frac{100}{\log^{-1} 0.869} \\ &= \frac{100}{7.4} \end{aligned}$$

In taking the antilog of 0.869 ($\log^{-1} 0.869$) note that the mantissa is 0.869, while the characteristic number is zero. Looking up the 0.869 on the L scale of the slide rule indicates the digits 74 on the D scale. Since the characteristic number is zero, the number which is the $\log^{-1} 0.869$ must have *one* digit to the left of its decimal point, giving 7.4.

$$E_2 = \frac{100}{7.4}$$
$$E_2 = 13.5 \text{ volts}$$

E_2 is the voltage across the resistor, or E_R. From Kirchhoff's voltage law, the applied voltage, E_T, is equal to the sum of the capacitor voltage, E_C, and the resistor voltage E_R. Therefore, the capacitor voltage, E_C, may be found as follows:

$$\begin{aligned} E_T &= E_R + E_C \\ E_T - E_R &= E_C \\ 100 - 13.5 &= E_C \\ 86.5 &= E_C \end{aligned}$$

Note that in this example, the *R-C* time constant is

$$\begin{aligned} RC &= 10 \text{ kilohms } (0.01\ \mu\text{f}) \\ &= 10 \times 10^3\ (0.01 \times 10^{-6}) \\ &= 0.1 \times 10^{-3} \\ &= 0.0001 \text{ sec} \\ &= 100\ \mu\text{sec} \end{aligned}$$

The time that the voltage was applied is given as 200 μsec. This is equal to two *R-C* time constants. Checking with the universal time-constant chart, Fig. 3·1, it can be seen from curve *A* that, in two time constants, E_C becomes approximately 86 per cent of the applied voltage, and from curve *B*, it can be seen that E_R becomes approximately 14 per cent. This agrees very closely with the results of Example 11·1 where E_C becomes 86.5 volts and E_R 13.5 volts.

In the previous example, the time, T, the voltage, E_1, and the *R-C* were known, requiring Eq. (11·13), to solve for E_2. In the following example, the time, T, is the unknown, and another form of the equation must be used.

Example 11·2 If the applied voltage of Fig. 11·1*a* and Example 11·1 were replaced by a short circuit after the capacitor had charged to 86.5 volts, how long would it take for the capacitor voltage, E_C, to decrease to 0.1 volt? The values for the capacitor and resistor are those of Example 11·1, 0.01 μf and 10 kilohms.

Solution. Equation (11·11) is used here to solve for the time, T. On the discharge of a capacitor, the capacitor voltage acts as the applied voltage, and this is also the voltage across the resistor. This can be seen during the period T_2 of Fig. 11·1*b*. When E_C becomes zero, E_R also becomes zero.

$$\begin{aligned}
T &= 2.3RC\left(\log \frac{E_1}{E_2}\right) \qquad (11\cdot 11)\\
&= 2.3(10 \text{ kilohms})\ 0.01\ \mu\text{f}\left(\log \frac{86.5}{0.1}\right)\\
&= 2.3(10 \times 10^3)\ (0.01 \times 10^{-6}) \log 865.\\
&= (0.23 \times 10^{-3})\ (2.937)\\
&= 0.675 \times 10^{-3}\\
&= 0.000675 \text{ sec}\\
&= 675.\ \mu\text{sec}
\end{aligned}$$

It takes a capacitor approximately five or six *R-C* time constants to practically fully charge or to fully discharge. Note that the *R-C* time constant here is 100 μsec (10 kilohms × 0.01 μf), and six *R-C*s would be 600 μsec, which is close to the answer in this problem.

Square waves in *R-C* circuits. Chapter 3 discusses the effect that an *R-C* circuit has upon a square wave, using the approximate values from the universal time-constant chart, Fig. 3·1. In this section, examples are given using the more exact values of Eqs. (11·11) and (11·13).

Example 11·3 As shown in Fig. 11·2*b*, a 100-volt square wave, having 75 μsec duration and 25 μsec between square waves, is applied to a 0.001-μf capacitor in series with a 50-kilohm resistor, shown in Fig. 11·2*a*. Calculate the capacitor voltage E_C and the resistor voltage E_R (Fig. 11·2*d* and *c* respectively) at each instant t_1, t_2, t_3, and t_4. Also determine how long, after t_4, it takes for E_R and E_C to decrease to 0.1 volt.

Solution. The calculations for this example will be shown as a series of steps, each step solving for E_R and E_C at each of the instants, t_1, t_2, t_3, and t_4. The solutions are shown in Fig. 11·2*c* and *d*.

1. *At instant* t_1, the circuit is effectively as shown in Fig. 11·2*e*, with +100 volts applied. At this first instant, current flows, but the capacitor has not had time yet to charge at all. As a result, E_C is zero volts, while E_R is +100 volts. These values are indicated at instant t_1 in Fig. 11·2*c* and *d*.

2. *Between instant* t_1 *and* t_2, *just before* t_2, the circuit is as shown in Fig. 11·2*f*, with the +100 volts still applied. Current flow has occurred for the 75-μsec period, and the .001-μf capacitor has charged up somewhat.

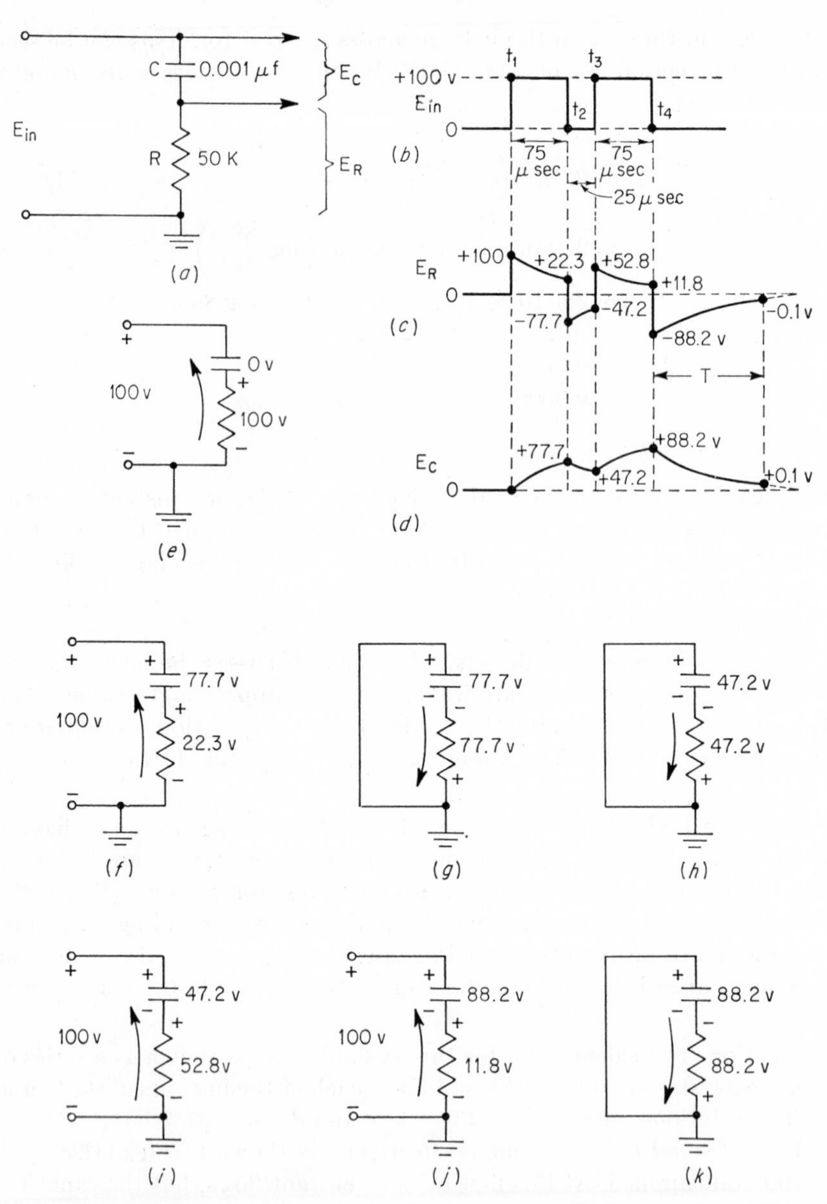

Figure 11·2 Square waves (d-c) in *R-C* circuit, Example 11·3. (*a*) Circuit; (*b*) E_{in}; (*c*) E_R; (*d*) E_C; (*e*) circuit at t_1; (*f*) circuit just before t_2; (*g*) circuit at t_2; (*h*) circuit just before t_3; (*i*) circuit at t_3; (*j*) circuit just before t_4; (*k*) circuit at t_4.

Voltage across the 50-kilohm resistor has therefore decreased. From Kirchhoff's voltage law $E_R + E_C$ must equal the applied 100 volts. E_R has started off at the beginning of this period with a voltage of 100. This is E_1 in Eq. (11·13). At the end of this period, just before t_2, E_R has decreased to a new value, E_2 in the equation. Solving for E_2 gives

$$E_2 = \frac{E_1}{\log^{-1} \dfrac{T}{2.3RC}} \tag{11·13}$$

$$= \frac{100}{\log^{-1} \left(\dfrac{75\ \mu\text{sec}}{2.3(50\ \text{kilohms})\ 0.001\ \mu\text{f}} \right)}$$

$$= \frac{100}{\log^{-1} \left(\dfrac{75 \times 10^{-6}}{2.3(50 \times 10^3)\ 0.001 \times 10^{-6}} \right)}$$

$$= \frac{100}{\log^{-1} \left(\dfrac{75 \times 10^{-6}}{115 \times 10^{-6}} \right)}$$

$$= \frac{100}{\log^{-1} 0.652}$$

$$= \frac{100}{4.49}$$

$$= 22.3\ \text{volts}$$

As shown in Fig. 11·2*f*, with 100 volts applied and with 22.3 volts across the resistor, the difference, from Kirchhoff's law, is across the capacitor. E_C, as shown, is 77.7 volts (100 − 22.3). E_R and E_C, just before t_2, are shown at this instant in Fig. 11·2*c* and *d*.

3. *At instant* t_2, the applied E_{in} signal becomes zero, and the circuit is now effectively that shown in Fig. 11·2*g* where the capacitor has not had time yet to discharge. E_C is still 77.7 volts and acts as if it were the applied voltage. Discharge current flows *down* through the resistor, and E_R at this instant is −77.7 volts. E_R and E_C values are indicated at instant t_2 in Fig. 11·2*c* and *d*.

4. *Between instants* t_2 *and* t_3*, just before* t_3, the circuit is as shown in Fig. 11·2*h*. The capacitor continues to discharge for this 25-μsec period. E_C decreases, acting as the applied voltage, and E_R, equal to E_C during the discharge period, also decreases. At the start of this period (instant t_2), E_R is −77.7 volts. This is E_1 in the equation. At the end of this 25-μsec period, just before t_3, the decreased value of E_R is E_2 in the following equation:

$$E_2 = \frac{E_1}{\log^{-1}\left(\dfrac{T}{2.3RC}\right)} \tag{11·13}$$

$$= \frac{-77.7}{\log^{-1}\left(\dfrac{25\ \mu\text{sec}}{2.3(50\text{ kilohms})\ 0.001\ \mu\text{f}}\right)}$$

$$= \frac{-77.7}{\log^{-1}\left(\dfrac{25 \times 10^{-6}}{2.3(50 \times 10^3)\ 0.001 \times 10^{-6}}\right)}$$

$$= \frac{-77.7}{\log^{-1}\left(\dfrac{25 \times 10^{-6}}{115 \times 10^{-6}}\right)}$$

$$= \frac{-77.7}{\log^{-1} .217}$$

$$= \frac{-77.7}{1.65}$$

$$= -47.2\text{ volts}$$

Since $E_R = E_C$ during this discharge period, then E_C is also equal to 47.2 volts, having discharged down to this value from its previous charge of 77.7 volts. Values of E_R and E_C just before instant t_3 are also indicated in Fig. 11·2*c* and *d*.

5. *At instant* t_3, the applied voltage, E_{in} (Fig. 11·2*b*), again becomes +100, and the circuit is effectively that shown in Fig. 11·2*i*. At this instant, the capacitor is still at 47.2 volts, not having been able to charge up any higher yet. The applied +100 volts and the voltage on the capacitor are bucking, leaving a net *equivalent* applied voltage of 52.8 volts (100 − 47.2). At this instant, t_3, when the voltage suddenly rises, this 52.8 volts appears immediately across the resistor. Note in Fig. 11·2*i*, from Kirchhoff's law, that $E_C + E_R = E_{\text{applied}}$ or $47.2 + 52.8 = 100$. The values of E_R and E_C, at instant t_3, are indicated in Fig. 11·2*c* and *d*.

6. *Between instants* t_3 *and* t_4, *just before* t_4, E_{in} (Fig. 11·2*b*) remains at a steady +100 volts, and the circuit is effectively that shown in Fig. 11·2*j*. During this 75-μsec period, t_3 to t_4, current flows, causing the capacitor to charge up to a higher voltage. As a result, E_R decreases. The starting voltage across the resistor at the beginning of this time period is the 52.8 volts shown in the previous diagram, Fig. 11·2*i*, and is E_1 in the equation. The voltage to which E_R decreases at the end of this period is

E_2 in the equation. Solving for E_2 gives

$$E_2 = \frac{E_1}{\log^{-1} \dfrac{T}{2.3RC}} \tag{11·13}$$

$$= \frac{52.8}{\log^{-1} \left(\dfrac{75\ \mu\text{sec}}{2.3(50\ \text{kilohms})\ 0.001\ \mu\text{f}} \right)}$$

$$= \frac{52.8}{\log^{-1} \left(\dfrac{75 \times 10^{-6}}{115 \times 10^{-6}} \right)}$$

$$= \frac{52.8}{\log^{-1} 0.652}$$

$$= \frac{52.8}{4.49}$$

$$= 11.8\ \text{volts}$$

From Kirchhoff's voltage law, with 100 volts applied (Fig. 11·2*j*) and $E_R = 11.8$ volts, then E_C is equal to the difference or 88.2 volts (100 − 11.8). These values for E_R and E_C are shown just before t_4 in Fig. 11·2*c* and *d*.

7. *At instant* t_4, E_{in} (Fig. 11·2*b*) has decreased to zero, and the circuit is effectively that shown in Fig. 11·2*k*. The capacitor voltage is still at 88.2 volts, not having had time to discharge at all. E_C acts as if it were the voltage applied. Discharge current now flows *down* through the resistor, and at this instant, t_4, E_R becomes −88.2 volts, as shown in Fig. 11·2*c*.

8. *After instant* t_4, E_{in} (Fig. 11·2*b*) remains at zero. The capacitor now starts discharging towards zero, and E_R, equal to E_C during the discharge period, likewise decreases toward zero. The time required for E_C and E_R to decrease to 0.1 volt (the last part of this example) is

$$T = 2.3RC \left(\log \frac{E_1}{E_2} \right) \tag{11·11}$$

$$= 2.3(50\ \text{kilohms})\ 0.001\ \mu\text{f} \left(\log \frac{-88.2}{-0.1} \right)$$

$$= 2.3(50 \times 10^3)\ 0.001 \times 10^{-6}\ (\log 882.)$$

$$= 115 \times 10^{-6}(2.946)$$

$$= 339. \times 10^{-6}\ \text{sec}$$

$$= 339.\ \mu\text{sec}$$

Note that the R-C time constant of the circuit of Fig. 11·2a is

$$\begin{aligned} R\text{-}C &= (50 \text{ kilohms})\ 0.001\ \mu\text{f} \\ &= 50 \times 10^3\ (0.001 \times 10^{-6}) \\ &= 0.05 \times 10^{-3} \\ &= 0.00005 \text{ sec} \\ &= 50\ \mu\text{sec} \end{aligned}$$

It takes about five or six R-C time constants for a capacitor to discharge practically completely. Therefore, the time required, T, of 339 μsec can be seen to be slightly more than six R-C, which would be $6 \times 50 = 300$ μsec.

In the previous Examples, 11·1 to 11·3, the applied voltages were d-c or fluctuating d-c which went from zero to some positive value. The next example and discussion become more involved since the input square wave is an a-c type, going positive and negative, as shown in the following. The capacitor voltage may first charge up with one polarity, and then, if given sufficient time, may discharge and charge up with the opposite polarity.

Example 11·4 An R-C circuit similar to that of Fig. 11·2a has a time constant of 200 μsec. The input square wave is that shown in Fig. 11·3a. Calculate the voltage across the resistor, E_R, and that across the capacitor, E_C, at each of the instants t_1, t_2, t_3, and t_4. Also calculate the length of time, after t_4, that is required for the resistor voltage to decrease to -1 volt.

Solution. The following series of steps, together with the diagrams of Fig. 11·3, are used to determine the values of E_R and E_C, shown in Fig. 11·3b and c.

1. *Before instant* t_1, E_{in} (Fig. 11·3a) is -100 volts for a long time, allowing the capacitor sufficient time to have charged fully to this voltage. E_R becomes, and remains, zero once the capacitor has fully charged. The values of E_R and E_C are shown, before t_1, in Fig. 11·3b and c.

2. *At instant* t_1, E_{in} (Fig. 11·3a) rises abruptly to $+200$ volts, and the circuit is effectively that shown in Fig. 11·3d. The 200 volts applied is in series aiding with the 100 volts on the capacitor, making an equivalent applied voltage of 300. Current flow is *up* through the resistor, and E_R becomes 300 volts. The sum of E_R ($+300$) and E_C (-100), from Kirchhoff's voltage law, is equal to the $+200$ volts applied. E_R and E_C at t_1 are shown in Fig. 11·3b and c.

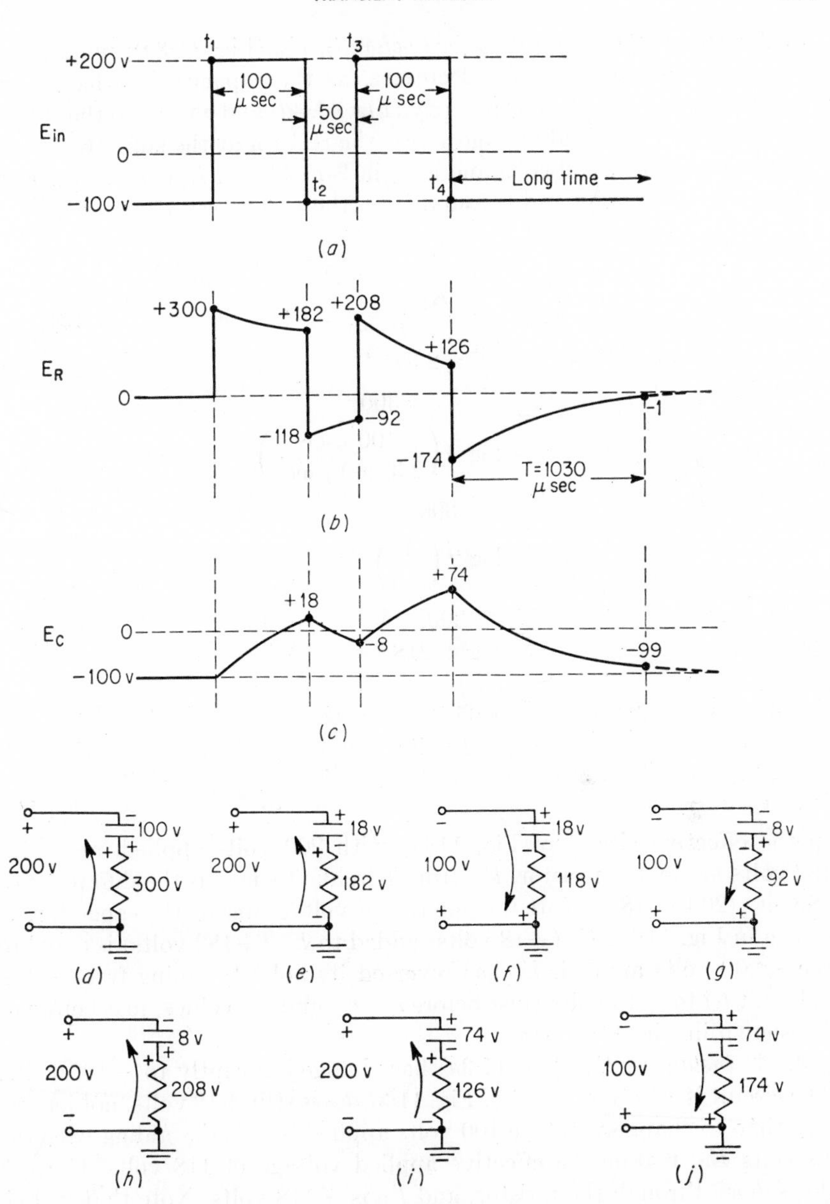

Figure 11·3 Square waves (a-c) in *R-C* circuit, Example 11·4. (*a*) E_{in}; (*b*) E_R; (*c*) E_C; (*d*) at t_1; (e) just before t_2; (*f*) at t_2; (*g*) just before t_3; (*h*) at t_3; (*i*) just before t_4; (*j*) at t_4.

3. *Between instants t_1 and t_2, just before t_2,* E_{in} (Fig. 11·3*a*), remains at +200 volts for 100 μsec. E_R decreases, as the capacitor discharges in trying to reverse its previous polarity (Fig. 11·3*d*) and charge to the +200 volts applied. The initial voltage across the resistor at the start (t_1) of the 100-μsec period is +300 volts and is E_1 in Eq. (11·13). E_2 is E_R at the end of this 100-μsec period and is found from the following:

$$E_2 = \frac{E_1}{\log^{-1} \dfrac{T}{2.3RC}} \tag{11·13}$$

$$= \frac{300}{\log^{-1}\left(\dfrac{100\ \mu\text{sec}}{2.3(200\ \mu\text{sec})}\right)}$$

$$= \frac{300}{\log^{-1}\left(\dfrac{1}{4.6}\right)}$$

$$= \frac{300}{\log^{-1} .218}$$

$$= \frac{300}{1.65}$$

$$= 182 \text{ volts}$$

In the effective circuit of Fig. 11·3*e*, with 200 volts applied, and with 182 volts across the resistor, E_C, from Kirchhoff's law, is the difference or 18 volts (200 − 182). Note the polarity of voltage across the capacitor, as shown in Fig. 11·3*e*. E_C (+18 volts) added to E_R (+182 volts) is equal to the +200 volts applied. E_C has reversed its polarity, going from −100 volts (at t_1) to +18 volts (just before t_2). E_R and E_C values, just before t_2, are shown in Fig. 11·3*b* and *c*.

4. *At instant t_2,* E_{in} (Fig. 11·3*a*) has changed abruptly to −100 volts. The circuit is effectively that of Fig. 11·3*f*. E_C is still +18 volts, not having had time to change yet. The 100 volts applied is in series aiding with the 18 volts E_C, making an effective applied voltage of 118 volts. Current flows *down* through the resistor, and E_R is −118 volts. Note that in Fig. 11·3*f*, the sum of E_R (−118) and E_C (+18) is equal to the −100 volts applied. The values of E_R and E_C, at instant t_2, are indicated in Fig. 11·3*b* and *c*.

5. *Between instants t_2 and t_3, just before t_3,* E_{in} (Fig. 11·3*a*) remains at −100 volts for 50 μsec. E_R at the start (t_2) of this period of time is −118 volts and is E_1 in Eq. (11·13). E_R decreases to a value, E_2, at the end (just before t_3) of this period. E_2 is

$$E_2 = \frac{E_1}{\log^{-1}\dfrac{T}{2.3RC}} \tag{11·13}$$

$$= \frac{-118}{\log^{-1}\left(\dfrac{50\ \mu\text{sec}}{2.3(200\ \mu\text{sec})}\right)}$$

$$= \frac{-118}{\log^{-1}\left(\dfrac{1}{9.2}\right)}$$

$$= \frac{-118}{\log^{-1} 0.109}$$

$$= \frac{-118}{1.285}$$

$$= -92.0 \text{ volts}$$

As shown in the effective diagram of Fig. 11·3*g*, with −100 volts applied and with $E_R = -92$ volts, E_C, from Kirchhoff's law, must be −8 volts. Values for E_R (−92) and E_C (−8) just before instant t_3 are shown in Fig. 11·3*b* and *c*.

6. *At instant t_3,* E_{in} (Fig. 11·3*a*) rises to +200 volts. The capacitor voltage has not had any time to change yet, and E_C is still −8 volts. The circuit is effectively that shown in Fig. 11·3*h*. The 200 volts applied is aiding the 8 volts across the capacitor, making it appear to be 208 volts applied. Current flows *up* through the resistor, and E_R therefore becomes +208 volts. Note that, in Fig. 11·3*h*, the sum of E_R (+208) and E_C (−8) equals the +200 volts applied, as required by Kirchhoff's law. These values of E_R and E_C at t_3 are indicated in Fig. 11·3*b* and *c*.

7. *Between instants t_3 and t_4, just before t_4,* E_{in} remains at a steady value of +200 volts. During this 100-μsec period, the capacitor, with a −8 volts charge at the start, will attempt to lose this charge and charge up to the reverse polarity of the +200 volts applied. E_R has a starting value at t_3 of +208 volts, and this is E_1 in Eq. (11·13). At the end of the 100-μsec period (just before t_4) E_R slumps to a lesser value, E_2 in the equation.

Solving for E_2:

$$E_2 = \frac{E_1}{\log^{-1} \dfrac{T}{2.3RC}} \tag{11·13}$$

$$= \frac{208}{\log^{-1}\left(\dfrac{100\ \mu\text{sec}}{2.3(200\ \mu\text{sec})}\right)}$$

$$= \frac{208}{\log^{-1}\left(\dfrac{1}{4.6}\right)}$$

$$= \frac{208}{\log^{-1} 0.218}$$

$$= \frac{208}{1.65}$$

$$= 126 \text{ volts}$$

As shown in the effective circuit of Fig. 11·3*i*, with +200 volts applied and E_R equal to +126 volts, then E_C, from Kirchhoff's law, must be the difference or 74 volts (200 − 126). Values of E_R and E_C just before t_4 are shown in Fig. 11·3*b* and *c*.

8. *At instant* t_4, E_{in} (Fig. 11·3*a*) changes abruptly from +200 volts to −100 volts. E_C has not had time to change yet and is still +74 volts. The circuit is effectively that of Fig. 11·3*j*, with an applied voltage of −100 volts, which is in series aiding with the 74 volts E_C, making it appear as if there were 174 volts applied. Current, as indicated, flows *down* through the resistor, and E_R becomes −174 volts. Note that in Fig. 11·3*j*, from Kirchhoff's voltage law, the sum of E_R (−174) and E_C (+74) is equal to the applied voltage (−100). Values of E_R and E_C at t_4 are shown in Fig. 11·3*b* and *c*.

9. *After instant* t_4, E_{in} (Fig. 11·3*a*) remains at −100 volts for a long time, giving the capacitor sufficient time to change from its previous value (at t_4) of +74 volts to practically the full −100 volts applied. E_R, with the capacitor fully charged to the applied voltage, then decreases to zero. The time, after t_4, that it takes for E_R to decrease from −174 volts to −1 volt is

$$T = 2.3RC\left(\log \frac{E_1}{E_2}\right) \tag{11·11}$$

$$= 2.3(200\ \mu\text{sec})\left(\log \frac{-174}{-1}\right)$$

$$= 2.3(200 \times 10^{-6}) \log 174.$$

$$= 460. \times 10^{-6}(2.24)$$

$$= 1030 \times 10^{-6} \text{ sec}$$

$$= 1030\ \mu\text{sec}$$

It requires approximately five or six *R-C* time constants for a capacitor to charge to practically the full applied voltage. During this time, E_R will change from some value of voltage to just about zero. In this example, the *R-C* is 200 μsec, and five *R-C*s is 1,000 μsec, which is quite close to the 1,030-μsec answer.

11·3 *L-R* Circuit

The discussion of *L-R* circuits given in Chap. 3 used the universal time-constant chart of Fig. 3·1. Referring back to this diagram, it can be seen that the graph for E_C (in the *R-C* circuit) is the I and E_R in the *L-R* circuit, and the E_R graph (in the *R-C* circuit) is the E_L graph in an *L-R* circuit. This can also be seen in Fig. 3·3 of that earlier chapter and Fig. 11·4*b*. Also, as has been shown earlier, the time constant in the *L-R* circuit consists of the inductance, L, in henrys, divided by the resistor, R, in ohms, or

$$\text{Time constant} = \frac{L}{R} \tag{11·14}$$

In one L/R time constant, the *current* rises to 63 per cent of its maximum value, as does E_R. Referring to the basic equation for the *R-C* circuit,

$$\log \frac{E_1}{E_2} = \frac{T}{2.3RC} \tag{11·10}$$

Replace the RC term in Eq. (11·10) by L/R (the time constant of an *L-R* circuit), giving the following basic equation for an *L-R* circuit:

$$\log \frac{E_1}{E_2} = \frac{T}{2.3\,\dfrac{L}{R}} \tag{11·15}$$

where E_1 is now E_L at the *beginning* of a time period, and E_2 is E_L at the *end* of that time period. Solving for T, the time period, in Eq. (11·15) gives

$$T = 2.3\,\frac{L}{R}\log \frac{E_1}{E_2} \tag{11·16}$$

Solving for E_2 (the voltage across the coil at the end of a time period) gives the following equation:

$$E_2 = \frac{E_1}{\log^{-1}\left(\dfrac{T}{2.3\,\dfrac{L}{R}}\right)} \tag{11·17}$$

The equations that will be used in the *L-R* example of this chapter are Eqs. (11·16) and (11·17). Note that they are practically identical to Eqs.

(11·11) and (11·13) except that the term L/R replaces RC, and E_1 and E_2 are now E_L voltages instead of E_R.

Example 11·5 As shown in Fig. 11·4, 100 volts d-c is applied to a 1-henry coil in series with a 10-kilohm resistor. (*a*) Find the time constant of the circuit. (*b*) Also, if the voltage is only applied for 200 μsec, find E_L and E_R at that instant. (*c*) Finally, if the d-c voltage source is now replaced by a short circuit, find the time required for E_L to decrease to −0.1 volts.

Solution. (*a*) The time constant is

$$\begin{aligned}\text{Time constant} &= \frac{L}{R} \qquad (11\cdot14)\\ &= \frac{1\text{ henry}}{10\text{ kilohms}}\\ &= \frac{1}{10 \times 10^3}\\ &= 0.1 \times 10^{-3}\text{ sec}\\ &= 0.0001\text{ sec}\\ &= 100\ \mu\text{sec}\end{aligned}$$

(*b*) As shown in Fig. 11·4*b*, E_L is maximum, being equal to the applied voltage, at the first instant. Therefore, $E_L = 100$ volts, and this is E_1 in Eq. (11·17). E_L decreases to a smaller value, E_2 in the equation, after a period of time, 200 μsec in this example. Solving for E_2 yields

$$\begin{aligned}E_2 &= \frac{E_1}{\log^{-1}\left(\dfrac{T}{2.3\,\dfrac{L}{R}}\right)} \qquad (11\cdot17)\\ &= \frac{100}{\log^{-1}\left(\dfrac{200\ \mu\text{sec}}{2.3(100\ \mu\text{sec})}\right)}\\ &= \frac{100}{\log^{-1} 0.869}\\ &= \frac{100}{7.4}\\ &= 13.5\text{ volts}\end{aligned}$$

With 100 volts applied, and with $E_L = 13.5$ volts, then from Kirchhoff's voltage law, E_R must be equal to the difference or 86.5 volts (100 − 13.5).

The current, incidentally, is zero at the instant when the voltage was first applied, producing zero volts across the resistor. Gradually, because of the opposition of the coil, current started to increase. Current rises to 63 per cent of its maximum, final value in the first *L-R* time constant, and, like E_C in the *R-C* circuit, repeats the process in each succeeding time constant, reaching practically its maximum value in about five or six time

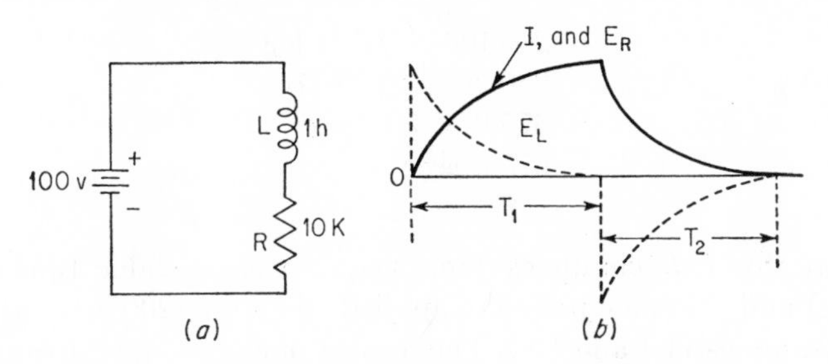

Figure 11·4 D-C applied to *L-R* circuit, Example 11·5.

constants. This is the period T_1 in Fig. 11·4*b*. The *maximum* value of current, from Ohm's law, is

$$I = \frac{E}{R} = \frac{100}{10 \text{ kilohms}} = \frac{100}{10{,}000} = 0.01 \text{ amp}$$

The 200-μsec period in which the voltage is applied is just twice the *L-R* time constant of 100 μsec. The current at this instant (the end of the 200-μsec time) is

$$I = \frac{E}{R} = \frac{86.5}{10{,}000} = 0.00865 \text{ amp}$$

where $E = E_R$

(*c*) When the d-c voltage source is removed and replaced by a short circuit, the current (0.00865 amp) tries to fall to zero. The magnetic field around the coil now starts collapsing, inducing a voltage in the coil. During the time T_2 in Fig. 11·4*b*, as I starts decreasing, E_R likewise decreases, and E_L reverses its polarity and becomes equal to E_R. The time that E_L (and E_R also) requires to decrease to -0.1 volt is

$$
\begin{aligned}
T &= 2.3 \frac{L}{R} \log \frac{E_1}{E_2} \qquad (11\cdot16)\\
&= 2.3(100\ \mu\text{sec}) \log \frac{-86.5}{-0.1}\\
&= 2.3(100 \times 10^{-6}) \log 865.\\
&= 230 \times 10^{-6}(2.937)\\
&= 675.0 \times 10^{-6}\ \text{sec}\\
&= 675.0\ \mu\text{sec}
\end{aligned}
$$

Note that this *L-R* circuit example uses the same value time constant (100 μsec) and the same period of applied 100 volts (200 μsec) as the *R-C* circuit Examples 11·1 and 11·2. The results, of course, are the same, except that the E_R and E_C values of Example 11·1 are now respectively the E_L and E_R values in Example 11·5.

To further show the similarity of the *L-R* circuit to the *R-C* circuit, the following example uses an *L-R* time constant and input square waves, which are identical to that of the *R-C* circuit in Fig. 11·2 of Example 11·3.

Example 11·6 A 5-henry coil in series with a 100-kilohm resistor, as shown in Fig. 11·5*a*, has an input signal applied, as shown in Fig. 11·5*b*. Find the time constant, and calculate the values of the voltages E_L and E_R at each of the indicated instants, t_1, t_2, t_3, and t_4. Also determine how much time is required, after the input has become zero at t_4, for E_L to decrease almost to zero, to 0.1 volt.

Solution. The voltages E_L and E_R are shown at the various instants in Fig. 11·5*c* and *d*. The calculations and discussions follow in the step-by-step procedures.

1. The time constant of the 5-henry coil and the 100 kilohm resistor is

$$
\begin{aligned}
\text{Time constant} &= \frac{L}{R} \qquad (11\cdot14)\\
&= \frac{5}{100\ \text{kilohm}}\\
&= \frac{5}{100 \times 10^{3}}\\
&= 0.05 \times 10^{-3}\ \text{sec}\\
&= 0.00005\ \text{sec}\\
&= 50\ \mu\text{sec}
\end{aligned}
$$

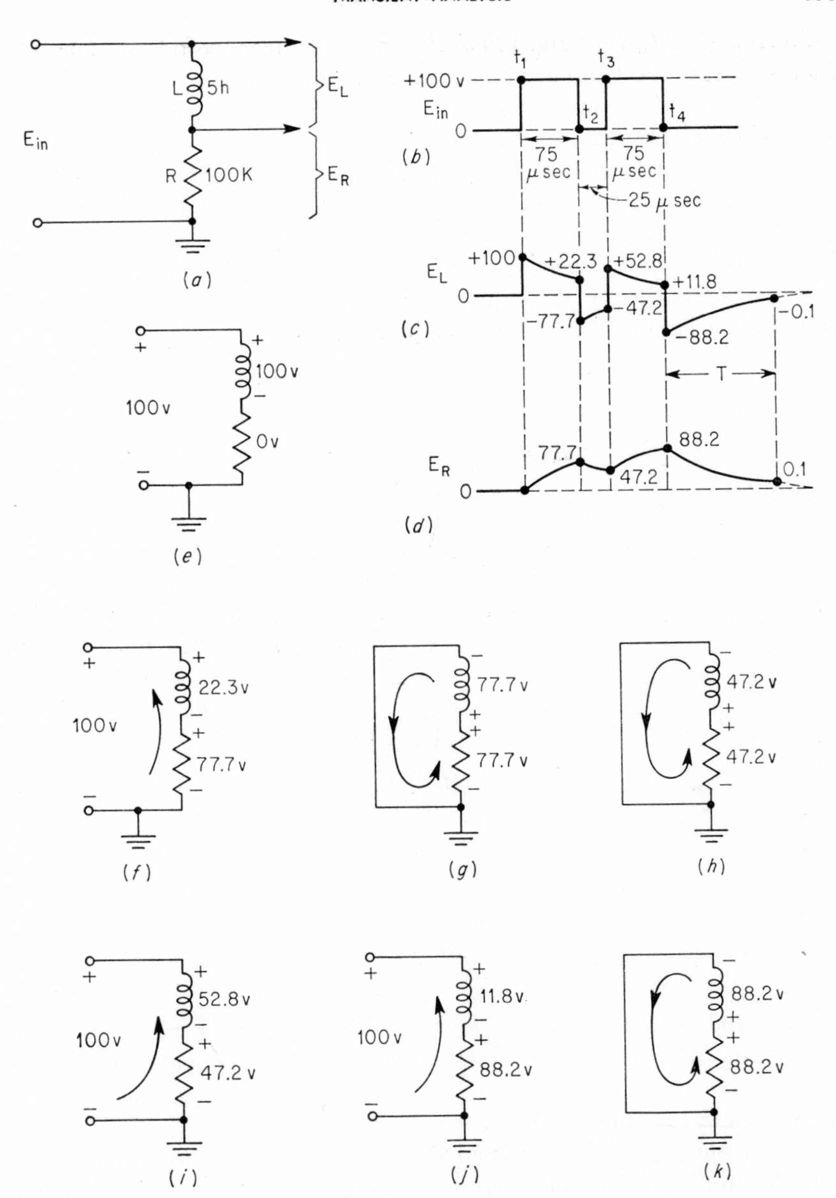

***Figure* 11·5** **Square waves in *L-R* circuit, Example 11·6. (*a*) Circuit; (*b*) E_{in}; (*c*) E_L; (*d*) E_R; (*e*) at t_1; (*f*) just before t_2; (*g*) at t_2; (*h*) just before t_3; (*i*) at t_3; (*j*) just before t_4; (*k*) at t_4.**

2. *At instant* t_1, E_{in} (Fig. 11·5*b*) rises abruptly from zero to +100 volts. At this first instant, no current flows yet, and E_R is zero volts while the full 100 volts of E_{in} appears across the coil. The effective circuit is that shown in Fig. 11·5*e*. E_L and E_R are shown in Fig. 11·5*c* and *d*.

3. *Between instants* t_1 *and* t_2, *just before* t_2, E_{in} (Fig. 11·5*b*) remains at +100 volts during this 75-μsec period of time. Current flow is *up* through the resistor, as shown in Fig. 11·5*f*, and current increases. E_R rises, and E_L decreases. The value of E_L at the start (t_1) of this period is +100 volts, and this is E_1 in Eq. (11·17). The decreased value of E_L at the end of the period (just before t_2) is E_2 in the equation. Solving for E_2 gives

$$E_2 = \frac{E_1}{\log^{-1}\left(\dfrac{T}{2.3\dfrac{L}{R}}\right)} \qquad (11\cdot17)$$

$$= \frac{100}{\log^{-1}\left(\dfrac{75\ \mu\text{sec}}{2.3(50\ \mu\text{sec})}\right)}$$

$$= \frac{100}{\log^{-1}\left(\dfrac{75}{115}\right)}$$

$$= \frac{100}{\log^{-1} 0.652}$$

$$= \frac{100}{4.49}$$

$$= 22.3 \text{ volts}$$

With 100 volts applied, as shown in Fig. 11·5*f*, and with $E_L = 22.3$ volts, then from Kirchhoff's voltage law, E_R must be equal to the difference or 77.7 volts (100 − 22.3). The values of E_L and E_R, just before t_2, are shown in Fig. 11·5*c* and *d*.

4. *At instant* t_2, E_{in} has dropped abruptly to zero. The circuit effectively is shown in Fig. 11·5*g*. The current tries to stop, but cannot do so immediately, and E_R, therefore, remains at the 77.7 volts. The collapsing magnetic field induces voltage in the coil which keeps the current flowing *up* through the resistor. Note that the voltage polarity across L (Fig. 11·5*g*) is opposite to that shown previously (Fig. 11·5*f*). At this instant, E_L is equal to E_R. The magnitude of each is 77.7 volts, as indicated in Fig. 11·5*c* and *d*.

5. *Between instants t_2 and t_3, just before t_3,* E_{in} (Fig. 11·5*b*) remains at zero for 25 μsec. The magnetic field which started to collapse at instant t_2 is still decreasing, and current is still flowing *up* through the resistor, but it is decreasing. The initial value of E_L at the start (t_2) of this 25-μsec period is -77.7 volts, and this is E_1 in Eq. (11·17). The value of E_L at the end of this period (just before t_3) is E_2 in the equation. Solving for E_2 gives

$$E_2 = \frac{E_1}{\log^{-1}\left(\dfrac{T}{2.3\dfrac{L}{R}}\right)} \qquad (11\cdot17)$$

$$= \frac{-77.7}{\log^{-1}\left(\dfrac{25\ \mu\text{sec}}{2.3(50\ \mu\text{sec})}\right)}$$

$$= \frac{-77.7}{\log^{-1}\left(\dfrac{1}{4.6}\right)}$$

$$= \frac{-77.7}{\log^{-1} .218}$$

$$= \frac{-77.7}{1.65}$$

$$= -47.2 \text{ volts}$$

During this time, $E_R = E_L$, as shown in the circuit of Fig. 11·5*h* and also in Fig. 11·5*c* and *d*.

6. *At instant t_3,* E_{in} (Fig. 11·5*b*) rises abruptly to $+100$ volts. The current cannot change yet, and E_R is still at its previous value of 47.2 volts. As shown in the equivalent diagram of Fig. 11·5*i*, with 100 volts applied and with $E_R = 47.2$ volts, then from Kirchhoff's voltage law, E_L must be equal to the difference, or 52.8 volts ($100 - 47.2$). Note that the polarity of voltage across the coil has again reversed, comparing Fig. 11·5*i* with the previous one of Fig. 11·5*h*, and current still flows *up* through the circuit. The values of E_L and E_R at t_3 are also shown in Fig. 11·5*c* and *d*.

7. *Between instants t_3 and t_4, just before t_4,* E_{in} (Fig. 11·5*b*) remains at $+100$ volts for 75 μsec. Current increases, and E_R therefore increases, while E_L decreases. The value of E_L at the start (t_3) of this 75-μsec period is 52.8 volts and is E_1 in Eq. (11·17). The decreased value of E_L at the end of the period (just before t_4) is E_2 in the equation, and E_2 is

$$E_2 = \frac{E_1}{\log^{-1}\left(\dfrac{T}{2.3\dfrac{L}{R}}\right)} \tag{11·17}$$

$$= \frac{52.8}{\log^{-1}\left(\dfrac{75\ \mu\text{sec}}{2.3(50\ \mu\text{sec})}\right)}$$

$$= \frac{52.8}{\log^{-1}\left(\dfrac{75}{115}\right)}$$

$$= \frac{52.8}{\log^{-1} 0.652}$$

$$= \frac{52.8}{4.49}$$

$$= 11.8 \text{ volts}$$

As shown in Fig. 11·5*j*, with 100 volts applied and with $E_L = 11.8$ volts, then from Kirchhoff's law, E_R must be equal to the difference, or 88.2 volts (100 − 11.8). Values of E_L and E_R, just before t_4, are shown in Fig. 11·5*c* and *d*.

8. *At instant* t_4, E_{in} drops sharply to zero. Since the current cannot change immediately, E_R remains at its previous value of 88.2 volts. The collapsing magnetic field induces a voltage in the coil which keeps the current flowing *up* through the circuit components. The polarity of this E_L is again reversed, as shown in the effective circuit of Fig. 11·5*k*, as compared to the previous one of Fig. 11·5*j*. As usual at this time, $E_L = E_R$ but is of opposite polarity. The values of E_L (−88.2 volts) and E_R (+88.2 volts) are shown in Fig. 11·5*c* and *d*.

9. *After instant* t_4, E_{in} (Fig. 11·5*b*) remains at zero. The magnetic field collapses completely after a time period, T, and the current decreases to zero. Both E_R and E_L likewise decrease to zero. The time, T, after instant t_4, that E_L requires to decrease from its value of −88.2 volts at t_4 [E_1 in Eq. (11·16)] to a value of −0.1 volt (E_2 in the equation) is

$$T = 2.3\frac{L}{R}\log\frac{E_1}{E_2} \tag{11·16}$$

$$= 2.3(50\ \mu\text{sec})\log\frac{-88.2}{-0.1}$$

$$= 115 \times 10^{-6} \log 882.$$

$$= 115 \times 10^{-6}(2.946)$$

$$= 339. \times 10^{-6} \text{ sec}$$

$$= 339.\ \mu\text{sec}$$

Note that all the answers in this Example 11·6 are exactly the same as that for the equal time constant *R-C* circuit of Example 11·3, Fig. 11·2.

11·4 Multivibrator *R-C* Circuits

Multivibrators are discussed in Chap. 5 using the charging and discharging of capacitors through resistors, with their resultant voltage waveshapes, to explain their intricate operation. In the present chapter, the amplitudes of these voltage waveshapes are calculated using the equations employed for *R-C* circuits in Sec. 11·2. The reader should be familiar first with the basic theory of multivibrators (Chap. 5), also with the *R-C* circuit transients (Sec. 11·2), and also with the discussion of Thévenin equivalent circuits (Chap. 1) before pursuing this present discussion.

The following example refers to a plate-coupled astable multivibrator employing two *R-C* circuits which operate simultaneously. That is, the capacitor in one *R-C* is charging, while the other is discharging. The circuit of Fig. 11·7*a* shows a plate-coupled astable multivibrator. The basic operation of this free-running oscillator consists of each tube (V_1 and V_2) conducting alternately. As one stage goes into conduction, its plate voltage falls, causing its coupling capacitor to start discharging down through the resistor which is connected to the other grid. This drives that grid negative, cutting off that stage. Cutoff time of a stage depends primarily on the size of the *R-C* circuit. When the cutoff stage begins conduction, it cuts off the other stage. A detailed explanation is given in Chap. 5.

In the solution of the following multivibrator example, approximations are used in order to show a simple but acceptable method of circuit analysis. Certain factors are purposely omitted so that the solution remains uncluttered. The internal tube resistance R_B is assumed, for simplification, to be a constant value when actually it is a variable, decreasing as the positive grid draws more current. Similarly, the grid-to-cathode resistance R_{GK} is assumed to be a constant, whereas it too fluctuates. The Miller effect, which could alter the equivalent circuit capacitance, is also disregarded. Another factor which is purposely omitted in the attempt for simplicity is the fact that the switching from one triode to the other in the multi does not occur at the precise grid cutoff voltage but at less negative values. As a result of the approximations, the derived voltage waveshapes do not show the true amplitudes of the undershoots at the plates nor those of the overshoots at the grids. Despite these shortcomings, the following example and solution should help the reader gain more insight of the multivibrator.

Example 11·7 In the plate-coupled astable multivibrator of Fig. 11·7*a*, using a 12AU7 duo-triode, calculate the amplitudes of the grid- and plate-voltage waveshapes of both V_1 and V_2, and determine the frequency of this oscillator.

Solution. Before the circuit can be analyzed, the internal resistance, R_B, of each tube must be known. These are derived from the load lines which are shown drawn on the tube characteristic curves of Fig. 11·6.

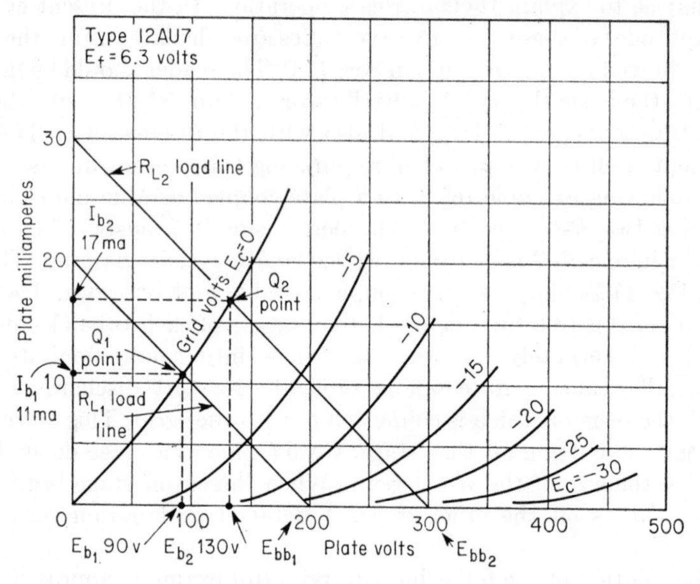

Figure 11·6 Characteristic curves and load lines, Example 11·7.

The load line for V_1, using E_{BB} (B+ supply) of +200 volts and a load resistor (R_{L_1}) of 10 kilohms, is shown connecting the +200 volts, E_{BB_1}, zero-ma point, and the 0 volts, 20 ma (I_B) point. The 20-ma value is from Ohm's law.

$$I_B = \frac{E_{BB_1}}{R_{L_1}} = \frac{200}{10 \text{ kilohms}} = 20 \text{ ma}$$

The intersection of this load line and the zero volts grid bias ($E_C = 0$) is called the Q point, in this example Q_1. I_{B_1} and E_{B_1}, from the Q_1 point of Fig. 11·6, are 11 ma and 90 volts respectively. The internal resistance of V_1, R_{B_1}, is then

$$R_{B_1} = \frac{E_{B_1}}{I_{B_1}}$$

$$= \frac{90}{11 \text{ ma}}$$

$$= \frac{90}{0.011}$$

$$= 8.2 \text{ kilohms}$$

Similarly, the load line for V_2, R_{L_2}, is drawn between that E_{BB} point (+300 volts) and an I_B value of 30 ma. The 30 ma is from

$$I_B = \frac{E_{BB_2}}{R_{L_2}} = \frac{300}{10 \text{ kilohms}} = 30 \text{ ma}$$

The Q_2 point, as shown in Fig. 11·6, gives a value I_{B_2} of 17 ma, and E_{B_2} of 130 volts. R_{B_2} is therefore

$$\begin{aligned} R_{B_2} &= \frac{E_{B_2}}{I_{B_2}} \\ &= \frac{130}{17 \text{ ma}} \\ &= \frac{130}{0.017} \\ &= 7.65 \text{ kilohms} \end{aligned}$$

Also available from the load lines and the characteristic curves of Fig. 11·6 are the grid cutoff voltages. The R_{L_1} load line intersects the zero plate current at somewhere between the -15 and the -20 E_C curves, with -17 volts a good estimate. Therefore, grid voltage cutoff for V_1, e_{Cco}, is -17 volts. Similarly, from the R_{L_2} load line, e_{Cco} for V_2 is -25 volts. Another tube characteristic that is required in the circuit analysis is the grid-to-cathode internal resistance (R_{G-K}) of each tube when the grid is *positive* with respect to the cathode. Positive-grid tube characteristic curves are available from the manufacturers but are not shown here. The resistance values, R_{G-K}, for a 12AU7 vary from a few hundred ohms to more than 1 kilohm, depending upon the plate and grid voltages. In this example, a value of 1 kilohm will be assumed for the R_{G-K} of each stage, causing no appreciable difference in the final results. The step-by-step discussion now follows.

1. The voltages at the plate of V_1, and at the grid of V_2, are found first by utilizing the shorter R-C circuit (0.0001 μf and 200 kilohms) before the larger one (0.002 μf and 500 kilohms). The circuit containing the shorter R-C is redrawn as the Thévenin circuit of Fig. 11·7*b*. P_1 is the plate of V_1, while G_2 is the grid of V_2. V_1 is replaced by its internal resistance, R_{B_1}, and the switch, S_1. The switch open denotes that V_1 is not conducting. At the same time, another switch, S_2, is shown closed, denoting that the grid of V_2 is positive with respect to its cathode and therefore drawing current. The 1-kilohm resistor R_{G-K_2} is the internal resistance existing between the grid and cathode of V_2 when grid current flows. The Thévenin equivalent of Fig. 11·7*b* is shown in *c*, where the 8.2-kilohm resistor, R_{B_1}, is omitted since switch S_1 is open. The 1-kilohm resistor, R_{G-K_2}, is in parallel with the 200-kilohm resistor, R_{G_2}, and the equivalent is just about 1 kilohm when G_2 is positive.

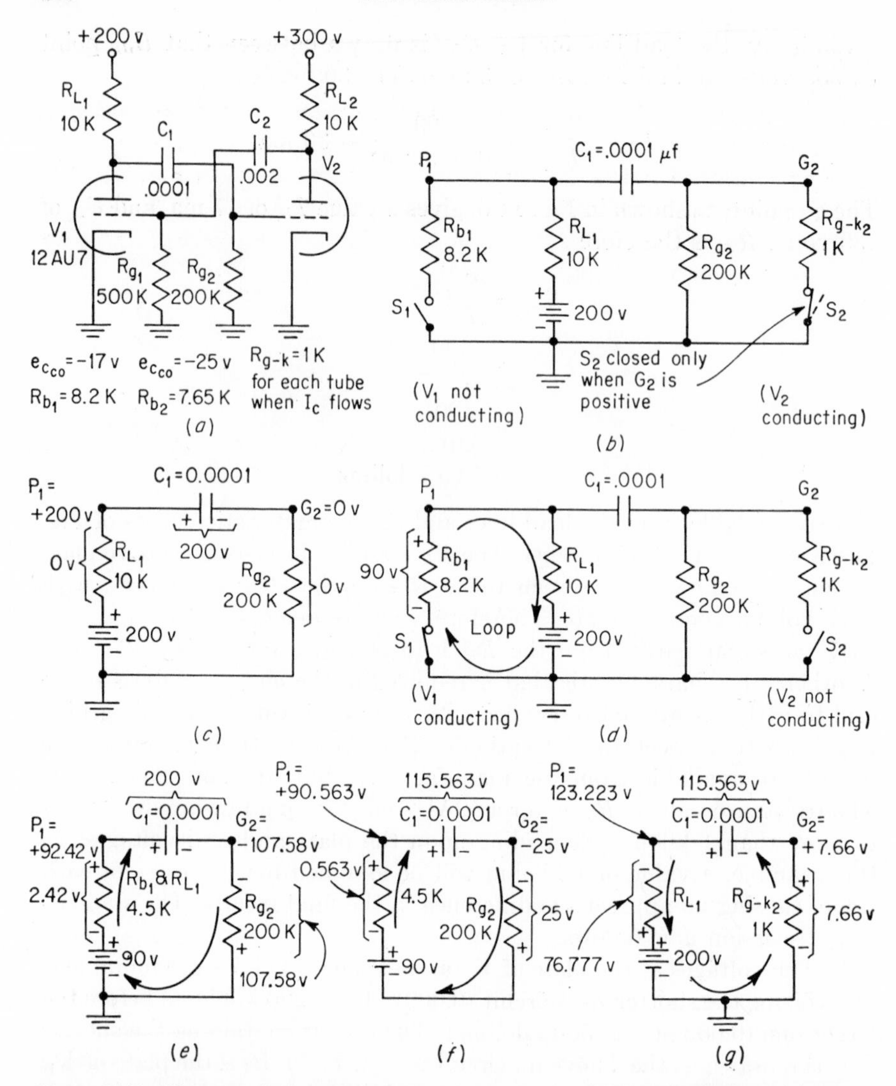

Figure 11·7 **Plate-coupled astable multivibrator, shorter *R-C*, Example 11·7. (*a*) Plate-coupled astable multivibrator; (*b*) circuit of short *R-C* of diagram (*a*), V_2 conducting; (*c*) Thévenin equivalent of (*b*) just before t_1; (*d*) circuit of short *R-C* of diagram (*a*), V_1 conducting; (e) Thévenin equivalent of (*d*) at t_1; (*f*) Thévenin equivalent just before t_2; (*g*) Thévenin equivalent at t_2.**

2. Just *before instant* t_1, in Fig. 11·8, it is assumed that V_1 is inoperative, while V_2 is conducting. This condition is as shown in Fig. 11·7*b* and *c*. Capacitor C_1 (0.0001 μf) in Fig. 11·7*c* will become fully charged to the +200 volts applied, and there is now zero volts across R_{L_1} (10 kilohms) and across the 200-kilohm grid resistor. Point P_1 (plate of V_1) is now +200 volts, while point G_2 (grid of V_2) is zero volts. These voltages, before t_1, are shown in the waveshapes of Fig. 11·8*a* and *b*.

3. *At instant* t_1, V_1 now starts conducting, and switch S_1 closes, as shown in Fig. 11·7*d*. At the same time, the grid of V_2 goes negative, and it no longer draws current. Switch S_2 is now shown open in Fig. 11·7*d* to signify this condition. The Thévenin equivalent of Fig. 11·7*d* is shown in *e*. The resistors R_{B_1} (8.2 kilohms) and R_{L_1} (10 kilohms) of Fig. 11·7*d* are considered to be in parallel, giving an equivalent of

$$\begin{aligned} R_{\text{equiv}} &= \frac{R_{B_1}R_{L_1}}{R_{B_1} + R_{L_1}} \\ &= \frac{(8.2 \text{ kilohms})(10 \text{ kilohms})}{8.2 \text{ kilohms} + 10 \text{ kilohms}} \\ &= \frac{82 \times 10^6}{18.2 \times 10^3} \\ &= 4.5 \times 10^3 \\ &= 4.5 \text{ kilohms} \end{aligned}$$

This 4.5-kilohm resistor is shown in the circuit of Fig. 11·7*e*. In Fig. 11·7*d*, the voltage across the 8.2-kilohm resistor, R_{B_1}, is found by considering the 8.2 kilohms to be in *series* with the 10 kilohms and the applied 200 volts and is

$$\begin{aligned} E_{R_{B_1}} &= \frac{R_B}{R_{B_1} + R_{L_1}} E_{\text{applied}} \\ &= \left(\frac{8.2 \text{ kilohms}}{8.2 \text{ kilohms} + 10 \text{ kilohms}}\right) 200 \\ &= \left(\frac{8.2 \text{ kilohms}}{18.2 \text{ kilohms}}\right) 200 \\ &= 90 \text{ volts} \end{aligned}$$

This 90 volts is the Thévenin equivalent voltage of the loop of Fig. 11·7*d* and is shown in Fig. 11·7*e*.

At instant t_1, the *net* applied voltage in Fig. 11·7*e* consists of the voltage charge of 200 volts on C_1, opposing the 90 volts (Thévenin equivalent) and is therefore 110 volts (200 − 90). Voltage across the 4.5-kilohm resistor may be found by considering this resistor to be in series with the 200-kilohm resistor, R_{G_2}, and the net applied 110 volts and is

$$\begin{aligned} E_{4.5 \text{ kilohms}} &= \left(\frac{4.5 \text{ kilohms}}{4.5 \text{ kilohms} + 200 \text{ kilohms}}\right) E_{\text{net applied}} \\ &= \left(\frac{4.5 \text{ kilohms}}{204.5 \text{ kilohms}}\right) 110 \\ &= 2.42 \text{ volts} \end{aligned}$$

Voltage across R_{G_2} (200 kilohms) is the remainder, or 107.58 volts (110 − 2.42). Current flow, as indicated by the arrows in Fig. 11·7*e*, is *down* through R_{G_2} and *up* through the 4.5-kilohm resistor, producing the voltage polarities shown. Point G_2 is now −107.58 volts, while point P_1 is +92.42 volts (2.42 + 90). These voltages, at instant t_1, are shown in Fig. 11·8*a* and *b*. The grid of V_2, Fig. 11·8*b*, is now beyond cutoff, and V_2 is inoperative.

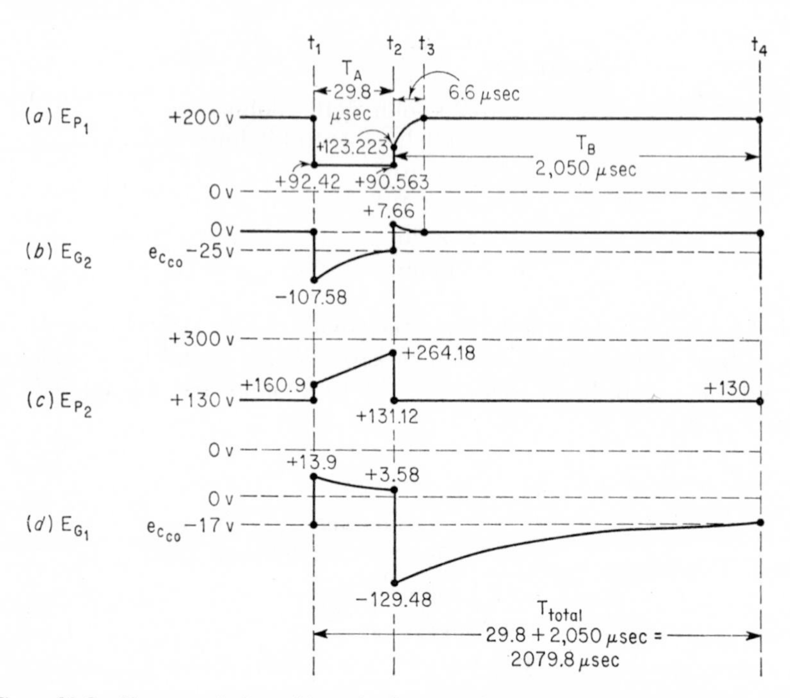

Figure 11·8 Plate-coupled astable multivibrator voltage waveshapes, Example 11·7.

4. *Between instants t_1 and t_2, just before t_2,* of Fig. 11·8, the capacitor of Fig. 11·7*e* is discharging. Voltage across R_{G_2} is decreasing, and point G_2 becomes less negative. Just before instant t_2, the grid of V_2 becomes −25 volts. This is the cutoff point of V_2, found previously from the load line. This stage is therefore cut off during the period between t_1 and t_2 when its grid goes from −107.58 volts to −25 volts. At the beginning (t_1) of this time period, the voltage across the resistors of Fig. 11·7*e* is 110 volts (2.42 + 107.58). This is E_1 in Eq. (11·11). At the end (just before t_2) of this time period, the voltage across the resistors of Fig. 11·7*f* is

$$E_{R_{G_2}} = 25 \text{ volts (given as cutoff of } V_2)$$

added to $E_{4.5\text{ kilohms}}$, which is

$$\frac{E_{4.5\text{ kilohms}}}{E_{R_{G_2}}} = \frac{4.5\text{ kilohms}}{200\text{ kilohms}}$$

$$\frac{E_{4.5\text{ kilohms}}}{25\text{ volts}} = \frac{4.5\text{ kilohms}}{200\text{ kilohms}}$$

$$E_{4.5\text{ kilohms}} = \left(\frac{4.5\text{ kilohms}}{200\text{ kilohms}}\right)(25)$$

$$= 0.563\text{ volt}$$

Therefore, the voltage across the resistors at the end of this period is 25.563 volts (25 + 0.563). This is E_2 in Eq. 11·11. To find the time from t_1 to just before t_2, called here T_A, solve the following equation for T.

$$T = 2.3RC\left(\log\frac{E_1}{E_2}\right) \qquad (11\cdot11)$$

$$= 2.3(204.5\text{ kilohms})\ 0.0001\text{ mf}\left(\log\frac{110.}{25.563}\right)$$

$$= 2.3(204.5 \times 10^3)\ 0.0001 \times 10^{-6}(\log 4.3)$$

$$= 0.047 \times 10^{-3}(0.634)$$

$$= 0.0298 \times 10^{-3}\text{ sec}$$

$$= 0.0000298\text{ sec}$$

$$= 29.8\ \mu\text{sec}$$

At this instant, *just before* t_2, the voltage at point P_1 of Fig. 11·7f is +90.563 volts (90 + 0.563) and at point G_2, of course, −25 volts. These are indicated in Fig. 11·8a and b. Voltage across the capacitor C_1 in Fig. 11·7f is now 115.563 volts (0.563 + 90 + 25).

5. *At instant* t_2, V_2 is no longer cut off and starts conducting. This, as in any astable multivibrator, now cuts off V_1. The circuit now returns to that of Fig. 11·7b, with S_1 open and S_2 closed. The Thévenin equivalent is shown in Fig. 11·7g. C_1 at t_2 still has its 115.563 volts charge. The applied voltage is 200 volts, as shown, which opposes E_{C_1}, producing a *net* applied voltage of 84.437 volts (200 − 115.563). C_1 now starts charging up again, trying to charge to the 200 volts. Current flows, as shown, *up* through R_{G-K_2} and *down* through R_{L_1}, producing the voltage polarities shown. Point G_2 now becomes positive. Voltage across R_{G-K_2}, the 1-kilohm resistor, is now

$$E_{R_{G-K_2}} = \left(\frac{R_{G-K_2}}{R_{L_1} + R_{G-K_2}}\right) E_{\text{net applied}}$$

$$= \left(\frac{1\text{ kilohm}}{10\text{ kilohms} + 1\text{ kilohm}}\right) 84.437$$

$$= \left(\frac{1}{11}\right) 84.437$$

$$= 7.66\text{ volts}$$

Point G_2 is now +7.66 volts. $E_{R_{L_1}}$ is the remainder or 76.777 volts (84.437 − 7.66). Voltage at point P_1 is now the 200 volts applied less $E_{R_{L_1}}$ or +123.223 volts (200 − 76.777). Voltages at P_1 and G_2, *at instant* t_2, are indicated in Fig. 11·8*a* and *b*.

6. With V_1 cut off, C_1, in Fig. 11·7*g*, continues to charge up toward the full applied 200 volts. It will require approximately five or six *R-C* time constants for C_1 to fully charge, and six *R-C*'s are

$$6R\text{-}C$$
$$6(10 \text{ kilohms} + 1 \text{ kilohm})\ 0.0001\ \mu\text{f}$$
$$6(11 \times 10^3)\ (0.0001 \times 10^{-6})$$
$$0.0066 \times 10^{-3} \text{ sec}$$
$$0.0000066 \text{ sec}$$
$$6.6\ \mu\text{sec}$$

When C_1 has fully charged, I no longer flows in Fig. 11·7*g*, and point P_1 is at +200 volts, while point G_2 is zero volts. This occurs at instant t_3, Fig. 11·8*a* and *b*, and time from t_2 to t_3 is approximately 6.6 μsec.

7. Now the larger *R-C* circuit in the multivibrator diagram of Fig. 11·7*a* is analyzed. This *R-C* is the 0.002-μf capacitor, C_2, and the 500-kilohm resistor, R_{G_1}, connecting the plate of V_2 and the grid of V_1. A Thévenin circuit from the plate of V_2, point P_2, to the grid of V_1, point G_1, is shown in Fig. 11·9*a*. Here, it is assumed that V_2 is conducting, and switch S_3 is shown closed, completing the loop circuit. V_1 is assumed to be nonconducting, and its grid is not drawing current at this time. Switch S_4 is shown open to signify that no grid current is flowing in V_1.

A Thévenin equivalent circuit of Fig. 11·9*a* is shown in *b*. R_{B_2} (7.65 kilohms) and R_{L_2} (10 kilohms) are first assumed to be in parallel, and the equivalent resistor is

$$\begin{aligned} R_{\text{equiv}} &= \frac{R_{B_2}R_{L_2}}{R_{B_2} + R_{L_2}} \\ &= \frac{7.65 \text{ kilohms (10 kilohms)}}{7.65 \text{ kilohms} + 10 \text{ kilohms}} \\ &= \frac{76.5 \times 10^6}{17.65 \times 10^3} \\ &= 4.34 \text{ kilohms} \end{aligned}$$

This 4.34-kilohm equivalent resistor is shown in Fig. 11·9*b*. The equivalent voltage is found by getting the voltage across R_{B_2}. For this, R_{B_2} is considered to be in *series* with R_{L_2} and the 300 volts applied. $E_{R_{B_2}}$ is then

$$E_{R_{B_2}} = \left(\frac{R_{B_2}}{R_{B_2} + R_{L_2}}\right) E_{\text{applied}}$$
$$= \left(\frac{7.65 \text{ kilohms}}{7.65 \text{ kilohms} + 10 \text{ kilohms}}\right) 300$$
$$= \left(\frac{7.65 \text{ kilohms}}{17.65 \text{ kilohms}}\right) 300$$
$$= 130 \text{ volts}$$

This 130 volts is the Thévenin equivalent voltage of the loop in Fig. 11·9*a* and is shown in Fig. 11·9*b*.

8. *Just before* t_1, V_2 is assumed to be conducting, and V_1 is not. As shown in the waveshapes, Fig. 11·8*d*, the grid of V_1 is at -17 volts. This is the cutoff value found previously from the load line. As shown in Fig. 11·9*b*, with point G_1 at -17 volts, the current must be flowing *down* through R_{G_1} and *up* through the 4.34-kilohm resistor (the equivalent of R_{B_2} and R_{L_2}). This resistor, 4.34 kilohms, is so small compared to the 500-kilohm resistor that with 17 volts across the larger resistor, the voltage across the smaller may be ignored. (Actually, it is about 0.1 volt.) As shown in Fig. 11·9*b*, point P_2 is then just about $+130$ volts, while point G_1 is -17 volts. These values for P_2 and G_1 just before t_1 are shown in Fig. 11·8*c* and *d*. With 130 volts applied and with 17 volts across R_{G_1}, E_{C_2}, as shown in Fig. 11·9*b*, must be 147 volts.

9. *At instant* t_1, V_1 starts conducting, and V_2 cuts off. The Thévenin circuit is that now shown in Fig. 11·9*c*, with switch S_3 open and S_4 closed. The Thévenin equivalent of Fig. 11·9*c* is shown in (*d*). R_{B_2} is now omitted since it is opened up by S_3. The closed switch S_4 places the 1-kilohm resistor (R_{G-K_1}) in parallel with the 500 kilohms (R_{G_1}), making the equivalent just about 1 kilohm. The 300 volts applied, in Fig. 11·9*d*, is opposing the 147-volt charge on C_2, producing a *net* applied voltage of 153 volts (300 − 147). C_2 starts charging, and current flows *up* through the 1-kilohm resistor and *down* through R_{L_2}, producing the voltage polarities shown. Voltage across the 1-kilohm resistor (R_{G_1} and R_{G-K_1}) is then

$$E = \left(\frac{1 \text{ kilohm}}{10 \text{ kilohms} + 1 \text{ kilohm}}\right) 153$$
$$= \left(\frac{1}{11}\right) 153$$
$$= 13.9 \text{ volts}$$

The remainder, 139.1 volts (153 − 13.9), is across R_{L_2} in Fig. 11·9*d*. Voltage at point P_2 is the 300 volts applied less the opposing-polarity voltage $E_{R_{L_2}}$ or 160.9 volts (300 − 139.1). As shown in Fig. 11·9*d*, voltage

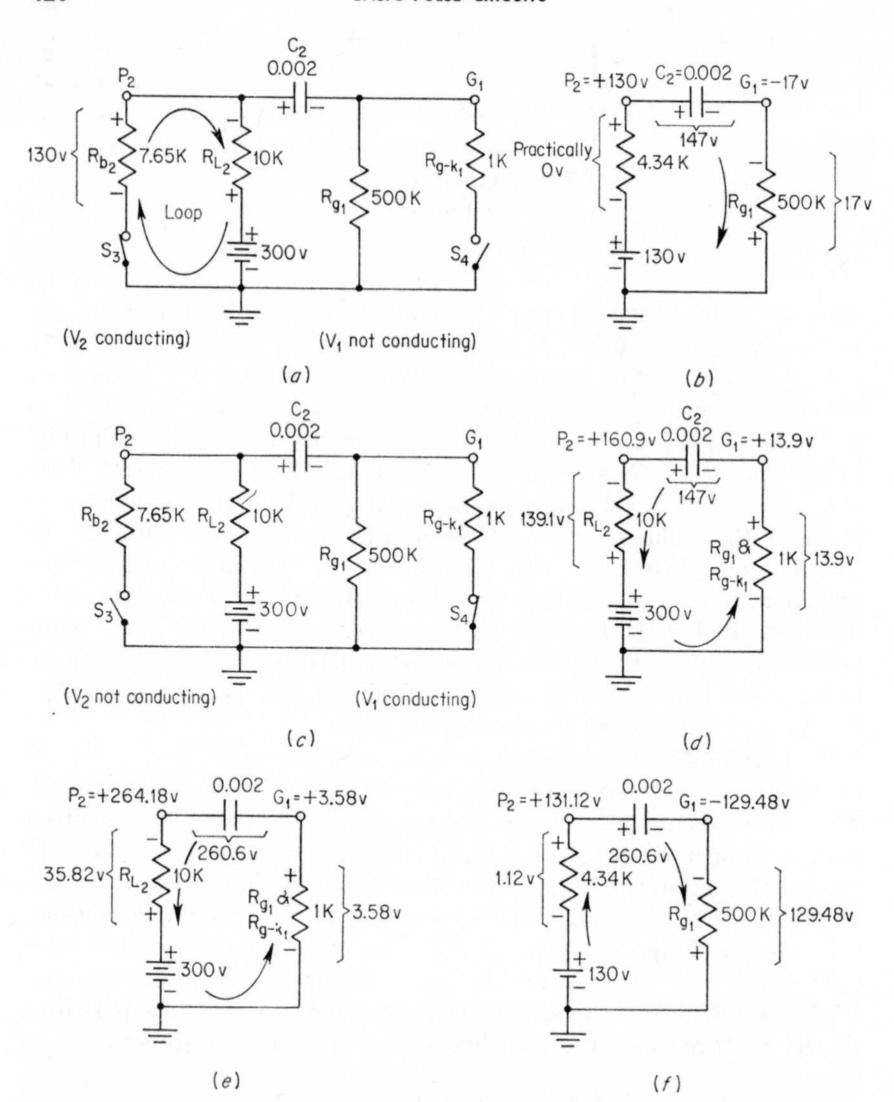

Figure 11·9 Large *R-C* in Example 11·7. (*a*) Circuit of large *R-C* of Fig. 11·7*a*, V_2 conducting; (*b*) Thévenin equivalent of (*a*) just before t_1; (*c*) circuit of large *R-C* of Fig. 11·7*a*, V_1 conducting; (*d*) Thévenin equivalent of (*c*) at t_1; (*e*) Thévenin equivalent just before t_2; (*f*) Thévenin equivalent at t_2.

at point P_2 is $+160.9$, and at point G_1 it is $+13.9$ volts. These values, at instant t_1, are also indicated in Fig. 11·8*c* and *d*.

10. *Between instants t_1 and t_2, just before t_2,* V_1 is still conducting, and V_2 is still cut off. C_2 is continuing to charge up, trying to reach the 300 volts applied, as shown in Fig. 11·9*e*. The time that C_2 is given in which to charge is the 29.8 μsec from t_1 to t_2, referred to as T_A, as shown in Fig. 11·8*a*, and found previously. The voltage that C_2 will reach in this period may be found by first solving for E_2, the voltage across the *total* resistance in the circuit, in Eq. (11·13). The difference between the applied voltage and E_2 is the charge on the capacitor.

$$E_2 = \frac{E_1}{\log^{-1} \dfrac{T}{2.3RC}} \tag{11·13}$$

where E_1 = *total* resistor voltage at the beginning (t_1) of the time period
T = 29.8 μsec
R-C = the circuit of Fig. 11·9*e*.

E_1 consists of the sum of 139.1 volts across R_{L_2} in Fig. 11·9*d* and the 13.9 volts across the 1-kilohm resistor in the same diagram. E_1 is 153 volts (139.1 + 13.9).

$$\begin{aligned} E_2 &= \frac{153}{\log^{-1}\left(\dfrac{29.8\ \mu\text{sec}}{2.3(10\text{ kilohms} + 1\text{ kilohm})\ 0.002\ \mu\text{f}}\right)} \\ &= \frac{153}{\log^{-1}\left(\dfrac{29.8 \times 10^{-6}}{2.3(11 \times 10^3)\ 0.002 \times 10^{-6}}\right)} \\ &= \frac{153}{\log^{-1}\left(\dfrac{29.8 \times 10^{-6}}{0.0506 \times 10^{-3}}\right)} \\ &= \frac{153}{\log^{-1} 0.589} \\ &= \frac{153}{3.88} \\ &= 39.4\text{ volts} \end{aligned}$$

This 39.4 volts, E_2, is the *sum* voltage across *both* resistors in Fig. 11·9*e*. Voltage across the 1-kilohm resistor, R_{G_1} and R_{G-K_1}, is therefore

$$\begin{aligned} E_{1\text{ kilohm}} &= \left(\frac{1\text{ kilohm}}{10\text{ kilohms} + 1\text{ kilohm}}\right) 39.4\text{ volts} \\ &= \left(\frac{1}{11}\right) 39.4 \\ &= 3.58\text{ volts} \end{aligned}$$

Voltage at point G_1 is +3.58 volts. The remainder of the voltage is across R_{L_2} and is 39.4 − 3.58 or 35.82 volts. Voltage at point P_2 is now the 300 volts applied less $E_{R_{L_2}}$ or 300 − 35.82 = 264.18 volts. The voltages at points P_2 and G_1 are shown in Fig. 11·9*e* and are also indicated at just before t_2 in the waveshapes of Fig. 11·8*c* and *d*. E_{C_2} is now at 260.6 (300 − E_2) or also the difference between points P_2 and G_1.

11. *At instant* t_2, as shown in the waveshape drawings of Fig. 11·8, V_2 again starts conducting, and V_1 becomes inoperative. The circuit again becomes that shown in Fig. 11·9*a*. The Thévenin equivalent is shown in Fig. 11·9*f*. C_2 is still at its 260.6-volt charge. The applied voltage, as shown, is 130 volts and is opposing E_{C_2}. C_2 now starts discharging, and current flows *down* through R_{G_1}, as shown, and *up* through the 4.34-kilohm equivalent resistor. The *net* applied voltage is 260.6 (E_{C_2}) − 130 = 130.6 volts. Voltage across the 4.34-kilohm resistor is

$$E_{4.34\text{ kilohms}} = \left(\frac{4.34\text{ kilohms}}{4.34\text{ kilohms} + 500\text{ kilohms}}\right) 130.6$$
$$= \left(\frac{4.34\text{ kilohms}}{504.34\text{ kilohms}}\right) 130.6$$
$$\cong 1.12\text{ volts}$$

Voltage across the 500-kilohm resistor (R_{G_1}) is the difference or

$$130.6 - 1.12 = 129.48\text{ volts}$$

Voltage at point P_2 in Fig. 11·9*f* is +131.12 volts (130 + 1.12). Voltage at point G_1 is −129.48. These are also indicated in Fig. 11·8*c* and *d*.

12. *After instant* t_2, Fig. 11·8, V_2 continues to conduct for some time. C_2 in Fig. 11·9*f* continues to discharge *down* through R_{G_1}, keeping the grid of V_1 negative. *Just before* t_4 of Fig. 11·8*d*, this grid finally reaches its cutoff voltage of −17 volts. At t_4, therefore, V_1 starts to conduct, and the next cycle begins. The time from t_2 to just before t_4, called here T_B, is therefore found from Eq. (11·11)

$$T = 2.3RC \log \frac{E_1}{E_2} \tag{11·11}$$

E_1 is the *total* resistor voltage at the beginning (t_2) of the T_B time period. E_1, from Fig. 11·9*f*, is 130.6 (129.48 + 1.12). E_2 at the end of T_B time period consists of the 17 volts (cutoff voltage of V_1) across the 500-kilohm resistor, R_{G_1}, plus the fraction (about 0.1 volt) of a volt across the small 4.34-kilohm equivalent resistor. This insignificant voltage is ignored here. T_B is then

$$T = 2.3(500\text{ kilohms} + 4.34\text{ kilohms})\ 0.002\ \mu\text{f} \left(\log \frac{130.6}{17}\right)$$
$$= 2.3(504.34 \times 10^3)\ (0.002 \times 10^{-6}) \log 7.67$$
$$= 2.32 \times 10^{-3}(0.885)$$
$$= 2.05 \times 10^{-3}\text{ sec}$$
$$= 0.00205\text{ sec}$$
$$= 2{,}050\ \mu\text{sec}$$

13. The total time of one cycle, as shown in Fig. 11·8, is

$$\begin{aligned} T_{\text{total}} &= T_A + T_B \\ &= 29.8\ \mu\text{sec} + 2{,}050\ \mu\text{sec} \\ &= 2{,}079.8\ \mu\text{sec} \\ &= 2{,}079.8 \times 10^{-6}\ \text{sec} \end{aligned}$$

The frequency of the waveshapes shown in Fig. 11·8 is therefore

$$\begin{aligned} \text{Frequency} &= \frac{1}{T_{\text{total}}} \\ &= \frac{1}{2{,}079.8 \times 10^{-6}} \\ &= 0.000481 \times 10^{6} \text{ pulses per second} \\ &= 481. \text{ pulses per second} \end{aligned}$$

PROBLEMS

11·1 In the circuit of Fig. 11·1*a*, if E_{applied} is 300 volts, $C = 0.05\ \mu$f, and $R = 2$ kilohms, find E_R and E_C 150 μsec after the voltage has been applied.

11·2 In the circuit of Fig. 11·4*a*, if E_{applied} is 150 volts, $L = 150$ mh, and $R = 30$ kilohms, find E_L and E_R, 20 μsec after the voltage has been applied.

MORE CHALLENGING PROBLEMS

11·3 For the diagram accompanying this problem, draw the E_L and E_R voltage waveshapes with values at (*a*), (*b*), (*c*), (*d*), (*e*), and (*f*).

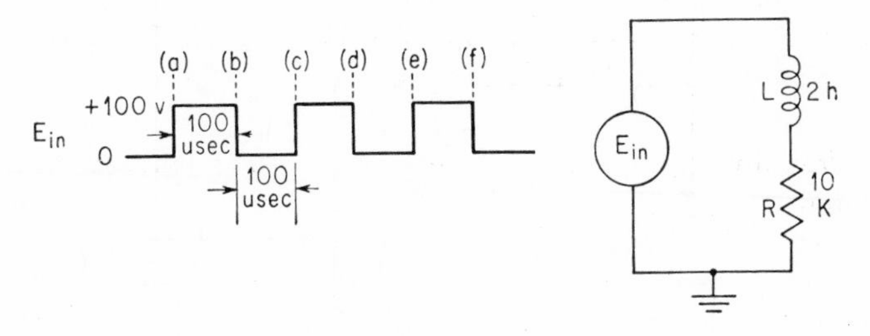

Problem 11·3

11·4 For the diagram given with this problem, draw E_R and E_C voltage waveshapes with values until a *steady-state* condition is reached. This occurs when the E_R (and E_C) waveshape values keep repeating.

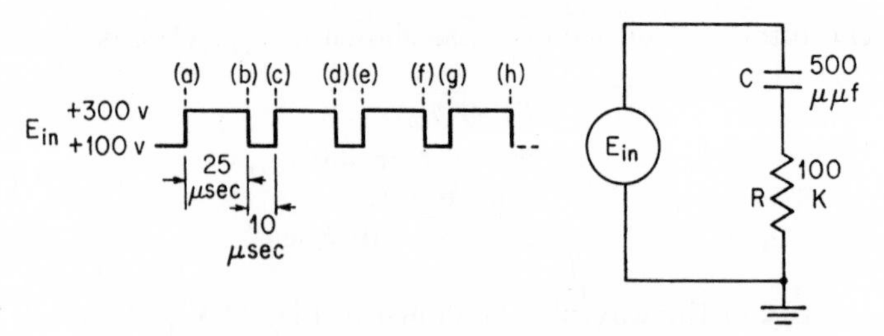

Problem 11·4

11·5 For the circuit given with this problem, the switch is first *open* for a long time, and then it *closes* and *opens* for 400-μsec periods. Draw the E_C and E_R waveshapes with values until a *steady-state condition* is reached. (Hint: First solve for the Thévenin equivalent resistance and voltage between point X and point Y with switch open and with switch closed.)

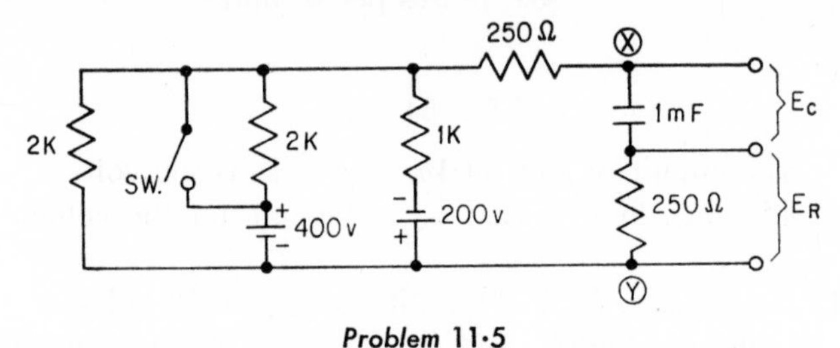

Problem 11·5

11·6 From the circuit given with this problem, draw grid and plate voltage waveshapes with values, and calculate the frequency of this

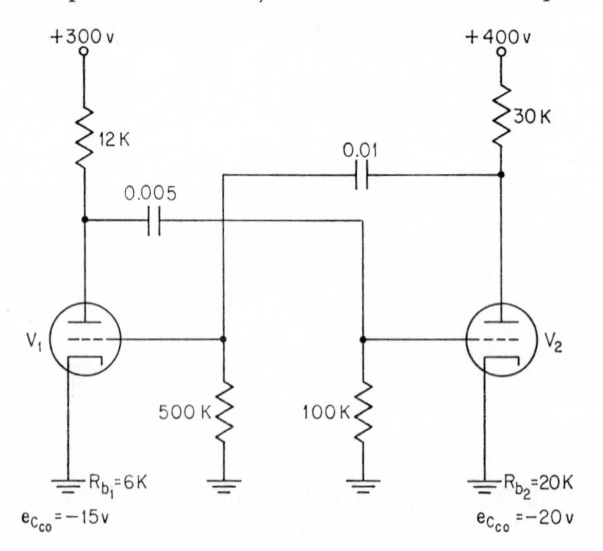

Problem 11·6

astable plate-coupled multivibrator. Start with V_2 conducting, just before instant (a) when V_1 starts to conduct, and solve first using the shorter R-C circuit.

11·7 From the circuit of the accompanying diagram and using the 6C5 tube characteristic curves, draw the load lines and determine the internal resistances of V_1 and V_2 and the required grid cutoff voltages. Draw the grid and plate voltage waveshapes with values, and determine the frequency of the generated signals.

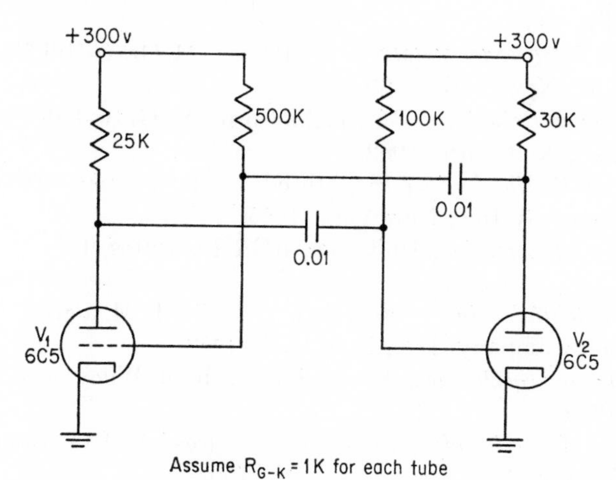

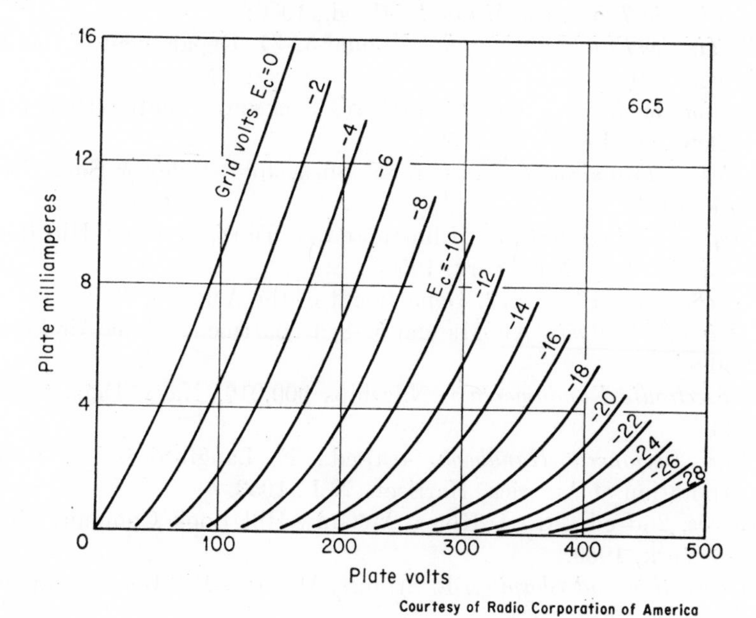

Courtesy of Radio Corporation of America

Problem 11·7

Book References

Basic Theory and Application of Transistors, TM11-690, Department of the Army, 1959.

Digital Computer Fundamentals, T. C. Bartee, McGraw-Hill Book Company, Inc., New York, 1960.

Digital Computer Principles, Burroughs Corp., McGraw-Hill Book Company, Inc., New York, 1962.

Electronic and Radio Engineering, 4th ed., F. E. Terman, McGraw-Hill Book Company Inc., New York, 1955.

Electronics for Automation, Unit One, RCA Institutes, Inc., New York, 1961.

Electronic Switching, Timing, and Pulse Circuits, J. M. Pettit, McGraw-Hill Book Company, Inc., New York, 1959.

Fundamentals of Electronics, E. N. Lurch, John Wiley & Sons, New York, 1960.

Fundamentals of Semiconductor and Tube Electronics, H. A. Romanowitz, John Wiley & Sons, Inc., New York, 1962.

General Electric Transistor Manual, 5th ed., 1960.

Guided Missiles Fundamentals, AF Manual 52-31, Department of the Air Force, 1959.

Introduction to Electric Circuits, H. W. Jackson, Prentice-Hall Inc., Englewood Cliffs, N.J., 1959.

Principles of Transistor Circuits, R. F. Shea, John Wiley & Sons, New York, 1953.

Pulse and Digital Circuits, J. Millman and H. Taub, McGraw-Hill Book Company, Inc., New York, 1956.

Pulse Techniques, TM11-672, Department of the Army, 1951.

Radar Circuit Analysis, AF Manual 52-8, Department of the Air Force, 1960.

Radar Electronic Fundamentals, Navships 900,016, Navy Department, 1944.

Radiotron Designer's Handbook, 4th ed., F. Langford-Smith, Radio Corporation of America, Harrison, N.J., 1953.

Transistors, 2nd ed., M. S. Kiver, McGraw-Hill Book Company, Inc., New York, 1959.

Wave Generation and Shaping, L. Strauss, McGraw-Hill Book Company, Inc., New York, 1960.

Answers to Problems

Chapter 1

1·1 2.14 volts **1·3** (*a*) 0.17 amp (*b*) 0.119 amp (*c*) 0.2 amp **1·5** 100 ohms, −100 volts **1·7** 18.6 volts **1·9** 50.9 ohms

Chapter 2

2·1 (*a*) C_T (*b*) C_S and C_K **2·3** (*a*) 3,980 ohms (*b*) 158 μh **2·5** (*a*) uncompensated (*b*) shunt series (*c*) uncompensated **2·7** (*a*) leading (*b*) lagging **2·9** (*a*) E_{R_1} (*b*) E_{R_2} (*c*) E_{R_T} (*d*) E_C (*e*) E_T (*f*) E_{out} **2·11** (*a*) 5 μf (*b*) 31,800 ohms

Chapter 3

3·1 (*a*) 0.0005 sec or 500 μsec (*b*) 189 volts (*c*) 111 volts **3·3** (*a*) E_C is square, and E_R consists of positive and negative spikes. (*b*) E_C is a small amplitude triangular waveshape, and E_R is square with slight downward tilts. **3·5** (*a*) 5 amp (*b*) 2 amp **3·7** (*a*) 430 microamps (*b*) 43 volts

Chapter 4

4·1 (*a*) +20 volts base, pulse extending to +25 volts (*b*) +20 volts d-c, no signal (*c*) +20 volts d-c, no signal (*d*) +20 volts base, pulse extending to +30 volts **4·3** (*a*) +20 volts base, pulse extending to +35 volts (*b*) +20 volts base, pulse extending down to +5 volts (*c*) +20 volts base, pulse extending to +25 volts (*d*) +20 volts base, pulse extending down to zero **4·4** (*a*) +10 volts (*b*) −15 volts (*c*) +5 volts (*d*) −30 **4·5** (*a*) +15 volts (*b*) −10 volts (*c*) +5 volts (*d*) −10 volts **4·7** (*a*) zero volts (*b*) −5 volts (*c*) zero volts (*d*) −20 volts **4·9** (*a*) +10 volts base, pulse extending to +25 volts (*b*) +10 volts base, pulse extending down to zero (*c*) +10 volts base, pulse extending to +15 volts (*d*) +10 volts base, pulse extending down to zero.

Chapter 5

5·1 See schematic Fig. 5·1*a*. **5·2** See Fig. 5·2. **5·4** See schematic Fig. 5·5*a*. **5·5** (*a*) 1 kc (*b*) 1 kc (*c*) 1 kc (*d*) free running at about 950 cps. **5·6** See schematic Fig. 5·7*a*. **5·8** See Figs. 5·8, and 5·9. **5·9** See schematic Fig. 5·10. **5·10** See Fig. 5·11. **5·12** See schematic Fig. 5·12. **5·13** See Fig. 5·13. **5·14** See schematic Fig. 5·18*a*. **5·15** negative-going pulse.

Chapter 6

6·1 See schematic Fig. 6·3. **6·3** See schematic Fig. 6·4. **6·5** (*a*) 1 kc (*b*) 1 kc (*c*) 3 kc **6·7** (*a*) See diagram Fig. 6·7. (*b*) 50 μsec **6·9** See diagram Fig. 6·11. **6·13** similar to Fig. 6·12*b*, but with reverse polarity d-c voltages **6·15** Just before instant (a): $E_{C_1} = 0$, $I = 0$, $E_{R_2} = 0$, and $E_{R_1} = 0$. At instant (*a*): $E_{C_1} = 0$, $I = 0.05$ amp, $E_{R_2} = 50$, and $E_{R_1} = 50$. At instant (*b*): $E_{C_1} = 50$, $I = 0.05$ amp, $E_{R_2} = 0$, and $E_{R_1} = 50$. At instant (*c*): $E_C = 100$, $I = 0$, $E_{R_2} = 0$, and $E_{R_1} = 0$.

During the (*a*) to (*b*) time, E_{C_1} rises *linearly*, I remains *constant*, E_{R_2} falls *linearly*, and E_{R_1} remains *constant*. During the (*b*) to (*c*) time, E_{C_1} rises *exponentially*, I decreases *exponentially*, E_{R_2} remains at zero, and E_{R_1} decreases *exponentially*. **6·17** larger sawtooth *amplitude*, and lower *frequency*

Chapter 7

7·1 (*a*) 101111 (*b*) 111101 (*c*) 10111101 (*d*) 1001101 **7·3** (*a*) 17 (*b*) 143 (*c*) 29 (*d*) 69 **7·5** (*a*) 326 (*b*) 620 (*c*) 177 (*d*) 254 **7·6** (*a*) 2212 (*b*) 1202 (*c*) 21000 **7·7** (*a*) Neons in plates of V_1 and V_2 (*b*) Neon in plate of V_1 only **7·9** (*a*) neons number 2 and 8 (*b*) neon number 8 **7·10** (*a*) neon number 8 (*b*) neon number 0.

Chapter 8

8·1 (*a*) +10-volt pulse (*b*) +10-volt pulse (*c*) zero (*d*) OR gate **8·3** (*a*) +15-volt base, no signal (*b*) +15-volt base, no signal (*c*) +15-volt base, pulse extending negatively down to almost zero (*d*) +15-volt base, pulse extending negatively down to almost zero **8·5** (*a*) $E_{out} = +10$ volts; $I_{P_{D_1}} = 1$ ma, $I_{P_{D_2}} = 0$ (*b*) E_{out} rises abruptly from +10 volts to +15 volts; $I_{P_{D_1}}$ falls abruptly from 1 ma to zero; $I_{P_{D_2}}$ rises sharply from 0 to 1.5 ma (*c*) E_{out} remains at +15 volts; $I_{P_{D_1}} = 0$; $I_{P_{D_2}}$ remains at 1.5 ma (*d*) E_{out} falls sharply from +15 volts to zero; $I_{P_{D_1}} = 0$; $I_{P_{D_2}}$ falls sharply from 1.5 ma to zero **8·7** Switch *open:* (*a*) zero (*b*) +3-volt pulse (*c*) +10-volt pulse (*d*) −5-volt pulse

Switch *closed:* (*a*) zero (*b*) +3-volt pulse (*c*) +10-volt pulse (*d*) zero.

Chapter 9

9·1 (*a*) A signal from X_2C or X_3B OR gates (*b*) The previous flip-flop stage, V_3, going into or through its zero or flop state **9·6** 1011111010

9·7 101110110 **9·8** 10 **9·9** 0000 **9·10** 010 **9·11** 1001 **9·12** 15,750 cps equalizing pulses only **9·13** 31,500 cps equalizing pulses only.

Chapter 10

10·1 (*a*) E_{C_1}, also E_{grid} (*b*) E_{R_G} (*c*) E_{tank}, also E_{plate} (*d*) I_{plate} (*e*) X_L tube **10·7** (*a*) V_1 (*b*) E_{R_1} (*c*) negative (*d*) decreases G_M, increasing C_{equiv} [see Eq. (10·51)] (*e*) with a larger C, oscillator frequency decreases, thwarting the attempt to drift higher in frequency **10·8** (*a*) more heavily (*b*) E_{R_K} goes positive, decreasing the bias on V_2. (*c*) A smaller bias on V_2 permits the cutoff grid to reach conduction sooner, resulting in a higher frequency, thwarting the attempt to drift lower in frequency. **10·9** +72 volts **10·10** +110 volts **10·11** 3.6 volts **10·13** 0.4 volts, positive at the top.

Chapter 11

11·1 E_R = 66.8 volts; E_C = 233.2 volts **11·3** (*a*) E_L = abrupt rise from zero to +100 volts; E_R = 0 volts (*b*) E_L = slumps to +61 volts and abruptly goes to −39 volts; E_R = rises to +39 volts (*c*) E_L = gradual rise to −24 volts, and abruptly goes to +76 volts; E_R = slumps to +24 volts. (*d*) E_L = slumps to +46 volts and abruptly goes to −54 volts; E_R = rises to +54 volts. (*e*) E_L = rises gradually to −33 volts and abruptly goes to +67 volts; E_R = slumps to +33 volts. (*f*) E_L = slumps to +41 volts and abruptly goes to −59 volts; E_R = rises to +59 volts. **11·6** E_{plate} V_1 = +300 volts at the start; at (*a*) it falls abruptly to +108, slumping down to +101 volts at instant (*b*). (*a*)-to-(*b*) time is 1,170 μsec. At (*b*), it rises abruptly to +135 volts and then rises exponentially to +300 volts.

E_{grid} V_2 = approximately zero volts at start; at (*a*) it abruptly becomes −192 volts; slumping exponentially to −20 volts ($e_{C\text{c.o.}}$) at (*b*) during the 1,170 μsec; it then rises abruptly to +14 volts and then decreases to zero.

E_{plate} V_2 = starts at +160 volts and rises sharply to +182 volts at (*a*); it then rises gradually to +395 volts at (*b*) and then falls sharply to +165 volts, decreasing gradually down to +160 volts.

E_{grid} V_1 = starts at −15 volts ($e_{C\text{c.o.}}$), rising abruptly to +7 volts at (*a*); it slumps down to zero at (*b*) and then changes abruptly to −230 volts. It rises exponentially to −15 volts ($e_{C\text{c.o.}}$) at (*c*). The (*b*)-to-(*c*) time is 14,000 μsec. *Frequency* is the *reciprocal* of the time of one cycle [(*a*)-to-(*c*) time]. One cycle time is 1,170 μsec [(*a*)-to-(*b*)] +14,000 μsec [(*b*)-to-(*c*)] and is 15,170 μsec. *Frequency* is therefore 66 cps.

Index